AF510075

IET TELECOMMUNICATIONS SERIES 81

Applications of Machine Learning in Wireless Communications

Other volumes in this series:

Volume 9 **Phase Noise in Signal Sources** W.P. Robins
Volume 12 **Spread Spectrum in Communications** R. Skaug and J.F. Hjelmstad
Volume 13 **Advanced Signal Processing** D.J. Creasey (Editor)
Volume 19 **Telecommunications Traffic, Tariffs and Costs** R.E. Farr
Volume 20 **An Introduction to Satellite Communications** D.I. Dalgleish
Volume 26 **Common-Channel Signalling** R.J. Manterfield
Volume 28 **Very Small Aperture Terminals (VSATs)** J.L. Everett (Editor)
Volume 29 **ATM: The broadband telecommunications solution** L.G. Cuthbert and J.C. Sapanel
Volume 31 **Data Communications and Networks, 3rd Edition** R.L. Brewster (Editor)
Volume 32 **Analogue Optical Fibre Communications** B. Wilson, Z. Ghassemlooy and I.Z. Darwazeh (Editors)
Volume 33 **Modern Personal Radio Systems** R.C.V. Macario (Editor)
Volume 34 **Digital Broadcasting** P. Dambacher
Volume 35 **Principles of Performance Engineering for Telecommunication and Information Systems** M. Ghanbari, C.J. Hughes, M.C. Sinclair and J.P. Eade
Volume 36 **Telecommunication Networks, 2nd Edition** J.E. Flood (Editor)
Volume 37 **Optical Communication Receiver Design** S.B. Alexander
Volume 38 **Satellite Communication Systems, 3rd Edition** B.G. Evans (Editor)
Volume 40 **Spread Spectrum in Mobile Communication** O. Berg, T. Berg, J.F. Hjelmstad, S. Haavik and R. Skaug
Volume 41 **World Telecommunications Economics** J.J. Wheatley
Volume 43 **Telecommunications Signalling** R.J. Manterfield
Volume 44 **Digital Signal Filtering, Analysis and Restoration** J. Jan
Volume 45 **Radio Spectrum Management, 2nd Edition** D.J. Withers
Volume 46 **Intelligent Networks: Principles and applications** J.R. Anderson
Volume 47 **Local Access Network Technologies** P. France
Volume 48 **Telecommunications Quality of Service Management** A.P. Oodan (Editor)
Volume 49 **Standard Codecs: Image compression to advanced video coding** M. Ghanbari
Volume 50 **Telecommunications Regulation** J. Buckley
Volume 51 **Security for Mobility** C. Mitchell (Editor)
Volume 52 **Understanding Telecommunications Networks** A. Valdar
Volume 53 **Video Compression Systems: From first principles to concatenated codecs** A. Bock
Volume 54 **Standard Codecs: Image compression to advanced video coding, 3rd Edition** M. Ghanbari
Volume 59 **Dynamic Ad Hoc Networks** H. Rashvand and H. Chao (Editors)
Volume 60 **Understanding Telecommunications Business** A. Valdar and I. Morfett
Volume 65 **Advances in Body-Centric Wireless Communication: Applications and state-of-the-art** Q.H. Abbasi, M.U. Rehman, K. Qaraqe and A. Alomainy (Editors)
Volume 67 **Managing the Internet of Things: Architectures, theories and applications** J. Huang and K. Hua (Editors)
Volume 68 **Advanced Relay Technologies in Next Generation Wireless Communications** I. Krikidis and G. Zheng
Volume 69 **5G Wireless Technologies** A. Alexiou (Editor)
Volume 70 **Cloud and Fog Computing in 5G Mobile Networks** E. Markakis, G. Mastorakis, C.X. Mavromoustakis and E. Pallis (Editors)
Volume 71 **Understanding Telecommunications Networks, 2nd Edition** A. Valdar
Volume 72 **Introduction to Digital Wireless Communications** Hong-Chuan Yang
Volume 73 **Network as a Service for Next Generation Internet** Q. Duan and S. Wang (Editors)
Volume 74 **Access, Fronthaul and Backhaul Networks for 5G & Beyond** M.A. Imran, S.A.R. Zaidi and M.Z. Shakir (Editors)
Volume 76 **Trusted Communications with Physical Layer Security for 5G and Beyond** T.Q. Duong, X. Zhou and H.V. Poor (Editors)
Volume 77 **Network Design, Modelling and Performance Evaluation** Q. Vien
Volume 78 **Principles and Applications of Free Space Optical Communications** A.K. Majumdar, Z. Ghassemlooy, A.A.B. Raj (Editors)
Volume 79 **Satellite Communications in the 5G Era** S.K. Sharma, S. Chatzinotas and D. Arapoglou
Volume 80 **Transceiver and System Design for Digital Communications, 5th Edition Scott** R. Bullock
Volume 905 **ISDN Applications in Education and Training** R. Mason and P.D. Bacsich

Applications of Machine Learning in Wireless Communications

Edited by
Ruisi He and Zhiguo Ding

The Institution of Engineering and Technology

Published by The Institution of Engineering and Technology, London, United Kingdom

The Institution of Engineering and Technology is registered as a Charity in England & Wales (no. 211014) and Scotland (no. SC038698).

© The Institution of Engineering and Technology 2019

First published 2019

This publication is copyright under the Berne Convention and the Universal Copyright Convention. All rights reserved. Apart from any fair dealing for the purposes of research or private study, or criticism or review, as permitted under the Copyright, Designs and Patents Act 1988, this publication may be reproduced, stored or transmitted, in any form or by any means, only with the prior permission in writing of the publishers, or in the case of reprographic reproduction in accordance with the terms of licences issued by the Copyright Licensing Agency. Enquiries concerning reproduction outside those terms should be sent to the publisher at the undermentioned address:

The Institution of Engineering and Technology
Michael Faraday House
Six Hills Way, Stevenage
Herts, SG1 2AY, United Kingdom

www.theiet.org

While the authors and publisher believe that the information and guidance given in this work are correct, all parties must rely upon their own skill and judgement when making use of them. Neither the authors nor publisher assumes any liability to anyone for any loss or damage caused by any error or omission in the work, whether such an error or omission is the result of negligence or any other cause. Any and all such liability is disclaimed.

The moral rights of the authors to be identified as authors of this work have been asserted by them in accordance with the Copyright, Designs and Patents Act 1988.

British Library Cataloguing in Publication Data
A catalogue record for this product is available from the British Library

ISBN 978-1-78561-657-0 (hardback)
ISBN 978-1-78561-658-7 (PDF)

Typeset in India by MPS Limited

Contents

Foreword

The technologies of wireless communications have been changed drastically in recent years. The rapidly growing wave of wireless data is pushing against the boundary of wireless communication system's performance. Such pervasive and exponentially increasing data present imminent challenges to all aspects of wireless communication system's design, and the future wireless communications will require robust intelligent algorithms for different services in different scenarios. Contributions are needed from multidisciplinary fields to enhance wireless system, such as computer science, mathematics, control and many other science disciplines. The combined efforts from scientists from different disciplines are important for the success of the wireless communication industry.

In such an era of big data where data mining and data analysis technologies are effective approaches for wireless system evaluation and design, the applications of machine learning in wireless communications have received a lot of attention recently. Machine learning provides feasible and new solutions for the complex wireless communication system design. It has been a powerful tool and popular research topic with many potential applications to enhance wireless communications, e.g. radio channel modelling, channel estimation and signal detection, network management and performance improvement, access control, resource allocation. However, most of the current researches are separated into different fields and have not been well organized and presented yet. It is therefore difficult for academic and industrial groups to see the potentialities of using machine learning in wireless communications. It is now appropriate to present a detailed guidance of how to combine the disciplines of wireless communications and machine learning.

In this book, present and future developments and trends of wireless communication technologies are depicted based on contributions from machine learning and other fields in artificial intelligence. The prime focus of this book is given in the physical layer and network layer with a special emphasis on machine-learning projects that are (or are close to) achieving improvements in wireless communications. A wide variety of research results are merged together to make this book useful for students and researchers. There are 13 chapters in this book, and we have organized them as follows:

- In *Chapter 1*, an overview of machine-learning algorithms and their applications are presented to provide advice and references to fundamental concepts accessible to the broad community of wireless communication practitioners. Specifically, the materials are organized into three sections following the three main branches of machine learning: supervised learning, unsupervised learning and reinforcement learning (RL). Each section starts with an overview to illustrate the major concerns and ideas of this branch. Then, classic algorithms and their last developments are reviewed with typical applications and useful references. Furthermore, pseudocodes are added to provide interpretations and details of algorithms. Each section ends by a summary in which the structure of this section is untangled and relevant applications in wireless communication are given.

- In *Chapter 2*, using machine learning in wireless channel modelling is presented. First of all, the background of the machine-learning-enabled channel modelling is introduced. Then, four related aspects are presented: (i) propagation scenario classification, (ii) machine-learning-based multipath component (MPC) clustering, (iii) automatic MPC tracking and (iv) deep-learning-based channel modelling. The results in this chapter can provide references to other real-world measurement data-based channel modelling.
- In *Chapter 3*, the wireless channel prediction is addressed, which is a key issue for wireless communication network planning and operation. Instead of the classic model-based methods, a survey of recent advances in machine-learning technique-based channel prediction algorithms is provided, including both batch and online methods. Experimental results are provided using the real data.
- In *Chapter 4*, new types of channel estimators based on machine learning are introduced, which are different from traditional pilot-aided channel estimators such as least squares and linear minimum mean square errors. Specifically, two newly designed channel estimators based on deep learning and one blind estimator based on expectation maximization algorithm are provided for wireless communication systems. The challenges and open problems for channel estimation aided by machine-learning theories are also suggested.
- In *Chapter 5*, cognitive radio is introduced as a promising paradigm to solve the spectrum scarcity and to improve the energy efficiency of the next generation mobile communication network. In the context of cognitive radios, the necessity of using signal identification techniques is first presented. A survey of signal identification techniques and recent advances in this field using machine learning are then provided. Finally, open problems and possible future directions for cognitive radio are briefly discussed.
- In *Chapter 6*, the fundamental concepts that are important in the study of compressive sensing (CS) are introduced. Three conditions are described, i.e. the null space property, the restricted isometry property and mutual coherence, that are used to evaluate the quality of sensing matrices and to demonstrate the feasibility of reconstruction. Some widely used numerical algorithms for sparse recovery are briefly reviewed, which are classified into two categories, i.e. convex optimization algorithms and greedy algorithms. Various examples are illustrated where the CS principle has been applied to WSNs.
- In *Chapter 7*, the enhancement of the proposed IEEE 802.11p Medium Access Control (MAC) layer is studied for vehicular use by applying RL. The purpose of this adaptive channel access control technique is enabling more reliable, high-throughput data exchanges among moving vehicles for cooperative awareness purposes. Some technical background for vehicular networks is presented, as well as some relevant existing solutions tackling similar channel sharing problems. Finally, some new findings from combining the IEEE 802.11p MAC with

RL-based adaptation and insight of the various challenges appearing when applying such mechanisms in a wireless vehicular network are presented.

- In *Chapter 8*, the advantage of applying machine-learning-based perceptual coding strategies in relieving bandwidth limitation is presented for wireless multimedia communications. Typical video coding standards, especially the state-of-the-art high efficiency video coding (HEVC) standard, as well as recent research progress on perceptual video coding, are included. An example that minimizes the overall perceptual distortion is further demonstrated by modelling subjective quality with machine-learning-based saliency detection. Several promising directions in learning-based perceptual video coding are presented to further enhance wireless multimedia communication experience.
- In *Chapter 9*, it is argued that the state-of-the-art HEVC standard can be used for saliency detection to generate the useful features in compressed domain. Therefore, this chapter proposes to learn the video saliency model, with regard to HEVC features. First, an eye-tracking database is established for video saliency detection. Through the statistical analysis on our eye-tracking database, we find out that human fixations tend to fall into the regions with large-valued HEVC features on splitting depth, bit allocation and motion vector (MV). In addition, three observations are obtained with the further analysis on our eye-tracking database. Accordingly, several features in HEVC domain are proposed on the basis of splitting depth, bit allocation and MV. Next, a kind of support vector machine is learned to integrate those HEVC features together, for video saliency detection. Since almost all video data are stored in the compressed form, the proposed method is able to avoid both the computational cost on decoding and the storage cost on raw data. More importantly, experimental results show that the proposed method is superior to other state-of-the-art saliency detection methods, either in compressed or uncompressed domain.
- In *Chapter 10*, deep learning is incorporated for indoor localization based on channel state information (CSI) with commodity 5GHz Wi-Fi. The state-of-the-art deep-learning techniques are first introduced, including deep autoencoder networks, convolutional neural networks and recurrent neural networks. The CSI preliminaries and three hypotheses are further introduced, which are validated with experiments. Then a deep-learning-based algorithm is presented to leverage bimodal CSI data, i.e. average amplitudes and estimated angle of arrivals, in both offline and online stages of fingerprinting. The proposed scheme is validated with extensive experiments. Finally, several open research problems are examined for indoor localization based on deep-learning techniques.
- In *Chapter 11*, the reinforcement-learning-based wireless resource allocation is presented. First the basic principle of stochastic approximation is introduced, which is the basis of the RL. Then how to formulate the wireless resource allocation problems via three forms of Markov decision process (MDP) is demonstrated, respectively, namely finite-horizon MDP, infinite-horizon MDP with discount cost and infinite-horizon MDP with average cost. One of the key knowledge to solve MDP problem is the system state transition probability, which might be unknown in practice. Hence, finally it is shown that when some system statistics are unknown, the MDP problems can still be solved via the method of RL.

- In *Chapter 12*, by integrating information theory with the principle of effective capacity, an energy efficiency optimization problem is formulated with statistical QoS guarantee in the uplink of two-tier small cell networks. To solve the problem, a Q-learning mechanism based on Stackelberg game framework is introduced, in which a macro-user acts as a leader, and knows all small-cell-users' transmit power strategies, while the small-cell-users are followers and only communicate with the microcell base station not with other small-cell base stations. In a formulated Stackelberg game procedure, the macro-user selects the transmit power level based on the best responses of the small-cell-users. Then those small-cell-users find their best responses. And in order to improve the self-organizing ability of femtocell, based on the non-cooperative game framework, a Boltzmann distribution-based weighted filter Q-learning algorithm (BDB-WFQA) based on Boltzmann distribution is proposed to realize power allocation. The simulation results show the proposed distributed Q-learning algorithm has a better performance in terms of convergence speed while providing delay QoS provisioning. The proposed BDB-WFQA algorithm increases the achievable effective capacity of macro-users and a better performance compared with other power-control algorithm.
- In *Chapter 13*, the open Jackson queuing network models are used to model the macroscopic level vehicular mobility. The proposed simple model can accurately describe the vehicular mobility and then further predict various measures of network-level performance like the vehicular distribution, and vehicular-level performance like average sojourn time in each area and the number of sojourned areas in the vehicular networks. Model validation based on two large-scale urban city vehicular motion traces reveals that such a simple model can accurately predict a number of system metrics interested by the vehicular network performance. Moreover, two applications are proposed to illustrate the proposed model is effective in the analysis of system-level performance and dimensioning for vehicular networks.

The goal of this book is to help communications system designers gain an overview of the pertinent applications of machine learning in wireless communications, and for researchers to assess where the most pressing needs for further work lie. This book can also be used as a textbook for the courses dedicated for machine-learning-enabled wireless communications. With contributions from an international panel of leading researchers, this book will find a place on the bookshelves of academic and industrial researchers and advanced students working in wireless communications and machine learning. We hope that the above contributions will form an interesting and useful compendium on applications of machine learning in wireless communications.

Prof. Ruisi He
State Key Laboratory of Rail Traffic Control and Safety
Beijing Jiaotong University, China
and
Prof. Zhiguo Ding
School of Electrical and Electronic Engineering
The University of Manchester, UK

Chapter 1

Introduction of machine learning

Yangli-ao Geng[1], Ming Liu[1], Qingyong Li[1], and Ruisi He[2]

Machine learning, as a subfield of artificial intelligence, is a category of algorithms that allow computers to learn knowledge from examples and experience (data), without being explicitly programmed [1]. Machine-learning algorithms can find natural patterns hidden in massive complex data, which humans can hardly deal with manually. The past two decades have witnessed tremendous growth in big data, which makes machine learning become a key technique for solving problems in many areas such as computer vision, computational finance, computational biology, business decision, automotive and natural language processing (NLP). Furthermore, our life has been significantly improved by various technologies based on machine learning [2]. Facial-recognition technology allows social media platforms to help users tag and share photos of friends. Optical character recognition technology converts images of text into movable type. Recommendation systems, powered by machine learning, suggest what films or television shows to watch next based on user preferences. Information retrieval technology supports a search engine to return most related records after users input some keywords. NLP technology makes it possible to filter out spam from massive e-mails automatically. Self-driving cars that rely on machine learning to navigate are around the corner to consumers.

In wireless communications, when you encounter a complex task or problem involving a large amount of data and lots of variables, but without existing formula or equation, machine learning can be a solution. Traditionally, machine-learning algorithms can be roughly divided into three categories: supervised learning, unsupervised learning and reinforcement learning (RL). In this chapter, we present an overview of machine-learning algorithms and list their applications, with a goal of providing useful advice and references to fundamental concepts accessible to the broad community of wireless communications practitioners.

1.1 Supervised learning

Let us begin with an example to explain the basic idea of supervised learning. Imagine that you are a weatherman and have access to historical meteorological data

[1]School of Computer and Information Technology, Beijing Jiaotong University, China
[2]State Key Laboratory of Rail Traffic Control and Safety, Beijing Jiaotong University, China

(e.g. temperature, wind speed and precipitation for the past days). Now, given today's meteorological data, how to predict the weather of the next day? A natural idea is to explore a rule from the historical meteorological data. Specifically, you need to observe and analyse what the weather was under the meteorological data of the last day. If you are fortunate enough to find a rule, then you will make a successful prediction. However, in most cases, the meteorological data is too big to analyse for humans. Supervised learning would be a solution to this challenge.

In fact, what you try to do in the above example is a typical supervised learning task. Formally, *supervised learning* is a procedure of learning a function $f(\cdot)$ that maps an input $\mathbf{x}$ (meteorological data of a day) to an output y (weather of the next day) based on a set of sample pairs $T = \{(\mathbf{x}_i, y_i)\}_{i=1}^n$ (historical data), where T is called a *training set* and y_i is called a *label*. If y is a categorical variable (e.g. sunny or rainy), then the task is called a *classification* task. If y is a continuous variable (e.g. probability of precipitation), then the task is called a *regression* task. Furthermore, for a new input $\mathbf{x}_0$, which is called a *test sample*, $f(\mathbf{x}_0)$ will give the prediction.

In wireless communications, an important problem is estimating the channel noise in a MIMO wireless network, since knowing these parameters are essential to many tasks of a wireless network such as network management, event detection, location-based service and routing [3]. This problem can be solved by using supervised learning approaches. Let us consider the circumstances for the linear channel with white added Gaussian noise MIMO environments with t transmitting antennas and r receiving antennas. Assume the channel model is $\mathbf{z} = \mathbf{Hs} + \mathbf{u}$, where $\mathbf{s} \in \mathbb{R}^t$, $\mathbf{u} \in \mathbb{R}^r$ and $\mathbf{z} \in \mathbb{R}^r$ denote signal vector, noise vector and received vector, respectively. The goal in the channel noise estimation problem is to estimate $\mathbf{u}$ given $\mathbf{s}$ and $\mathbf{z}$. This problem can be formulated as r regression tasks, and the target of the kth regression task is to predict u_k for $1 \leq k \leq r$. In the kth regression task, a training pair is represented as $\{[\mathbf{s}^T, z_k]^T, u_k\}$. We can complete these tasks using any regression model (will be introduced later in this chapter). Once the model is well trained, u_k can be predicted when a new sample $[\bar{\mathbf{s}}^T, \bar{z}_k]^T$ comes. In this section, we will discuss three practical technologies of supervised learning.

1.1.1 k-Nearest neighbours method

The k-nearest neighbours (k-NNs) method is a basic supervised learning method which is applicable to both classification and regression tasks. Here, we will focus on the classification since the regression shares similar steps with the classification. Given a training set $T = \{(\mathbf{x}_i, y_i)\}_{i=1}^n$ and a test sample $\mathbf{x}_0$, the task is to predict the category of $\mathbf{x}_0$ under the instruction of T. The main idea of k-NN is that first search k-NNs of $\mathbf{x}_0$ in the training set and then classify $\mathbf{x}_0$ into the category which is most common among the k-NNs (the majority principle). Particularly, if $k = 1$, $\mathbf{x}_0$ is simply assigned to the class of its nearest neighbour.

Figure 1.1(a) shows an illustration for the main idea of k-NN. From Figure 1.1(a), we observe that there are seven samples in the training set, four of which are labelled as the first class (denoted by squares) and the others are labelled as the second class (denoted by triangles). We intend to predict the category of a test sample (denoted by a circle) using the k-NN method. When $k = 3$, as shown in Figure 1.1(b), the test sample will be assigned to the first class according to the majority principle. When $k = 1$, as shown in Figure 1.1(c), the test sample will be assigned to the second class since its nearest neighbour belongs to the second class. A formal description of k-NN is presented in Algorithm 1.1.

The output of the k-NN algorithm is related to two things. One is the distance function, which measures how near two samples are. Different distance functions will lead to different k-NN sets and thus different classification results. The most commonly used distance function is the L_p distance. Given two vectors $\mathbf{x} = (x_1, \ldots, x_d)^T$ and $\mathbf{z} = (z_1, \ldots, z_d)^T$, the L_p distance between them is defined as

$$L_p(\mathbf{x}, \mathbf{z}) = \left(\sum_{i=1}^{d} |x_i - z_i|^p \right)^{1/p}. \tag{1.1}$$

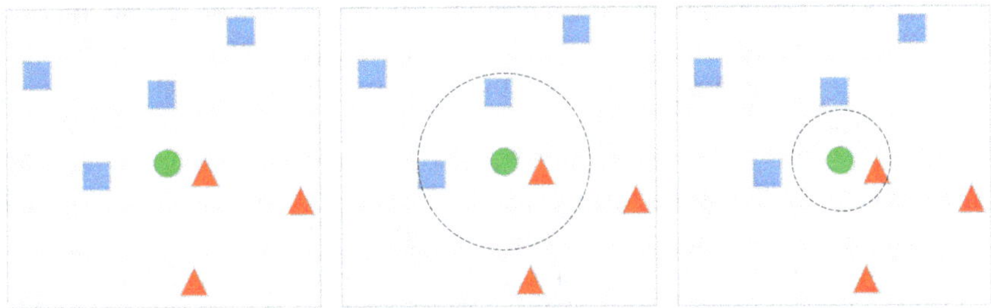

(a) (b) (c)

Figure 1.1 An illustration for the main idea of k-NN. (a) A training set there consists of seven samples, four of which are labelled as the first class (denoted by squares) and the others are labelled as the second class (denoted by triangles). A test sample is denoted as a circle. (b) When k = 3, the test sample is classified as the first class. (c) When k = 1, the test sample is assigned as the second class

---

Algorithm 1.1: k-NN method

Input: number of neighbours k, training set $T = \{(\mathbf{x}_i, y_i)\}_{i=1}^{n}$, test sample $\mathbf{x}_0$

Output: label of test sample y_0

1 Find the k-NN set of $\mathbf{x}$ in T and denote the set by $N_k(\mathbf{x}_0)$;

2 Determine the label of $\mathbf{x}$ according to the majority principle, i.e.

$$y_0 = \arg \max_{1 \le c \le m} \sum_{\mathbf{x}_i \in N_k(\mathbf{x}_0)} I(y_i = c),$$

where $I(y_i = c)$ equals 1 if $y_i = c$ and 0 otherwise;

When p equals 2, the L_p distance becomes the Euclidean distance. When p equals 1, the L_p distance is also called the Manhattan distance. When p goes to ∞, it can be shown that

$$L_\infty(\mathbf{x}, \mathbf{z}) = \max_i |x_i - z_i|. \tag{1.2}$$

Another useful distance is the angular distance, which is defined as

$$D_A(\mathbf{x}, \mathbf{z}) = \frac{\arccos\left(\mathbf{x}^T\mathbf{z}/(\|\mathbf{x}\|\,\|\mathbf{z}\|)\right)}{\pi}. \tag{1.3}$$

As its name suggests, the angular distance measures the included angle between two vectors, and thus it is independent of the length of the vectors. This property makes the angular distance useful in the situation that we only concern the proportion of each component of features. Readers can refer to [4] for more information about distances.

The other factor affecting the result of the algorithm is the value of k. As shown in Figure 1.1, different values of k may lead to different results. The best choice of k depends upon the data. Generally, smaller values of k can generate more accurate result for a high-quality training set, but it is sensitive to noises. In other words, the output for a test sample may be severely affected by the noise samples near to it. In contrast, larger values of k reduce the effect of noise on the classification but make boundaries between classes less distinct [5]. In practice, one popular way of choosing the empirically optimal k is via cross validation [6].

The k-NN algorithm is easy to implement by computing the distance between the test sample and all training samples, but it is computationally intensive for a big training set. The acceleration strategy for searching k-NNs can be found in [7]. Some theoretical results about k-NN have been presented in [8,9]. References [10,11] demonstrate two applications of the k-NN method to fall detection via wireless sensor network data and energy enhancements for smart mobile devices, respectively.

1.1.2 Decision tree

Decision tree is a supervised learning model based on a tree structure, which is used for both classification and regression tasks. It is one of the most popular models in supervised learning due to its effectiveness and strong interpretability. As shown in Figure 1.2, a decision tree consists of three parts: internal nodes, leaf nodes and branches. Among them, each internal node defines a set of if–then rules; each leaf node defines a category (or a target value for a regression task), and branches determine the topology structure of the tree. To predict the category (or target value) for a given test sample, we should find a path from the root node to a leaf node following the steps below. Starting from the root node, chose a branch according to that which rules the test sample meet in each internal node. Then go to the next node along the branch. Repeat the above two steps until arriving at a leaf node, then the category (target value) is given by this leaf node.

For a given training set $\{(\mathbf{x}_i, y_i)\}_{i=1}^n$, we say a decision tree affirms it if the tree outputs a correct prediction y_i for any $\mathbf{x}_i$ $(i = 1, \ldots, n)$. Given a training set, there may exist tremendous trees affirming it. However, only a few of them will achieve good performance on test samples (we call these trees effective trees), but we cannot afford

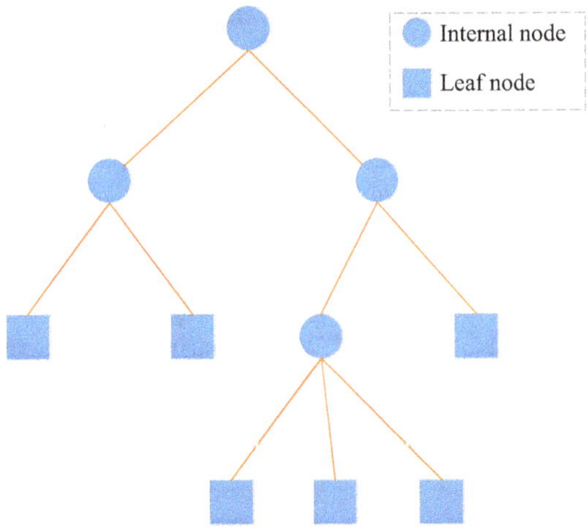

Figure 1.2 Decision tree structure

to enumerate all trees to find an effective tree. Thus, the key problem is how to construct an effective tree in a reasonable span of time. A variety of methods have been developed for learning an effective tree, such as ID3 [12], C4.5 [13] and classification and regression tree (CART) [14]. Most of them share a similar core idea that employs a top-down, greedy strategy to search through the space of possible decision trees. In this section, we will focus on the common-used CART method and its two improvements, random forest (RF) and gradient boosting decision tree (GBDT).

1.1.2.1 Classification and regression tree

CART [14] is a recursive partitioning method to build a classification or regression tree. Different from other methods, CART constraints the tree as a binary tree, which means there are only two branches in an internal node. In each internal node, a test sample will go down the left or right branch according to whether it meets the rule defined in the node or not.

Constructing CARTs share a similar process. The only difference between them is the partition criterion in each internal node. For a classification tree, the partition criterion is to minimize the Gini coefficient. Specifically, given a training set T of n samples and k categories, with n_i samples in the ith category, the Gini coefficient of T is defined as

$$\text{Gini}(T) = \sum_{i=1}^{k} \frac{n_i}{n}\left(1 - \frac{n_i}{n}\right) = 1 - \sum_{i=1}^{k}\left(\frac{n_i}{n}\right)^2, \tag{1.4}$$

where $n = \sum_{i=1}^{k} n_i$. In the root node, CART will find a partition rule to divide a training set T into two partitions, say T_1 and T_2, which minimizes the following function:

$$\frac{|T_1|}{|T|}\text{Gini}(|T_1|) + \frac{|T_2|}{|T|}\text{Gini}(T_2). \tag{1.5}$$

The similar steps will be carried out recursively for T_1 and T_2, respectively, until a stop condition meets.

In contrast, a regression tree is to predict continuous variables and its partition criteria is usually chosen as the minimum mean square error. Specifically, given a training set $T=\{(\mathbf{x}_i, y_i)\}_{i=1}^n$, a regression tree will divide T into T_1 and T_2 such that the following equation is minimized:

$$\sum_{(\mathbf{x}_i,y_i)\in T_1} (y_i - m_1)^2 + \sum_{(\mathbf{x}_j,y_j)\in T_2} (y_j - m_2)^2, \tag{1.6}$$

where $m_j = (1/|T_j|)\sum_{(\mathbf{x}_i,y_i)\in T_j} y_i$ $(j = 1, 2)$. For clarity, we summarize the constructing process and the predicting process in Algorithms 1.2 and 1.3, respectively.

By using Algorithm 1.2, we can construct a decision tree. However, this tree is so fine that it may cause *overfitting* (i.e. it achieves perfect performance on a training set but bad predictions for test samples). An extra pruning step can improve this situation. The pruning step consists of two main phases. First, iteratively prune the tree from the leaf nodes to the root node and thus acquire a tree sequence $Tree_0, Tree_1, \ldots, Tree_n$, where $Tree_0$ denotes the entire tree and $Tree_n$ denotes the tree which only contain

---

Algorithm 1.2: CART constructing tree

Input: training set $T=\{(\mathbf{x}_i, y_i) \in \mathbb{R}^{d+1}\}_{i=1}^n$, stop number s

if $|T| < s$ **then**
- *node.left* = *NULL*, *node.right* = *NULL*;
- **if** *the task is classification* **then**
  - set *node.y* as the category which is most common among T;
- **if** *the task is regression* **then**
  - $node.y = \frac{1}{|T|}\sum_{y_i\in T} y_i$
- **return** node;

$\hat{v} = \infty$;

for $j = 1$ **to** d **do**
- **if** *the task is classification* **then**
  - $\bar{v}, \bar{p} = \min_p \frac{|T_1(p)|}{|T|}\mathrm{Gini}(|T_1(p)|) + \frac{|T_2(p)|}{|T|}\mathrm{Gini}(T_2(p))$;
- **if** *the task is regression* **then**
  - $\bar{v}, \bar{p} = \min_p \sum_{(\mathbf{x}_i,y_i)\in T_1(p)} (y_i - m_1)^2 + \sum_{(\mathbf{x}_j,y_j)\in T_2(p)} (y_j - m_2)^2$;
- where $T_1(p) = \{(\mathbf{x}_i, y_i)|\mathbf{x}_i[j] \le p\}$, $T_2(p) = \{(\mathbf{x}_i, y_i)|\mathbf{x}_i[j] > p\}$;
- **if** $\bar{v} < \hat{v}$ **then**
  - $\hat{v} = \bar{v}$, $node.dim = j$, $node.p = \bar{p}$;

node.left = Constructing Tree(T_1(*node.p*));

node.right = Constructing Tree(T_2(*node.p*));

return *node*;

Algorithm 1.3: CART predicting

Input: test sample $\mathbf{x}_0$, root node *node*
Output: prediction y_0
if *node.left* = *NULL and node.right* = *NULL* **then**
  return *node.y*;
if $\mathbf{x}_0[node.dim] \le node.p$ **then**
  return Predicting($\mathbf{x}_0$, *node.left*);
else
  return Predicting($\mathbf{x}_0$, *node.right*);

the root node. Second, select the optimal tree from the sequence by using the cross validation. For more details, readers can refer to [14]. References [15–17] demonstrate three applications of CART in wireless communications.

1.1.2.2 Random forest

As discussed in Section 1.1.2.1, a tree constructed by the CART method has a risk of overfitting. To meet this challenge, Breiman proposed the RF model in [18]. As its name suggests, RF consists of many trees and introduces a random step in its constructing process to prevent overfitting.

Suppose we are given a training set T. To construct a tree $Tree_j$, RF first generates a training subset T_j by sampling from T uniformly and with replacement (T_j has the same size with T). Then, a construction algorithm will be carried out on T_j. The construction algorithm is similar to the CART method but introduces an extra random step. Specifically, in each internal node, CART chooses the optimal feature from all d features, but RF first randomly select l features from all d features and then chooses the optimal feature from the l features. The above construction process will be repeated m times and thus a forest (which contains m trees) is constructed. For a classification task, the output is determined by taking the majority vote in m trees. For a regression task, the output is the mean of m outputs. The whole process is summarized in Algorithm 1.4. In wireless communications, RF has been applied in many fields, such as indoor localization [19] and device-free fall detection [20].

1.1.2.3 Gradient boosting decision tree

GBDT [21] is a special case of the famous boosting method [22] based on a tree structure. Specifically, the model of GBDT is represented as a sum of some CART trees, i.e.:

$$f_m(\mathbf{x}) = \sum_{j=1}^{m} Tree_j(\mathbf{x}; \Theta_j), \tag{1.7}$$

Algorithm 1.4: Random forest

Input: training set $T = \{(\mathbf{x}_i, y_i) \in \mathbb{R}^{d+1}\}_{i=1}^{n}$, number of trees m, number of categories k, test sample $\mathbf{x}_0$
Output: prediction y_0

```
/* training */
for j = 1, ..., m do
    T_j ← Ø;
    for i = 1, ..., n do
        randomly select a train sample (x, y) from T;
        T_j ← T_j ∪ {(x, y)};
    based on T_j, construct a decision tree Tree_j using randomized CART;
/* testing */
if the task is classification then
    y_0 = arg max_{1≤c≤k} Σ_{j=1}^{m} I(Tree_j(x_0) = c);
if the task is regression then
    y_0 = (1/m) Σ_{j=1}^{m} Tree_j(x_0);
```

where $Tree_j(\mathbf{x}; \Theta_j)$ denotes the jth tree with parameter of Θ_j. Given a training set $\{(\mathbf{x}_1, y_1), \ldots, (\mathbf{x}_n, y_n)\}$, the goal of GBDT is to minimize:

$$\sum_{i=1}^{n} L(f_m(\mathbf{x}_i), y_i), \tag{1.8}$$

where $L(\cdot, \cdot)$ is a differentiable function which measures the difference between $f_m(\mathbf{x}_i)$ and y_i and is chosen according to the task.

However, it is often difficult to find an optimal solution to minimize (1.8). As a trade-off, GBDT uses a greedy strategy to yield an approximate solution. First, notice that (1.7) can be written as a recursive form:

$$f_j(\mathbf{x}) = f_{j-1}(\mathbf{x}) + Tree_j(\mathbf{x}; \Theta_j) \quad (j = 1, \ldots, m), \tag{1.9}$$

where we have defined $f_0(\mathbf{x}) = 0$. Then, by fixing the parameters of f_{j-1}, GBDT finds the parameter set Θ_j by solving:

$$\min_{\Theta_j} \sum_{i=1}^{n} L(f_{j-1}(\mathbf{x}_i) + Tree_j(\mathbf{x}_i; \Theta_j), y_i). \tag{1.10}$$

Replacing the loss function $L(u, v)$ by its first-order Taylor series approximation with respect to u at $u = f_{j-1}(\mathbf{x}_i)$, we have

$$\sum_{i=1}^{n} L(f_{j-1}(\mathbf{x}_i) + Tree_j(\mathbf{x}_i; \Theta_j), y_i) \approx \sum_{i=1}^{n} \left[L(f_{j-1}(\mathbf{x}_i), y_i) + \frac{\partial L(f_{j-1}(\mathbf{x}_i), y_i)}{\partial f_{j-1}(\mathbf{x}_i)} Tree(\mathbf{x}_i; \Theta_j) \right]. \tag{1.11}$$

Notice that the right side is a linear function with respect to $Tree(\mathbf{x}_i; \Theta_j)$ and its value can decrease by letting $Tree(\mathbf{x}_i; \Theta_j) = -(\partial L(f_{j-1}(\mathbf{x}_i), y_i)/\partial f_{j-1}(\mathbf{x}_i))$. Thus, GBDT trains $Tree(\cdot, \Theta_j)$ by using a new training set $\left\{(\mathbf{x}_i, -(\partial L(f_{j-1}(\mathbf{x}_i), y_i)/(\partial f_{j-1}(\mathbf{x}_i))))\right\}_{i=1}^{n}$. The above steps will be repeated for $j = 1, \ldots, m$ and thus a gradient boosting tree is generated.

GBDT is known as one of the best methods in supervised learning and has been widely applied in many tasks. There are many tricks in its implementation. Two popular implementations, XGboost and LightGBM, can be found in [23] and [24], respectively. References [25] and [26] demonstrate two applications of GBDT in obstacle detection and quality of experience (QoE) prediction, respectively.

1.1.3 Perceptron

A perceptron is a linear model for a binary-classification task and is the foundation of the famous support vector machine (SVM) and deep neural networks (DNNs). Intuitively, it tries to find a hyperplane to separate the input space (feature space) into two half-spaces such that the samples of different classes lie in the different half-spaces. An illustration is shown in Figure 1.3(a). A hyperplane in $\mathbb{R}^d$ can be described by an equation $\mathbf{w}^T\mathbf{x} + b = 0$, where $\mathbf{w} \in \mathbb{R}^D$ is the normal vector. Correspondingly,

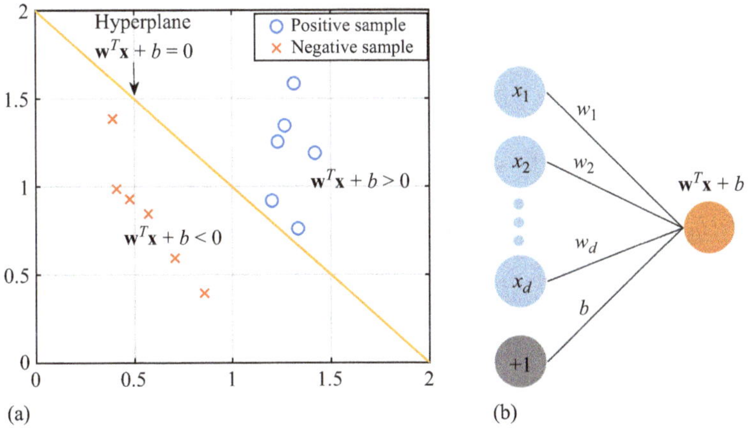

Figure 1.3 (a) An illustration of a perceptron and (b) the graph representation of a perceptron

Algorithm 1.5: Perceptron learning

Input: training set $\{(\mathbf{x}_1, y_1), \ldots, (\mathbf{x}_n, y_n)\}$, learning rate $\eta \in (0, 1]$
Output: parameters of perceptron $\mathbf{w}, b$

randomly initialize $\mathbf{w}, b$;
flag = *True*;
while *flag* **do**
    flag = *False*;
    for $i = 1, \ldots, n$ **do**
        if $y_i(\mathbf{w}^T\mathbf{x}_i + b) < 0$ **then**
            $\mathbf{w} = \mathbf{w} + \eta y_i \mathbf{x}_i$;
            $b = b + \eta y_i$;
            flag = *True*;

$\mathbf{w}^T\mathbf{x} + b > 0$ and $\mathbf{w}^T\mathbf{x} + b < 0$ represent the two half-spaces separated by the hyperplane $\mathbf{w}^T\mathbf{x} + b = 0$. For a sample $\mathbf{x}_0$, if $\mathbf{w}^T\mathbf{x}_0 + b$ is larger than 0, we say $\mathbf{x}_0$ is in the positive direction of the hyperplane, and if $\mathbf{w}^T\mathbf{x}_0 + b$ is less than 0, we say it is in the negative direction.

In addition, by writing $\mathbf{w}^T\mathbf{x} + b = [\mathbf{x}^T, 1] \cdot [\mathbf{w}^T, b]^T = \sum_{i=1}^{d} x_i w_i + b$, we can view $[\mathbf{x}^T, 1]^T$, $[\mathbf{w}^T, b]^T$ and $\mathbf{w}^T\mathbf{x} + b$ as the inputs, parameters and output of a perceptron, respectively. Their relation can be described by a graph, where the inputs and output are represented by nodes, and the parameters are represented by edges, as shown in Figure 1.3(b). This graph representation is convenient for describing the multilayer perceptron (neural networks) which will be introduced in Section 1.1.3.3.

Suppose we have a training set $T = \{(\mathbf{x}_1, y_1), \ldots, (\mathbf{x}_n, y_n)\}$, where $\mathbf{x}_i \in \mathbb{R}^d$ and $y_i \in \{+1, -1\}$ is the ground truth. The perceptron algorithm can be formulated as

$$\min_{\mathbf{w}\in\mathbb{R}^d, b\in\mathbb{R}} L(\mathbf{w}, b) \triangleq -\sum_{i=1}^{N} y_i(\mathbf{w}^T\mathbf{x}_i + b), \tag{1.12}$$

where $\mathbf{w}^T\mathbf{x} + b = 0$ is the classification hyperplane, and $y_i(\mathbf{w}^T\mathbf{x}_i + b) > 0$ implies that the ith sample lie in the correct half-space. Generally, the stochastic gradient descent algorithm is used to obtain a solution to (1.12) and its convergence has been shown in [27]. The learning process is summarized in Algorithm 1.5.

1.1.3.1 Support vector machine

SVM is a binary-classification model. SVM shares a similar idea with the perceptron model, i.e. find a hyperplane to separate two classes of training samples. In general, there may be several hyperplanes meeting the requirement. A perceptron finds any one of them as the classification hyperplane. In contrast, SVM will seek the one that maximizes the classification margin, which is defined as the distance from the hyperplane to the nearest training sample. As shown in Figure 1.4(a), three hyperplanes which can separate the two classes of training samples are drawn in three different

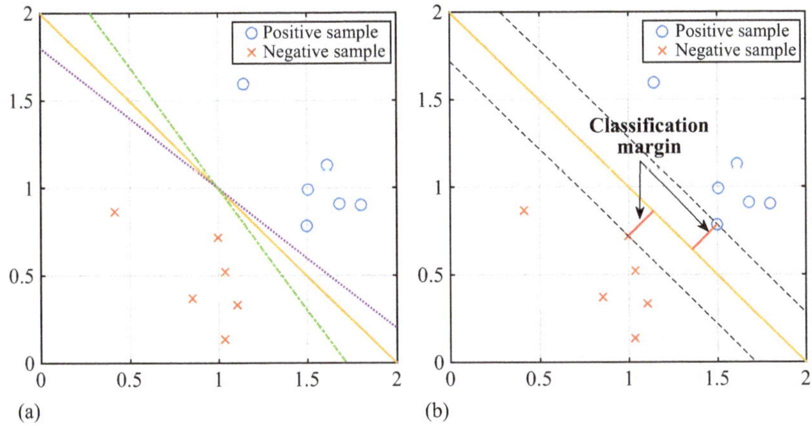

Figure 1.4 *(a) Three hyperplanes which can separate the two classes of training samples and (b) the hyperplane which maximizes the classification margin*

styles, and each of them can serve as a solution to the perceptron. However, as shown in Figure 1.4(b), only the hyperplane which maximizes the classification margin can serve as the solution to SVM.

According to the definition of the classification margin, distance from any training sample to the classification hyperplane should be not less than it. Thus, given the training set $\{(\mathbf{x}_i, y_i)\}_{i=1}^n$, the learning of SVM can be formulated as the following optimization problem:

$$\begin{aligned} &\max_{\mathbf{w}\in\mathbb{R}^d, b\in\mathbb{R}} \quad \gamma \\ &\quad \text{s.t.} \quad y_i\left(\frac{\mathbf{w}^T\mathbf{x}_i}{\|w\|} + \frac{b}{\|w\|}\right) \geq \gamma \quad (i = 1, \ldots, n) \end{aligned} \tag{1.13}$$

where $\left((\mathbf{w}^T\mathbf{x}_i/\|w\|) + (b/\|w\|)\right)$ can be viewed as the signed distance from $\mathbf{x}_i$ to the classification hyperplane $\mathbf{w}^T\mathbf{x} + b = 0$, and the sign of $y_i\left((\mathbf{w}^T\mathbf{x}_i/\|w\|) + (b/\|w\|)\right)$ denotes whether $\mathbf{x}_i$ lies in the right half-space. It can be shown that problem (1.13) is equivalent to

$$\begin{aligned} &\min_{\mathbf{w}\in\mathbb{R}^d, b\in\mathbb{R}} \quad \frac{1}{2}\|w\|^2 \\ &\quad \text{s.t.} \quad y_i\left(\mathbf{w}^T\mathbf{x}_i + b\right) - 1 \geq 0 \quad (i = 1, \ldots, n). \end{aligned} \tag{1.14}$$

Problem (1.14) is a quadratic programming problem [28] and can be efficiently solved by several optimization tools [29,30].

Note that both the perceptron and SVM suppose that the training set can be separated linearly. However, this supposition is not always correct. Correspondingly, the soft-margin hyperplane and the kernel trick have been introduced to deal with the non-linear situation. Please refer to [31] and [32] for more details. In addition, SVM can also be used to handle regression tasks, which is also known as support vector regression [33]. SVM has been widely applied in many fields of wireless communications, such as superimposed transmission mode identification [34], selective forwarding attacks detection [35], localization [36] and MIMO channel learning [3].

1.1.3.2 Logistic regression

First of all, we should clarify that logistic regression is a classification model and its name is borrowed from the field of statistics. Logistic regression can be binomial or multinomial depending on whether the classification task is binary or multi-class. We will introduce the binomial case followed by the multinomial case.

Recall that in a perceptron, a hyperplane $\mathbf{w}^T\mathbf{x} + b = 0$ is learned to discriminate the two classes. A test sample $\mathbf{x}_0$ will be classified as positive or negative according to whether $\mathbf{w}^T\mathbf{x}_0 + b$ is larger than 0 or $\mathbf{w}^T\mathbf{x}_0 + b$ is less than 0. Sometimes, however, we want to know the probability of a test sample belonging to a class. In the example shown in Figure 1.5(a), though both $\mathbf{x}_1$ and $\mathbf{x}_2$ are classified as the positive class, we are more confident that $\mathbf{x}_2$ is the positive sample since $\mathbf{x}_2$ is farther from the decision hyperplane than $\mathbf{x}_1$ (i.e. $\mathbf{w}^T\mathbf{x}_2 + b$ is larger than $\mathbf{w}^T\mathbf{x}_1 + b$). The sample $\mathbf{x}_3$ is classified as the negative sample since $\mathbf{w}^T\mathbf{x}_3 + b$ is less than 0.

From the above example, we can infer that the sign of $\mathbf{w}^T\mathbf{x} + b$ decides the class of $\mathbf{x}$, and $|\mathbf{w}^T\mathbf{x} + b|$ gives the confidence level of the decision. However, the value of $\mathbf{w}^T\mathbf{x} + b$ can be any one of $(-\infty, \infty)$ but we want a probability value. A nature idea is that find a monotone increasing function $g : (-\infty, \infty) \mapsto (0, 1)$ such that

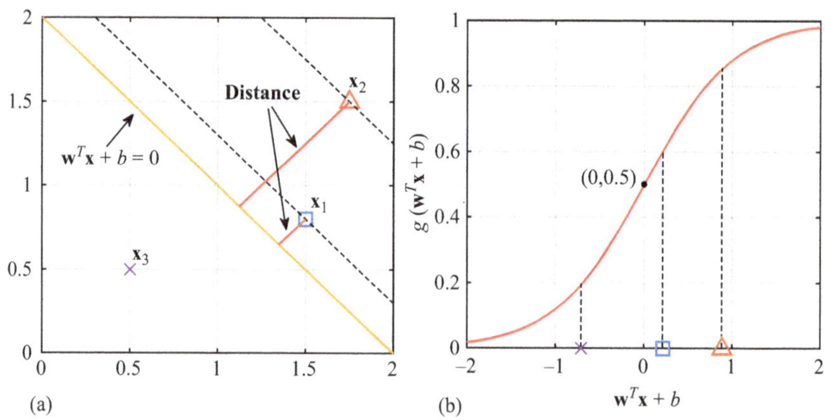

Figure 1.5 (a) A decision hyperplane $\mathbf{w}^T\mathbf{x} + b = 0$ and three samples: $\mathbf{x}_1$ (blue square), $\mathbf{x}_2$ (red triangle) and $\mathbf{x}_3$ (purple cross). (b) The graph of function $g(t) \triangleq (\exp(t)/(1 + \exp(t)))$ and the probability of $\mathbf{x}_1$, $\mathbf{x}_2$ and $\mathbf{x}_3$ belong to the positive class

$g(\mathbf{w}^T\mathbf{x}+b)$ represents the probability of $\mathbf{x}$ belonging to the positive class. Thus, by choosing $g(t) \triangleq (\exp(t)/(1+\exp(t)))$, we get the model of logistic regression:

$$P\{y=1|\mathbf{x}\} = \frac{\exp\left(\mathbf{w}^T\mathbf{x}+b\right)}{1+\exp\left(\mathbf{w}^T\mathbf{x}+b\right)}. \tag{1.15}$$

$P\{y=1|\mathbf{x}\}$ denotes the probability of $\mathbf{x}$ belonging to the positive class, and correspondingly $1-P\{y=1|\mathbf{x}\}$ is the probability of $\mathbf{x}$ belonging to the negative class. As shown in Figure 1.5(b), by the above transformation, the probability of $\mathbf{x}_1$, $\mathbf{x}_2$ and $\mathbf{x}_3$ belonging to the positive class are around 0.6, 0.85 and 0.2, respectively.

Given a training set $\{(\mathbf{x}_i, y_i)\}_{i=1}^n$, the parameters $\mathbf{w}$ and b of the binomial logistic regression model can be estimated by the maximum likelihood estimation. Specifically, its log-likelihood is written as

$$\begin{aligned} L(\mathbf{w},b) &= \log\left(\prod_{y_i=1}\frac{\exp\left(\mathbf{w}^T\mathbf{x}_i+b\right)}{1+\exp\left(\mathbf{w}^T\mathbf{x}_i+b\right)}\cdot\prod_{y_i=-1}\left(1-\frac{\exp\left(\mathbf{w}^T\mathbf{x}+b\right)}{1+\exp\left(\mathbf{w}^T\mathbf{x}+b\right)}\right)\right) \\ &= \sum_{y_i=1}\left(\mathbf{w}^T\mathbf{x}_i+b\right)-\sum_{i=1}^{n}\log\left(1+\exp\left(\mathbf{w}^T\mathbf{x}_i+b\right)\right). \end{aligned} \tag{1.16}$$

Because there is no closed-form solution to the problem of maximizing (1.16), the gradient descent method and the quasi-Newton method are generally used to obtain a numerical solution.

Following the binomial logistic regression, we can deduce the multinomial case. For a k-classification task, the multinomial logistic regression model are given by

$$P(y=j|\mathbf{x}) = \frac{\exp\left(\mathbf{w}_j^T\mathbf{x}+b_j\right)}{\sum_{i=1}^{k}\exp\left(\mathbf{w}_i^T\mathbf{x}+b_i\right)} \quad (j=1,\ldots,k), \tag{1.17}$$

where $P(y=j|\mathbf{x})$ denotes the probability of $\mathbf{x}$ belonging to the jth class. The parameter set $\{(\mathbf{w}_i, b_i)\}_{j=1}^k$ can also be estimated by using the maximum likelihood estimation. Another name of multinomial logistic regression is softmax regression, which is often used as the last layer of a multilayer perceptron that will be introduced in the next section. Logistic regression has been applied to predict device wireless data and location interface configurations that can optimize energy consumption in mobile devices [11]. References [37], [38] and [39] demonstrate three applications of logistic regression to home wireless security, reliability evaluation and patient anomaly detection in medical wireless sensor networks, respectively.

1.1.3.3 Multilayer perceptron and deep learning

A multilayer perceptron is also known as a multilayer neural network, which is also called a DNN when the number of layers is large enough. As shown in Figure 1.6(a), a multilayer perceptron is a perceptron with three or more layers rather than two layers in the original perceptron. The leftmost and rightmost layers are called an input layer and an output layer, respectively. Notice that the output layer can have more than one

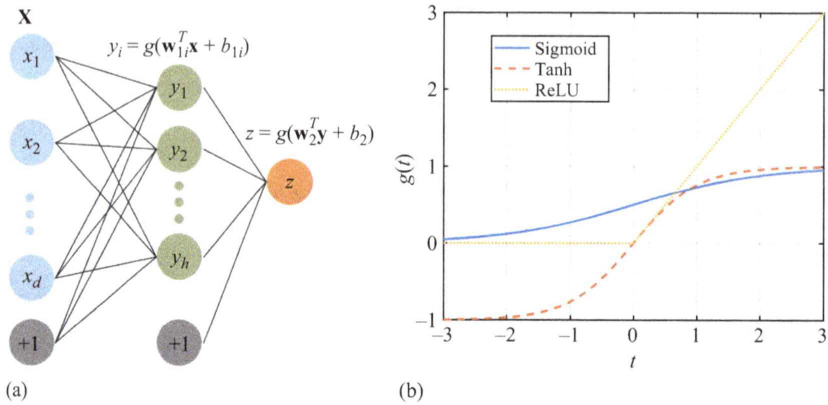

Figure 1.6 (a) A simple neural network with three layers, where g(·) is a non-linear activation function and (b) the curves of three commonly used activation functions

node though we only use one for simplicity in this example. The middle layer is called a hidden layer, because its nodes are not observed in the training process. Similar to a Markov chain, the node values of each layer are computed only depending on the node values of its previous layer.

Because the original perceptron is just a linear function that maps the weighted inputs to the output of each layer, the linear algebra shows that any number of layers can be reduced to a two-layer input–output model. Thus, a non-linear activation function $g : \mathbb{R} \mapsto \mathbb{R}$, which is usually monotonously increasing and differentiable almost everywhere [40], is introduced to achieve a non-linear mapping. Here, we list some commonly used activation functions in Table 1.1, and their curves are shown in Figure 1.6(b). Notice that the sigmoid function is just a modification of the probability mapping (1.15) used in logistic regression. As shown in Figure 1.6(b), the hyperbolic tangent (tanh) function shares similar curve trace with the sigmoid function except the output range being $(-1,1)$ instead of $(0, 1)$. Their good mathematical properties make them popular in early research [41]. However, they encounter difficulties in DNNs. It is easy to verify that $\lim_{t\to\infty} g'(t) = 0$ and $|g'(t)|$ is a small value in most areas of the domain for both of them. This property restricts their use in DNNs since training DNNs require that the gradient of the activation function is around 1. To meet this challenge, a rectified linear unit (ReLU) activation function is proposed. As shown in Table 1.1, the ReLU function is piece-wise linear function and saturates at exactly 0 whenever the input t is less than 0. Though it is simple enough, the ReLU function has achieved great success and became the default choice in DNNs [40].

Now, we will have a brief discussion about the training process of the multi-layer perceptron. For simplicity, let us consider the performance of a regression task

Table 1.1 Three commonly used activation functions

Name	Abbreviation	Formula	Range
Sigmoid	Sigmoid	$g(t) = 1/(1 + e^{-t})$	$(0, 1)$
Hyperbolic tangent	Tanh	$g(t) = ((e^t - e^{-t})/(e^t + e^{-t}))$	$(-1, 1)$
Rectified linear unit	ReLU	$g(t) = \max(0, t)$	$[0, +\infty)$

by using the model shown in Figure 1.6(a). For convenience, denote the parameters between the input and hidden layers as a matrix $\mathbf{W}_1 = (\hat{\mathbf{w}}_{11}, \ldots, \hat{\mathbf{w}}_{1h}, \hat{\mathbf{w}}_{1h+1}) \in \mathbb{R}^{(d+1)\times(h+1)}$, where $\hat{\mathbf{w}}_{1i} = (\mathbf{w}_{1i}^T, b_{1i})^T \in \mathbb{R}^{d+1}$ and $\hat{\mathbf{w}}_{1h+1} = (0, 0, \ldots, 1)^T$. Similarly, the parameters between the hidden and output layers are denoted as a vector $\mathbf{w}_2 = (w_{21}, \ldots, w_{2h}, w_{2h+1})^T \in \mathbb{R}^{h+1}$, where $w_{2h+1} = b_2$. Let $\mathbf{x} = (x_1, \ldots, x_d, 1)^T \in \mathbb{R}^{d+1}$, $\mathbf{y} = (y_1, \ldots, y_h, 1)^T \in \mathbb{R}^{h+1}$ and z denote the input vector, the hidden vector, and the output scalar, respectively. Then the relations among $\mathbf{x} = (x_1, \ldots, x_d, 1)^T \in \mathbb{R}^{d+1}$, $\mathbf{y} = (y_1, \ldots, y_h, 1)^T \in \mathbb{R}^{h+1}$ and z can be presented as

$$
\begin{aligned}
\mathbf{y} &= g(\mathbf{W}_1^T \mathbf{x}), \\
z &= g(\mathbf{w}_2^T \mathbf{y}),
\end{aligned}
\tag{1.18}
$$

where the activation function g will act on each element for a vector as input. Suppose we expect that the model outputs $\bar{z}$ for the input $\mathbf{x}$, and thus the square error is given by $e = (1/2)(z - \bar{z})^2$. We decrease this error by using the gradient descent method. This means that $\partial e/\partial \mathbf{W}_1$ and $\partial e/\partial \mathbf{w}_2$ need to be computed. By the gradient chain rule, we have

$$
\begin{aligned}
\frac{\partial e}{\partial z} &= (z - \bar{z}), \\
\frac{\partial e}{\partial \mathbf{w}_2} &= \frac{\partial e}{\partial z}\frac{\partial z}{\partial \mathbf{w}_2} = (z - \bar{z})g'(\mathbf{w}_2^T \mathbf{y} + b_2)\mathbf{y},
\end{aligned}
\tag{1.19}
$$

and

$$
\frac{\partial e}{\partial \mathbf{W}_1} = \frac{\partial e}{\partial z}\frac{\partial z}{\partial \mathbf{y}}\frac{\partial \mathbf{y}}{\partial \mathbf{W}_1},
\tag{1.20}
$$

where we have omitted the dimensions for simplicity. Thus, to compute $\partial e/\partial \mathbf{W}_1$, we first need to compute:

$$
\begin{aligned}
\frac{\partial z}{\partial \mathbf{y}} &= g'(\mathbf{w}_2^T \mathbf{y})\mathbf{w}_2, \\
\frac{\partial \mathbf{y}}{\partial \mathbf{W}_1} &= \left\{\frac{\partial y_i}{\partial \mathbf{W}_1}\right\}_{i=1}^{h+1} = \left\{g'(\mathbf{w}_{1i}^T \mathbf{x}) \cdot \mathbf{x}\mathbf{e}_i^T\right\}_{i=1}^{h+1},
\end{aligned}
\tag{1.21}
$$

where $\mathbf{e}_i \in \mathbb{R}^{h+1}$ denotes the unit vector with its ith element being 1. By plugging (1.21) into (1.20), we have

$$\begin{aligned}\frac{\partial e}{\partial \mathbf{W}_1} = \frac{\partial e}{\partial z}\frac{\partial z}{\partial \mathbf{y}}\frac{\partial \mathbf{y}}{\partial \mathbf{W}_1} &= (z-\bar{z})g'(\mathbf{w}_2^T\mathbf{y}+b_2)\mathbf{w}_2 \odot \left\{g'(\mathbf{w}_{1i}^T\mathbf{x})\cdot \mathbf{x}\mathbf{e}_i^T\right\}_{i=1}^{h+1} \\ &= (z-\bar{z})g'(\mathbf{w}_2^T\mathbf{y}+b_2)\sum_{i=1}^{h+1} w_{2i}g'(\mathbf{w}_{1i}^T\mathbf{x})\cdot \mathbf{x}\mathbf{e}_i^T\end{aligned} \tag{1.22}$$

Thus, we can update the parameters by using the gradient descent method to reduce the error. In the above deduction, what we really need to calculate are just $\partial e/\partial z$, $\partial z/\partial \mathbf{w}_2$, $\partial z/\partial \mathbf{y}$ and $\partial \mathbf{y}/\partial \mathbf{W}_1$. As shown in Figure 1.7(a), we find these terms are nothing but the derivatives of the node values or the parameters of each layer with respect to the node values of the next layer. Beginning from the output layer, ‘multiply’ them layer by layer according to the chain rule, and then we obtain the derivatives of the square error with respect to the parameters of each layer. The above strategy is the so-called backpropagation (BP) algorithm [42]. Equipped with the ReLU activation function, the BP algorithm can train the neural networks with dozens or even hundreds of layers, which constitutes the foundation of deep learning.

In the model shown in Figure 1.6(a), we can observe that every node of the input layer is connected to every node of the hidden layer. This connection structure is called fully connection, and the layer which is fully connected (FC) to the previous layer is called the *FC layer* [42]. Supposing that the number of nodes of two layers are m and n, then the number of parameters of fully connection will be $m \times n$, which will be a large number even when m and n are moderate. Excessive parameters will slow down the training process and increase the risk of overfitting, which is especially serious in DNNs. Parameter sharing is an effective technology to meet this challenge. A representative example using the parameter sharing technology is convolution neural networks (CNNs), which are a specialized kind of neural networks for processing

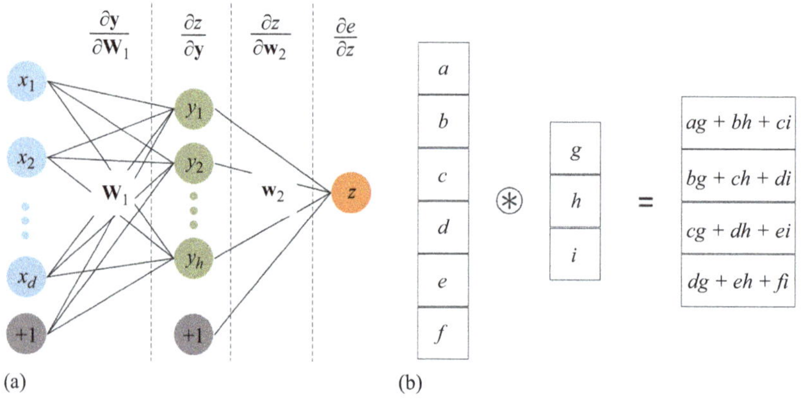

Figure 1.7 (a) The derivatives of each layer with respect to its previous layer and (b) an example of the convolution operation performed on vectors

data that has a known grid-like topology [43], such as time-series data and matrix data. The name of CNNs comes from its basic operation called convolution (which has a little difference from the convolution in mathematics). Though convolution operation can be performed on vectors, matrices and even tensors with arbitrary order, we will introduce the vector case here for simplicity.

To perform the convolution operation on a vector $\mathbf{x} \in \mathbb{R}^d$, we first need a kernel, which is also a vector $\mathbf{k} \in \mathbb{R}^l$ with $l \leq d$. Let $\mathbf{x}[i:j]$ denotes the vector generated by extracting the elements from the ith position to the jth position of $\mathbf{x}$, i.e. $\mathbf{x}[i:j] \triangleq (x_i, x_{i+1}, \ldots, x_j)^T$. Then the convolution of $\mathbf{x}$ and $\mathbf{k}$ is defined as

$$\mathbf{x} \circledast \mathbf{k} \triangleq \begin{pmatrix} \langle \mathbf{x}[1:l], \mathbf{k} \rangle \\ \langle \mathbf{x}[2:l+1], \mathbf{k} \rangle \\ \vdots \\ \langle \mathbf{x}[d-l+1:d], \mathbf{k} \rangle \end{pmatrix} = \hat{\mathbf{y}} \in \mathbb{R}^{d-l+1}, \tag{1.23}$$

where $\langle \cdot, \cdot \rangle$ denotes the inner product of two vectors. See Figure 1.7(b) for an example of convolution. The convolution operation for matrices and tensors can be defined similarly by using a matrix kernel and a tensor kernel, respectively. Based on the convolution operation, a new transformation structure, as distinct from the fully connection, can be built as

$$\mathbf{y} = g\,(\mathbf{x} \circledast \mathbf{k}), \tag{1.24}$$

where $\mathbf{k}$ is the parameter that needs to be trained. The layer with this kind of transformation structure is called the convolution layer. Compared with the FC layer, the number of parameters has dramatically decreased for the convolution layer. Furthermore, the size of the kernel is independent to the number of nodes of the previous layer.

It should be noted that we can set several kernels in a convolution layer to generate richer features. For example, if we choose the convolution layer with M kernels as the hidden layer in example shown in Figure 1.6(a), then m features will be generated as

$$g\,(\mathbf{x} \circledast (\mathbf{k}_1, \ldots, \mathbf{k}_m)) = (g\,(\mathbf{x} \circledast \mathbf{k}_1), \ldots, g\,(\mathbf{x} \circledast \mathbf{k}_m)) = (\mathbf{y}_1, \ldots, \mathbf{y}_m). \tag{1.25}$$

Similarly, we can continue to transform $(\mathbf{y}_1, \ldots, \mathbf{y}_m)$ by using more convolution layers.

In addition to the convolution operation, another operation widely used in CNNs is the max-pooling operation. As a kernel is needed in the convolution operation, the max-pooling operation needs a window to determine the scope of the operation. Specifically, given a vector $\mathbf{x} \in \mathbb{R}^d$ and a window of size l with d being divisible by l, the max-pooling operation to $\mathbf{x}$ with the window is defined as

$$\text{max-pooling}(\mathbf{x}, l) \triangleq \begin{pmatrix} \max\{\mathbf{x}[1:l]\} \\ \max\{\mathbf{x}[l+1:2l]\} \\ \vdots \\ \max\{\mathbf{x}[d-l+1:d]\} \end{pmatrix} = \hat{\mathbf{y}} \in \mathbb{R}^{\frac{d}{l}}. \tag{1.26}$$

See Figure 1.8(a) for an example of max-pooling. The max-pooling operation for matrices and tensors can be defined similarly by using windows with different dimensions. Other layers such as normalization layers and average-pooling layers [44], we do not detail here to stay focused.

Generally, a neural network may be constructed by using several kinds of layers. For example, in a classical architecture, the first few layers are usually composed alternate of the convolution layer and the max-pooling layer, and the FC layer is often used as the last few layers. A simple example of architecture for classification with a convolution network is shown in Figure 1.8(b).

The recent 10 years have witnessed earthshaking development of deep learning. The state-of-the-art of many applications has been dramatically improved due to its development. In particular, CNNs have brought about breakthroughs in processing multidimensional data such as image and video. In addition, recurrent neural networks [42] have shone light on sequential data such as text and speech; generative anniversary networks [45] are known as a class of models which can learn a mimic distribution from the true data distribution to generate high-quality artificial samples, such as images and speeches; deep RL (DRL) [46] is a kind of tool to solve control and decision-making problems with high-dimensional inputs, such as board game, robot navigation and smart transportation. Reference [42] is an excellent introduction to deep learning. More details about the theory and the implementation of deep learning can be found in [43]. For a historical survey of deep learning, readers can refer to [47]. Many open-source deep-learning frameworks, such as TensorFlow and Caffe, make neural networks easy to implement. Readers can find abundant user-friendly tutorials from the Internet. Deep learning has been widely applied in many fields of wireless communications, such as network prediction [48,49], traffic classification [50,51], modulation recognition [52,53], localization [54,55] and anomaly detection [56–58]. Readers can refer to [59] for a comprehensive survey of deep learning in mobile and wireless networks.

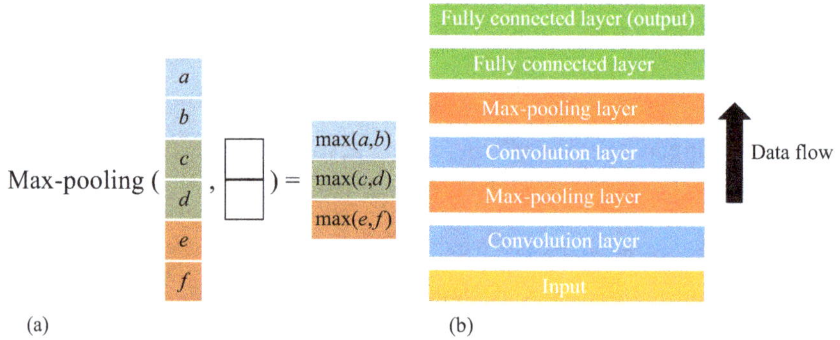

Figure 1.8 (a) An example of max-pooling to $\mathbf{x} \in \mathbb{R}^6$ *with a window of size 2 and (b) an simple example of architecture for classification with convolutional network*

1.1.4 Summary of supervised learning

In this section, we have discussed supervised learning. The main task of supervised learning is to learn a function that maps an input to an output based on a training set, which consists of examples of input–output pairs. According to whether the predicted variable is categorical or continuous, a supervised learning task is called a classification or regression task.

As shown in Figure 1.9, three popular technologies and their improvements have been introduced in this section. Among them, the k-NNs method has the simplest form. The k-NN method does not need explicit training steps and is very easy to implement. In most cases, it can give a not-bad result. However, if your target is high accuracy, then the latter two technologies will be better choices. Decision tree is a kind of supervised learning algorithms based on tree structures. CART is known as an effective method to construct a single tree. To acquire a better performance, RF constructs many trees by using a randomized CART, and then a final prediction will be given by integrating predictions of all trees. GBDT is a boosting method based on CART, and it is known as one of the best methods in supervised learning. A perceptron is a linear classification model as a foundation of SVM, logistic regression and multilayer perceptron. SVM improves the performance of the perceptron by maximizing the classification margin. Logistic regression is more robust to outliers than SVM and can give the probability of a sample belongs to a category. Multilayer perceptron is a perceptron with multilayers, which is also known as deep learning when the number of layers is large enough. Deep learning, as a heavy tool in machine learning, can have millions of parameters and

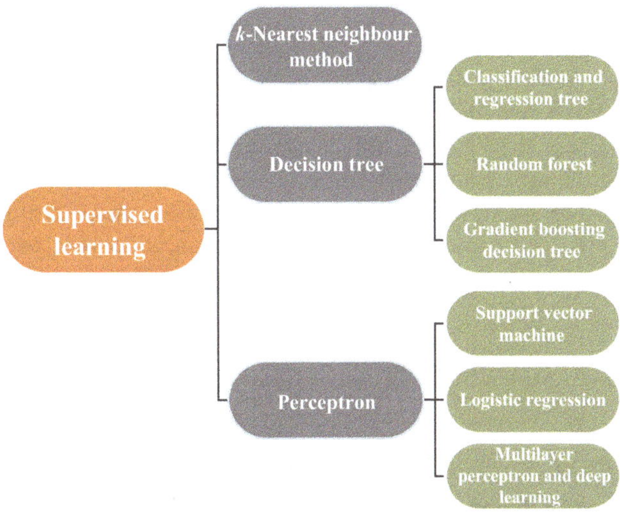

Figure 1.9 Structure chart for supervised learning technologies discussed in this chapter

Table 1.2 Summary of applications of supervised learning in wireless communications

Method	Function	Application in wireless communications
k-Nearest neighbours method	Classification/Regression	Fall detection [10] Energy enhancements [11]
Classification and regression tree	Classification/Regression	Improving congestion control [15] Monitoring animal behaviour [16] Intrusion detection [17]
Gradient boosting decision tree	Classification/Regression	Indoor localization [19] Fall detection [20]
Random forest	Classification/Regression	Obstacle detection [25] QoE prediction [26]
Support vector machine	Classification/Regression	Transmission mode identification [34] Attack detection [35] Localization [36] MIMO channel learning [3]
Logistic regression	Classification	Home wireless security [37] Reliability evaluation [36] Patient anomaly detection [39] Energy enhancements [11]
Multilayer perceptron and deep learning	Classification/Regression	Network prediction [48,49] Traffic classification [50,51] Modulation recognition [52,53] Localization [54,55] Anomaly detection [56–58]

cost huge amount of computing resources for training. As a reward, it is state-of-the-art method for most machine-learning tasks. Furthermore, many technologies from hardware to software have been applied to accelerate its training stage. In Table 1.2, we summarize the applications of supervised learning in wireless communications. Zhang *et al.* [59] conduct a comprehensive survey regarding deep learning in wireless communications. Readers can refer to [8] for more advanced approaches of supervised learning.

1.2 Unsupervised learning

Unsupervised learning is a process of discovering and exploring for investigating inherent and hidden structures from data without labels [60]. Unlike supervised learning where a training set $\{(\mathbf{x}_i, y_i)\}_{i=1}^n$ is provided, we have to work with an unlabelled data set $\{\mathbf{x}_i\}_{i=1}^n$ (there is no y_i) in unsupervised learning. Three common unsupervised learning tasks are clustering, density estimation and dimension reduction. The goal of clustering is to divide samples into groups (called *clusters*) such that the samples in the same cluster are more similar to each other than to those in different clusters.

Rather than defining classes before observing the test data, clustering allows us to find and analyse the undiscovered groups hidden in data. From Sections 1.2.1–1.2.4, we will discuss four representative clustering algorithms. Density estimation aims to estimate the distribution density of data in the feature space, and thus we can find the high-density regions which usually reveal some important characteristics of the data. In Section 1.2.5, we will introduce a popular density-estimation method: the Gaussian mixture model (GMM). Dimension reduction pursues to transform the data in a high-dimensional space to a low-dimensional space, and the low-dimensional representation should reserve principal structures of the data. In Sections 1.2.6 and 1.2.7, we will discuss two practical dimension-reduction technologies: principal component analysis (PCA) and autoencoder.

1.2.1 k-Means

k-Means [61] is one of the simplest unsupervised learning algorithms which solve a clustering problem. This method only needs one input parameter k, which is the number of clusters we expect to output. The main idea of k-means is to find k optimal points (in the feature space) as the representatives of k clusters according to an evaluation function, and each point in the data set will be assigned to a cluster based on the distance between the point to each representative. Given a data set $X = \{\mathbf{x}_i \in \mathbb{R}^d\}_{i=1}^n$, let X_i and $\mathbf{r}_i$ denote the ith cluster and the corresponding representative, respectively. Then, k-means aims to find the solution of the following problem:

$$\begin{aligned} \min_{\mathbf{r}_i, X_i} \quad & \sum_{i=1}^{k} \sum_{\mathbf{x} \in X_i} \|\mathbf{x} - \mathbf{r}_i\|^2 \\ \text{s.t.} \quad & \bigcup_{i=1}^{k} X_i = X \\ & X_i \bigcap X_j = \emptyset \quad (i \neq j). \end{aligned} \tag{1.27}$$

Notice that $\sum_{\mathbf{x} \in X_i} \|\mathbf{x} - \mathbf{r}_i\|^2$ measures how dissimilar the points in ith cluster to the corresponding representative, and thus the object is to minimize the sum of these dissimilarities.

However, the above problem has been shown to be an NP-hard problem [62], which means the global optimal solution cannot be found efficiently in general cases. As an alternative, k-means provides an iterative process to obtain an approximate solution. Initially, it randomly selects k points as initial representative. Then it alternately conducts two steps as follows. First, partitions the all points into k clusters in which each point is assigned to the cluster with the nearest representative. Second, take the mean points of each cluster as the k new representatives, which reveals the origin of the name of k-means. The above steps will be repeated until the clusters remain stable. The whole process is summarized in Algorithm 1.6.

Let us checkout the correctness of k-means step by step. In the first step, when fixing k representatives, each point is assigned to the nearest representative and thus the object value of (1.27) will decrease or remain unchanged. In the second step,

Algorithm 1.6: K-means algorithm

Input: dataset $\left\{\mathbf{x}_i \in \mathbb{R}^d\right\}_{i=1}^n$, number of clusters k
Output: clusters $X_1, \ldots, X_k$
randomly select k points $\mathbf{r}_1, \ldots, \mathbf{r}_k$ as representatives;
repeat
    `/* the first step */`
    for $i = 1$ **to** k **do**
        $X_i \leftarrow \emptyset$;
    for $j = 1$ **to** n **do**
        $\hat{i} = \arg\min_{1 \le i \le k} \|\mathbf{x}_j - \mathbf{r}_i\|^2$;
        $X_{\hat{i}} \leftarrow X_{\hat{i}} \cup \{\mathbf{x}_j\}$;
    `/* the second step */`
    for $i = 1$ **to** k **do**
        $\mathbf{r}_i \leftarrow \frac{\sum_{\mathbf{x} \in X_i} \mathbf{x}}{|X_i|}$;
until $X_1, \ldots, X_k$ *do not change*;

by fixing the k clusters, we can find the optimal solutions to the sub-problems of (1.27), i.e.:

$$\arg\min_{\mathbf{r} \in \mathbb{R}^d} \sum_{\mathbf{x} \in X_i} \|\mathbf{x} - \mathbf{r}\|^2 = \frac{\sum_{\mathbf{x} \in X_i} \mathbf{x}}{|X_i|} \quad (i = 1, \ldots, k). \tag{1.28}$$

Thus, the value of object function will also decrease or remain unchanged in the second step. In summary, the two steps in k-means will decrease the object value or reach convergence.

Because k-means is easy to implement and has short running time for low-dimensional data, it has been widely used in various topics and as a preprocessing step for other algorithms [63–66]. However, three major shortcomings are known for the original k-means algorithm. The first one is that choosing an appropriate k is a non-trivial problem. Accordingly, X-means [67] and G-means [68] have been proposed based on the Bayesian information criterion [69] and Gaussian distribution. They can estimate k automatically by using model-selection criteria from statistics. The second one is that an inappropriate choice for the k initial representatives may lead to poor performance. As a solution, the k-means++ algorithm [70] augmented k-means with a simple randomized seeding technique and is guaranteed to find a solution that is $O(\log k)$ competitive to the optimal k-means solution. The third one is that k-means fails to discover clusters with complex shapes [71]. Accordingly, kernel k-means [72] was proposed to detect arbitrary-shaped clusters, with an appropriate choice of the kernel function. References [73], [74] and [75] present three applications of k-means in collaborative signal processing, wireless surveillance systems and wireless hybrid networks, respectively.

1.2.2 Density-based spatial clustering of applications with noise

k-Means is designed to discover spherical shaped clusters. Though kernel k-means can find arbitrary-shaped clusters with an appropriate kernel function, there is no general guidelines for how to choose an appropriate kernel function. In contrast, density-based clustering algorithms are known for the advantage of discovering clusters with arbitrary shapes. In this section, we will introduce the most famous density-based clustering algorithm named DBSCAN (density-based spatial clustering of applications with noise).

DBSCAN [76] was proposed as the first density-based clustering algorithm in 1996. The main idea of DBSCAN can be summarized as three steps. First, DBSCAN estimates the density of each point $\mathbf{x}$ by counting the number of points which belong to the neighbourhood of $\mathbf{x}$. Second, it finds *core points* as points with high density. Third, it connects core points that are very close and their neighbourhood to form dense regions as clusters. Next, we will detail the three steps.

To define density, DBSCAN introduces the concept of ε-neighbourhood, where ε is a user-specified parameter. Given a data set $X = \{\mathbf{x}_i\}_{i=1}^n$, the ε-neighbourhood of a point $\mathbf{x}$ denoted by $N_\varepsilon(\mathbf{x})$ is defined as

$$N_\varepsilon(\mathbf{x}) = \{\mathbf{x}_i | \mathbf{x}_i \in X, \mathrm{dist}(\mathbf{x}, \mathbf{x}_i) \leq \varepsilon\}, \tag{1.29}$$

where $\mathrm{dist}(\cdot,\cdot)$ can choose any distance function according to the application. Then, the density of $\mathbf{x}$, denoted by $\rho(\mathbf{x})$, is defined as the number of points belonging to the neighbourhood of $\mathbf{x}$, i.e.:

$$\rho(\mathbf{x}) = |N_\varepsilon(\mathbf{x})|. \tag{1.30}$$

After the density being defined, DBSCAN introduces another user-specified parameter *MinPts* to find core points. Specifically, if the density of a point $\mathbf{x}$ is not less than *MinPts*, then $\mathbf{x}$ is called a core point. Furthermore, the set consisting of all core points is denoted by $O \triangleq \{\mathbf{x} \in X | \rho(\mathbf{x}) \geq MinPts\}$.

To form clusters, DBSCAN defines the connected relation between core points. For two core points $\mathbf{x}$ and $\mathbf{y}$, we say they are connected if there exists a core-point sequence $\langle \mathbf{x} \equiv \mathbf{z}_1, \ldots, \mathbf{z}_t \equiv \mathbf{y} \rangle$ such that $\mathbf{z}_{i+1} \in N_\varepsilon(\mathbf{z}_i)$ for $1 \leq i < t$ and $\{\mathbf{z}_i\}_{i=1}^t \subset O$. An illustration is shown in Figure 1.10(a). Notice that the connected relation can be verified as an *equivalence relation* on the set O. Thus, DBSCAN uses this equivalence relation to divide core points into equivalence classes. Suppose that core points are divided into k equivalence classes $O_1, \ldots, O_k$, where $\bigcup_{i=1}^k O_i = O$ and $O_i \bigcap O_j = \emptyset$ for $i \neq j$. These equivalence classes constitute the skeleton of clusters. Then, DBSCAN forms the clusters $C_1, \ldots, C_k$ by letting:

$$C_i = N_\varepsilon(O_i) \triangleq \bigcup_{\mathbf{x} \in O_i} N_\varepsilon(\mathbf{x}) \quad (i = 1, \ldots, k). \tag{1.31}$$

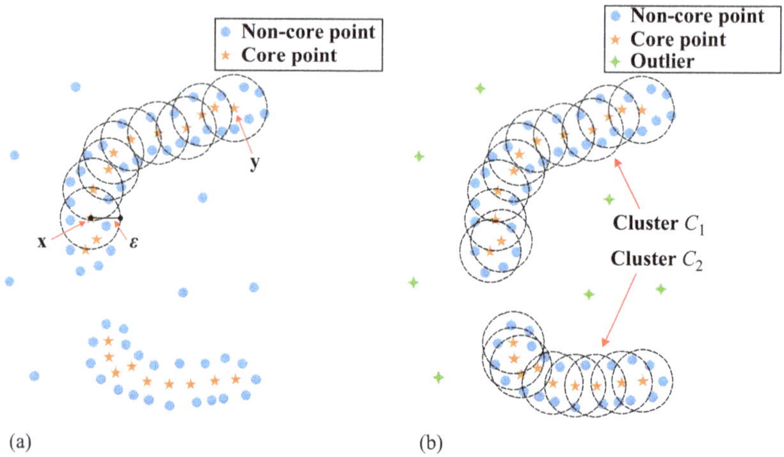

Figure 1.10 (a) An illustration for the connected relation, where the two core points **x** *and* **y** *are connected and (b) an illustration for clusters and outliers. There are two clusters and seven outliers denoted by four-pointed stars*

Notice that there may exist some points which do not belong to any clusters, that is, $X \backslash \left(\bigcup_{i=1}^{k} C_i \right) \neq \emptyset$. DBSCAN assigns these points as outliers because they are far from any normal points. An illustration is shown in Figure 1.10(b).

To this end, we have presented the three main steps of DBSCAN. Algorithm 1.7 summarizes the details of DBSCAN. Let the number of points be $n = |X|$. Finding the ε-neighbourhood for points is the most time-consuming step with computational complexity of $O(n^2)$. The other steps can be implemented within nearly linear computational complexity. Thus, the computation complexity of DBSCAN is $O(n^2)$. The neighbour searching step of DBSCAN can be accelerated by using spatial index technology [60] and groups method [77]. DBSCAN can find arbitrary-shaped clusters and is robust to outliers. However, the clustering quality of DBSCAN highly depends on the parameter ε and it is non-trivial to find an appropriate value for ε. Accordingly, the OPTICS algorithm [78] provides a visual tool to help users find the hierarchical cluster structure and determine the parameters. Some applications of DBSCAN in wireless sensor networks can be found in [79–83].

1.2.3 Clustering by fast search and find of density peaks

In 2014, Rodriguez and Laio proposed a novel-density-based clustering method, named Fast search-and-find of Density Peaks (FDP) [84]. FDP has received extensive attention due to its brilliant idea and the capacity to detect clusters with complex point distribution. As a density-based clustering algorithm, FDP shares similar steps with

Algorithm 1.7: DBSCAN

Input: dataset $X = \{\mathbf{x}_i\}_{i=1}^n$, ε, $MinPts$
Output: clusters $C_1, \ldots, C_k$, outliers set A
for $i = 1$ **to** n **do**
  find $N_\varepsilon(\mathbf{x}_i)$;
  $\rho(\mathbf{x}_i) = |N_\varepsilon(\mathbf{x}_i)|$;
define $O \triangleq \{\mathbf{x} \in X \,|\, \rho(\mathbf{x}) \geq MinPts\}$;
$k = 0$;
repeat
  $k = k + 1$;
  randomly select a core point $\mathbf{o}$ from O;
  use the depth-first-search algorithm to find the set
   $O_k \triangleq \{\mathbf{x} \in O | \mathbf{x} \text{ is connected to } \mathbf{o}\}$;
  define $C_k \triangleq N_\varepsilon(O_k)$;
  $O = O \backslash O_k$;
until $O = \emptyset$;
define $A \triangleq X \backslash \left(\bigcup_{i=1}^k C_i\right)$;

DBSCAN, that is, estimating density, finding core points and forming clusters. However, there are two differences between them. First, FDP detects core points based on a novel criterion, named delta-distance, other than the density. Second, FDP forms cluster by using a novel concept, named higher density nearest neighbour (HDN), rather than the neighbourhood in DBSCAN. Next, we will introduce the two novel concepts followed by the details of FDP.

To begin with, FDP shares the same density definition with DBSCAN. Specifically, given a data set $X = \{\mathbf{x}_i\}_{i=1}^n$, the density of a point $\mathbf{x}$ is computed as

$$\rho(\mathbf{x}) = |N_\varepsilon(\mathbf{x})|, \tag{1.32}$$

where $N_\varepsilon(\mathbf{x})$ denotes the ε-neighbourhood of $\mathbf{x}$ (see (1.29)). After computing the density, FDP defines the HDN of a point $\mathbf{x}$, denoted by $\pi(\mathbf{x})$, as the nearest point whose density is higher than $\mathbf{x}$, i.e.:

$$\pi(\mathbf{x}) \triangleq \arg \min_{\mathbf{y} \in X, \rho(\mathbf{y}) > \rho(\mathbf{x})} \text{dist}(\mathbf{y}, \mathbf{x}). \tag{1.33}$$

Specially, for the point with the highest density, its HDN is defined as the farthest point in X. Then, FDP defines the delta-distance of a point $\mathbf{x}$ as

$$\delta(\mathbf{x}) = \text{dist}(\mathbf{x}, \pi(\mathbf{x})). \tag{1.34}$$

Note that the delta-distance is small for most points and only much larger for a point being either a local maxima in the density or an outlier because the HDN of a outlier may be far from it. In FDP, a local maxima in the density is called a core point.

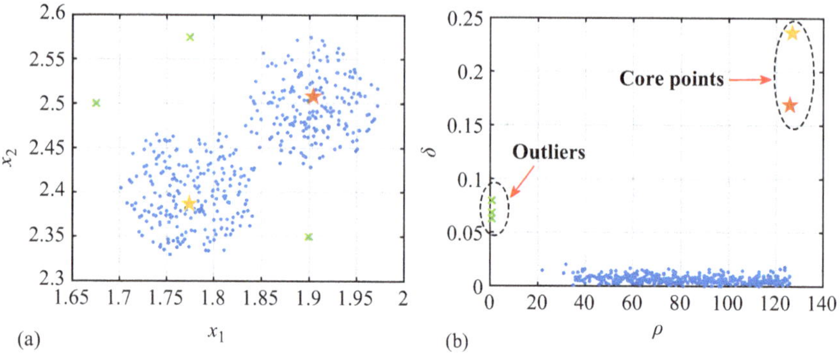

Figure 1.11 (a) A simple data set distributed in a two-dimensional space. There are two clusters and three outliers (denoted by green 'x'). The core point of each cluster is denoted by a pentagram. (b) The decision graph corresponding to the example shown in (a), which is the plot of δ as a function of ρ for all points

This observation is illustrated by a simple example shown in Figure 1.11. Figure 1.11(a) shows a data set distributed in a two-dimensional space. We can find that there are two clusters, and the core point of each cluster is denoted by a pentagram. In addition, three outliers are shown by green '×'. Figure 1.11(b) shows the plot of δ as a function of ρ for all points. This representation is called a decision graph in FDP. As shown in the decision graph, though the density of the ordinary points (the blue dots) fluctuates from around 20 to 120, the low δ values situate them in the bottom of the graph. The outliers have higher δ values than the ordinary points, but they locate in the left of the graph due to their low densities. In contrast, the two core points, which have higher densities and much larger δ values, locate in the top-right area of the ρ–δ plane.

Thus, by using the decision graph, core points and outliers can be found visually. FDP recognizes the cluster number as the number of core points. After that, each remaining point is assigned to the same cluster as its HDN. An illustration is shown in Figure 1.12. There are three core points and thus three clusters, and different clusters are distinguished by different colours.

To this end, we have presented the main steps of FDP. The detail of FDP is summarized in Algorithm 1.8. The clustering quality of FDP is largely dependent on the parameter ε, and it is non-trivial to choose an appropriate ε. To reduce this dependence, one can use the kernel density [85] instead of (1.32). Compared with DBSCAN, FDP is more robust to the clusters with various densities, and they have the same time complexity of $O(n^2)$. In addition, FDP provides a brilliant idea, the HDN. Many algorithms have been proposed inspired by this idea, such as [86–90]. References [91] and [92] present two applications of FDP to optimal energy replenishment and balancing the energy consumption in wireless sensor networks, respectively.

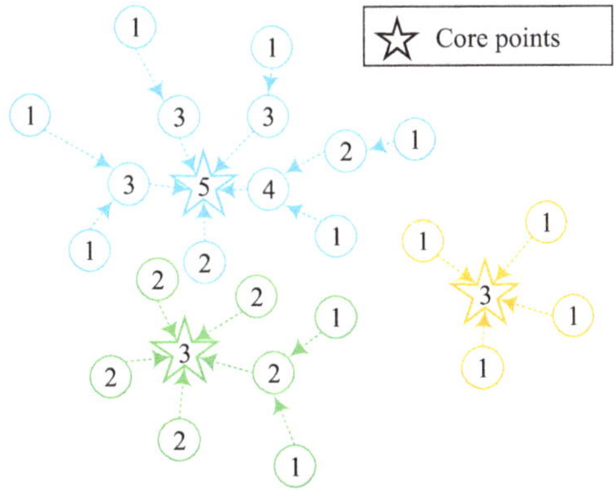

Figure 1.12 An illustration for assigning remaining points. The number in each point denotes its density. The HDN of each point is specified by an arrow. The three core points are denoted by pentagrams

---

Algorithm 1.8: FDP

Input: dataset $X = \{\mathbf{x}_i\}_{i=1}^n$, ε
Output: clusters $C_1, \ldots, C_k$, outliers set A
for $i = 1$ **to** n **do**
  find $N_\varepsilon(\mathbf{x}_i)$;
  $\rho(\mathbf{x}_i) = |N_\varepsilon(\mathbf{x}_i)|$;
for $i = 1$ **to** n **do**
  find $\boldsymbol{\pi}(\mathbf{x}_i) = \arg\min_{\mathbf{y}\in X, \rho(\mathbf{y})>\rho(\mathbf{x})} \mathrm{dist}(\mathbf{y}, \mathbf{x})$;
  $\delta(\mathbf{x}_i) = \mathrm{dist}(\mathbf{x}_i, \boldsymbol{\pi}(\mathbf{x}_i))$;
draw the decision graph, that is, the plot of δ as a function of ρ for all points;
Find core point O set and outlier set A by using the decision graph visually;
Suppose $O = \{\mathbf{o}_1, \ldots, \mathbf{o}_k\}$;
Create clusters $C_1 = \{\mathbf{o}_1\}, \ldots, C_k = \{\mathbf{o}_k\}$;
for $\mathbf{x} \in X \backslash (O \cup A)$ **do**
  assign $\mathbf{x}$ to a cluster according to $\boldsymbol{\pi}(\mathbf{x})$;

1.2.4 Relative core merge clustering algorithm

In real applications, clusters within a data set often have various shapes, densities and scales. To detect clusters with various distributions, Geng *et al.* proposed a RElative COre MErge (RECOME) clustering algorithm [90]. The core of RECOME is a novel density measure, i.e. relative k-NN kernel density (NKD) (RNKD). RECOME recognizes core samples with unit RNKD and divides noncore samples into atom clusters by using the HDN relation as mentioned in Section 1.2.3. Core samples and their corresponding atom clusters are then merged through α-reachable paths on a k-NNs graph. Next, we will introduce RNKD followed by the details of RECOME.

The RNKD is based on an NKD. For a sample $\mathbf{x}$ in a data set $X = \{\mathbf{x}_i\}_{i=1}^{n}$, the NKD of $\mathbf{x}$ is defined as

$$\rho(\mathbf{x}) = \sum_{\mathbf{z}\in N_k(\mathbf{x})} \exp\left\{-\frac{\text{dist}(\mathbf{x},\mathbf{z})}{\sigma}\right\}, \tag{1.35}$$

where $N_k(\mathbf{x})$ denotes the k-NNs set of $\mathbf{x}$ in X, and σ is a constant which can be estimated from the data set. NKD enjoys some good properties and allows easy discrimination of outliers. However, it fails to reveal clusters with various densities. To overcome this shortcoming, RNKD is proposed with the definition of:

$$\rho^*(\mathbf{x}) = \frac{\rho(\mathbf{x})}{\max\limits_{\mathbf{z}\in N_k(\mathbf{x})\cup\{\mathbf{x}\}} \{\rho(\mathbf{z})\}}. \tag{1.36}$$

Intuitively, RNKD is a ratio of densities of two neighbouring samples, and thus it is robust to the change of density since the densities of two neighbouring samples are always at the same level. Figure 1.13 shows a comparison between NKD and RNKD for a simple data set. We can observe that RNKD successfully detects three dense clusters and five sparse clusters which NKD fails to reveal. For more detailed discussions about RNKD, readers can refer to [90].

From the definition of RNKD, we know that a sample with unite RNKD has a local maximal density. Thus, the samples with unite RNKD are good candidates for cluster centres and are called core samples in RECOME. In particular, the set of core samples is denoted by $O \triangleq \{\mathbf{x}|\mathbf{x} \in X, \rho^*(\mathbf{x}) = 1\}$.

Inspired by the idea from FDP, RECOME defines a directed graph $G = (X, A)$ with the arc set $A = \{\langle \mathbf{x}, \boldsymbol{\pi}(\mathbf{x})\rangle | \mathbf{x} \in X\backslash O\}$, where $\boldsymbol{\pi}(\mathbf{x})$ was defined in (1.33). It can be shown that, starting from any noncore sample and following the arcs, a core sample will be reached eventually. In fact, G consists of many trees with disjoint samples,

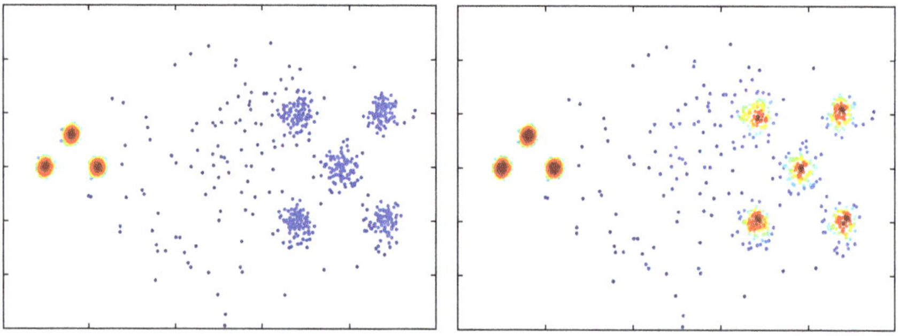

(a) (b)

Figure 1.13 (a) The heat map of NKD for a two-dimensional data set. (b) The heat map of RNKD for a two-dimensional data set. Figures are quoted from [90] and each tree is rooted at a core sample (similar to the relation shown in Figure 1.12). Furthermore, a core sample and its descendants in the tree are called an atom cluster in RECOME. Atom clusters form the basis of final clusters; however, a true cluster may consist of several atom clusters. This happens when many local maximal exist in one true cluster. Thus, a merging step is introduced to combine atom clusters into true clusters.

RECOME treats each core sample as the representative of the atom cluster that it belongs to and merges atom clusters by merging core samples. To do that, it defines another graph with undirected edges, k-NN graph, as

$$G_k = (X, E), E = \{\langle \mathbf{x}, \mathbf{z} \rangle | \mathbf{x} \in N_k(\mathbf{z}) \wedge \mathbf{z} \in N_k(\mathbf{x})\}. \tag{1.37}$$

Furthermore, on the k-NN graph, two samples $\mathbf{x}$ and $\mathbf{z}$ are called α-connected if there exists a path $\langle \mathbf{x}, \mathbf{w}_1, \ldots, \mathbf{w}_s, \mathbf{z} \rangle$ in G_k such that $\rho^*(\mathbf{w}_i) > \alpha$ for $i = 1, \ldots, s$, where α is a user-specified parameter. It can be verified that the α-connected relation is an equivalence relation on the core sample set. RECOME divides core samples into equivalence classes by using this relation. Correspondingly, atom clusters associated with core samples in the same equivalent class are merged into a final cluster. For clarity, we summarize the details of RECOME in Algorithm 1.9.

In RECOME, there are two user-specified parameters k and α. As discussed in [90], the clustering quality of RECOME is not sensitive to k, and it is recommended to tune k in the range $[\sqrt{n}/2, \sqrt{n}]$. On the other hand, the clustering result of RECOME largely depends on parameter α. In particular, as α increases, cluster granularity (i.e. the volume of clusters) decreases and cluster purity increases. In [90], authors also provide an auxiliary algorithm to help users to tune α fast. RECOME has been shown to be effective on detecting clusters with different shapes, densities and scales. Furthermore, it has nearly linear computational complexity if the k-NNs of each sample are computed in advance. In addition, readers can refer to [93] for an application to channel modelling in wireless communications.

1.2.5 Gaussian mixture models and EM algorithm

In this section, we will introduce GMM, which is used for density estimation. In machine learning, the goal of density estimation is to estimate an unobservable underlying probability density function, based on a finite observed data set. Once a probability density function is obtained, we can learn a lot of valuable information based on it. GMM has been widely used for density estimation due to its simple form but strong capacity in data representation. Next, we will introduce the formulation of GMM followed by the estimation for parameters of GMM.

As the name suggests, the distribution of GMM can be written as a weighted linear combination of several Gaussian distributions. Specifically, the probability density of a vector $\mathbf{x} \in \mathbb{R}^d$ is given by

$$f(\mathbf{x}) = \sum_{i=1}^{k} \phi_i N(\mathbf{x} | \boldsymbol{\mu}_i, \boldsymbol{\Sigma}_i), \tag{1.38}$$

Algorithm 1.9: RECOME clustering

Input: dataset $X = \{\mathbf{x}_i\}_{i=1}^n$, parameters k, α
Output: clusters $C_1, \ldots, C_t$

for $i = 1$ **to** n **do**
  find $N_k(\mathbf{x}_i)$;
  $\rho(\mathbf{x}_i) = \sum_{\mathbf{z} \in N_k(\mathbf{x}_i)} \exp\left\{-\frac{\text{dist}(\mathbf{x}_i,\mathbf{z})}{\sigma}\right\}$;
for $i = 1$ **to** n **do**
  $\rho^*(\mathbf{x}_i) = \frac{\rho(\mathbf{x}_i)}{\max_{\mathbf{z} \in N_k(\mathbf{x}_i) \cup \{\mathbf{x}_i\}} \{\rho(\mathbf{z})\}}$;
find $O \triangleq \{\mathbf{x} | \mathbf{x} \in X, \rho^*(\mathbf{x}) = 1\}$;
for $\mathbf{x} \in X \backslash O$ **do**
  find $\boldsymbol{\pi}(\mathbf{x}) \triangleq \arg \min_{\mathbf{y} \in X, \rho(\mathbf{y}) > \rho(\mathbf{x})} \text{dist}(\mathbf{y}, \mathbf{x})$;
construct directed graph $G = (X, A)$, where $A = \{\langle \mathbf{x}, \boldsymbol{\pi}(\mathbf{x}) \rangle | \mathbf{x} \in X \backslash O\}$;
for $\mathbf{o} \in O$ **do**
  find atom cluster $C_{\mathbf{o}} = \{\mathbf{o}\} \cup \{\mathbf{x} | \mathbf{x}$ is connected to $\mathbf{o}$ in $G\}$;
construct k-NN graph $G_k = (X, E)$, where $E = \{\langle \mathbf{x}, \mathbf{z} \rangle | \mathbf{x} \in N_k(\mathbf{z}) \wedge \mathbf{z} \in N_k(\mathbf{x})\}$;
$t = 0$;
repeat
  $t = t + 1$;
  randomly select a core sample $\mathbf{o}$ from O;
  use the depth-first-search algorithm to find set
   $O_t \triangleq \{\mathbf{x} \in O | \mathbf{x}$ is α-connected to $\mathbf{o}$ in $G_k\}$;
  define $C_t = \bigcup_{\mathbf{x} \in O_t} C_{\mathbf{x}}$;
  $O = O \backslash O_t$;
until $O = \emptyset$;

where $\phi_1, \ldots, \phi_k$ are non-negative with $\sum_{i=1}^k \phi_i = 1$, and

$$N(\mathbf{x}|\boldsymbol{\mu}_i, \boldsymbol{\Sigma}_i) = \frac{1}{(2\pi)^{d/2}} \frac{1}{|\Sigma_i|^{1/2}} \exp\left\{-\frac{1}{2}(\mathbf{x} - \boldsymbol{\mu}_i)^T \boldsymbol{\Sigma}_i^{-1} (\mathbf{x} - \boldsymbol{\mu}_i)\right\} \tag{1.39}$$

is the Gaussian distribution with mean vector $\boldsymbol{\mu}_i$ and covariance matrix $\boldsymbol{\Sigma}_i$. Parameter k controls the capacity and the complexity of GMM. Considering a two-dimensional data set shown in Figure 1.14(a), Figure 1.14(b) shows the fitted distribution by using GMM with $k = 5$, where we observe that only a fuzzy outline is reserved and most details are lost. In contrast, Figure 1.14(c) shows the case for $k = 20$, where we find the fitted distribution reflects the main characteristic of the data set. In fact, by increasing k, GMM can approximate any continuous distribution to some desired degree of accuracy. However, it does not mean that the larger k is better, because a large k may lead to overfitting and a huge time cost for the parameter estimation. In most cases, k is inferred experientially from the data.

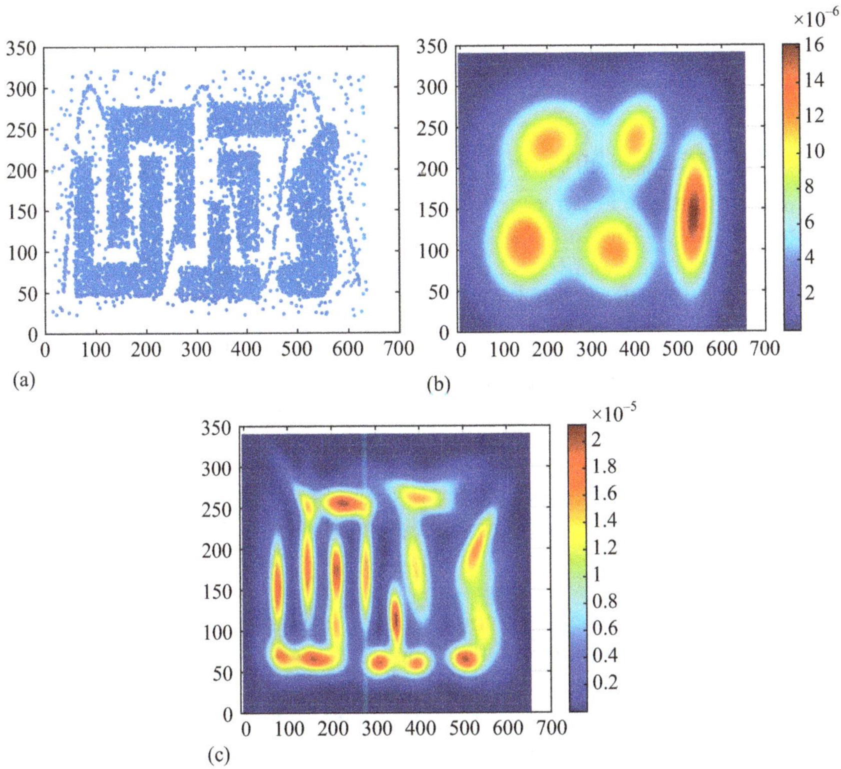

Figure 1.14 *(a) A two-dimensional data set, (b) the fitted distribution by using GMM with k = 5 and (c) the fitted distribution by using GMM with k = 20*

Now, we discuss the parameter estimation for GMM. Given a data set $\{\mathbf{x}_j\}_{j=1}^n$, the log-likelihood of GMM is given by

$$L = \sum_{j=1}^{n} \ln f(\mathbf{x}_j) = \sum_{j=1}^{n} \ln \left(\sum_{i=1}^{k} \phi_i N(\mathbf{x}_j | \boldsymbol{\mu}_i, \boldsymbol{\Sigma}_i) \right). \tag{1.40}$$

Thus, the maximum likelihood estimation is to solve:

$$\begin{aligned} \max_{\{\phi_i\},\{\boldsymbol{\mu}_i\},\{\boldsymbol{\Sigma}_i\}} \quad & L \\ \text{s.t.} \quad & \boldsymbol{\Sigma}_i \succ \mathbf{0} \quad (i = 1, \ldots, k), \\ & \phi_i \geq 0 \quad (i = 1, \ldots, k), \\ & \sum_{i=1}^{k} \phi_i = 1. \end{aligned} \tag{1.41}$$

where $\boldsymbol{\Sigma}_i \succ \mathbf{0}$ means that $\boldsymbol{\Sigma}_i$ should be a positive semi-definite matrix. Unfortunately, there is no effective ways to find the global optimal solution to (1.41). Next, we will introduce the expectation–maximization (EM) algorithm to find an approximate solution to (1.41).

1.2.5.1 The EM algorithm

The EM algorithm is an iterative method to solve the problem of maximum likelihood estimation for the model which depends on unobserved *latent variables*. To explain the meaning of *latent variables*, let us consider the aforementioned GMM. Samples obeying the distribution of GMM can be generated by the following two steps:

1. Randomly select one from the k Gaussian models with the probability that $P\{\text{the } i\text{th model being chosen}\} = \phi_i \quad (i = 1, \dots, k)$.
2. Supposing that the zth model is selected in the first step, then generate a sample $\mathbf{x}$ by using the zth Gaussian model.

By the above steps, we know that the joint probability distribution of $(z, \mathbf{x})$ is given by $p(z, \mathbf{x}) = \phi_z N(\mathbf{x}|\boldsymbol{\mu}_z, \boldsymbol{\Sigma}_z)$. However, in the parameter-estimation of GMM, what we are given is only the value of $\mathbf{x}$, and the value of z is unobservable to us. Here, z is called a latent variable.

Now, let us consider the maximum likelihood estimation of a model with latent variables. Supposing that an observable sample set $X = \{\mathbf{x}_j\}_{j=1}^n$ is given and the corresponding latent sample set $Z = \{\mathbf{z}_j\}_{j=1}^n$ is unobservable to us, the goal is to estimate the parameter set θ of a probability distribution $p(\mathbf{x}|\theta)$. Notice that we regard Z as a random variable since it is unknown to us. By denoting $p(X|\theta) = \prod_{j=1}^n p(\mathbf{x}_j|\theta)$, the problem of maximum likelihood estimation can be written as

$$\max_{\theta} \quad \ln p(X|\theta) \equiv \ln \sum_{Z} p(X, Z|\theta). \tag{1.42}$$

However, solving (1.42) is usually a difficult problem since the unknown Z. As an alternative, the EM algorithm can generate an approximate solution by using an iterative process. Initially, it initializes the parameter set θ to some random values $\bar{\theta}$. Then, the following two steps will be alternately executed until convergence.

1. E-step: Compute the expectation of $\ln p(X, Z|\theta)$ with respect to Z with the distribution $p(Z|X, \bar{\theta})$, that is,

$$\mathbb{E}_{Z|X,\bar{\theta}} \ln p(X, Z|\theta) = \sum_{Z} p(Z|X, \bar{\theta}) \ln p(X, Z|\theta). \tag{1.43}$$

2. M-step: Update the parameter $\bar{\theta}$ by the solution of maximizing (1.43), that is,

$$\bar{\theta} = \arg\max_{\theta} \sum_{Z} p(Z|X, \bar{\theta}) \ln p(X, Z|\theta). \tag{1.44}$$

Finally, the resulting $\bar{\theta}$ will be the estimated parameter set. Here, we do not discuss the correctness of the EM algorithm to stay focused. Readers can refer to [94] for more details.

1.2.5.2 The EM algorithm for GMM

We are ready to apply the EM algorithm to solve the parameter-estimation problem for GMM. As mentioned in Section 1.2.5.1, let z_j be a latent variable which denotes the index of the Gaussian model that generates $\mathbf{x}_j$. To begin with, initialize the parameters $\phi_i, \boldsymbol{\mu}_i, \boldsymbol{\Sigma}_i$ to some random values $\bar{\phi}_i, \bar{\boldsymbol{\mu}}_i, \bar{\boldsymbol{\Sigma}}_i$ for $i = 1, \ldots, k$.

In the E-step, we first need to compute:

$$p(z_j = l|\mathbf{x}_j, \bar{\theta}) = \frac{\bar{\phi}_l N(\mathbf{x}_j|\bar{\boldsymbol{\mu}}_l, \bar{\boldsymbol{\Sigma}}_l)}{\sum_{i=1}^{k} \bar{\phi}_i N(\mathbf{x}_j|\bar{\boldsymbol{\mu}}_i, \bar{\boldsymbol{\Sigma}}_i)}. \tag{1.45}$$

Since $p(z_j = l|\mathbf{x}_j, \bar{\theta})$ is a constant in an iteration, we denote it as γ_{jl} for simplicity. Then, compute the exception:

$$\begin{aligned}
\mathbb{E}_{Z|X,\bar{\theta}} \ln p(X, Z|\theta) &= \sum_{Z} p(Z|X, \bar{\theta}) \ln p(X, Z|\theta) \\
&= \sum_{j=1}^{n} \sum_{z_j} p(z_j|\mathbf{x}_j, \bar{\theta}) \ln p(\mathbf{x}_j, z_j|\theta) \\
&= \sum_{j=1}^{n} \sum_{l=1}^{k} p(z_j = l|\mathbf{x}_j, \bar{\theta}) \ln (\phi_l N(\mathbf{x}_j|\boldsymbol{\mu}_l, \boldsymbol{\Sigma}_l)) \\
&= \sum_{j=1}^{n} \sum_{l=1}^{k} \gamma_{jl} \ln (\phi_l N(\mathbf{x}_j|\boldsymbol{\mu}_l, \boldsymbol{\Sigma}_l)).
\end{aligned} \tag{1.46}$$

In the M-step, we need to solve:

$$\begin{aligned}
\max_{\{\phi_l\},\{\boldsymbol{\mu}_l\},\{\boldsymbol{\Sigma}_l\}} \quad & \sum_{j=1}^{n} \sum_{l=1}^{k} \gamma_{jl} \left(\ln \phi_l - \frac{1}{2} \ln |\boldsymbol{\Sigma}_l| - \frac{1}{2} (\mathbf{x}_j - \boldsymbol{\mu}_l)^T \boldsymbol{\Sigma}_l^{-1} (\mathbf{x}_j - \boldsymbol{\mu}_l) \right) \\
\text{s.t.} \quad & \boldsymbol{\Sigma}_l \succ \mathbf{0} \quad (l = 1, \ldots, k), \\
& \phi_l \geq 0 \quad (l = 1, \ldots, k), \\
& \sum_{l=1}^{k} \phi_l = 1,
\end{aligned} \tag{1.47}$$

where we have omitted unrelated constants. By using the Karush–Kuhn–Tucker conditions [28], the optimal solution to (1.47) is given by

$$\begin{aligned}
\phi_l &= \frac{n_l}{n}, \\
\boldsymbol{\mu}_l &= \frac{1}{n_l} \sum_{j=1}^{n} \gamma_{jl} \mathbf{x}_j, \\
\boldsymbol{\Sigma}_l &= \frac{1}{n_l} \sum_{j=1}^{n} \gamma_{jl} (\mathbf{x}_j - \boldsymbol{\mu}_l)(\mathbf{x}_j - \boldsymbol{\mu}_l)^T,
\end{aligned} \tag{1.48}$$

Algorithm 1.10: The EM algorithm for GMM

Input: dataset $X = \{\mathbf{x}_j\}_{j=1}^n$, number of Gaussian models k

Output: estimated parameters $\{\bar{\phi}_l\}_{l=1}^k, \{\bar{\boldsymbol{\mu}}_l\}_{l=1}^k, \{\bar{\boldsymbol{\Sigma}}_l\}_{l=1}^k$

randomly initialize $\{\bar{\phi}_l\}_{l=1}^k, \{\bar{\boldsymbol{\mu}}_l\}_{l=1}^k, \{\bar{\boldsymbol{\Sigma}}_l\}_{l=1}^k$;

repeat

  /* E-step */

  for $l = 1$ **to** k **do**

    for $i = 1$ **to** n **do**

      $\gamma_{jl} = \frac{\bar{\phi}_l N(\mathbf{x}_j|\bar{\boldsymbol{\mu}}_l,\bar{\boldsymbol{\Sigma}}_l)}{\sum_{i=1}^k \bar{\phi}_i N(\mathbf{x}_j|\bar{\boldsymbol{\mu}}_i,\bar{\boldsymbol{\Sigma}}_i)}$;

    $n_l = \sum_{j=1}^n \gamma_{jl}$;

  /* M-step */

  for $l = 1$ **to** k **do**

    $\bar{\phi}_l = \frac{n_l}{n}$;

    $\bar{\boldsymbol{\mu}}_l = \frac{1}{n_l}\sum_{j=1}^n \gamma_{jl}\mathbf{x}_j$;

  for $l = 1$ **to** k **do**

    $\bar{\boldsymbol{\Sigma}}_l = \frac{1}{n_l}\sum_{j=1}^n \gamma_{jl}(\mathbf{x}_j - \bar{\boldsymbol{\mu}}_l)(\mathbf{x}_j - \bar{\boldsymbol{\mu}}_l)^T$;

until *convergence*;

---

where we have defined $n_l = \sum_{j=1}^n \gamma_{jl}$. We conclude the whole procedure in Algorithm 1.10.

In addition to fitting the distribution density of data, GMM can also be used for clustering data. Specifically, if we regard the k Gaussian models as the 'patterns' of k clusters, then the probability that a sample $\mathbf{x}_i$ comes from the lth pattern is given by

$$p(z_j = l|\mathbf{x}_j) = \frac{\bar{\phi}_l N(\mathbf{x}_j|\bar{\boldsymbol{\mu}}_l, \bar{\boldsymbol{\Sigma}}_l)}{\sum_{i=1}^k \bar{\phi}_i N(\mathbf{x}_j|\bar{\boldsymbol{\mu}}_i, \bar{\boldsymbol{\Sigma}}_i)}. \tag{1.49}$$

Thus, $l^* \triangleq \arg\max_l p(z_j = l|\mathbf{x}_j)$ gives the index of the cluster that $\mathbf{x}_j$ most likely belongs to. Furthermore, low $p(z_j = l^*|\mathbf{x}_j)$ may imply that $\mathbf{x}_j$ is an outlier. Reference [95] shows a diffusion-based EM algorithm for distributed estimation of GMM in wireless sensor networks. References [96] and [97] present two applications of GMM in target tracking and signal-strength prediction, respectively.

1.2.6 Principal component analysis

In real applications, we usually encounter data with high dimensionality, such as speech signals or digital images. In order to handle such data adequately, we need to reduce its dimensionality. *Dimension reduction* is a process of transforming original high-dimensional data into low-dimensional representation which reserve meaningful characteristics of the data. Among all dimension-reduction methods, PCA is the most

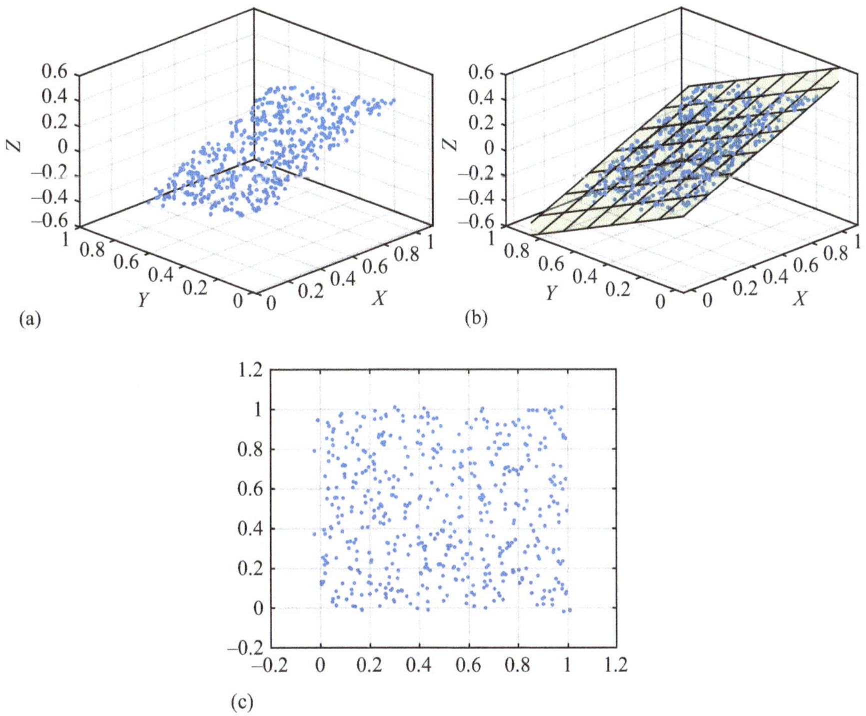

Figure 1.15 *(a) A three-dimensional data set, (b) the points are distributed near a two-dimensional plane and (c) projection of the points onto the plane*

popular one. PCA can be derived from the perspectives of both geometry and statistics. Here we will focus on the former perspective since it meets our intuition better.

To begin with, let us consider a three-dimensional data set as shown in Figure 1.15(a). Though all points lie in a three-dimensional space, as shown in Figure 1.15(b), they are distributed near a two-dimensional plane. As shown in Figure 1.15(c), after projecting all points onto the plane, we can observe that, in fact, they are distributed in a rectangle. In this example, we find that the low-dimensional representation captures the key characteristic of the data set.

Reviewing the above example, the key step is to find a low-dimensional plane near all points. In PCA, this problem is formalized as an optimization problem by using linear algebra. Specifically, given a data set $\{\mathbf{x}_i \in \mathbb{R}^d\}_{i=1}^n$, PCA intends to find a t-dimensional ($t < d$) plane[1] that minimizes the sum of the square of the distance between each point and its projection onto the plane. Formally, a t-dimensional plane can be described by a semi-orthogonal matrix $\mathbf{B} = (\mathbf{b}_1, \ldots, \mathbf{b}_t) \in \mathbb{R}^{d\times t}$ (i.e. $\mathbf{B}^T\mathbf{B} = \mathbf{I}$) and a shift vector $\mathbf{s} \in \mathbb{R}^d$. By linear algebra, $\{\mathbf{B}\mathbf{z}|\mathbf{z} \in \mathbb{R}^t\}$ is a t-dimensional subspace.

[1]A formal name should be affine subspace. Here we use 'plane' for simplicity.

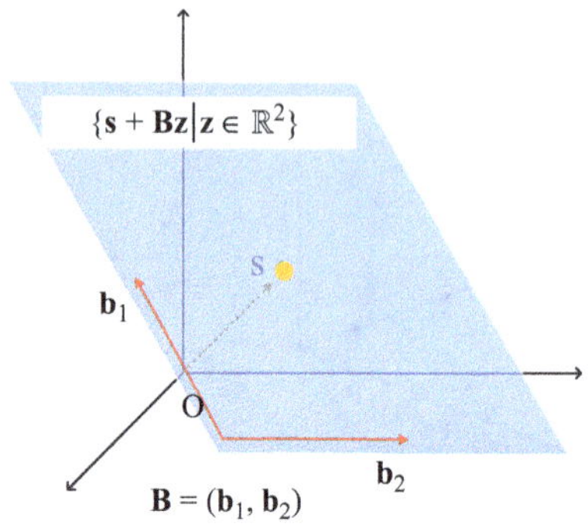

Figure 1.16 A two-dimensional plane $\{s + Bz | z \in \mathbb{R}^2\}$ in $\mathbb{R}^3$

Then, we shift the subspace by $\mathbf{s}$ and get a plane $\{\mathbf{s} + \mathbf{B}\mathbf{z} | \mathbf{z} \in \mathbb{R}^t\}$. An illustration is shown in Figure 1.16. Furthermore, for any point $\mathbf{x} \in \mathbb{R}^d$, the projection of it to the plane is given by

$$\mathbf{s} + \mathbf{B}\mathbf{B}^T(\mathbf{x} - \mathbf{s}). \tag{1.50}$$

Correspondingly, the distance from $\mathbf{x}$ to its projection is given by $\left\|(\mathbf{I} - \mathbf{B}\mathbf{B}^T)(\mathbf{x} - \mathbf{s})\right\|$.

Thus, denoting the data matrix as $\mathbf{X} = (\mathbf{x}_1, \ldots, \mathbf{x}_n) \in \mathbb{R}^{d \times n}$, the goal of PCA is to find the optimal solution to

$$\begin{aligned} \min_{\mathbf{s},\mathbf{B}} \sum_{i=1}^{n} \quad & \left\|(\mathbf{I} - \mathbf{B}\mathbf{B}^T)(\mathbf{x}_i - \mathbf{s})\right\|^2 \\ \text{s.t.} \quad & \mathbf{B}^T\mathbf{B} = \mathbf{I}. \end{aligned} \tag{1.51}$$

By taking gradient with respect to $\mathbf{s}$, the optimal $\mathbf{s}$ is given by $\boldsymbol{\mu} = (1/n)\sum_{i=1}^{n} \mathbf{x}_i$. Replacing $\mathbf{s}$ by $\boldsymbol{\mu}$, we still need to solve:

$$\begin{aligned} \min_{\mathbf{B}} \sum_{i=1}^{n} \quad & \left\|(\mathbf{I} - \mathbf{B}\mathbf{B}^T)(\mathbf{x}_i - \boldsymbol{\mu})\right\|^2 \\ \text{s.t.} \quad & \mathbf{B}^T\mathbf{B} = \mathbf{I}. \end{aligned} \tag{1.52}$$

Noticing that

$$\left\|(\mathbf{I} - \mathbf{B}\mathbf{B}^T)(\mathbf{x}_i - \boldsymbol{\mu})\right\|^2 = \|\mathbf{x}_i - \boldsymbol{\mu}\|^2 - (\mathbf{x}_i - \boldsymbol{\mu})^T\mathbf{B}\mathbf{B}^T(\mathbf{x}_i - \boldsymbol{\mu}), \tag{1.53}$$

(1.52) is equivalent to

$$\begin{aligned} \max_{\mathbf{B}} \quad & \sum_{i=1}^{n} (\mathbf{x}_i - \boldsymbol{\mu})^T\mathbf{B}\mathbf{B}^T(\mathbf{x}_i - \boldsymbol{\mu}) = \mathrm{Tr}\left(\bar{\mathbf{X}}^T\mathbf{B}\mathbf{B}^T\bar{\mathbf{X}}\right) = \mathrm{Tr}\left(\mathbf{B}^T\bar{\mathbf{X}}\bar{\mathbf{X}}^T\mathbf{B}\right) \\ \text{s.t.} \quad & \mathbf{B}^T\mathbf{B} = \mathbf{I}. \end{aligned} \tag{1.54}$$

Algorithm 1.11: Principal component analysis

Input: data set $X = \{\mathbf{x}_i \in \mathbb{R}^d\}_{i=1}^n$, target dimension $t \quad (t < d)$
Output: coordinate matrix in dimension-reduction space
$\mathbf{Y} = (\mathbf{y}_1, \ldots, \mathbf{y}_n) \in \mathbb{R}^{t \times n}$

compute $\mu = \frac{1}{n}\sum_{i=1}^{n} \mathbf{x}_i$;
define $\bar{\mathbf{X}} = (\mathbf{x}_1 - \boldsymbol{\mu}, \ldots, \mathbf{x}_n - \boldsymbol{\mu}) \in \mathbb{R}^{d \times n}$;
compute the first t orthogonal eigenvectors $\mathbf{p}_1, \ldots, \mathbf{p}_t$ of $\bar{\mathbf{X}}\bar{\mathbf{X}}^T$;
define $\mathbf{P} = (\mathbf{p}_1, \ldots, \mathbf{p}_t) \in \mathbb{R}^{d \times t}$;
$\mathbf{Y} = \mathbf{P}^T\bar{\mathbf{X}}$;

---

where $\bar{\mathbf{X}} = (\mathbf{x}_1 - \boldsymbol{\mu}, \ldots, \mathbf{x}_n - \boldsymbol{\mu})$. Suppose that $\mathbf{p}_1, \ldots, \mathbf{p}_t$ are t orthogonal eigenvectors corresponding to the t largest eigenvalues of $\bar{\mathbf{X}}\bar{\mathbf{X}}^T$. By denoting $\mathbf{P} = (\mathbf{p}_1, \ldots, \mathbf{p}_t) \in \mathbb{R}^{d \times t}$, the optimal solution to (1.54) is given by

$$\mathbf{B}^* = \mathbf{P} \tag{1.55}$$

according to Corollary 4.3.39 in [98]. Thus, the coordinate of $\mathbf{x}_i$ in the dimension reduction space is given by $\mathbf{P}^T(\mathbf{x}_i - \boldsymbol{\mu}) = \mathbf{y}_i \in \mathbb{R}^t$. Finally, we conclude the whole process in Algorithm 1.11.

PCA is effective when data distribute near a low-dimensional plane. However, sometimes data may be approximately embedded in a non-linear structure, such as an ellipsoid or a hyperboloid. To handle the non-linear case, kernel PCA [99] is proposed by using the kernel trick. In addition, Generalized PCA [100] is proposed to deal with the case where data distribute in multi low-dimensional planes. When data is contaminated by a few noises, Robust PCA [101] has been shown to be effective in this case. References [102] and [103] demonstrate two applications of PCA to multivariate sampling for wireless sensor networks and wireless capsule endoscopy, respectively.

1.2.7 Autoencoder

In this section, we will introduce another dimension-reduction method *autoencoder*. An autoencoder is a neural network (see Section 1.1.3.3) used to learn an effective representation (encoded mode) for a data set, where the transformed code has lower dimensions compared with the original data.

As shown in Figure 1.17, an autoencoder consists of two parts, i.e. an encoder $f(\cdot|\eta)$ and a decoder $g(\cdot|\boldsymbol{\theta})$. Each of them is a neural network, where η and $\boldsymbol{\theta}$ denote the parameter sets of the encoder and the decoder, respectively. Given an input $\mathbf{x} \in \mathbb{R}^d$, the encoder is in charge of transforming $\mathbf{x}$ into a code $\mathbf{z} \in \mathbb{R}^t$, i.e. $f(\mathbf{x}|\eta) = \mathbf{z}$, where t is the length of the code with $t < d$. In contrast, the decoder tries to recover the original feature $\mathbf{x}$ from the code $\mathbf{z}$, i.e. $g(\mathbf{z}|\boldsymbol{\theta}) = \bar{\mathbf{x}} \in \mathbb{R}^d$ such that $\bar{\mathbf{x}} \approx \mathbf{x}$. Thus, given

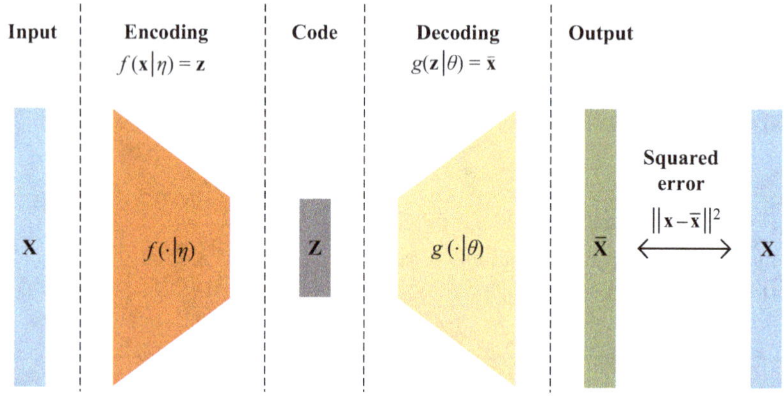

Figure 1.17 The flowchart of an autoencoder. First, the encoder $f(\cdot|\boldsymbol{\eta})$ encodes an input $\mathbf{x}$ into a code $\mathbf{z}$ with short length. Then the code $\mathbf{z}$ is transformed into an output $\bar{\mathbf{x}}$ being the same size as $\mathbf{x}$, by the decoder $g(\cdot|\boldsymbol{\theta})$. Given a data set $\{\boldsymbol{x}_i\}_{i=i}^n$, the object of training is to learn the parameter set such that minimize the sum of squared error $\sum_{i=1}^n \|\boldsymbol{x}_i - \bar{\boldsymbol{x}}_i\|^2$

a data set $\{\mathbf{x}_i\}_{i=1}^n$, the training process of the autoencoder can be formulated as the following optimization problem:

$$\min_{\eta,\theta} \sum_{i=1}^{n} \|\mathbf{x}_i - g(f(\mathbf{x}_i|\boldsymbol{\eta})|\boldsymbol{\theta})\|^2. \tag{1.56}$$

By limiting the length of code, minimizing the object function will force the code to capture critical structure of input features and ignore trivial details such as sparse noises. Thus, besides dimension reduction, an autoencoder can also be used to de-noise.

In Figure 1.18, we show a specific implementation for the dimension-reduction task on the MNIST data set of handwritten digits [41], where each sample is a grey-scale image with the size of 28 × 28. For simplicity, stacking the columns of each image into a vector, thus the input has a dimension of $28 \times 28 = 784$. As we see, the encoder consists of three FC layers, where each layer is equipped with a sigmoid activation function. The first layer non-linearly transforms an input vector with 784 dimensions into a hidden vector with 256 dimensions, and the second layer continues to reduce the dimension of the hidden vector from 256 to 128. Finally, after the transformation of the third layer, we get a code of 64 dimensions, which is far less than the dimension of the input vector. On the other hand, the decoder shares a same structure with the encoder except that each FC layer transforms a low-dimensional vector into a high-dimensional vector. The encoder tries to reconstruct the original input vector with 784 dimensions from the code with 64 dimensions. In addition, a sparse constraint on the parameters should be added as a regularization term to achieve better performance. After training the autoencoder using the BP algorithm

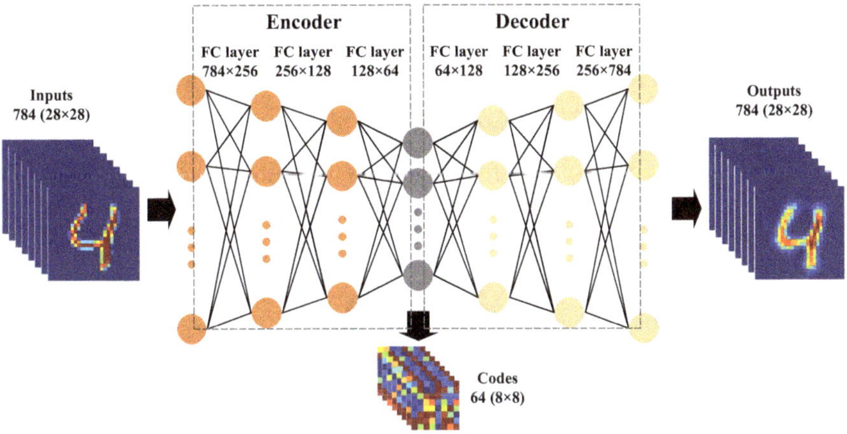

Figure 1.18 A specific implementation for the dimension-reduction task on the MNIST data set. The encoder consists of three fully connected layers (FC layers), where each layer is equipped with a sigmoid activation function. The decoder shares a same structure with the encoder except that each FC layer transforms a low-dimensional vector into a high-dimension vector

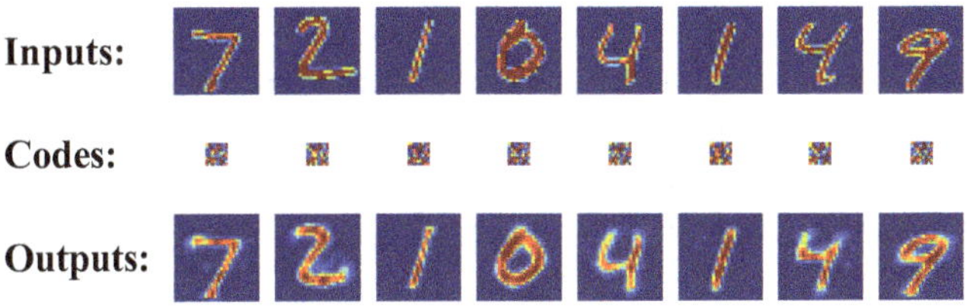

Figure 1.19 Partial results on the MNIST data set

(see Section 1.1.3.3), we can obtain the result shown in Figure 1.19. From this figure, we observe that an image can be reconstructed with high quality from a small-size code, which indicates that the main feature of the original image is encoded into the code.

We can use an encoder to compress a high-dimensional sample into a low-dimensional code and then use a decoder to reconstruct the sample from the code. An interesting problem is that, can we feed a randomly generated code into the decoder to obtain a new sample. Unfortunately, in most cases, the generated samples either are very similar to the original data or become meaningless things. Inspired by this idea, a variational autoencoder [104] is proposed. Different from the autoencoder, the variational autoencoder tries to learn an encoder to encode the distribution of original data rather than the data itself. By using a well-designed object function, the distribution of original data can be encoded into some low-dimensional normal distributions through an encoder. Correspondingly, a decoder is trained to transform the normal

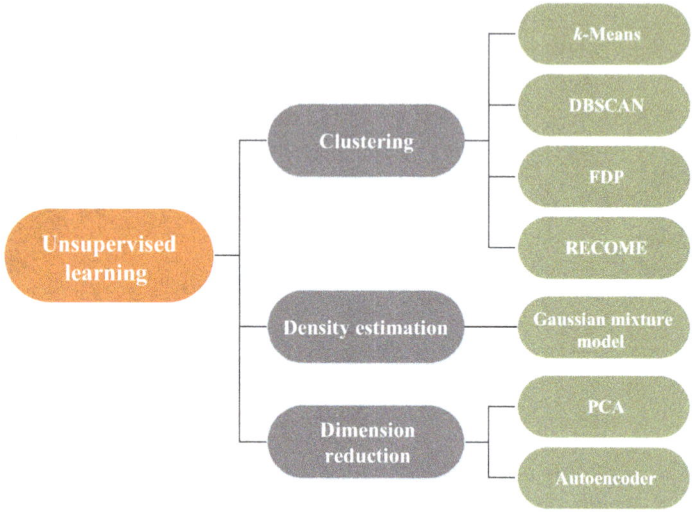

Figure 1.20 Structure chart for unsupervised learning technologies discussed in this chapter

distributions into real data distribution. Thus, one can first sample a code from the normal distributions and then feed it to the decoder to obtain a new sample. For more details about a variational autoencoder, please refer to [104]. In wireless communications and sensor networks, autoencoder has been applied in many fields, such as data compression [105], sparse data representation [106], wireless localization [107] and anomaly detection [108].

1.2.8 Summary of unsupervised learning

In this section, we have discussed unsupervised learning. In contrast to supervised learning, unsupervised learning needs to discover and explore the inherent and hidden structures of a data set without labels. As shown in Figure 1.20, unsupervised learning tasks can be mainly divided into three categories, i.e. clustering, density estimation and dimension reduction.

Clustering is the area studied most in unsupervised learning. We have introduced four practical clustering methods, i.e. k-means, DBSCAN, FDP and RECOME. k-means has the simplest form among four methods. It is very easy to implement and has short running time for low-dimensional data. But the original k-means prefers to divide data into clusters with convex shapes. DBSCAN is a density-based clustering algorithm. It is apt at detecting clusters with different shapes. But the clustering result of DBSCAN is sensitive to the choice of its parameters. FDP introduces a brilliant idea, HDN, and based on it, a clustering algorithm is constructed. FDP is easy to implement and can generate satisfactory results with cluster centres being selected visually. RECOME is based on a novel density measure, relative NKD. It has been

Table 1.3 Summary of applications of unsupervised learning in wireless communications

Method	Function	Application in wireless communications
k-Means	Clustering	Collaborative signal processing [73] Wireless surveillance systems [74] Wireless hybrid networks [75]
DBSCAN	Clustering	Localization [79] Wireless agriculture [80] Anomaly detection [81,83] Power backup [82]
FDP	Clustering	Wireless energy strategy [91,92]
RECOME	Clustering	MIMO channel learning [93]
Gaussian mixture models	Density estimation	Target tracking [96] Signal-strength prediction [97]
Principal component analysis	Dimension reduction	Multivariate sampling [37] Wireless capsule endoscopy [103]
Autoencoder	Dimension reduction	Wireless data representation [105,106] Localization [107] Anomaly detection [108]

shown to be effective on detecting clusters with various shapes, densities and scales. In addition, it also provides an auxiliary algorithm to help users selecting parameters.

Density estimation is a basic problem in unsupervised learning. We have presented the GMM, which is one of most popular models for density estimation. The GMM can approximate any continuous distribution to some desired degree of accuracy as soon as the parameter k is large enough, but accordingly, time cost for the estimation of its parameters will increase. Dimension reduction plays an important role in the compression, comprehension and visualization of data. We have introduced two dimension-reduction technologies, PCA and autoencoder. PCA can be deduced from an intuitive geometry view and has been shown to be highly effective for data distributed in a linear structure. However, it may destroy non-linear topological relations in original data. Autoencoder is a dimension-reduction method based on neural networks. Compared with PCA, autoencoder has great potential to reserve the non-linear structure in original data, but it needs more time to adjust parameters for a given data set. In Table 1.3, we summarize the applications of supervised learning in wireless communications. For more technologies of unsupervised learning, readers can refer to [8].

1.3 Reinforcement learning

So far, we have discussed two kinds of machine-learning methods: supervised learning, which is adapted to handle a classification or regression task, and unsupervised

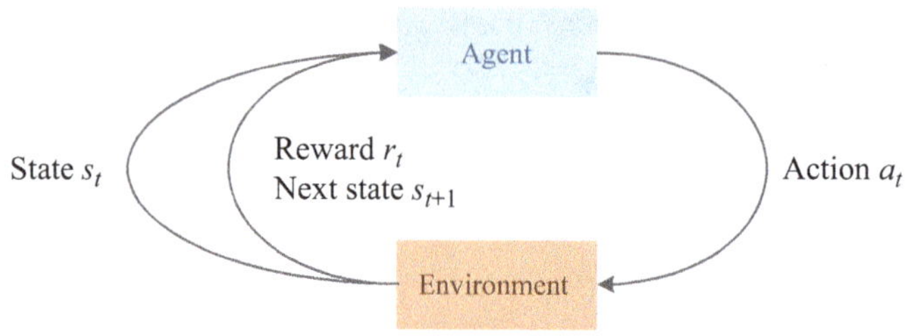

Figure 1.21 Markov decision process

learning, which is used to learn underlying hidden structure of data. However, in wireless communications, we sometimes encounter some problems of real-time controlling, which is hard to be solved by supervised or unsupervised learning methods. For example, in radio access networks, we need to dynamically turn on/off some base stations according to the traffic load variations so as to improve energy efficiency [109]. As a solution, RL is a powerful tool to deal with these real-time control problems. In this section, we will introduce the main idea and classic approaches of RL.

1.3.1 Markov decision process

In RL, a real-time control problem is simplified as a system where an agent and an environment interact over time. As illustrated in Figure 1.21, at time step t, the environment is in a state s_t (e.g. the traffic load variations in radio access networks). Then, the agent takes an action a_t (e.g. turn on/off some base stations) according to the state s_t. After that, the environment will return a reward r_t (e.g. the saved energy cost) to the agent and turn into the next state s_{t+1} on the basis of s_t and a_t. Since the rule of states and rewards is determined by the environment, what the agent controls is choosing actions in accordance with states to maximize total rewards in a long period.

The above idea can be formulated as a Markov decision process (MDP). Formally, an MDP, represented by a tuple $\langle \mathfrak{S}, \mathfrak{A}, P, R, \gamma \rangle$, consists of five parts:

- $\mathfrak{S}$ is a finite set of states.
- $\mathfrak{A}$ is a finite set of actions.
- P is a state transition probability function. $P(\cdot|s,a)$ gives the distribution over next state given a pair (s,a), where $s \in \mathfrak{S}$ and $a \in \mathfrak{A}$.
- $R : \mathfrak{S} \times \mathfrak{A} \to \mathbb{R}$ is a reward function.[2] $R(s,a)$ gives the reward after the agent takes an action a in a state s.
- $\gamma \in [0,1]$ is a discount factor, which is introduced to discount the long-period reward.[3]

[2]Here we suppose the reward function is deterministic for simplicity though it can be a random function.
[3]People often pay more attention to the short-term reward.

In addition, the strategy of the agent taking actions is defined as a policy π, where $\pi(a|s)$ gives the probability of the agent taking an action a in a state s. In other words, a policy fully defines the behaviour of an agent. Given an initial state s_0 and a policy π, an MDP can 'run' as follows:

$$\begin{aligned} &\text{For } t = 0, 1, 2, \ldots \\ &\quad a_t \sim \pi(\cdot|s_t); \\ &\quad r_t = R(s_t, a_t); \\ &\quad s_{t+1} \sim P(\cdot|s_t, a_t). \end{aligned} \tag{1.57}$$

Our objective is to find a policy π^* that maximizes cumulative discounted award $\sum_{t=0}^{\infty} \gamma^t r_t$ on average.

To smooth the ensuing discussion, we need to introduce two functions, i.e. a value function and a Q-value function. The value function with the policy π is defined as the expectation of cumulative discounted reward, i.e.:

$$V^\pi(s) \triangleq \mathbb{E}\left[\sum_{t=0}^{\infty} \gamma^t r_t \middle| s_0 = s\right], \quad s \in \mathfrak{S}. \tag{1.58}$$

The Q-value function (also called action-value function) is defined as

$$Q^\pi(s,a) \triangleq \mathbb{E}\left[\sum_{t=1}^{\infty} \gamma^t r_t \middle| s_0 = s, a_0 = a\right], \quad s \in \mathfrak{S}, a \in \mathfrak{A}. \tag{1.59}$$

Intuitively, the value function and the Q-value function evaluate how good a state and a state-action pair are under a policy π, respectively. If we expand the summations in the value function, we have

$$\begin{aligned} V^\pi(s) &= \mathbb{E}\left[\sum_{t=0}^{\infty} \gamma^t r_t \middle| s_0 = s\right] \\ &= \mathbb{E}\left[r_0 + \sum_{t=1}^{\infty} \gamma^t r_t \middle| s_0 = s\right] \\ &= \sum_{a \in \mathfrak{A}} \pi(a|s) \left(R(s,a) + \gamma \sum_{s' \in \mathfrak{S}} P(s'|s,a) \mathbb{E}\left[\sum_{t=1}^{\infty} \gamma^{t-1} r_t \middle| s_1 = s'\right]\right) \\ &= \sum_{a \in \mathfrak{A}} \pi(a|s) \left(R(s,a) + \gamma \sum_{s' \in \mathfrak{S}} P(s'|s,a) V^\pi(s')\right) \end{aligned} \tag{1.60}$$

Similarly, for Q-value function, we have

$$Q^\pi(s,a) = R(s,a) + \gamma \sum_{s' \in \mathfrak{S}} P(s'|s,a) \sum_{a' \in \mathfrak{A}} \pi(a'|s') Q^\pi(s',a'). \tag{1.61}$$

Equations (1.60) and (1.61) are so-called Bellman equations, which are foundations of RL.

On the other hand, if we fix s and a, $V^{\pi}(s)$ and $Q^{\pi}(s,a)$ in fact evaluate how good a policy π is. Thus, a policy that maximizes $V^{\pi}(s)$ ($Q^{\pi}(s,a)$) will be a good candidate for π^* though s and a are fixed. An arising problem is that there exists a policy that maximizes $V^{\pi}(s)$ ($Q^{\pi}(s,a)$) for any $s \in \mathfrak{S}$ (and $a \in \mathfrak{A}$). The following theorem gives a positive answer:

Theorem 1.1. *[110] For any MDP, there exists an optimal policy π^* such that*

$$V^{\pi^*}(s) = \max_{\pi} V^{\pi}(s) \quad \forall s \in \mathfrak{S}$$

and

$$Q^{\pi^*}(s,a) = \max_{\pi} Q^{\pi}(s,a) \quad \forall s \in \mathfrak{S} \text{ and } \forall a \in \mathfrak{A}.$$

According to Theorem 1.1, we can define the optimal value function and the optimal Q-value function as

$$V^*(\cdot) \triangleq V^{\pi^*}(\cdot) \text{ and } Q^*(\cdot,\cdot) \triangleq Q^{\pi^*}(\cdot,\cdot) \tag{1.62}$$

respectively, which are useful in finding the optimal policy. Furthermore, if $V^*(\cdot)$ and $Q^*(\cdot,\cdot)$ have been obtained, we can construct the optimal policy π^* by letting:

$$\pi^*(a|s) = \begin{cases} 1 \ \text{if } a = \arg\max_{a \in \mathfrak{A}} Q^*(s,a) \\ \qquad\quad = \arg\max_{a \in \mathfrak{A}} R(s,a) + \gamma \sum_{s' \in \mathfrak{S}} P(s'|s,a) V^*(s) \\ 0 \ \text{otherwise.} \end{cases} \tag{1.63}$$

In other word, there always exists a deterministic optimal policy for any MDP. In addition, we have the Bellman optimality equations as follows:

$$V^*(s) = \max_{a} R(s,a) + \gamma \sum_{s' \in \mathfrak{S}} P(s'|s,a) V^*(s) \tag{1.64}$$

and

$$Q^*(s,a) = R(s,a) + \gamma \sum_{s' \in \mathfrak{S}} P(s'|s,a) \max_{a' \in \mathfrak{A}} Q^*(s',a'). \tag{1.65}$$

MDP and the Bellman equations are theoretical cornerstones of RL, based on which many RL algorithms have been derived as we will show below. For more results regarding MDP, readers can refer to [111].

1.3.2 Model-based methods

In this subsection, we will discuss model-based methods, where the term 'model-based' means that the model of MDP (i.e. $\langle \mathfrak{S}, \mathfrak{A}, P, R, \gamma \rangle$) have been given as known information. There are two typical model-based algorithms. The one is policy iteration, and the other is value iteration. We will introduce the former followed by the latter.

Algorithm 1.12: Computing value function

Input: MDP $\mathfrak{M} = \langle \mathfrak{S}, \mathfrak{A}, P, R, \gamma \rangle$, policy π
Output: value function V^π
Initialize V_0^π randomly;
for $i = 1, 2, \ldots$ **do**
  for $s \in \mathfrak{S}$ **do**

$$V_i^\pi(s) \leftarrow \sum_{a \in \mathfrak{A}} \pi(a|s) \left(R(s,a) + \gamma \sum_{s' \in \mathfrak{S}} P(s'|s,a) V_{i-1}^\pi(s') \right);$$

  if V_i^π *converges* **then**
    break;
$V^\pi \leftarrow V_i^\pi$

The policy iteration takes an iterative strategy to find the optimal policy π^*. Given an MDP $\mathfrak{M} = \langle \mathfrak{S}, \mathfrak{A}, P, R, \gamma \rangle$ and an initial policy π, the policy iteration alternatively executes the following two steps:

1. Computing the value function V^π based on $\mathfrak{M}$ and π.
2. Improve the policy π according to V^π.

How to compute a value function? Given an MDP and a policy, the corresponding value function can be evaluated by Bellman equation (1.60). This process is described in Algorithm 1.12. The function sequence in Algorithm 1.12 can be proved to converge to V^π.

How to improve a policy? Given an MDP and the value function of a policy, the policy can be improved by using (1.63). As a result, we conclude the policy iteration algorithm in Algorithm 1.13.

The value iteration, as its name suggests, iteratively updates a value function until it achieves the optimal value function. It has a very brief form since it just iterates according to the optimal Bellman equation (1.64). We present the value iteration algorithm in Algorithm 1.14. After obtaining the optimal value function, we can construct the optimal policy by using (1.63).

In summary, when an MDP is given as known information, we can use either of the policy and the value iteration to find the optimal policy. The value iteration has more simple form, but the policy iteration usually converges more quickly in practice [111]. In wireless communications, the policy and the value iteration methods have been applied to many tasks, such as heterogeneous wireless networks [112], energy-efficient communications [113] and energy harvesting [114].

Algorithm 1.13: Policy iteration

Input: MDP $\mathfrak{M} = \langle \mathfrak{S}, \mathfrak{A}, P, R, \gamma \rangle$
Output: optimal policy π^*
Initialize π randomly;
repeat
  Compute value function V^{π} by using Algorithm 1.12;

$$\bar{\pi}(a|s) = \begin{cases} 1 & \text{if } a = \arg\max_{a \in \mathfrak{A}} R(s,a) + \gamma \sum_{s' \in \mathfrak{S}} P(s'|s,a) V^{\pi}(s') \\ 0 & \text{otherwise.} \end{cases}$$

  $\pi \leftarrow \bar{\pi}$;
until π *converges*;

Algorithm 1.14: Value iteration

Input: MDP $\mathfrak{M} = \langle \mathfrak{S}, \mathfrak{A}, P, R, \gamma \rangle$
Output: optimal policy V^*
Initialize V^* randomly;
repeat
  $\bar{V}^*(s) = \max_{a \in \mathfrak{A}} R(s,a) + \gamma \sum_{s' \in \mathfrak{S}} P(s'|s,a) V^*(s')$;
  $V^* \leftarrow \bar{V}^*$;
until V^* *not change*;

1.3.3 Model-free methods

In practice, we often encounter a problem where the MDP model behind it is unknown to us, and thus model-based algorithms are prohibited in this situation. In this subsection, we will discuss two kinds of model-free methods: Monte Carlo (MC) methods, temporal-difference (TD) learning, which can be applied when the MDP model is unobservable.

1.3.3.1 Monte Carlo methods

The MC approach is a general idea derived from the law of large numbers, i.e. the average of the results obtained from numerous samples will be close to the expected value. It performs estimations of the value or Q-value function based on the experience of the agent (samples).

For example, consider the value function with a policy π:

$$V^{\pi}(s) = \mathbb{E}\left[\sum_{t=0}^{\infty} \gamma^t r_t \middle| s_0 = s \right].$$

Algorithm 1.15: Incremental Monte Carlo estimation

Input: experience trajectory $s_0, a_0, r_0, \ldots$, policy π, discount factor γ, truncation number n, iteration number m

Output: Q-value function Q^π

for $s \in \mathfrak{S}$ **do**
  for $a \in \mathfrak{A}$ **do**
    $N(s, a) \leftarrow 0$;
    $Q^\pi(s, a) \leftarrow 0$;
set $R = \sum_{t=0}^{n} \gamma^t r_t$;
for $t = 0, 1, \ldots, m$ **do**
  $N(s_t, a_t) \leftarrow N(s_t, a_t) + 1$;
  $Q^\pi(s_t, a_t) \leftarrow Q^\pi(s_t, a_t) + \frac{1}{N(s_t, a_t)}(R - Q^\pi(s_t, a_t))$;
  $R \leftarrow \frac{R - r_t}{\gamma} + \gamma^n r_{t+1+n}$;
  if Q^π *converges* **then**
    break;

It is defined by the expectation of all trails starting from s. Now, suppose that we independently conduct l experiments by applying the policy π, and thus we obtain l trajectories $\{\tau_i\}_{i=1}^{l}$, where $\tau_i = \left\langle s \equiv s_0^{(i)}, a_0^{(i)}, r_0^{(i)}, s_1^{(i)}, a_1^{(i)}, r_1^{(i)}, \ldots, r_{n_i}^{(i)} \right\rangle$. Let $R^{(i)} = \sum_{t=0}^{n_i} \gamma^t r_t^{(i)}$. Then, according to the law of large numbers, we have

$$
\begin{aligned}
V^\pi(s) &= \mathbb{E}\left[\sum_{t=0}^{\infty} \gamma^t r_t \middle| s_0 = s\right] \approx \frac{1}{l}\sum_{i=1}^{l} R^{(i)} \\
&= \frac{1}{l-1}\sum_{i=1}^{l-1} R^{(i)} + \frac{1}{l}\left(R^{(l)} - \frac{1}{l-1}\sum_{i=1}^{l-1} R^{(i)}\right)
\end{aligned}
$$

when l is large enough. Therefore, the value function can be estimated if we afford numerous experiments. Similarly, the Q-value function can also be estimated by using the MC method. However, in practice, we often face an online infinity trajectory: $s_0, a_0, r_0, \ldots$. In this situation, we can update the Q-value (or value) function incrementally as shown in Algorithm 1.15. The truncation number n in Algorithm 1.15 is used to shield negligible remainders.

Once we estimate the Q-value function for a given policy π, the optimal policy can be obtained iteratively as presented in Algorithm 1.16. Here ε is introduced to take a chance on small probability events.

MC methods are unbiased and easy to implement. However, they often suffer from high variance in practice since the MDP model in real world may be so complicated that a huge amount of samples are required to achieve a stable estimation. This restricts the usage of MC methods when the cost of experiments is high.

Algorithm 1.16: Incremental Monte Carlo policy iteration

Input: discount factor γ, truncation number n, iteration number m, ε
Output: optimal policy π^*
Initialize π randomly;
repeat
  Apply π to generate a trail $s_0, a_0, r_0, \ldots, r_{n+m}$;
  Estimate Q-value function Q^π by using Algorithm 1.15;

$$\bar{\pi}(a|s) = \begin{cases} (1-\varepsilon) & \text{if } a = \arg\max_{a\in\mathfrak{A}} Q^\pi(s,a) \\ \varepsilon/(|\mathfrak{S}|-1) & \text{otherwise.} \end{cases}$$

  $\pi \leftarrow \bar{\pi}$;
until π *converges*;

1.3.3.2 Temporal-difference learning

Like the MC approach, TD learning also tries to estimate the value or Q-value function from the experience of the agent. But it performs an incremental estimation based on the Bellman equations besides MC sampling.

To begin with, suppose that we obtain a sample set $\left\{\langle s, a^{(i)}, r^{(i)}, s^{(i)}\rangle\right\}_{i=1}^{l}$ by applying a policy π. Then, by applying MC sampling to the Bellman equation of the value function, we have

$$\begin{aligned} V^\pi(s) &= \mathbb{E}\left[R(s,a) + \gamma V^\pi(s')|\pi\right] \\ &\approx \frac{1}{l}\sum_{i=1}^{l} r^{(i)} + \gamma V^\pi(s^{(i)}) \\ &= \mu_{l-1} + \frac{1}{l}\left(r^{(l)} + \gamma V^\pi(s^{(l)}) - \mu_{l-1}\right), \\ &\approx V^\pi(s) + \frac{1}{l}\left(r^{(l)} + \gamma V^\pi(s^{(l)}) - V^\pi(s)\right), \end{aligned} \tag{1.66}$$

where $\mu_{l-1} = (1/(l-1))\sum_{i=1}^{l-1} r^{(i)} + \gamma V^\pi(s^{(i)})$. Therefore, to acquire an estimation of $V^\pi(s)$, we can update it by the fixed point iteration [111]:

$$V^\pi(s) \leftarrow V^\pi(s) + \frac{1}{l}\left(r^{(l)} + \gamma V^\pi(s^{(l)}) - V^\pi(s)\right). \tag{1.67}$$

In practice, $1/l$ in (1.67) is usually replaced by a monotonically decreasing sequence. So far, we have presented the main idea of the TD learning. The detailed steps of TD learning are concluded in Algorithm 1.17, where the learning rate sequence should satisfy $\sum_{t=0}^{\infty}\alpha_t = \infty$ and $\sum_{t=0}^{\infty}\alpha_t^2 < \infty$.

Similarly, the Q-value function w.r.t. a policy can be estimated by using Algorithm 1.18, which is also known as the Sarsa algorithm. Based on the Sarsa algorithm,

Algorithm 1.17: TD learning

Input: experience trajectory $s_0, a_0, r_0, s_1 \ldots$ w.r.t. policy π, discount factor γ, learning rate sequence $\alpha_0, \alpha_1, \ldots$
Output: value function V^π

for $s \in \mathfrak{S}$ **do**
  $V^\pi(s) \leftarrow 0$;
for $t = 0, 1, \ldots$ **do**
  $V^\pi(s_t) \leftarrow V^\pi(s_t) + \alpha_t\,(r_t + \gamma V^\pi(s_{t+1}) - V^\pi(s_t))$;

Algorithm 1.18: Sarsa algorithm

Input: experience trajectory $s_0, a_0, r_0, s_1 \ldots$ w.r.t. policy π, discount factor γ, learning rate sequence $\alpha_0, \alpha_1, \ldots$
Output: Q-value function V^π

for $s \in \mathfrak{S}$ **do**
  for $a \in \mathfrak{A}$ **do**
    $Q^\pi(s, a) \leftarrow 0$;
for $t = 0, 1, \ldots$ **do**
  $Q^\pi(s_t, a_t) \leftarrow Q^\pi(s_t, a_t) + \alpha_t\,(r_t + \gamma Q^\pi(s_{t+1}, a_{t+1}) - Q^\pi(s_t, a_t))$;

we can improve the policy alternatively by using Algorithm 1.16, where the Q-value function is estimated by the Sarsa algorithm.

On the other hand, if choosing the optimal Bellman equation (1.65) as the iteration strategy, we can derive the famous Q-learning algorithm as presented in Algorithm 1.19.

In summary, TD learning, Sarsa and Q-learning are all algorithms based on the Bellman equations and MC sampling. Among them, the goal of TD learning and Sarsa is to estimate the value or Q-value function for a given policy, while Q-learning aims at learning the optimal Q-value function directly. It should be noted that, by using TD learning, we can only estimate the value function, which is not enough to determine a policy because the state transition probability is unknown to us. In contrast, a policy can be derived from the Q-value function, which is estimated by Sarsa and Q-learning. In practice, Sarsa often demonstrates better performance than Q-learning. Furthermore, all of the three methods can be improved to converge more quickly by introducing the eligibility trace. Readers can refer to [111] for more details.

Moreover, TD learning, Sarsa and Q-learning have been widely applied in wireless communications. References [115] and [116] demonstrate two applications ofTD learning in energy-aware sensor communications and detection of spectral resources,

Algorithm 1.19: Q-learning

Input: discount factor γ, learning rate sequence $\alpha_0, \alpha_1, \ldots$
Output: optimal Q-value function Q^*

for $s \in \mathfrak{S}$ **do**
  for $a \in \mathfrak{A}$ **do**
    $Q^*(s, a) \leftarrow 0$;
Initialize s_0;
for $t = 0, 1, \ldots$ **do**
  $a_t \sim \pi(\cdot|s_t)$, where

$$\pi(a|s) = \begin{cases} (1-\varepsilon) & \text{if } a = \arg\max_{a\in\mathfrak{A}} Q^*(s,a) \\ \varepsilon/(|\mathfrak{S}|-1) & \text{otherwise.} \end{cases}$$

  Take action a_t, observe r_t and s_{t+1};
  $Q^*(s_t, a_t) \leftarrow Q^*(s_t, a_t) + \alpha_t\,(r_t + \gamma \max_{a\in\mathfrak{A}} Q^*(s_{t+1}, a) - Q^*(s_t, a_t))$;

respectively. References [117], [118] and [119] show three applications of Sarsa in channel allocation, interference mitigation and energy harvesting, respectively. References [120], [121] and [122] present three applications of Q-learning in routing protocols, power allocation and caching policy, respectively.

1.3.4 Deep reinforcement learning

So far we have discussed both model-based and model-free methods in RL. All of these methods need to store one or two tables with size $|\mathfrak{S}|$ (for the value function) or $|\mathfrak{S}| \times |\mathfrak{A}|$ (for the Q-value function and the policy). In practice, however, we often encounter the situation where $|\mathfrak{S}|$ is very large or even infinite. In this case, it is impractical to store a table whose size is proportional to $|\mathfrak{S}|$. DNNs, as discussed in Section 1.1.3.3, have a strong ability in representation and can be used to approximate a complex function. As a result, DNNs have been applied to approximate the value function, the Q-value function and the policy. In this subsection, we will discuss these approximation ideas.

1.3.4.1 Value function approximation

First of all, we consider the problem of approximating the value function by using a DNN. For simplicity, let $\hat{V}(s, \mathbf{W})$ denote a DNN which is with parameter $\mathbf{W}$ and receives an input $s \in \mathfrak{S}$. For a given policy π, our goal is turning $\mathbf{W}$ so as to $\hat{V}(s, \mathbf{W}) \approx V^\pi(s), \forall s \in \mathfrak{S}$. Unfortunately, since the true $V^\pi(\cdot)$ is unknown to us, we cannot learn $\mathbf{W}$ directly. Alternatively, let us consider minimizing the difference between $\hat{V}(\cdot, \mathbf{W})$ and $V^\pi(\cdot)$ in expectation, i.e.:

$$\min_{\mathbf{W}} \frac{1}{2} \mathop{\mathbb{E}}_{s} \left(\hat{V}(s, \mathbf{W}) - V^\pi(s) \right)^2. \tag{1.68}$$

Algorithm 1.20: Value function approximation

Input: a sample set $\mathfrak{D} = \{(s^{(i)}, r^{(i)}, s'^{(i)})\}_{i=1}^{l}$, batch size m, learning rate α
Output: approximate value function $\hat{V}(\cdot, \mathbf{W})$

Initialize $\mathbf{W}$;
repeat
  Randomly sample a subset $\{(s^{(j)}, r^{(j)}, s'^{(j)})\}_{j=1}^{m}$ from $\mathfrak{D}$;
  $\mathbf{W} \leftarrow \mathbf{W} + \frac{\alpha}{m}\sum_{j=1}^{m}\left(\hat{V}(s^{(j)}, \mathbf{W}) - (r^{(j)} + \hat{V}(s'^{(j)}, \mathbf{W})\right)\nabla_{\mathbf{W}}\hat{V}(s^{(j)}, \mathbf{W})$;
until *convergence*;

As mentioned in Section 1.1.3.3, we use the gradient descent method to update $\mathbf{W}$. Taking the gradient of (1.68) w.r.t. $\mathbf{W}$, we have

$$\nabla_{\mathbf{W}}\frac{1}{2}\mathop{\mathbb{E}}_{s}\left(\hat{V}(s, \mathbf{W}) - V^{\pi}(s)\right)^2 = \mathop{\mathbb{E}}_{s}\left(\hat{V}(s, \mathbf{W}) - V^{\pi}(s)\right)\nabla_{\mathbf{W}}\hat{V}(s, \mathbf{W}). \tag{1.69}$$

By applying the Bellman equation, (1.69) can be transformed into

$$\mathop{\mathbb{E}}_{s,r,s'}\left(\hat{V}(s, \mathbf{W}) - (r + V^{\pi}(s'))\right)\nabla_{\mathbf{W}}\hat{V}(s, \mathbf{W}). \tag{1.70}$$

However, since the true $V^{\pi}(s')$ is unknown, we substitute $\hat{V}(s', \mathbf{W})$ for $V^{\pi}(s')$ and get

$$\mathop{\mathbb{E}}_{s,r,s'}\left(\hat{V}(s, \mathbf{W}) - (r + \hat{V}(s', \mathbf{W})\right)\nabla_{\mathbf{W}}\hat{V}(s, \mathbf{W}). \tag{1.71}$$

Now, if we have obtained a finite sample set $\{(s^{(i)}, r^{(i)}, s'^{(i)})\}_{i=1}^{l}$ from the experience, (1.71) can be estimated as

$$\frac{1}{l}\sum_{i=1}^{l}\left(\hat{V}(s^{(i)}, \mathbf{W}) - (r^{(i)} + \hat{V}(s'^{(i)}, \mathbf{W})\right)\nabla_{\mathbf{W}}\hat{V}(s^{(i)}, \mathbf{W}). \tag{1.72}$$

Thus, we can use (1.72) to update $\mathbf{W}$ until convergence. The value function approximation via DNNs is summarized in Algorithm 1.20.

On the other hand, for the Q-value function, we can approximate it with a similar way as described in Algorithm 1.21. After the value function or the Q-value function is approximated, we can work out the optimal policy by using the policy iteration (Algorithms 1.13 or 1.16). However, given a large size problem, a more smarter way is to parametrize the policy by using another DNN, which will be discussed in the following part.

1.3.4.2 Policy gradient methods

Similar to value or Q-value functions, a policy can be parametrized by using DNNs too. However, it is non-trivial to estimate a gradient to improve the parametrized policy. Accordingly, a policy gradient has been proposed to solve this problem. In this part, we will discuss the policy gradient and its derivation.

Algorithm 1.21: Q-value function approximation

Input: a sample set $\mathfrak{D} = \{(s^{(i)}, a^{(i)}, r^{(i)}, s'^{(i)}, a'^{(i)})\}_{i=1}^{l}$, batch size m, learning rate α

Output: approximate value function $\hat{Q}(\cdot, \cdot, \mathbf{U})$

Initialize $\mathbf{U}$;

repeat

Randomly sample a subset $\{(s^{(j)}, a^{(j)}, r^{(j)}, s'^{(j)}, a'^{(j)})\}_{j=1}^{m}$ from $\mathfrak{D}$;

$\mathbf{U} \leftarrow$

$\mathbf{U} + \frac{\alpha}{m}\sum_{j=1}^{m}\left(\hat{Q}(s^{(j)}, a^{(j)}, \mathbf{U}) - (r^{(j)} + \hat{Q}(s'^{(j)}, a'^{(j)}, \mathbf{U})\right)\nabla_{\mathbf{U}}\hat{Q}(s^{(j)}, a^{(j)}, \mathbf{U})$;

until *convergence*;

To begin with, let $\hat{\pi}(s, a, \theta)$ denote a DNN which is with parameter θ and receive two inputs $s \in \mathfrak{S}$ and $a \in \mathfrak{A}$. Our goal is to learn the parameter θ such that the expectation of the total reward is maximized, i.e.:

$$\max_{\theta} J(\theta) \triangleq \mathbb{E}\left[\sum_{t=0}^{\infty}\gamma^t r_t \middle| \hat{\pi}(\cdot, \cdot, \theta)\right] = \int_{\tau} g(\tau)\mathbb{P}\{\tau | \hat{\pi}(\cdot, \cdot, \theta)\}\mathrm{d}\tau, \tag{1.73}$$

where $\tau = \langle s_0, a_0, r_0, s_1, a_1, r_1, \ldots \rangle$ and $g(\tau) = \sum_{t=0}^{\infty}\gamma^t r_t$ denote a trajectory and its reward, respectively. To update θ, we need to take the gradient w.r.t. θ, that is

$$\nabla_\theta J(\theta) = \int_{\tau} g(\tau)\nabla_\theta\mathbb{P}\{\tau | \hat{\pi}(\cdot, \cdot, \theta)\}\mathrm{d}\tau. \tag{1.74}$$

But the gradient in (1.74) is hard to be estimated since it relies on the probability. Fortunately, this difficulty can be resolved by using a nice trick as follows:

$$\begin{aligned}
\nabla_\theta J(\theta) &= \int_{\tau} g(\tau)\nabla_\theta\mathbb{P}\{\tau | \hat{\pi}(\cdot, \cdot, \theta)\}\mathrm{d}\tau \\
&= \int_{\tau} g(\tau)\mathbb{P}\{\tau | \hat{\pi}(\cdot, \cdot, \theta)\}\frac{\nabla_\theta\mathbb{P}\{\tau | \hat{\pi}(\cdot, \cdot, \theta)\}}{\mathbb{P}\{\tau | \hat{\pi}(\cdot, \cdot, \theta)\}}\mathrm{d}\tau \\
&= \int_{\tau} g(\tau)\mathbb{P}\{\tau | \hat{\pi}(\cdot, \cdot, \theta)\}\nabla_\theta \log\mathbb{P}\{\tau | \hat{\pi}(\cdot, \cdot, \theta)\}\mathrm{d}\tau \\
&= \mathbb{E}\left[g(\tau)\nabla_\theta \log\mathbb{P}\{\tau | \hat{\pi}(\cdot, \cdot, \theta)\}\middle| \hat{\pi}(\cdot, \cdot, \theta)\right].
\end{aligned} \tag{1.75}$$

Moreover, we have

$$\begin{aligned}\nabla_\theta \log \mathbb{P}\{\tau|\hat{\pi}(\cdot,\cdot,\theta)\} &= \nabla_\theta \log \left(\mathbb{P}(s_0) \prod_{t=0}^{\infty} \hat{\pi}(s_t, a_t, \theta) P(s_{t+1}|s_t, a_t) \right) \\ &= \nabla_\theta \left(\log \mathbb{P}(s_0) + \sum_{t=0}^{\infty} \log P(s_{t+1}|s_t, a_t) + \sum_{t=0}^{\infty} \log \hat{\pi}(s_t, a_t, \theta) \right) \\ &= \sum_{t=0}^{\infty} \nabla_\theta \log \hat{\pi}(s_t, a_t, \theta). \end{aligned} \tag{1.76}$$

Plugging (1.76) into (1.75), we have

$$\nabla_\theta J(\theta) = \mathbb{E}\left[g(\tau) \sum_{t=0}^{\infty} \nabla_\theta \log \hat{\pi}(s_t, a_t, \theta) \Big| \hat{\pi}(\cdot,\cdot,\theta) \right]. \tag{1.77}$$

Equation (1.77) can be estimated by the MC approach in principle. In practice, however, it suffers from high variance because credit assignment is really hard [123]. A way to reduce the variance is to replace (1.77) by the following equation [124]:

$$\begin{aligned}\nabla_\theta J(\theta) &\approx \mathbb{E}\left[\sum_{t=0}^{\infty} (Q^{\hat{\pi}(\cdot,\cdot,\theta)}(s_t, a_t) - V^{\hat{\pi}(\cdot,\cdot,\theta)}(s_t)) \nabla_\theta \log \hat{\pi}(s_t, a_t, \theta) \Big| \hat{\pi}(\cdot,\cdot,\theta) \right] \\ &\approx \frac{1}{l} \sum_{i=1}^{l} (Q^{\hat{\pi}(\cdot,\cdot,\theta)}(s^{(i)}, a^{(i)}) - V^{\hat{\pi}(\cdot,\cdot,\theta)}(s^{(i)})) \nabla_\theta \log \hat{\pi}(s^{(i)}, a^{(i)}, \theta), \end{aligned} \tag{1.78}$$

where $\{(s^{(i)}, a^{(i)})\}_{i=1}^{l}$ is a sample set from the experience under the policy $\hat{\pi}(\cdot,\cdot,\theta)$. So far, a remaining problem is that $Q^{\hat{\pi}(\cdot,\cdot,\theta)}$ and $V^{\hat{\pi}(\cdot,\cdot,\theta)}$ are unknown to us. The answer would be using the value and Q-value function approximations as described in Section 1.3.4.1. We summary the whole process in Algorithm 1.22. This algorithm is known as the famous actor–critic (AC) algorithm (also known as the A3C algorithm), where actor and critic refer to the policy DNN and the value (Q-value) DNN, respectively.

It is worth mentioning that the AC algorithm has an extension named asynchronous advantage AC algorithm [125]. The A3C algorithm has better convergence and became a standard starting point in many recent works [126].

DRL is popular with current wireless communications research. For example, Q-value function approximation has been applied in mobile edge computing [127], resource allocation [128] and base station control [129]. In addition, [130], [131] and [132] demonstrate three applications of actor–critic algorithm in quality of service (QoS) driven scheduling, bandwidth intrusion detection and spectrum management, respectively.

1.3.5 Summary of reinforcement learning

In this section, we have discussed RL, which is an effective tool to solve real-time control problems in various fields. As a theoretical basis of RL, the MDP theory has provided essential concepts for the RL algorithm design, such as Bellman equations, (optimal) value function, (optimal) Q-value function and optimal policy. As shown in

Algorithm 1.22: Actor–critic algorithm

Input: sampling sizes l, m, learning rates $\alpha_1, \alpha_2, \alpha_3$

Output: approximate optimal policy $\hat{\pi}(\cdot, \cdot, \theta)$, approximate optimal value function $\hat{V}(\cdot, \mathbf{W})$, approximate Q-value function $\hat{Q}(\cdot, \cdot, \mathbf{U})$

Initialize $\theta, \mathbf{U}, \hat{Q}(\cdot, \cdot, \mathbf{U})$;

repeat

  Generate a sample set $\mathfrak{D} = \{(s^{(i)}, a^{(i)}, r^{(i)}, s'^{(i)}, a'^{(i)})\}_{i=1}^{l}$ by using policy $\hat{\pi}(\cdot, \cdot, \theta)$;

  Update $\mathbf{W}$ by using Algorithm 1.20 with $\mathfrak{D}, m, \alpha_1, \mathbf{W}$ (without parameter initialization);

  Update $\mathbf{U}$ by using Algorithm 1.21 with $\mathfrak{D}, m, \alpha_2, \mathbf{U}$ (without parameter initialization);

$$\theta \leftarrow \theta + \frac{\alpha_3}{l} \sum_{i=1}^{l} (Q(s^{(i)}, a^{(i)}, \mathbf{U}) - V(s^{(i)}, \mathbf{W})) \nabla_\theta \log \hat{\pi}(s^{(i)}, a^{(i)}, \theta);$$

until *convergence*;

Figure 1.22, we have introduced three parts of RL: model-based methods, model-free methods and DRL.

Model-based methods assume that the MDP model is given as prior information. Based on the model information and Bellman equations, this kind of algorithms try to learn the (optimal) value function, the (optimal) Q-value function and the optimal policy. In general, model-based algorithms have a better effect and a faster convergence than model-free algorithms provided that the given MDP model is accurate. However, model-based algorithms are rarely used in practice, since MDP models in real world are usually too complicated to be estimated accurately.

Model-free methods are designed for the case where information of hidden MDP is unknown. Model-free algorithms can be further divided into two subclasses: MC methods and TD learning. Based on the law of large numbers, MC methods try to estimate the value or Q-value function from an appropriate number of samples generated from experiments. MC methods are unbiased, but they suffer from high variance in practice since MDP models in real world are usually complex such that it needs massive samples to achieve a stable result. On the other hand, TD learning integrates the Bellman equations and MC sampling in its algorithms design. By introducing the Bellman equations, TD learning reduces the estimation variance compared with MC methods, though its estimation may be biased. TD learning has shown a decent performance in practice and provides basic ideas for many subsequent RL algorithms.

DRL is proposed to deal with the condition where the number of states is extremely large or even infinite. DRL applies DNNs to approximate the value function, the Q-value function and the policy. Among them, the update rule of the value

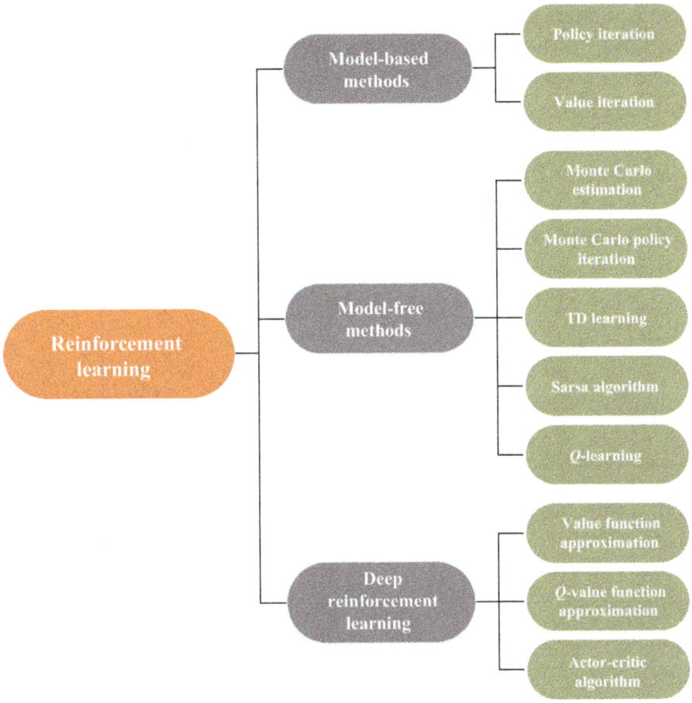

Figure 1.22 Structure chart for reinforcement-learning technologies discussed in this chapter

Table 1.4 Summary of applications of reinforcement learning in wireless communications

Method	**Application in wireless communications**
Policy iteration	Energy harvesting [114]
Value iteration	Heterogeneous wireless networks [112]
	Energy-efficient communications [113]
TD learning	Energy-aware sensor communications [115]
	Detection of spectral resources [116]
Sarsa	Channel allocation [117]
	Interference mitigation [118]
	Energy harvesting [119]
Q-learning	Routing protocols [120]
	Power allocation [121]
	Caching policy [122]
Q-value function approximation	Mobile edge computing [127]
	Resource allocation [128]
	Base station control [129]
Actor–critic algorithm	QoS-driven scheduling [130]
	Bandwidth intrusion detection [131]
	Spectrum management [132]

(Q-value) approximation is very similar to that of TD learning except replacing a table with a DNN in the learning process. In this case, the parameter of the DNN can be trained conveniently by using the gradient descent method. In contrast, the policy approximation is more difficult since its gradient cannot be estimated directly. Accordingly, the policy gradient is introduced which provides an approximate gradient to update parameters. Based on the policy gradient, actor–critic (AC) algorithm is proposed where both the actor and the critic are realized by DNNs. AC algorithm is very practical and has become a framework of many cutting-edge RL techniques.

In Table 1.4, we summarize the applications of RL in wireless communications. A historical survey of RL can be seen in [133], and the new developments in DRL can be seen in [126].

1.4 Summary

In this chapter, we have reviewed three main branches of machine learning: supervised learning, unsupervised learning and RL. Supervised learning tries to learn a function that maps an input to an output by referring to a training set. A supervised learning task is called a classification task or a regression task according to whether the predicted variable is categorical or continuous. In contrast, unsupervised learning aims at discovering and exploring the inherent and hidden structures of a data set without labels. Unsupervised learning has three main functions: clustering, density estimation and dimension reduction. RL is commonly employed to deal with the optimal decision-making in a dynamic system. By modelling the problem as the MDP, RL seeks to find an optimal policy. An RL algorithm is called a model-based algorithm or a model-free algorithm depending on whether the MDP model parameter is required or not. Furthermore, if an RL algorithm applies DNNs to approximate a function, it also called a deep RL method.

There is no doubt that machine learning is achieving increasingly promising results in wireless communications. However, there are several essential open-research issues that are noteworthy in the future [59]:

1. In general, supervised models require massive training data to gain satisfying performance, especially for deep models. Unfortunately, unlike some popular research areas such as computer vision and NLP, there still lacks high-quality and large-volume-labelled data sets for wireless applications. Moreover, due to limitations of sensors and network equipment, wireless data collected are usually subjected to loss, redundancy, mislabelling and class imbalance. How to implement supervised learning with limited low-quality training data is a significant and urgent problem in the research of wireless learning.
2. On the other hand, wireless networks generate large amounts of data every day. However, data labelling is an expensive and time-consuming process. To facilitate the analysis of raw wireless data, unsupervised learning is increasingly essential in extracting insights from unlabelled data [134]. Furthermore, recent success in generative models (e.g. variational autoencoder and generative adversarial networks) greatly boosts the development of unsupervised learning.

It will be worthwhile and beneficial to employ these new technologies to handle unsupervised tasks in wireless communications.

3. Currently, many wireless network control problems have been solved by constrained optimization, dynamic programming and game theory approaches. These methods either make strong assumptions about the objective functions (e.g. linearity or convexity) or sample distribution (e.g. Gaussian or Poisson distributed), or endure high time and space complexity. Unfortunately, as wireless networks are getting increasingly complex, such assumptions become unrealistic sometimes. As a solution, DRL is a powerful tool to handle complex control problems. Inspired by its remarkable achievements in self-driving [135] and the game of Go [136], a few researchers start to apply DRL to solve the wireless network control problems. However, this work only demonstrates a small part of DRL's advantages, and its potential in wireless communications remains largely unexplored.

Acknowledgement

This work is supported in part by the National Natural Science Foundation of China (Grant No. 61501022).

References

[1] Samuel AL. Some studies in machine learning using the game of checkers. IBM Journal of Research and Development. 1959;3(3):210–229.
[2] Tagliaferri L. An Introduction to Machine Learning; 2017. https://www.digitalocean.com/community/tutorials/an-introduction-to-machine-learning.
[3] Feng Vs, and Chang SY. Determination of wireless networks parameters through parallel hierarchical support vector machines. IEEE Transactions on Parallel and Distributed Systems. 2012;23(3):505–512.
[4] Deza E, and Deza MM. Dictionary of Distances. Elsevier; 2006. Available from: https://www.sciencedirect.com/science/article/pii/B9780444520876500007.
[5] Everitt BS, Landau S, Leese M, *et al.* Miscellaneous Clustering Methods. Hoboken, NJ: John Wiley & Sons, Ltd; 2011.
[6] Kohavi R. A study of cross-validation and bootstrap for accuracy estimation and model selection. In: International Joint Conference on Artificial Intelligence; 1995. p. 1137–1143.
[7] Samet H. The Design and Analysis of Spatial Data Structures. Boston, MA: Addison-Wesley; 1990.
[8] Hastie T, Tibshirani R, and Friedman J. The Elements of Statistical Learning: Data Mining, Inference and Prediction. 2nd ed. Berlin: Springer; 2008.
[9] Friedman JH. Flexible Metric Nearest Neighbor Classification; 1994. Technical report. Available from: https://statistics.stanford.edu/research/flexible-metric-nearest- neighbor-classification.

[10] Erdogan SZ, and Bilgin TT. A data mining approach for fall detection by using k-nearest neighbour algorithm on wireless sensor network data. IET Communications. 2012;6(18):3281–3287.

[11] Donohoo BK, Ohlsen C, Pasricha S, *et al.* Context-aware energy enhancements for smart mobile devices. IEEE Transactions on Mobile Computing. 2014;13(8):1720–1732.

[12] Quinlan JR. Induction of decision trees. Machine Learning. 1986;1(1): 81–106. Available from: https://doi.org/10.1007/BF00116251.

[13] Quinlan JR. C4.5: Programs for Machine Learning. San Francisco, CA, USA: Morgan Kaufmann Publishers Inc.; 1993.

[14] Breiman L, Friedman J, Stone CJ, *et al.* Classification and Regression Trees. The Wadsworth and Brooks-Cole statistics-probability series. Taylor & Francis; 1984. Available from: https://books.google.com/books?id=JwQx-WOmSyQC.

[15] Geurts P, El Khayat I, and Leduc G. A machine learning approach to improve congestion control over wireless computer networks. In: International Conference on Data Mining. IEEE; 2004. p. 383–386.

[16] Nadimi ES, Søgaard HT, and Bak T. ZigBee-based wireless sensor networks for classifying the behaviour of a herd of animals using classification trees. Biosystems Engineering. 2008;100(2):167–176.

[17] Coppolino L, D'Antonio S, Garofalo A, *et al.* Applying data mining techniques to intrusion detection in wireless sensor networks. In: International Conference on P2P, Parallel, Grid, Cloud and Internet Computing. IEEE; 2013. p. 247–254.

[18] Breiman L. Random forests. Machine Learning. 2001;45(1):5–32. Available from: https://doi.org/10.1023/A:1010933404324.

[19] Calderoni L, Ferrara M, Franco A, *et al.* Indoor localization in a hospital environment using random forest classifiers. Expert Systems with Applications. 2015;42(1):125–134.

[20] Wang Y, Wu K, and Ni LM. WiFall: Device-free fall detection by wireless networks. IEEE Transactions on Mobile Computing. 2017;16(2):581–594.

[21] Friedman JH. Greedy function approximation: a gradient boosting machine. Annals of Statistics. 2001;29:1189–1232.

[22] Freund Y, and Schapire RE. A decision-theoretic generalization of on-line learning and an application to boosting. Journal of Computer and System Sciences. 1997;55(1):119–139.

[23] Chen T, and Guestrin C. XGBoost: a scalable tree boosting system. In: SIGKDD International Conference on Knowledge Discovery and Data Mining. ACM; 2016. p. 785–794.

[24] Ke G, Meng Q, Finley T, *et al.* LightGBM: a highly efficient gradient boosting decision tree. In: Advances in Neural Information Processing Systems; 2017. p. 3149–3157.

[25] Yu X, Chen H, Zhao W, *et al.* No-reference QoE prediction model for video streaming service in 3G networks. In: International Conference on Wireless Communications, Networking and Mobile Computing. IEEE; 2012. p. 1–4.

[26] Sattiraju R, Kochems J, and Schotten HD. Machine learning based obstacle detection for Automatic Train Pairing. In: International Workshop on Factory Communication Systems. IEEE; 2017. p. 1–4.
[27] Novikoff AB. On convergence proofs on perceptrons. In: Proceedings of the Symposium on the Mathematical Theory of Automata. vol. 12. New York, NY, USA: Polytechnic Institute of Brooklyn; 1962. p. 615–622.
[28] Chi CY, Li WC, and Lin CH. Convex Optimization for Signal Processing and Communications: From Fundamentals to Applications. Boca Raton, FL: CRC Press; 2017.
[29] Grant M, Boyd S, and Ye Y. CVX: Matlab Software for Disciplined Convex Programming; 2008. Available from: http://cvxr.com/cvx.
[30] Platt J. Sequential Minimal Optimization: A Fast Algorithm for Training Support Vector Machines; 1998. Technical report. Available from: https://www.microsoft.com/en-us/research/publication/sequential-minimal-optimization-a-fast-algorithm-for-training-support-vector-machines/.
[31] Cortes C, and Vapnik V. Support-vector networks. Machine Learning. 1995; 20(3):273–297.
[32] Schölkopf B, and Smola AJ. Learning with Kernels: Support Vector Machines, Regularization, Optimization, and Beyond. Cambridge, MA: MIT Press; 2002.
[33] Smola AJ, and Schölkopf B. A tutorial on support vector regression. Statistics and Computing. 2004;14(3):199–222.
[34] Gandetto M, Guainazzo M, and Regazzoni CS. Use of time-frequency analysis and neural networks for mode identification in a wireless software-defined radio approach. EURASIP Journal on Applied Signal Processing. 2004;2004:1778–1790.
[35] Kaplantzis S, Shilton A, Mani N, *et al.* Detecting selective forwarding attacks in wireless sensor networks using support vector machines. In: International Conference on Intelligent Sensors, Sensor Networks and Information. IEEE; 2007. p. 335–340.
[36] Huan R, Chen Q, Mao K, *et al.* A three-dimension localization algorithm for wireless sensor network nodes based on SVM. In: International Conference on Green Circuits and Systems. IEEE; 2010. p. 651–654.
[37] Woon I, Tan GW, and Low R. Association for Information Systems. A protection motivation theory approach to home wireless security. In: International Conference on Information Systems; 2005. p. 31.
[38] Huang F, Jiang Z, Zhang S, *et al.* Reliability evaluation of wireless sensor networks using logistic regression. In: International Conference on Communications and Mobile Computing. vol. 3. IEEE; 2010. p. 334–338.
[39] Salem O, Guerassimov A, Mehaoua A, *et al.* Sensor fault and patient anomaly detection and classification in medical wireless sensor networks. In: International Conference on Communications. IEEE; 2013. p. 4373–4378.
[40] Gulcehre C, Moczulski M, Denil M, *et al.* Noisy activation functions. In: International Conference on Machine Learning; 2016. p. 3059–3068.
[41] LeCun Y, Bottou L, Bengio Y, *et al.* Gradient-based learning applied to document recognition. Proceedings of the IEEE. 1998;86(11):2278–2324.

[42] LeCun Y, Bengio Y, and Hinton G. Deep learning. Nature. 2015; 521(7553):436.

[43] Goodfellow I, Bengio Y, Courville A, *et al.* Deep learning. vol. 1. Cambridge: MIT Press; 2016.

[44] Krizhevsky A, Sutskever I, and Hinton GE. ImageNet classification with deep convolutional neural networks. In: Advances in Neural Information Processing Systems; 2012. p. 1097–1105.

[45] Goodfellow I, Pouget-Abadie J, Mirza M, *et al.* Generative adversarial nets. In: Advances in Neural Information Processing Systems; 2014. p. 2672–2680.

[46] Mnih V, Kavukcuoglu K, Silver D, *et al.* Human-level control through deep reinforcement learning. Nature. 2015;518(7540):529.

[47] Schmidhuber J. Deep learning in neural networks: an overview. Neural Networks. 2015;61:85–117.

[48] Pierucci L, and Micheli D. A neural network for quality of experience estimation in mobile communications. IEEE MultiMedia. 2016;23(4):42–49.

[49] Nie L, Jiang D, Yu S, *et al.* Network traffic prediction based on deep belief network in wireless mesh backbone networks. In: Wireless Communications and Networking Conference. IEEE; 2017. p. 1–5.

[50] Wang W, Zhu M, Wang J, *et al.* End-to-end encrypted traffic classification with one-dimensional convolution neural networks. In: International Conference on Intelligence and Security Informatics. IEEE; 2017. p. 43–48.

[51] Wang W, Zhu M, Zeng X, *et al.* Malware traffic classification using convolutional neural network for representation learning. In: International Conference on Information Networking. IEEE; 2017. p. 712–717.

[52] O'Shea TJ, Pemula L, Batra D, *et al.* Radio transformer networks: attention models for learning to synchronize in wireless systems. In: Signals, Systems and Computers, 2016 50th Asilomar Conference on. IEEE; 2016. p. 662–666.

[53] West NE, and O'Shea T. Deep architectures for modulation recognition. In: International Symposium on Dynamic Spectrum Access Networks. IEEE; 2017. p. 1–6.

[54] Yun S, Lee J, Chung W, *et al.* A soft computing approach to localization in wireless sensor networks. Expert Systems with Applications. 2009; 36(4):7552–7561.

[55] Chagas SH, Martins JB, and de Oliveira LL. An approach to localization scheme of wireless sensor networks based on artificial neural networks and genetic algorithms. In: International New Circuits and Systems Conference. IEEE; 2012. p. 137–140.

[56] Alsheikh MA, Lin S, Niyato D, *et al.* Machine learning in wireless sensor networks: Algorithms, strategies, and applications. IEEE Communications Surveys and Tutorials. 2014;16(4):1996–2018.

[57] Thing VL. IEEE 802.11 network anomaly detection and attack classification: a deep learning approach. In: Wireless Communications and Networking Conference. IEEE; 2017. p. 1–6.

[58] Yuan Z, Lu Y, Wang Z, *et al.* Droid-Sec: deep learning in android malware detection. In: ACM SIGCOMM Computer Communication Review. vol. 44. ACM; 2014. p. 371–372.

[59] Zhang C, Patras P, and Haddadi H. Deep learning in mobile and wireless networking: a survey. arXiv preprint arXiv:180304311. 2018.

[60] Han J, Pei J, and Kamber M. Data Mining: Concepts and Techniques. Singapore: Elsevier; 2011.

[61] MacQueen J. Some methods for classification and analysis of multivariate observations. In: The Fifth Berkeley Symposium on Mathematical Statistics and Probability. vol. 1. Oakland, CA, USA; 1967. p. 281–297.

[62] Mahajan M, Nimbhorkar P, and Varadarajan K. The planar k-means problem is NP-hard. Theoretical Computer Science. 2012;442:13–21.

[63] Ding C, and He X. K-means clustering via principal component analysis. In: International Conference on Machine Learning. ACM; 2004. p. 29.

[64] Csurka G, Dance C, Fan L, *et al.* Visual categorization with bags of keypoints. In: Workshop on Statistical Learning in Computer Vision. vol. 1. Prague; 2004. p. 1–2.

[65] Sivic J, and Zisserman A. Efficient visual search of videos cast as text retrieval. IEEE Transactions on Pattern Analysis and Machine Intelligence. 2009;31(4):591–606.

[66] Coates A, and Ng AY. Learning feature representations with k-means. In: Neural Networks: Tricks of the Trade. Springer; 2012. p. 561–580.

[67] Pelleg D, and Moore AW. X-means: extending k-means with efficient estimation of the number of clusters. In: International Conference on Machine Learning. vol. 1; 2000. p. 727–734.

[68] Hamerly G, and Elkan C. Learning the k in k-means. In: Advances in Neural Information Processing Systems; 2004. p. 281–288.

[69] Kass RE, and Wasserman L. A reference Bayesian test for nested hypotheses and its relationship to the Schwarz criterion. Journal of the American Statistical Association. 1995;90(431):928–934.

[70] Arthur D, and Vassilvitskii S. k-Means++: the advantages of careful seeding. In: Proceedings of the Eighteenth Annual ACM-SIAM Symposium on Discrete Algorithms. Society for Industrial and Applied Mathematics; 2007. p. 1027–1035.

[71] Jain AK. Data clustering: 50 years beyond K-means. Pattern Recognition Letters. 2010;31(8):651–666.

[72] Dhillon IS, Guan Y, and Kulis B. Kernel k-means: spectral clustering and normalized cuts. In: SIGKDD Conference on Knowledge Discovery and Data Mining. ACM; 2004. p. 551–556.

[73] Li D, Wong KD, Hu YH, *et al.* Detection, classification, and tracking of targets. IEEE Signal Processing Magazine. 2002;19(2):17–29.

[74] Tseng YC, Wang YC, Cheng KY, *et al.* iMouse: an integrated mobile surveillance and wireless sensor system. Computer. 2007;40(6):60–66.

[75] Xia M, Owada Y, Inoue M, *et al.* Optical and wireless hybrid access networks: design and optimization. Journal of Optical Communications and Networking. 2012;4(10):749–759.

[76] Ester M, Kriegel HP, Sander J, *et al.* A density-based algorithm for discovering clusters in large spatial databases with noise. In: SIGKDD Conference on Knowledge Discovery and Data Mining. vol. 96; 1996. p. 226–231.

[77] Kumar KM, and Reddy ARM. A fast DBSCAN clustering algorithm by accelerating neighbor searching using Groups method. Pattern Recognition. 2016;58:39–48.

[78] Kriegel HP, Kröger P, Sander J, *et al.* Density-based clustering. Wiley Interdisciplinary Reviews: Data Mining and Knowledge Discovery. 2011;1(3): 231–240.

[79] Zhao F, Luo Hy, and Quan L. A mobile beacon-assisted localization algorithm based on network-density clustering for wireless sensor networks. In: International Conference on Mobile Ad-hoc and Sensor Networks. IEEE; 2009. p. 304–310.

[80] Faiçal BS, Costa FG, Pessin G, *et al.* The use of unmanned aerial vehicles and wireless sensor networks for spraying pesticides. Journal of Systems Architecture. 2014;60(4):393–404.

[81] Shamshirband S, Amini A, Anuar NB, *et al.* D-FICCA: a density-based fuzzy imperialist competitive clustering algorithm for intrusion detection in wireless sensor networks. Measurement. 2014;55:212–226.

[82] Wagh S, and Prasad R. Power backup density based clustering algorithm for maximizing lifetime of wireless sensor networks. In: International Conference on Wireless Communications, Vehicular Technology, Information Theory and Aerospace & Electronic Systems (VITAE). IEEE; 2014. p. 1–5.

[83] Abid A, Kachouri A, and Mahfoudhi A. Outlier detection for wireless sensor networks using density-based clustering approach. IET Wireless Sensor Systems. 2017;7(4):83–90.

[84] Rodriguez A, and Laio A. Clustering by fast search and find of density peaks. Science. 2014;344(6191):1492–1496.

[85] Botev ZI, Grotowski JF, Kroese DP, *et al.* Kernel density estimation via diffusion. The Annals of Statistics. 2010;38(5):2916–2957.

[86] Xie J, Gao H, Xie W, *et al.* Robust clustering by detecting density peaks and assigning points based on fuzzy weighted k-nearest neighbors. Information Sciences. 2016;354:19–40.

[87] Liang Z, and Chen P. Delta-density based clustering with a divide-and-conquer strategy: 3DC clustering. Pattern Recognition Letters. 2016;73: 52–59.

[88] Wang G, and Song Q. Automatic clustering via outward statistical testing on density metrics. IEEE Transactions on Knowledge and Data Engineering. 2016;28(8):1971–1985.

[89] Yaohui L, Zhengming M, and Fang Y. Adaptive density peak clustering based on K-nearest neighbors with aggregating strategy. Knowledge-Based Systems. 2017;133:208–220.

[90] Geng Ya, Li Q, Zheng R, *et al.* RECOME: a new density-based clustering algorithm using relative KNN kernel density. Information Sciences. 2018;436:13–30.

[91] Gu X, Peng J, Zhang X, *et al.* A density-based clustering approach for optimal energy replenishment in WRSNs. In: International Symposium on Parallel and Distributed Processing with Applications. IEEE; 2017. p. 1018–1023.

[92] Zhang Y, Liu M, and Liu Q. An energy-balanced clustering protocol based on an improved CFSFDP algorithm for wireless sensor networks. Sensors. 2018;18(3):881.

[93] He R, Li Q, Ai B, *et al.* A kernel-power-density-based algorithm for channel multipath components clustering. IEEE Transactions on Wireless Communications. 2017;16(11):7138–7151.

[94] Borman S. The expectation maximization algorithm – a short tutorial. Submitted for publication. 2004:1–9.

[95] Weng Y, Xiao W, and Xie L. Diffusion-based EM algorithm for distributed estimation of Gaussian mixtures in wireless sensor networks. Sensors. 2011;11(6):6297–6316.

[96] Zuo L, Mehrotra K, Varshney PK, *et al.* Bandwidth-efficient target tracking in distributed sensor networks using particle filters. In: International Conference on Information Fusion. IEEE; 2006. p. 1–4.

[97] Wali PK, Prasad M, Shreyas N, *et al.* Gaussian mixture model-expectation maximization based signal strength prediction for seamless connectivity in hybrid wireless networks. In: International Conference on Advances in Mobile Computing and Multimedia. ACM; 2009. p. 493–497.

[98] Horn RA, and Johnson CR. Matrix Analysis. 2nd ed. Cambridge: Cambridge University Press; 2012.

[99] Schölkopf B, Smola A, and Müller KR. Nonlinear component analysis as a kernel eigenvalue problem. Neural Computation. 1998;10(5): 1299–1319.

[100] Vidal R, Ma Y, and Sastry S. Generalized principal component analysis (GPCA). IEEE Transactions on Pattern Analysis and Machine Intelligence. 2005;27(12):1945–1959.

[101] Candès EJ, Li X, Ma Y, *et al.* Robust principal component analysis?. Journal of the ACM (JACM). 2011;58(3):11.

[102] Aquino AL, Junior OS, Frery AC, *et al.* MuSA: multivariate sampling algorithm for wireless sensor networks. IEEE Transactions on Computers. 2014;63(4):968–978.

[103] Ghosh T, Fattah SA, and Wahid KA. CHOBS: color histogram of block statistics for automatic bleeding detection in wireless capsule endoscopy video. IEEE Journal of Translational Engineering in Health and Medicine. 2018;6:1–12.

[104] Kingma DP, and Welling M. Auto-encoding variational Bayes. arXiv preprint arXiv:13126114. 2013.

[105] Abu Alsheikh M, Poh PK, Lin S, *et al.* Efficient data compression with error bound guarantee in wireless sensor networks. In: International Conference on Modeling, Analysis and Simulation of Wireless and Mobile Systems. ACM; 2014. p. 307–311.

[106] Alsheikh MA, Lin S, Tan HP, *et al.* Toward a robust sparse data representation for wireless sensor networks. In: Conference on Local Computer Networks. IEEE; 2015. p. 117–124.

[107] Zhang W, Liu K, Zhang W, *et al.* Deep neural networks for wireless localization in indoor and outdoor environments. Neurocomputing. 2016;194: 279–287.

[108] Feng Q, Zhang Y, Li C, *et al.* Anomaly detection of spectrum in wireless communication via deep auto-encoders. The Journal of Supercomputing. 2017;73(7):3161–3178.

[109] Li R, Zhao Z, Chen X, Palicot J, and Zhang H. A Transfer Actor-Critic Learning Framework for Energy Saving in Cellular Radio Access Networks. IEEE Transactions on Wireless Communications. 2014;13(4):2000–2011.

[110] Puterman ML. Markov Decision Processes: Discrete Stochastic Dynamic Programming. Hoboken, NJ: John Wiley & Sons; 2014.

[111] Sigaud O, and Buffet O. Markov Decision Processes in Artificial Intelligence. Hoboken, NJ: John Wiley & Sons; 2013.

[112] Stevens-Navarro E, Lin Y, and Wong VW. An MDP-based vertical handoff decision algorithm for heterogeneous wireless networks. IEEE Transactions on Vehicular Technology. 2008;57(2):1243–1254.

[113] Mastronarde N, and van der Schaar M. Fast reinforcement learning for energy-efficient wireless communication. IEEE Transactions on Signal Processing. 2011;59(12):6262–6266.

[114] Blasco P, Gunduz D, and Dohler M. A learning theoretic approach to energy harvesting communication system optimization. IEEE Transactions on Wireless Communications. 2013;12(4):1872–1882.

[115] Pandana C, and Liu KR. Near-optimal reinforcement learning framework for energy-aware sensor communications. IEEE Journal on Selected Areas in Communications. 2005;23(4):788–797.

[116] Berthold U, Fu F, van der Schaar M, *et al.* Detection of spectral resources in cognitive radios using reinforcement learning. In: Symposium on New Frontiers in Dynamic Spectrum Access Networks. IEEE; 2008. p. 1–5.

[117] Lilith N, and Dogançay K. Distributed dynamic call admission control and channel allocation using SARSA. In: Communications, 2005 Asia-Pacific Conference on. IEEE; 2005. p. 376–380.

[118] Kazemi R, Vesilo R, Dutkiewicz E, *et al.* Reinforcement learning in power control games for internetwork interference mitigation in wireless body area networks. In: International Symposium on Communications and Information Technologies. IEEE; 2012. p. 256–262.

[119] Ortiz A, Al-Shatri H, Li X, *et al.* Reinforcement learning for energy harvesting point-to-point communications. In: Communications (ICC), 2016 IEEE International Conference on. IEEE; 2016. p. 1–6.

[120] Saleem Y, Yau KLA, Mohamad H, *et al.* Clustering and reinforcement-learning-based routing for cognitive radio networks. IEEE Wireless Communications. 2017;24(4):146–151.

[121] Xiao L, Li Y, Dai C, *et al.* Reinforcement learning-based NOMA power allocation in the presence of smart jamming. IEEE Transactions on Vehicular Technology. 2018;67(4):3377–3389.

[122] Sadeghi A, Sheikholeslami F, and Giannakis GB. Optimal and scalable caching for 5G using reinforcement learning of space-time popularities. IEEE Journal of Selected Topics in Signal Processing. 2018;12(1):180–190.

[123] Dam G, Kording K, and Wei K. Credit assignment during movement reinforcement learning. PLoS One. 2013;8(2):e55352.

[124] Sutton RS, McAllester DA, Singh SP, *et al.* Policy gradient methods for reinforcement learning with function approximation. In: Advances in Neural Information Processing Systems; 2000. p. 1057–1063.

[125] Mnih V, Badia AP, Mirza M, *et al.* Asynchronous methods for deep reinforcement learning. In: International Conference on Machine Learning; 2016. p. 1928–1937.

[126] Arulkumaran K, Deisenroth MP, Brundage M, *et al.* A brief survey of deep reinforcement learning. arXiv preprint arXiv:170805866. 2017.

[127] He Y, Yu FR, Zhao N, *et al.* Software-defined networks with mobile edge computing and caching for smart cities: a big data deep reinforcement learning approach. IEEE Communications Magazine. 2017;55(12):31–37.

[128] Li J, Gao H, Lv T, *et al.* Deep reinforcement learning based computation offloading and resource allocation for MEC. In: Wireless Communications and Networking Conference. IEEE; 2018. p. 1–6.

[129] Liu J, Krishnamachari B, Zhou S, *et al.* DeepNap: data-driven base station sleeping operations through deep reinforcement learning. IEEE Internet of Things Journal. 2018;5(6):4273–4282.

[130] Comsa IS, De-Domenico A, and Ktenas D. QoS-driven scheduling in 5G radio access networks – a reinforcement learning approach. In: IEEE Global Communications Conference. IEEE; 2017. p. 1–7.

[131] Gupta A, Jha RK, Gandotra P, Jain S. and Supply HE. Bandwidth spoofing and intrusion detection system for multistage 5G wireless communication network. IEEE Wireless Communications. 2018;67(1):618–632.

[132] Koushik A, Hu F, and Kumar S. Intelligent spectrum management based on transfer actor-critic learning for rateless transmissions in cognitive radio networks. IEEE Transactions on Mobile Computing. 2018;17(5):1204–1215.

[133] Kaelbling LP, Littman ML, and Moore AW. Reinforcement learning: a survey. Journal of Artificial Intelligence Research. 1996;4:237–285.

[134] Usama M, Qadir J, Raza A, *et al.* Unsupervised machine learning for networking: techniques, applications and research challenges. arXiv preprint arXiv:170906599. 2017.

[135] Chen Z and Huang X. End-to-end learning for lane keeping of self-driving cars. IEEE Intelligent Vehicles Symposium (IV). 2017: 1856–1860.

[136] Silver D, Schrittwieser J, Simonyan K, *et al.* Mastering the game of Go without human knowledge. Nature. 2017;550(7676):354.

Chapter 2
Machine-learning-enabled channel modeling

Chen Huang[1], Ruisi He[2], Andreas F. Molisch[3], Zhangdui Zhong[2], and Bo Ai[2]

In this chapter, we present an introduction to the use of machine learning in wireless propagation channel modeling. We also present a survey of some current research topics that have become important issues for 5G communications.

2.1 Introduction

Channel modeling is one of the most important research topics for wireless communications, since the propagation channel determines the performance of any communication system operating in it. Specifically, channel modeling is a process of exploring and representing channel features in real environments, which reveals how radio waves propagate in different scenarios. The fundamental physical propagation processes of the radio waves, such as reflections, diffractions, are hard to observe directly, since radio waves typically experience multiple such fundamental interactions on their way from the transmitter to the receiver. In this case, channel modeling is developed to characterize some *effective* channel parameters, e.g., delay dispersion or attenuation, which can provide guidelines for the design and optimization of the communication system.

Most channel models are based on measurements in representative scenarios. Data collected during such measurement campaigns usually are the impulse response or transfer function for specific transmit and receive antenna configurations. With the emergence of multiple-input–multiple-output (MIMO) systems, directional characteristics of the channels can be extracted as well. In particular for such MIMO measurements, high-resolution parameter estimation (HRPE) techniques can be applied to obtain high-accuracy characteristics of the multipath components (MPCs). Examples for HRPE include space-alternating generalized expectation-maximization (SAGE) [1], clean [2], or joint maximum likelihood estimation (RiMAX) [3].

[1]School of Computer and Information Technology, Beijing Jiaotong University, China
[2]State Key Lab of Rail Traffic Control and Safety, Beijing Jiaotong University, China
[3]Department of Electrical Engineering, University of Southern California, USA

Machine learning, as an important branch of artificial intelligence, is considered to be a powerful tool to analyze measurement data, understand propagation processes, and create models. This is especially true for learning the principles and properties in channel measurement data from modern measurement campaigns, since the data volume and dimensionality of the measurement data have increased rapidly with the advent of massive MIMO systems. Therefore, machine-learning-based channel modeling has become a popular research topic. Typical applications of machine learning in channel modeling include

- **Propagation scenario classification.** Classification of the scenarios is an important part for the channel modeling and communication system deployment, since the channel models, or at a minimum their parameterizations, depend on the considered environment and scenario. For example, most models show a difference between line-of-sight (LOS) and non-line-of-sight (NLOS) scenarios. Current solutions for LOS/NLOS classification are generally based on different metrics, e.g., the K-factor [4], the root-mean-square delay spread and mean excess delay [5], or the Kurtosis of the channel state information [6]. However, the classification of the scenarios by using a binary hypothesis test based on single metric is not accurate enough for variable environments in wireless communications. On the other hand, some machine-learning techniques, e.g., support vector machines (SVM) and deep learning, which have a great advantage for extracting data features, can be used for scenario classification as well. In this case, learning and extracting the difference of channel properties in the different scenarios helps to automatically separate the measured data into different scenarios and discover the scenario features for resource allocation, system optimization, or localization. In this chapter, we present some first result on the machine-learning-based LOS/NLOS scenarios identification.
- **Machine-learning-based MPC clustering.** A large body of MIMO measurements has shown that the MPCs occur in groups, also known as clusters, such that the MPCs within one group have similar characteristics but have significantly different characteristics from the MPCs in other clusters. Separately characterizing the intra-cluster and intercluster properties can allow to significantly simplify channel models without major loss of accuracy. Therefore, many channel models have been proposed and developed based on the concept of clusters, e.g., Saleh–Valenzuela (SV) [7], COST 259 [8,9] COST 2100 [10], 3GPP spatial channel model [11], and WINNER model [12]. In the past, visual inspection has been widely used for cluster identification, which is inapplicable for extensive measurement data, and also subjective. Automated algorithms, in particular the KPowerMeans algorithm [13], have gained popularity in recent years but still suffer from the use of arbitrary thresholds and/or a priori assumption of the number of clusters. Moreover, many clustering approaches require to extract the MPCs via HRPE methods before clustering. These algorithms have generally high computational complexity, which makes real-time operation in time-varying channels difficult. Hence, automatic clustering for MPCs based on machine-learning algorithms has drawn a lot of attention.

- **Automatic MPC tracking.** Time-variation of the propagation is relevant for many applications such as high-speed railways and vehicular communications. However, in time-varying channels, the MPCs need to be not only clustered but also tracked to characterize their dynamic features. Machine-learning-based tracking algorithms can be adopted or developed to track the MPCs during consecutive snapshots, e.g., Kalman filters [14,15], or the Kuhn–Munkres algorithm [16]. However, the MPCs' behaviors during the time, e.g., split, merge, and lifetime, are still not fully utilized in the current tracking procedures. Hence, how to track the MPCs more accurately and efficiently is still an open question.
- **Deep-learning-based channel modeling approach.** The main goal of channel modeling is to find the interconnections among the transmitted signals, environments, and the received signals and model them by appropriate functions. Meanwhile, with the dramatic development of artificial intelligence, neural-network-based deep learning has shown great performance on characterizing data and extracting the mapping relationship between system input and output [17]. Therefore, many studies pay attention to modeling the channels by using neural network. For example, a back-propagation (BP) network is used for modeling the amplitude in propagation channels in [18], a radial-basis-function (RBF)-based neural network is used for modeling the Doppler frequency shift in [19]. Moreover, some other data-mining approaches can be adopted to pre-process the measured data, which makes the data easier to be analyzed and processed.

In this chapter, we introduce recent progress of the above applications for machine learning in channel modeling. The results in this chapter can provide references to other real-world measurement data-based channel modeling.

2.2 Propagation scenarios classification

In this section, machine-learning-based propagation scenario classification is introduced. Generally, different channel models are used for different typical scenarios. In particular, most models are developed based on different propagation assumptions and parameter settings. At the same time, some machine-learning algorithms are able to learn the interconnections and features of different training data and then refine them into classification principles, which can automatically classify the input data in applications. Due to the good accuracy of the classification, the machine-learning approaches are expected to extract the features and properties of the different channels and automatically distinguish the propagation scenarios. There are many machine-learning algorithms that have been widely used for classification, e.g., SVM or deep learning.

A particularly important example for classification problems is the identification of LOS/NLOS scenarios, which is a binary classification problem. The SVM is one of the promising solutions for such binary classification problems and offers a good trade-off between accuracy and complexity compared to deep learning. Therefore, we investigate in the following in more detail the application of the SVM to distinguish LOS/NLOS scenarios based on the channel properties.

The main goal of the algorithm described in the following is to use the machine-learning tool, i.e., the SVM, to learn the internal features of the LOS/NLOS parameters, which can be obtained by using parameters estimation algorithms, e.g., beamformers, and build an automatic classifier based on the extracted features. Consequently, there are two main steps of the proposed algorithm: (i) develop the input vector for the SVM method and (ii) adjust the parameters of the SVM method to achieve a better accuracy of classification.

2.2.1 Design of input vector

In the described algorithm, the power-angle-spectrums (PASs) of the LOS/NLOS scenarios obtained by using the Bartlett beamformer [20] are used as the training database. Figure 2.1(a) and (b) depicts example PASs of the LOS and NLOS scenario, respectively, where the data are collected from 30×4 MIMO measurements at a carrier frequency of 5.3 GHz and estimated by using the Bartlett beamformer [20]. Since the SVM can only use a vector as the input for training, the design of the input vector is

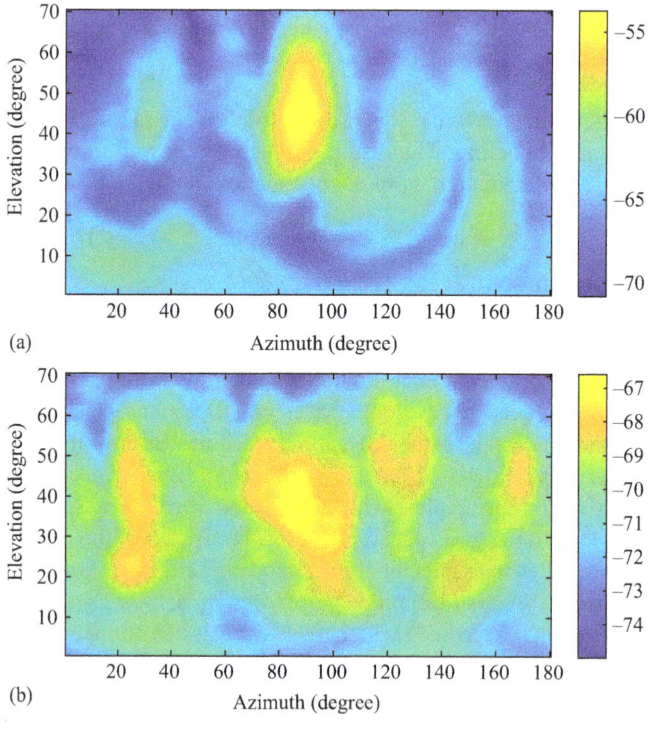

Figure 2.1 Power angle spectrum of (a) LOS and (b) NLOS scenarios, which are estimated by using Bartlett beamformer

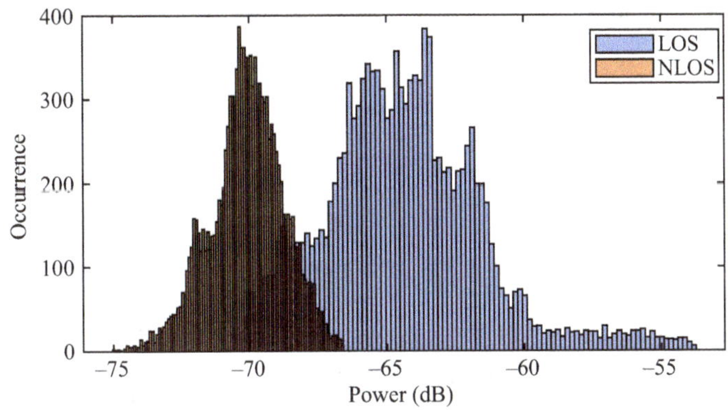

Figure 2.2 Histograms of the power distribution of the LOS and NLOS scenarios, respectively

crucial for the performance of the SVM. In the described algorithm, the SVM is used to learn the difference between the LOS and NLOS from the classified data (training data) and distinguish the LOS and NLOS condition of the unclassified data (test data). In this case, an input vector that is able to most clearly present the physical features of the LOS/NLOS data can achieve the best classification accuracy.

In order to design an appropriate input vector, we first consider the main difference of physical features between the MPCs in the LOS and NLOS scenarios. First, the average power is usually different, where the LOS scenario usually has higher power. Second, the power distribution is another noteworthy difference between the LOS and NLOS scenarios. Since the LOS path is blocked in the NLOS scenario, the impact of MPCs undergoing reflections, scatterings, and diffusions is more significant in the NLOS case. In other words, even if all such indirect MPCs are exactly the same in LOS and NLOS scenarios, the existence of the direct MPC changes the power distribution.

From the above, it follows that the histogram of the power is a characteristic that can be used to distinguish the LOS/NLOS scenarios, where the abscissa represents different power intervals, and the ordinate represents how many elements in the PAS distribute in the different power intervals. Furthermore, to simplify the feature vector, the number of power intervals is set at 100, with a uniform distribution in the range of the power of the PAS, as shown in Figure 2.2. In this case, the histogram of the power is considered as the input vector **X**, which can be expressed as

$$\mathbf{X} = \{x_1, x_2, \ldots, x_{100}\} \tag{2.1}$$

2.2.2 Training and adjustment

The thus-obtained input vector can now be fed into the SVM method. Nevertheless, the typical linear-kernel-function-based SVM cannot achieve the best accuracy of classification, considering that the physical features of the LOS/NLOS scenario are generally complicated to characterize. Consequently, we use here the RBF as the kernel function, which can be expressed as

$$k(x, x_i) = \exp\left(-\frac{\|x - x_i\|^2}{\delta^2}\right) \tag{2.2}$$

By using the RBF kernel function, the training data are projected to a higher dimension, in which the difference between the LOS and NLOS data can be observed more easily.

In this case, the histogram of the power in each PAS is considered as the feature vector for the input of the SVM to distinguish the LOS and NLOS scenarios. Based on our experiments, the described solution achieves nearly 94% accuracy on the classification.

In addition, the angle (azimuth/elevation) distribution of the power is also generally considered to be different in the LOS and NLOS scenarios. Since there is no LOS component in the NLOS scenario, it will more concentrate on reflections and scatterings in the environment, which leads to a lower average power and smaller power spread in the histograms. Therefore, utilizing the angle distribution in the feature vector may also increase the classification accuracy of the solution.

2.3 Machine-learning-based MPC clustering

As outlined in the introduction, modeling of inter- and intra-cluster properties, instead of the properties of the individual MPCs, offers an attractive trade-off between accuracy and complexity of the models and is thus widely used in the literature [21,22]. The basic requirement for such models is to identify clusters in measured data. In the past, visual inspection has been widely used to recognize the clusters. However, it is inapplicable to the analysis of large amounts of high-dimensional measurement data, which commonly are encountered in particular in MIMO measurement campaigns. Besides, visual inspection is a subjective approach, thus different inspectors may provide different clustering results, which makes comparisons between results from different groups difficult.

On the other hand, clustering is one of the most fundamental applications for machine learning. Therefore, machine-learning-based clustering algorithms have become a hot topic and are expected to be able to automatically cluster MPCs with high accuracy. The main challenges of automatic clustering of MPCs include the following: (i) the definition of MPCs' cluster has not been addressed clearly; (ii) the ground truth of MPCs' clusters is generally unknown, which makes it difficult to validate the clustering result; (iii) the number of clusters, which is required for many machine-learning-clustering methods, is usually unknown; and (iv) the dynamic changes of MPCs that occur in time-varying channels are difficult to utilize in many clustering algorithm. To provide a benchmark, in the following we describe some widely used classical MPC clustering algorithms.

2.3.1 *K*PowerMeans-based clustering

The *K*PowerMeans algorithm described in [13] is one of the most popular clustering approaches for MPCs in the radio channels. The key idea of *K*PowerMeans is based on the conventional *K*Means method, which is a typical hard partition approach and clusters data objects based on the distance among each other. Similar to *K*Means, the *K*PowerMeans requires the number of clusters as prior information, an indeterminate cluster number may have an impact on the performance of clustering. While a number of different methods have been described, the most straightforward way is to compute results with different cluster numbers and compare the results. The main idea of *K*PowerMeans is summarized in the following subsection.

2.3.1.1 Clustering

Figure 2.3(a)–(d) shows the four stages in the iteration of clustering. The dots and blocks in (a) present the input MPCs and initialed cluster-centroids, respectively, whereas the different colors of the dots in (b)–(d) represent different categories of clusters. The *K*PowerMeans algorithm requires the number of clusters as prior

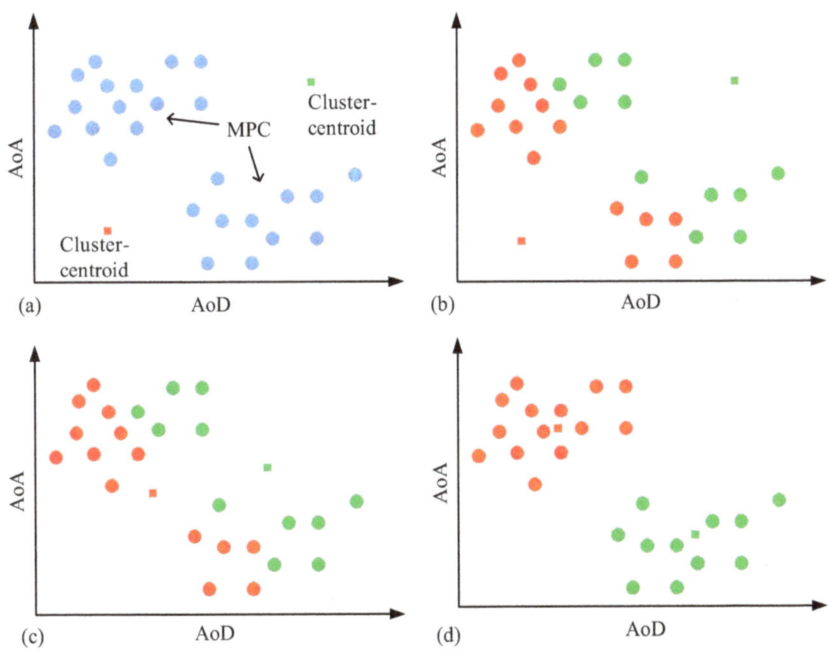

Figure 2.3 The clustering framework of the KPowerMeans algorithm, where (a)–(d) is four stage in the iteration of clustering. The dots and blocks in (a) present input objects and initialed cluster-centroids, respectively, whereas the different colors of the dots in (b)–(d) represent different categories of clusters

information, e.g., the blue and red blocks in Figure 2.3(a) and then clusters the MPCs preliminarily to the closest cluster-centroid, as shown in Figure 2.3(b). To accurately measure the similarity between MPCs/clusters, the MPCs distance (MCD) is used to measure the distance between MPCs and cluster-centroids, where the angle of arrival (AoA), angle of departure (AoD) and delay of the MPCs/cluster-centroids are considered. The MCD between the ith MPC and the jth MPC can be obtained as

$$\mathrm{MCD}_{ij} = \sqrt{\|\mathrm{MCD}_{\mathrm{AoA},ij}\|^2 + \|\mathrm{MCD}_{\mathrm{AoD},ij}\|^2 + \mathrm{MCD}_{\tau,ij}^2} \tag{2.3}$$

where

$$\mathrm{MCD}_{\mathrm{AoA/AoD},ij} = \frac{1}{2}\left| \begin{pmatrix} \sin(\theta_i)\cos(\varphi_i) \\ \sin(\theta_i)\sin(\varphi_i) \\ \cos(\theta_i) \end{pmatrix} - \begin{pmatrix} \sin(\theta_j)\cos(\varphi_j) \\ \sin(\theta_j)\sin(\varphi_j) \\ \cos(\theta_j) \end{pmatrix} \right|, \tag{2.4}$$

$$\mathrm{MCD}_{\tau,ij} = \zeta \cdot \frac{|\tau_i - \tau_j|}{\Delta\tau_{\max}} \cdot \frac{\tau_{std}}{\Delta\tau_{\max}}, \tag{2.5}$$

with $\Delta\tau_{\max} = \max_{i,j}\{|\tau_i - \tau_j|\}$ and ζ as an opportune delay scaling factor; various ways to select this scaling factor have been described in the literature. After the MPCs are clustered preliminarily, the cluster-centroids are recomputed, as shown in Figure 2.3(c). Then, the cluster members and the cluster-centroids are alternately recomputed in each iteration, until the data converge to stable clusters or reach the restraint of a preset running time.

2.3.1.2 Validation

To avoid the impact of an indeterminate number of the clusters, [13] develops the CombinedValidate method based on the combination of the Calinski–Harabasz (CH) index and the Davies–Bouldin criterion (DB). The basic idea of CombinedValidate is to restrict valid choices of the optimum number of clusters by a threshold set in the DB index. Subsequently, the CH index is used to decide on the optimum number out of the restricted set of possibilities.

2.3.1.3 Cluster pruning—ShapePrune

After successfully finding the optimum number of the clusters, the ShapePrune cluster pruning algorithm is adopted for discarding outliers. The basic idea of ShapePrune is to remove data points that have the largest distance from their own cluster-centroid with the constraint that cluster power and cluster spreads must not be changed significantly. In this case, the features of clusters can be more easily observed, where the clusters' properties can also be preserved as much as possible. Figure 2.4 shows the extracted MPCs from the MIMO measurement data in [13], where the power of MPCs is color coded. By applying the *K*PowerMeans, the MPCs can be automatically clustered without human interaction, as shown in Figure 2.5. Compared with

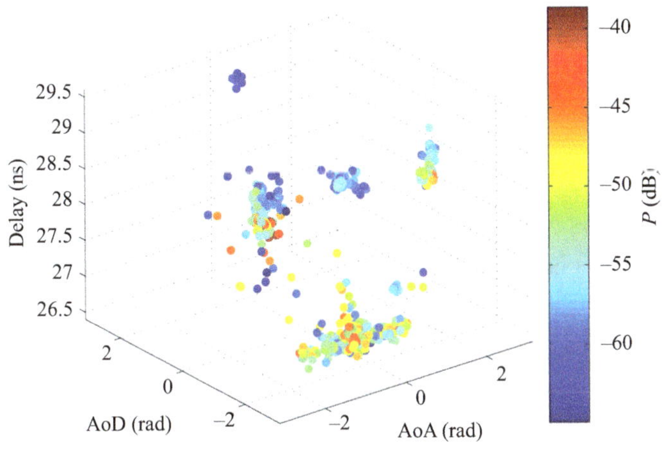

Figure 2.4 The unclustered MIMO measurement data in LOS scenario from [13], where the power of MPCs is color coded

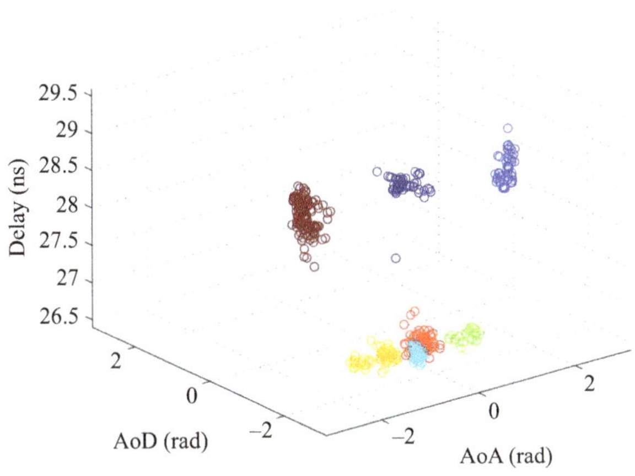

Figure 2.5 The result of clustering from [13], where the weak MPCs are removed

the visual inspection, the *K*PowerMeans can well identify the clusters closest to each other, e.g., the clusters in red, yellow, blue, and green in Figure 2.4.

2.3.1.4 Development

It is noteworthy that the initial parameters, e.g., cluster number and position of initial cluster-centroid, have a great impact on the performance of *K*PowerMeans. In *K*PowerMeans, the validation method is applied to select the best estimation of the number of the clusters; thus the performance of the validation method also affects the performance and efficiency of clustering. In [23], a performance assessment of several cluster validation methods is presented. There it was found that the Xie–Beni index and generalized Dunnes index reach the best performance, although the result also shows that none of the indices is able to always predict correctly the desired number of clusters. Moreover, to improve the efficiency of clustering, [24] develops the *K*PowerMeans by using the MPCs that have the highest power as the initial cluster-centroids. On the other hand, the study in [25] claims that as a hard partition approach, *K*Means is not the best choice for clustering the MPCs, considering that some MPCs are located near the middle of more than one cluster, and thus cannot be directly associated with a single cluster. Therefore, instead of using hard decisions as the *K*PowerMeans, a Fuzzy-*c*-means-based MPC clustering algorithm is described in [25], where soft information regarding the association of multipaths to a centroid are considered. As the result, the Fuzzy-*c*-means-based MPC clustering algorithm performs a robust and automatic clustering.

2.3.2 Sparsity-based clustering

In this subsection, the framework of a sparsity-based MPC clustering algorithm [26] is introduced, which was described to cluster channel impulse responses (CIRs) consisting of multiple groups of clusters. The key idea of the described algorithm is to use a sparsity-based optimization to recover the CIRs from measured data and then use a heuristic approach to separate the clusters from the recovered CIRs. The main idea can be summarized as follows [26,27]:

The CIRs are assumed to follow the SV model [7], i.e., the power of MPCs generally decreases with the delays as follows:

$$|\overline{\alpha}_{l,k}|^2 = |\overline{\alpha}_{0,0}|^2 \cdot \underbrace{\exp\left(-\frac{T_l}{\Gamma}\right)}_{A_1} \cdot \underbrace{\exp\left(-\frac{\tau_{l,k}}{\Lambda_l}\right)}_{A_2} \tag{2.6}$$

where A_1 and A_2 denote the intercluster and intra-cluster power decay, respectively; $|\overline{\alpha}_{0,0}|^2$ denotes the average power of the first MPC in the first cluster. Γ and Λ_l are the cluster and MPC power decay constants, respectively.

Then, the measured power delay profile (PDP) vector $\mathbf{P}$ is considered as the given signal, and the convex optimization is used to recover an original signal vector $\hat{\mathbf{P}}$, which is assumed to have the formulation (2.6). Furthermore, re-weighted l_1 minimization [28], which employed the weighted norm and iterations, is performed to enhance the sparsity of the solution.

Finally, based on the enhanced sparsity of $\hat{\mathbf{P}}$, clusters are identified from the curve of $\hat{\mathbf{P}}$. Generally, each cluster appears as a sharp onset followed by a linear decay, in the curve of the $\hat{\mathbf{P}}$ on a dB-scale. Hence, the clusters can be identified based on this feature, which can be formulated as the following optimization problem:

$$\min_{\hat{\mathbf{P}}} \|\mathbf{P} - \hat{\mathbf{P}}\|_2^2 + \lambda \|\Omega_2 \cdot \Omega_1 \cdot \hat{\mathbf{P}}\|_0 \tag{2.7}$$

where $\|\cdot\|_x$ denotes the l_x norm operation, and the l_0 norm operation returns the number of nonzero coefficients. λ is a regularization parameter, and Ω_1 is the finite-difference operator, which can be expressed as

$$\Omega_1 = \begin{pmatrix} \frac{\Delta\tau}{|\tau_1-\tau_2|} & \frac{\Delta\tau}{|\tau_1-\tau_2|} & 0 & \cdots & \cdots & 0 \\ 0 & \frac{\Delta\tau}{|\tau_2-\tau_3|} & -\frac{\Delta\tau}{|\tau_2-\tau_3|} & \cdots & \cdots & 0 \\ \vdots & \ddots & \ddots & \ddots & \ddots & \vdots \\ 0 & 0 & \ddots & \frac{\Delta\tau}{|\tau_{N-2}-\tau_{N-1}|} & -\frac{\Delta\tau}{|\tau_{N-2}-\tau_{N-1}|} & 0 \\ 0 & 0 & \cdots & \cdots & \frac{\Delta\tau}{|\tau_{N-1}-\tau_N|} & -\frac{\Delta\tau}{|\tau_{N-1}-\tau_N|} \end{pmatrix}_{(N-1)\times N} \tag{2.8}$$

where N is the dimension of $\mathbf{P}$ and $\hat{\mathbf{P}}$, $\Delta\tau$ is the minimum resolvable delay difference of data. Ω_2 is used to obtain the turning point at which the slope changes significantly and can be expressed as

$$\Omega_2 = \begin{pmatrix} 1 & -1 & 0 & \cdots & \cdots & 0 \\ 0 & 1 & -1 & \cdots & \cdots & 0 \\ \vdots & \ddots & \ddots & \ddots & \ddots & \vdots \\ 0 & 0 & \vdots & 1 & -1 & 0 \\ 0 & 0 & \cdots & \cdots & 1 & -1 \end{pmatrix}_{(N-2)\times(N-1)} \tag{2.9}$$

Note that $\lambda\|\Omega_2 \cdot \Omega_1 \cdot \hat{\mathbf{P}}\|_0$ in (2.7) is used to ensure that the recovered $\hat{\mathbf{P}}$ conform with the anticipated behavior of A_2 in (2.6). In this case, even a small number of clusters can be well identified by using the described algorithm. Moreover, [26] also incorporates the anticipated behavior of A_1 in (2.6) into $\hat{\mathbf{P}}$ by using a clustering-enhancement approach.

The details of the implementation of the sparsity-based clustering algorithm can be found in [26]. To evaluate the performance, Figure 2.6(a) gives the cluster identification result by using the sparsity-based algorithm, while (b) and (c) give the results by using *K*Means and *K*PowerMeans approaches, respectively. It can be seen that the clusters identified by the sparsity-based algorithm show more distinct features, where each cluster begins with a sharp power peak and ends with a low power valley before the next cluster. This feature well conforms to the assumption of the cluster in the SV model. On the other hand, as shown in Figure 2.6(b) and (c), the *K*Means and *K*PowerMeans tend to group the tail of one cluster into the next cluster, which may lead to the parameterized intra-cluster PDP model having a larger delay spread. More details and further analysis can be found in [26].

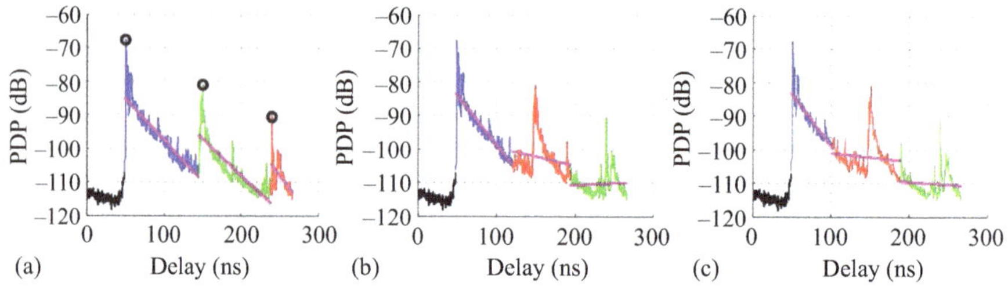

Figure 2.6 Example plots of PDP clustering in [26], where (a) gives the cluster identification result by using the sparsity-based algorithm, (b) and (c) give the results by using KMeans and KPowerMeans approaches, respectively. Different clusters are identified by using different colors, where the magenta lines represent the least squared regression of PDPs within clusters

2.3.3 Kernel-power-density-based clustering

In this section, the framework of the Kernel-power-density (KPD)-based algorithm is introduced, which was described in [29,30] to cluster the MPCs in MIMO channels. In this algorithm, the Kernel density of MPCs is adopted to characterize the modeled behavior of MPCs, where the powers of MPCs are also considered in the clustering process. Moreover, the relative density is considered, using a threshold to determine whether two clusters are density reachable.

To better elaborate, an example [29] is given in Figure 2.7, where (a) shows the measured MPCs, (b) shows the estimated density ρ, (c) shows the estimated density ρ^*, and (d) gives the final clustering results by using the KPD algorithm. The details are introduced as follows:

1. The KPD-based algorithm identifies clusters by using the Kernel density to identify the clusters; therefore, the density needs to be calculated first. For each MPC x, the density ρ with the K nearest MPCs can be obtained as follows:

$$\rho_x = \sum_{y \in K_x} \exp(\alpha_y) \times \exp\left(-\frac{|\tau_x - \tau_y|^2}{(\sigma_\tau)^2}\right) \times \exp\left(-\frac{|\Omega_{T,x} - \Omega_{T,y}|}{\sigma_{\Omega_T}}\right)$$
$$\times \exp\left(-\frac{|\Theta_{T,x} - \Theta_{T,y}|}{\sigma_{\Theta_T}}\right) \times \exp\left(-\frac{|\Theta_{R,x} - \Theta_{R,y}|}{\sigma_{\Theta_R}}\right) \quad (2.10)$$

where y is an arbitrary MPC ($y \neq x$). K_x is the set of the K nearest MPCs for the MPC x. $\sigma_{(\cdot)}$ is the standard deviation of the MPCs in the domain of $(\cdot)$. Specifically, past studies have modeled with good accuracy the intra-cluster power angle distribution as Laplacian distribution [31]; therefore, the Laplacian Kernel density is also used for the angular domain in (2.10).

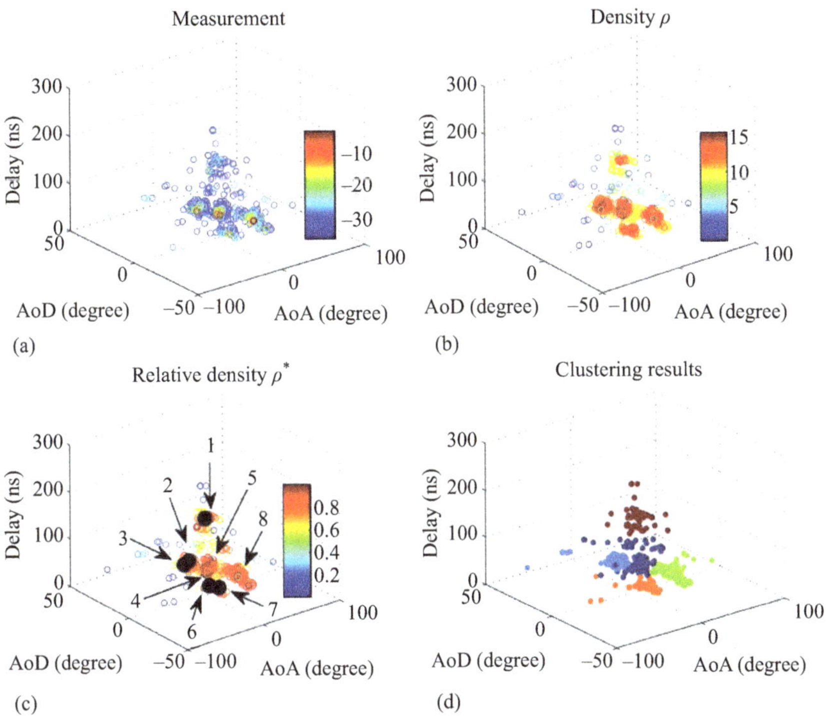

Figure 2.7 Illustration of KPD clustering using the measured MPCs: Part (a) shows the measured MPCs, where the color bar indicates the power of an MPC. Part (b) plots the estimated density ρ, where the color bar indicates the level of ρ. Part (c) plots the estimated density ρ, where the color bar indicates the level of ρ*. The eight solid black points are the key MPCs with ρ* = 1. Part (d) shows the clustering results by using the KPD algorithm, where the clusters are plotted with different colors*

2. In the next step, the relative density ρ^* also needs to be calculated based on the obtained density ρ_x, which can be expressed as

$$\rho_x^* = \frac{\rho_x}{\max_{y \in K_x \cup \{x\}} \{\rho_y\}}. \tag{2.11}$$

Figure 2.7 shows an example plot of the relative density ρ^*. Specifically, the relative density ρ^* in (2.11) can be used to identify the clusters with relatively weak power.

3. Next, the key MPCs need to be obtained. An MPC x will be labeled as key MPC $\hat{x}$ if $\rho^* = 1$:

$$\hat{\Phi} = \{x | x \in \Phi, \rho_x^* = 1\}. \tag{2.12}$$

In the described algorithm, the obtained key MPCs are selected as the initial cluster-centroids. Figure 2.7(c) gives an example of the key MPCs, which are plotted as solid black points.

4. The main goal of the KPD algorithm is to cluster MPCs based on the Kernel density, therefore, for each non-key MPC x, we define its high-density-neighboring MPC $\tilde{x}$ as

$$\tilde{x} = \arg\min_{\substack{y \in \Phi, \\ \rho_y > \rho_x}} d(x, y) \tag{2.13}$$

where d represents the Euclidean distance. Then, the MPCs are connected based on their own high-density-neighboring $\tilde{x}$ and the connection is defined as

$$p_x = \{x \rightarrow \tilde{x}\} \tag{2.14}$$

and thus a connection map ζ_1 can be obtained as follows:

$$\zeta_1 = \{p_x | x \in \Phi\}. \tag{2.15}$$

In this case, the MPCs that are connected to the same key MPC in ζ_1 are grouped as one cluster.

5. For each MPC, the connection between itself and its K nearest MPCs can be expressed as follows:

$$q_x = \{x \rightarrow y, y \in K_x\} \tag{2.16}$$

where another connectedness map ζ_2 can be obtained, as follows:

$$\zeta_2 = \{q_x | x \in \Phi\}. \tag{2.17}$$

In this case, two key MPCs clusters will be merged into a new cluster if the following criteria are met:

- The two key MPCs are included in ζ_2
- Any MPC belonging to the two key MPCs' clusters has $\rho^* > \chi$

where χ is a density threshold. As shown in Figure 2.7(c), clusters 2 and 3, 6 and 7 meet the conditions and are merged into new clusters, respectively.

To validate the performance of the clustering result, the F-measure is used in [29], where the precision and recall of each cluster are considered. It is noteworthy that the validation by using F-measure requires the ground truth of the cluster members. Generally, the ground truth is unavailable in measured channels; hence, the F-measure can be only applied for the clustering result of simulated channel, for which the (clustered) MPC generation mechanism, and thus the ground truth, is known. The 3GPP 3D MIMO channel model is used to simulate the channels in [29], and 300 random channels are simulated to validate the performance of the KPD-based algorithm, where the conventional KPowerMeans [13] and DBSCAN [32] are shown as comparisons. Figure 2.8 depicts the impact of the cluster number on the F-measure, where the described algorithm shows better performance than the others, especially in the scenarios containing more clusters, and the clustering performances of all three reduce with the increasing number of clusters.

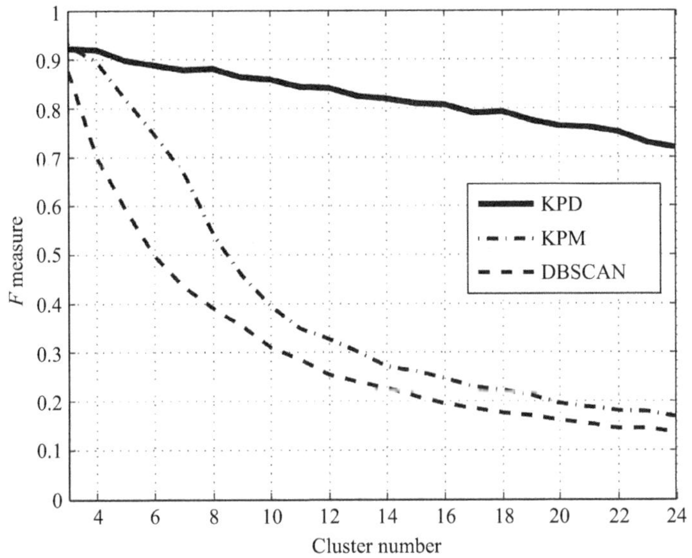

Figure 2.8 Impact of cluster number on the F measure in [29]

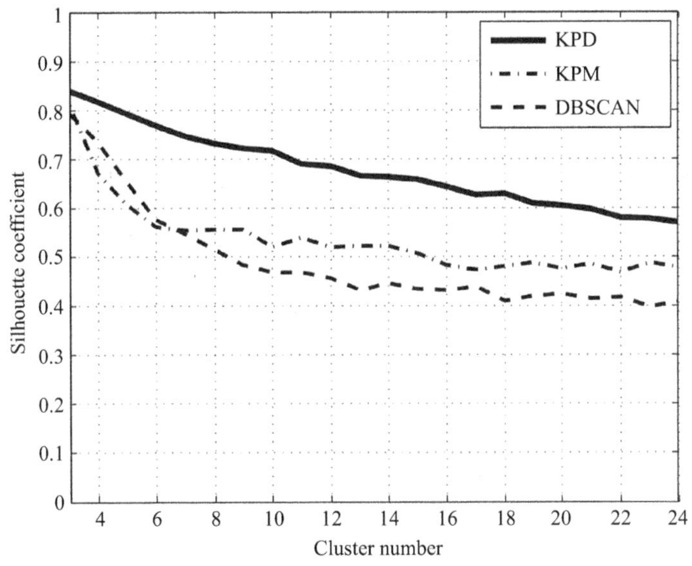

Figure 2.9 Impact of cluster angular spread on the F measure in [29]

Figure 2.9 shows the impact of the cluster angular spread on the F-measure of the three algorithms. It is found that the F-measure generally decreases with the increasing cluster angular spread, where the KPD-based algorithm shows the best performance among the three candidates. Further validation and analysis can be found in [29].

2.3.4 Time-cluster-spatial-lobe (TCSL)-based clustering

This section describes the time-cluster (TC)-spatial-lobe (SL) (TCSL) algorithm described in [33] for 3D-millimeter-wave statistical channel models, where the NYUSIM channel simulator [34] is implemented. A comparison to measured field data yielded a fit to 2D and 3D measurements using the TCSL algorithm, and it is found that TCSL can well fit the measurement data from urban NYC at mmWave using directional antennas with a lower complexity structure compared to other classical joint time-space modeling approaches [33,35–38]. The TCSL approach uses a fixed intercluster void interval representing the minimum propagation time between likely reflection or scattering objects. The framework of the TCSL algorithm is described in the following subsection.

2.3.4.1 TC clustering

In [33], the TCs are defined as a group of MPCs that have similar runtime and separated from other MPCs by a minimum interval, but which may arrive from different directions. Specifically, the minimum intercluster void interval is set to 25 ns. In other words, the MPCs whose inter-arrival time is less than 25 ns are considered as one TC; otherwise, they are considered as different TCs. Besides, the propagation phases of each MPC can be uniform between 0 and 2π. The choice of different intercluster voids results in different number of clusters in the delay domain; to be physically meaningful, this parameter needs to be adapted to the environment of observation.

2.3.4.2 SL clustering

Meanwhile, SLs are defined by the main directions of arrival/departure of the signal. Since the TCSL is based on measurements without HRPE of the MPCs, the angular width of an SL is determined by the beamwidth of the antenna (horn or lens or phased array) and measured over several hundred nanoseconds. A -10 dB power threshold with respect to the maximum received angle power is set in [33] to obtain the SLs (again, different thresholds might lead to different clusterings).

By applying the TCs and SLs, the MPCs in the time-space domain are decoupled into temporal and spatial statistics. Since the SLs and TCs are obtained individually, it is possible that a TC contains MPCs which belong to different SLs. On the contrary, an SL may contain many MPCs which belong to different TCs. These cases have been observed in real-world measurements [35–37], where the MPCs in the same TC may be observed in different SLs, or the MPCs in the same SL may be observed in different TCs.

The TCSL-clustering approach has low complexity, and some of its parameters can be related to the physical propagation environment [33]. However, it requires some prior parameters, such as the threshold to obtain the SLs, the delays and power levels of TC.

2.3.5 Target-recognition-based clustering

As we mentioned before, many current clustering algorithms for channel modeling are based on the characteristics of the MPCs, which are extracted by using HRPE algorithms, e.g., SAGE or CLEAN. However, performing the high-resolution estimation is time-consuming, and usually requires selection of algorithm parameters such as model order. Hence, some research focuses on the alternative approaches that require much lower computational effort.

For example, a power angle spectrum-based clustering and tracking algorithm (PASCT) is described in [39]. The PASCT algorithm first obtains the PAS by using a Bartlett (Fourier) beamformer [20], as shown in Figure 2.10. In this case, the cluster is defined as an "energy group," which has obvious higher power than the background, in the azimuth-elevation domain.

Generally, to recognize the clusters from the PAS, we need to distinguish between clusters and background. Clusters close to each other tend to be identified as one big target (called a target object), which contains one or more clusters. In this case, to further identify clusters in the big target objects, a density-peak-searching method is developed to divide the clusters. The details of the clustering process are as follows.

To recognize the target objects in PAS, the maximum-between-class-variance method [40] is applied to automatically determine a selection threshold of power for the elements in the PAS. This can separate the target objects from background noise at first stage. The between-class-variance of the average power levels of background noise and target objects can be expressed by

$$\delta^2(\alpha'_T) = p_B(\alpha'_T)(e_B(\alpha'_T) - E(\alpha'_i))^2 + p_O(\alpha'_T)(e_O(\alpha'_T) - E(\alpha'_i))^2, \tag{2.18}$$

where α'_T is the separation threshold between the clusters and background noise, $p_B(\alpha'_T)$ and $p_O(\alpha'_T)$ are the probabilities of the background noise and target objects occurrence in the current PAS, respectively, $e_B(\alpha'_T)$ and $e_O(\alpha'_T)$ are the average power levels of background noise and target objects, respectively, and $E(\alpha'_i)$ is the total mean power level of all the elements in the PAS.

The difference between background noise and groups of clusters can be maximized by maximizing the between-class-variance, and the best selection threshold α'^*_T can be therefore expressed as

$$\alpha'^*_T = \arg\{\max \delta^2(\alpha'_T) | \alpha'_1 \le \alpha'_T < \alpha'_L\}. \tag{2.19}$$

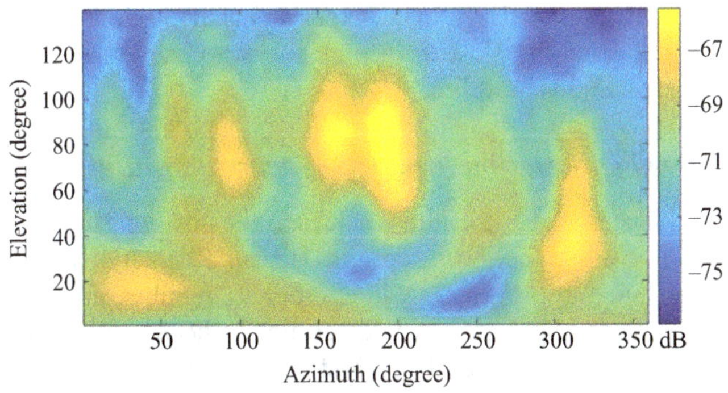

Figure 2.10 PAS obtained by using Bartlett beamformer in [39]

Since the number of the power levels is limited, ${\alpha'_T}^*$ can be easily found by a sequential search.

Nevertheless, the signal to interference plus noise ratio of the PAS has a strong impact on the performance of the target recognition. In an LOS scenario, the clusters are generally easily discernible, with strong power and low background noise, i.e., the targets can be easily recognized and detected. However, in many NLOS scenarios, the power distribution of the PAS is more complicated with high background noise, and many small clusters caused by reflections and scatterings interfere with the recognition process. In this case, the targets that contain many small clusters are difficult to be separated. To avoid this effect, an observation window $\Delta\alpha'_W$ is set so that only the elements having the power level between $[\alpha'_L - \Delta\alpha'_W, \ldots, \alpha'_L]$ are processed in the target recognition approach. In this case, the best selection threshold ${\alpha'_T}^*$ is obtained by

$$ {\alpha'_T}^* = \arg\{\max \delta^2(\alpha'_T) | \alpha'_L - \Delta\alpha'_W \leq \alpha'_T < \alpha'_L\}. \tag{2.20} $$

By using the observation window, the recognition process can focus on the elements with stronger power compared to the noise background. Moreover, a heuristic sequential search is used to select an appropriate observation window size $\Delta\alpha'_W$ as follows. Parameter $\Delta\alpha'_W$ is initialized to $0.1\alpha'_L$ at the beginning of the searching process and keeps increasing until the following constraints are no longer satisfied:

- Size of recognized targets: $S < S_{\max}$
- Power gap of each single target: $\Delta A < \Delta A_{\max}$

where S is the size of the recognized targets indicating how many elements the target consists of and $S_{\max}$ is the upper limit of size. Specifically, to avoid the interference caused by the small and fragmental targets, the lower limit of the size is also considered: only a target bigger than $S_{\min}$ is counted, whereas the target smaller than $S_{\min}$ is considered as noise rather than clusters. Parameter ΔA is the gap between the highest power and the mean power of each target. In each iteration, S and ΔA are updated based on the recognized target objects by using the new ${\alpha'_T}^*$ from (2.20), until the above constraints are no longer satisfied.

Examples for the clustering results in an LOS and NLOS scenarios are given in Figure 2.11(a) and (b), respectively. In the experiments in [39], the PASCT algorithm is able to well recognize the clusters in time-varying channels without using any high-resolution estimation algorithm.

2.3.6 Improved subtraction for cluster-centroid initialization

As mentioned in Section 2.3.1, the initial values of the positions of cluster-centroids have a great impact on the clustering results. Hence, a density-based initialization algorithm is developed in [41] to find an appropriate number and positions of the initial cluster-centroids. Once those are determined, the *K*PowerMeans algorithm [13] can be initiated to cluster the MPCs, and the position of the cluster-centroid is updated in each iteration. It is noteworthy that the MPC closest to an initial cluster-centroid is considered the initial power weighted centroid position in *K*PowerMeans.

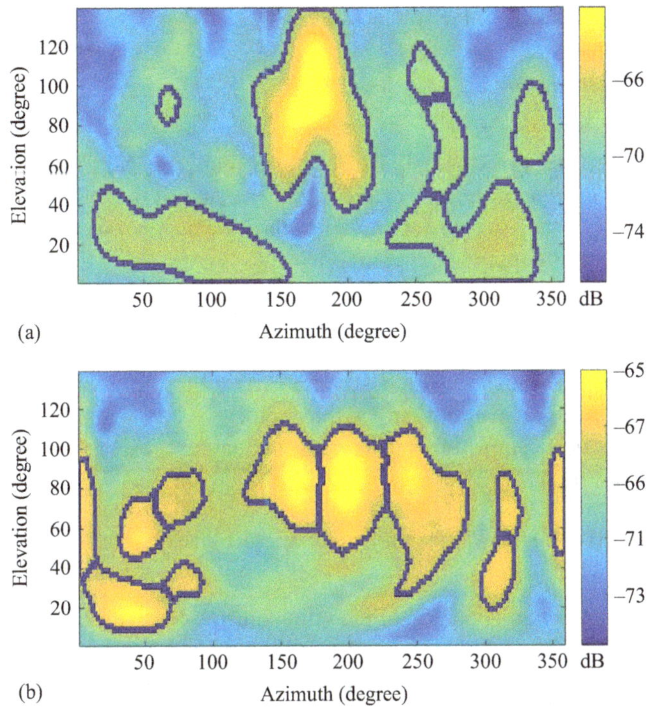

Figure 2.11 Cluster recognition results of the (a) LOS scenario and (b) NLOS scenario, respectively, in [39]

To accurately measure the distance between MPCs, the BMCD (balanced multipath component distance) [41] is used here. The main difference between the BMCD and MCD is that the BMCD introduces *additional normalization factors* for the angular domains. The normalization factors are calculated as

$$\delta_{AoD/AoA} = 2 \cdot \frac{\text{std}_j(d_{MCD,AoD/AoA}(\mathbf{x}_j, \bar{\mathbf{x}}))}{\max_j^2(d_{MCD,AoD/AoA}(\mathbf{x}_j, \bar{\mathbf{x}}))}, \tag{2.21}$$

where std_j is the standard deviation of the MCD between all MPC positions $\mathbf{x}_j$ and the center of data space $\bar{\mathbf{x}}$, and $\max_j$ is the corresponding maximum.

The concrete steps of the improved subtraction are expressed as follows:

1. Calculate the normalized parameter β:

$$\beta = \frac{N}{\sum_{j=i}^{N} d_{MPC}(\mathbf{x}_j, \bar{\mathbf{x}})}, \tag{2.22}$$

where N is the total number of MPCs and $d_{MPC}(\mathbf{x}_j, \bar{\mathbf{x}})$ is the BMCD between $\mathbf{x}_j$ and $\bar{\mathbf{x}}$.

2. Calculate the density value for each MPC $\mathbf{x}_i$:

$$P_j^{\mathbf{m}} = \sum_{j=1}^{N} \exp(-\mathbf{m}^T \cdot \beta \cdot \mathbf{d}_{MPC}(\mathbf{x}_j, \bar{\mathbf{x}})) \tag{2.23}$$

where $\mathbf{m}^T \cdot \beta$ scales the actual influence of neighboring MPCs and its inverse is called *neighborhood radius*. For measurement data, it is more practical to find the appropriate radii for DoA, DoD, and delay dimension separately. Hence, both $\mathbf{m}$ and $\mathbf{d}$ vectors contain three components:

$$\mathbf{d}_{MPC}(\mathbf{x}_i, \mathbf{x}_j) = [d_{MPC,DoA}((\mathbf{x}_i, \mathbf{x}_j)), d_{MPC,DoD}((\mathbf{x}_i, \mathbf{x}_j)), d_{MPC,delay}((\mathbf{x}_i, \mathbf{x}_j))]^T. \tag{2.24}$$

3. The points $\mathbf{x}_k$ with the highest density value are selected as the new cluster-centroids if their density value is above a certain threshold. Stop the iteration if all density values are lower than the threshold.
4. Subtract the new centroid from the data by updating the density values:

$$P_i^{\mathbf{m}} = P_i^{\mathbf{m}} - P_k^{\mathbf{m}} \cdot \exp(-\eta \cdot \mathbf{m}^T \cdot \beta \cdot \mathbf{d}_{MPC}(\mathbf{x}_i, \mathbf{x}_k)), \tag{2.25}$$

where $\eta \in (0, 1]$ is a weight parameter for the density subtraction. Return to step 3.

Then, the number and position of the initial cluster-centroids can be determined, and the *K*PowerMeans can be initialized with these values.

Specifically, to find a proper neighborhood radius, the correlation self-comparison method [41] is used. The detailed steps are

1. Calculate the set of density values for all MPCs P^{m_l} for an increasing m_l, where $m_l \in \{1, 5, 10, 15, \ldots\}$, and the other components in $\mathbf{m}$ are set to be 1.
2. Calculate the correlation between P^{m_l} and $P^{m_{l+1}}$. If the correlation increases above a preset threshold, m_l here is selected as the value for $\mathbf{m}$ in this dimension.

2.3.7 MR-DMS clustering

MR-DMS (multi-reference detection of maximum separation) [42] is developed based on the hierarchical cluster method, which first clusters all elements into one single cluster and then further separates the cluster into more smaller clusters. Specifically, the distances between all MPCs of a cluster seen from multiple reference points are measured, and the MPC group with the biggest distance is separated into two clusters. Besides, the BMCD introduced in Section 2.3.6 is used in [42] to measure the distance between MPCs and reference points. In this study, the optimum number of the clusters can be obtained by different ways: (i) using cluster validation indices, e.g., Xie–Beni index [43], to validate different clustering results or (ii) predefine a threshold for the separation process.

2.3.7.1 Cluster the MPCs

The concrete steps of the MR-DMS are as follows:

1. Spread N reference points over the data space. (e.g., $N = 16$).
2. Cluster all MPCs as single one cluster C_1.

3. Compare the current cluster number C_N and the maximum cluster number $C_{N,\max}$, if $C_N < C_{N,\max}$: for each recent cluster C_k, and calculate the BMCDs between all MPCs $\mathbf{x}_i$ in the current cluster and the reference points $\mathbf{r}_n$ according to

$$\mathbf{d}_n^k(i) = d_{\mathrm{MPC}}(\mathbf{x}_i, \mathbf{r}_n). \tag{2.26}$$

4. Sort the obtained BMCDs $\mathbf{d}_n^k(i)$ in ascending order.
5. Calculate the derivative $(\mathbf{d}_n^k(i))'$, which is the real distance between MPCs in the kth cluster and the nth reference point.
6. Separate the MPCs in the current cluster, which have maximum BMCD over all MPCs and the reference $\max_{k,n,i}(\mathbf{d}_n^k(i))'$.
7. Update the number of current clusters C_N and return to step 3.

2.3.7.2 Obtaining the optimum cluster number

The optimum cluster number can be determined by using the cluster validation indices or a predefined threshold during the separation stage. The implementation of the cluster validation indices is introduced in Section 2.3.1; thus, the threshold to automatically detect a proper cluster number is explained in the below.

During the separation of MPCs in step 6, only the MPCs whose maximum derivative $(\mathbf{d}_n^k(i))'$ exceeds a certain threshold for at least one reference point are considered for separation. In [42], an example of a dynamic threshold is defined by considering the distribution of $(\mathbf{d}_n^k(i))'$:

$$th_n^k = \mathrm{mean}((\mathbf{d}_n^k(i))') + \alpha \cdot \mathrm{std}(((\mathbf{d}_n^k(i))')), \tag{2.27}$$

where α is a weight parameter. Consequently, for each cluster, only the MPCs which have a BMCD significantly larger than the others in the same cluster are considered to be separated. The separation is stopped if all MPCs in the clusters are below the threshold.

Figure 2.12 compares the accuracy of detecting the cluster number by using the improved subtraction algorithm in [41] and the MR-DMS in [42]. In the validation, over 500 drops of the WINNER channel model scenario "urban macro cell" (C2) are tested. In addition, two different scenarios are used where the cluster angular spread of arrival (ASA) is varied (ASA $=\{6°, 15°\}$). From the results, the MR-DMS achieves better performance in detecting the correct cluster number than the improved subtraction algorithm.

Moreover, Figure 2.13 gives azimuth of AoA/AoD and delay domain clustering results based on a MIMO measurement campaign in Bonn, where Figure 2.13(a) and (b) is obtained by using the improved subtraction algorithm together with a run of KPowerMeans and the MR-DMS algorithm, respectively. Details of the measurement campaign can be found in [44], and the MPCs are extracted by using the RiMAX algorithm.

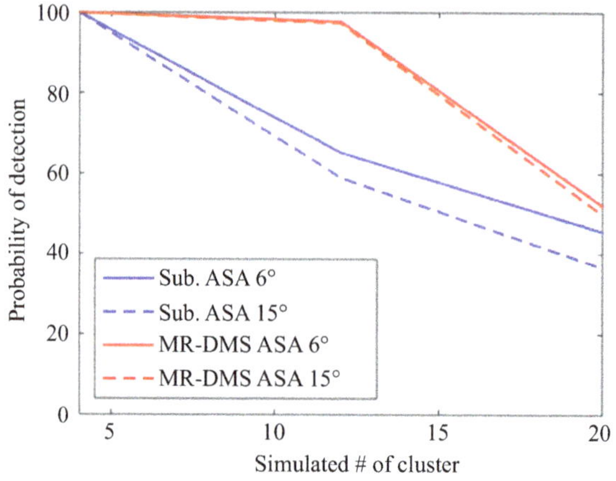

Figure 2.12 *Probability of correctly detecting the number of clusters by using the improved subtraction algorithm in [41] and the MR-DMS in [42], vs. cluster angular spread of arrival (ASA), where ASA = {6°, 15°}*

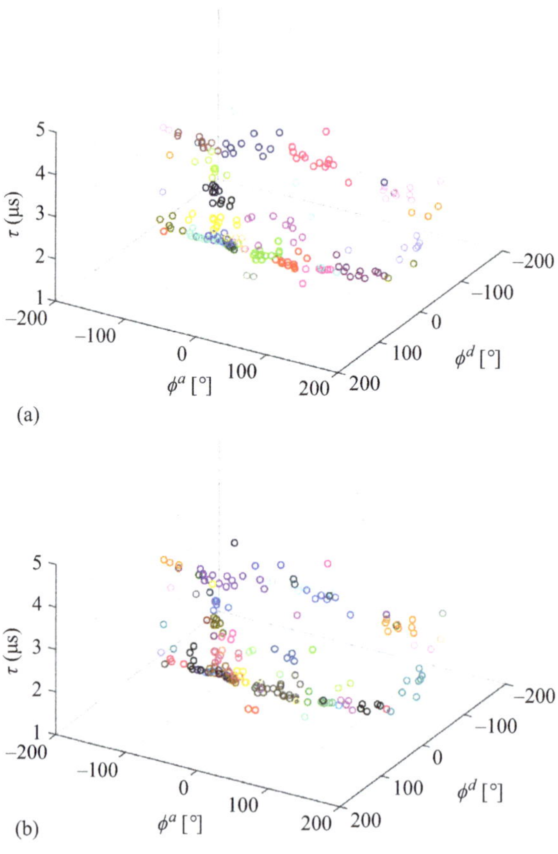

Figure 2.13 *Clustering results based on a MIMO measurement campaign in Bonn, where (a) and (b) are clustered by using the improved subtraction algorithm, together with one run of KPowerMeans and the MR-DMS algorithm, respectively*

2.4 Automatic MPC tracking algorithms

To accurately model time-varying channels, the dynamic changes of the MPCs and clusters need to be characterized. To achieve this, the dynamic MPCs need to be not only clustered but also tracked over time to model the channel [45,46]. Unfortunately, the true moving paths of the MPCs can never be obtained since the MPCs are individually extracted from each snapshot if single-snapshot evaluation algorithms such as SAGE or CLEAN are used. Therefore, many automatic tracking algorithm have been described for searching the most likely moving paths among MPCs/clusters. In this section, we present some machine-learning-based tracking algorithms used for channel modeling.

2.4.1 MCD-based tracking

A tracking method for MPCs needs to capture the moving feature of the clusters/MPCs considering the trade-off between tracking accuracy and computational complexity. The MCD-based tracking method [47,48] aims to track the MPCs in time-varying channels by measuring the MCD between each combination of MPC/cluster in two consecutive snapshots. Note that the MCD between different clusters is defined as the MCD between the cluster-centroids. The basic idea of the MCD-based tracking algorithm is expressed as follows:

1. Preset a threshold P_{T} based on measured data.
2. Measure the distance between MPCs:

$$D_{\mathrm{MPC}_i,\mathrm{MPC}_j} = \mathrm{MCD}(\mathrm{MPC}_i, \mathrm{MPC}_j), \quad \mathrm{MPC}_i \in \mathbf{S}_t, \mathrm{MPC}_j \in \mathbf{S}_{t+1} \tag{2.28}$$

   where $\mathbf{S}_t$ is the set of MPCs in the snapshot s_t, and $\mathbf{S}_{t+1}$ is the set of MPCs in the next snapshot.
3. Associate MPCs in the snapshots s_t and s_{t+1} based on the $D_{\mathrm{MPC}_i,\mathrm{MPC}_j}$ and P_{T}, where
   - if $D_{\mathrm{MPC}_i,\mathrm{MPC}_j} < P_{\mathrm{T}}$, the two MPCs are considered as the same MPC, as shown as the MPC_i and MPC_j in Figure 2.14(a);
   - if $D_{\mathrm{MPC}_i,\mathrm{MPC}_j} > P_{\mathrm{T}}$, the two MPC clusters are considered as different MPCs, as shown as the MPC_i and MPC_k in Figure 2.14(a);
   - if there are more than two MPCs in $\mathbf{S}_{t+1}$ that are close to MPC_i in $\mathbf{S}_t$, MPC_i is considered to be split in the next snapshot, as shown in Figure 2.14(b);
   - if there are more than two MPCs in $\mathbf{S}_t$ close to MPC_i in $\mathbf{S}_t$, MPC_i and MPC_j are considered to be merged in the next snapshot, as shown in Figure 2.14(c);

One of the advantages of the MCD-based tracking algorithm is its low computational complexity, which can be used for complicated scenarios containing many MPCs. Moreover, the behavior of the dynamic MPCs is properly considered in the MCD-based tracking algorithm, including split, merge, and birth–death of MPCs, which all correspond to realistic physical behavior. On the other hand, the value of the preset threshold has a great impact on the tracking results of the algorithm. Hence, the subjective threshold may cause unreliable performance.

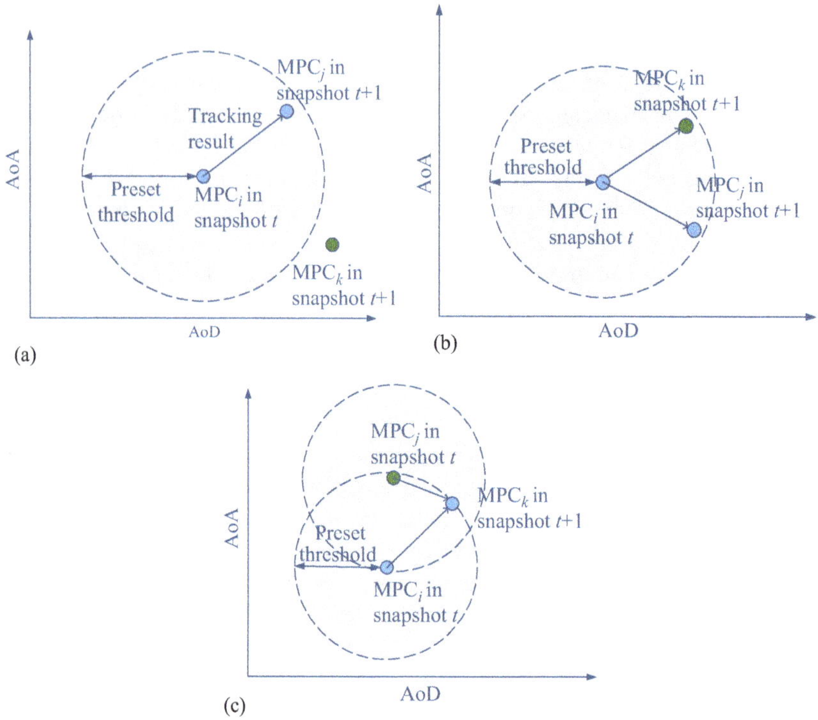

Figure 2.14 MCD-based tracking algorithm (a) gives the principle of tracking process, whereas (b) and (c) shows the cases of split and mergence, respectively

2.4.2 Two-way matching tracking

The two-way matching tracking algorithm is proposed in [45,46] for time-varying channels. It requires the estimated MPCs, which can be extracted by using an HRPE algorithm, and uses the MCD to measure the difference between MPSs. In addition, this tracking algorithm introduces a two-way matching process between two consecutive snapshots to improve the tracking accuracy. The main steps of the two way matching can be expressed as follows:

1. Obtain the MCD matrix **D** by calculating the MCD between each MPC during two consecutive snapshots. For two snapshots s and $s+1$, the MCD matrix **D** can be expressed as

$$\mathbf{D} = \begin{bmatrix} D_{1,1} & \cdots & D_{1,N(s+1)} \\ \vdots & \ddots & \vdots \\ D_{N(s),1} & \cdots & D_{N(s),N(s+1)} \end{bmatrix} \tag{2.29}$$

where $N(s)$ and $N(s+1)$ is the number of the MPCs in snapshots s and $s+1$, respectively.

2. MPCs x in snapshot s and y in snapshot $s+1$ are considered as the same MPC if the following condition is satisfied:

$$D_{x,y} \leq \varepsilon \tag{2.30}$$

$$x = \arg\min_{x}(D_{x\in s,y}) \tag{2.31}$$

$$y = \arg\min_{y}(D_{x,y\in s+1}) \tag{2.32}$$

   where ε is a preset threshold to determine whether the two MPCs in consecutive snapshots could be the same MPC.
3. Matching all MPCs between snapshots s and $s+1$, the matched MPC pairs are considered as the same MPC; the remaining MPCs in snapshots s and $s+1$ are considered as dead and new born MPC, respectively.
4. Repeat the preceding steps 1, 2, and 3 for the current and next snapshots.

One of the advantages of the two-way matching is low computation complexity, which makes it easy to implement for massive data, e.g., V2V channel measurement data. The described algorithm is implemented in [45] for V2V measurements, where it is found that only the MPCs with similar delay and angular characteristic are considered as the same MPCs.

2.4.3 Kalman filter-based tracking

Kalman filtering is one of the most popular machine-learning methods used for target tracking. Therefore, [15] described a cluster-tracking algorithm based on Kalman filters. It is noteworthy that the Kalman filter-based tracking algorithm is used for tracking cluster-centroids, instead of MPCs. The framework of the Kalman filtering is given in Figure 2.15, where $\mu_c^{(n)}$ is the cluster-centroid position in the angle domain or angle-delay domains, $x_c^{(n)}$ are the tracked objects in the input data: angle-delay vector ($\mathbf{X}^{(n)}$), power ($\mathbf{P}^{(n)}$), and n is the index of the snapshot.

For each iteration, the position of the cluster-centroid in the next snapshot is predicted by the Kalman filter based on the current position, and the predicted cluster-centroids are used for the clustering method, e.g., *K*PowerMeans, for the next snapshot.

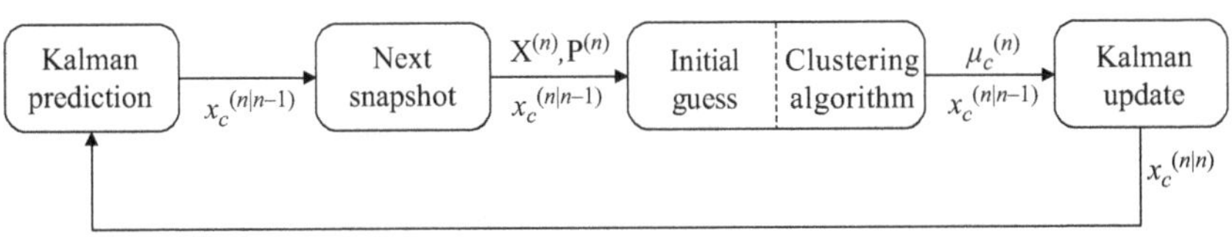

Figure 2.15 Framework of the Kalman filter, where $x_c^{(n)}$ are tracked objects in the input data ($\boldsymbol{X}^{(n)}, \boldsymbol{P}^{(n)}$)

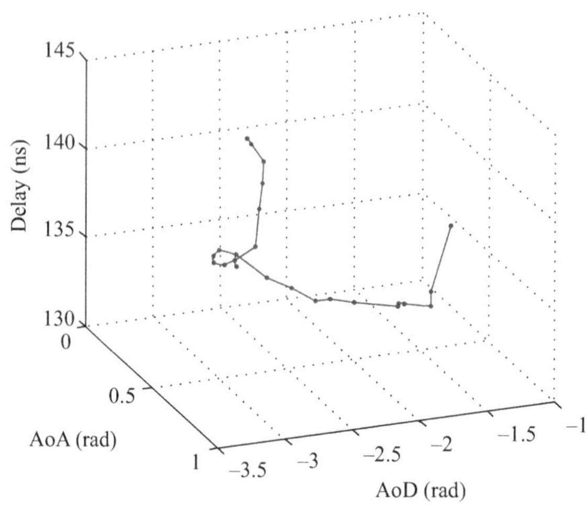

Figure 2.16 Illustration of the tracking result in [15]

Figure 2.16 gives the tracking result in [15]. For a single target, the Kalman filter-based tracking algorithm can achieve high tracking accuracy. To track multiple targets, the Kalman filter can be replaced by a particle filter [49].

2.4.4 Extended Kalman filter-based parameters estimation and tracking

In this section, we discuss an extended Kalman filter-based parameters estimation and tracking algorithm [14,50,51] to capture the dynamics of the channel parameters in time with a low computational complexity. A state-space model is proposed in [51], which is based on the assumption that the parameters are evolved slowly over time and correlated in consecutive time instances.

In the state-space model, the state vector consists of normalized delay $\mu^{\tau}_{k,p}$, the normalized AoA $\mu^{\varphi}_{k,p}$, and the path weight including the real part $\gamma^{Re}_{k,p}$ and imaginary part $\gamma^{Im}_{k,p}$. Consequently, the state model of the pth path at time k can expressed as

$$\theta_{k,p} = \left[\mu^{\tau}_{k,p}, \mu^{\varphi}_{k,p}, \gamma^{Re}_{k,p}, \gamma^{Im}_{k,p}\right]^T. \tag{2.33}$$

It is noteworthy that this model can be extended to contain additional parameters. Let the θ_k denote the state model of all MPCs at time k, then the state model can be rewritten as

$$\theta_k = \Phi\theta_k + \nu_k \tag{2.34}$$

$$\mathbf{y}_k = \mathbf{s}(\theta_k) + \mathbf{n}_{y,k} \tag{2.35}$$

where $\mathbf{y}_k$ is the observation vector, $\mathbf{n}_{y,k}$ is the complex vector containing dense MPCs and noise, $\mathbf{s}(\theta_k)$ is the mapping function between θ_k and $\mathbf{y}_k$, Φ is the state transition matrix. Specifically, all the parameters in the state model are assumed to be uncorrelated with each other. In other words, each parameter evolves independently in time and the state noise, which is additive real white Gaussian while the observation noise is circular complex white Gaussian, is also assumed to be uncorrelated with each other and the state. For each path, the covariance matrix of the state noise is represented by $\mathbf{Q}_{\theta,p} = diag\{\sigma^2_{\mu(\tau)}, \sigma^2_{\mu(\varphi)}, \sigma^2_{\gamma Re}, \sigma^2_{\gamma Im}\}$, whereas the covariance matrix of the observation noise is denoted by $\mathbf{R}_y$.

Considering the estimated parameters are real, the EKF equations can be expressed as

$$\hat{\theta}_{(k|k-1)} = \Phi\hat{\theta}_{(k-1|k-1)} \tag{2.36}$$

$$\mathbf{P}_{(k|k-1)} = \Phi\mathbf{P}_{(k-1|k-1)}\Phi^{\mathrm{T}} + \mathbf{Q}_\theta \tag{2.37}$$

$$\mathbf{P}_{(k|k)} = \left(\mathbf{J}(\hat{\theta}, \mathbf{R}_d) + \mathbf{P}^{-1}_{(k|k-1)}\right)^{-1} \tag{2.38}$$

$$\mathbf{K}_{(k)} = \mathbf{P}_{(k|k-1)}\left[\mathbf{I} - \mathbf{J}(\hat{\theta}, \mathbf{R}_d)\mathbf{P}_{(k|k)}\right] \cdot \begin{bmatrix} \mathscr{R}\{\mathbf{R}_y^{-1}\mathbf{D}_{(k)}\} \\ \mathscr{I}\{\mathbf{R}_y^{-1}\mathbf{D}_{(k)}\} \end{bmatrix}^{\mathrm{T}} \tag{2.39}$$

$$\hat{\theta}_{(k|k)} = \hat{\theta}_{(k|k-1)} + \mathbf{K}_{(k)} \begin{bmatrix} \mathscr{R}\{\mathbf{y}_k - \mathbf{s}(\hat{\theta}_{(k|k-1)})\} \\ \mathscr{I}\{\mathbf{y}_k - \mathbf{s}(\hat{\theta}_{(k|k-1)})\} \end{bmatrix}^{\mathrm{T}} \tag{2.40}$$

where $\mathscr{R}(\bullet)$ and $\mathscr{I}(\bullet)$ denotes the real and imaginary parts of $\bullet$, respectively, $\mathbf{P}_{(k|k)}$ is the estimated error covariance matrix, $\mathbf{J}(\hat{\theta}, \mathbf{R}_d) = \mathscr{R}\{\mathbf{D}^H_{(k)}\mathbf{R}_y^{-1}\mathbf{D}_{(k)}\}$, and $\mathbf{D}$ is the Jacobian matrix. For P paths containing L parameters, $\mathbf{D}$ can be expressed as

$$\mathbf{D}(\theta) = \frac{\partial}{\partial\theta^{\mathrm{T}}}\mathbf{s}(\theta) = \left[\frac{\partial}{\partial\theta_1^{\mathrm{T}}}\mathbf{s}(\theta) \quad \cdots \quad \frac{\partial}{\partial\theta_{LP}^{\mathrm{T}}}\mathbf{s}(\theta)\right]. \tag{2.41}$$

Apparently, the initialization value of parameters for the state transition matrix Φ and the covariance matrix of the state noise $\mathbf{Q}_\theta$ are crucial to the performance of the following tracking/estimation for the EKF. Therefore, it is suggested in [51] to employ another HRPE algorithm, e.g., SAGE [1], RiMAX [3], for this purpose.

2.4.5 Probability-based tracking

In this section, we discuss a probability-based tracking algorithm [16] to track the dynamic MPCs in time-varying channels. The algorithm aims to (i) identify the moving paths of the MPCs in consecutive snapshots and (ii) cluster these MPCs based on the relationships of moving paths. To track the MPCs, a novel-probability-based tracking process is used, which is conducted by maximizing the total sum probability of all moving paths.

In this algorithm, the number of MPCs is assumed to be time-invariant to reduce the complexity. For each MPC, four parameters are considered: AoD ϕ^{D}, AoA ϕ^{A}, delay τ and power α. Let $A_1, \ldots, A_m$ and $B_1, \ldots, B_m$ represent the MPCs in the snapshots S_i and S_{i+1}, respectively. l represents an ordered pair of the MPCs in consecutive snapshots, i.e., l_{A_x,B_y} is the moving path from A_x to B_y, between S_i and S_{i+1}, as shown in Figure 2.17(a). In the probability-based tracking algorithm, each moving path $l_{A,B}$ is weighed by a moving probability $P(A_x, B_y)$, as shown in Figure 2.17(b).

In the probability-based tracking algorithm, the moving paths are identified by maximizing the total probabilities of all selected moving paths, which can be expressed as

$$L^* = \arg\max_{L \subset \mathbf{L}} \sum_{(A_x,B_y)\in L} P(A_x, B_y) \tag{2.42}$$

where L is the selected set of the moving paths and $\mathbf{L}$ is the set of all moving paths. Then, the moving probability $P(A_x, B_y)$ is obtained by using the normalized

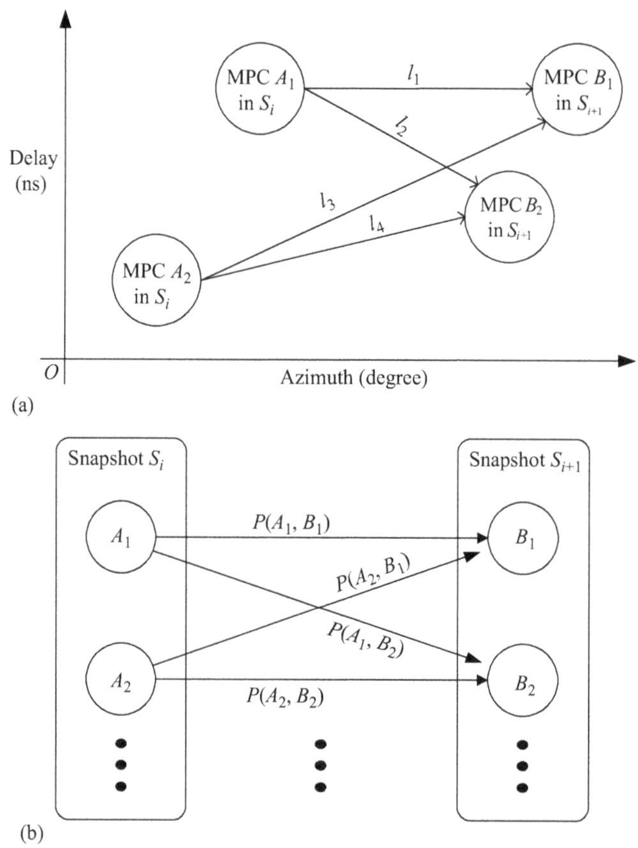

Figure 2.17 Illustration of the moving paths between two consecutive snapshots in [16], where (a) is delay and azimuth domain and (b) is bipartite graph domain

Euclidean distance D_{A_x,B_y} of the vector of parameters $[\phi^{\mathrm{D}}, \phi^{\mathrm{A}}, \tau, \alpha]$, which can be expressed as

$$P(A_x, B_y) = \begin{cases} 1 & D_{A_x,B_y} = 0, \\ 0 & D_{A_x,B_z} = 0, y \neq z, \\ \dfrac{1}{D_{A_x,B_y} \sum_{z=1}^{M} D_{A_x,B_z}^{-1}} & \text{others.} \end{cases} \tag{2.43}$$

To identify the sets of true moving paths L^*, the Kuhn–Munkres algorithm is executed, which is usually used to find the maximum weighted perfect-matching in a bipartite graph of a general assignment problem. In the bipartite graph, every node in two subsets links to each other and every link has its own weight. In this algorithm, the MPCs in two successive snapshots are considered as the two subsets in the bipartite graph, and the moving paths between each snapshot are considered as the links between two subsets. which is weighted by $P(A_x, B_y)$, as shown in Figure 2.17(b). In this case, the true moving paths can be recognized.

After obtaining the moving paths of all MPCs, a heuristic approach is developed to cluster these MPCs with the purpose of comparing the moving probability of the MPCs in the same snapshot with a preset threshold P_T. The basic idea of the clustering process is to group the MPCs with similar moving probabilities, which means their moving patterns are close to each other, e.g., if $P(A_x, B_y)$ and $P(A_x, B_z)$ are greater than P_T, it indicates that the MPCs B_y and B_z are fairly similar and estimated to belong to the same cluster. The clustering process can be expressed as

$$\mathbf{K}_x = \{B_y | P(A_x, B_y) > P_{\mathrm{T}}, A \in S_i, B \in S_{i+1}\}. \tag{2.44}$$

According to the simulations in [16], P_T is suggested to be set to 0.8. From (2.44), different A_x in S_i may lead to different clustering results. In this case, the result with the most occurrences is selected, e.g., if $\mathbf{K}_1 = \{B_1, B_2\}, \mathbf{K}_2 = \{B_1, B_2, B_3\}, \mathbf{K}_3 = \{B_1, B_2, B_3\}$, then $\mathbf{K} = \{B_1, B_2, B_3\}$.

2.5 Deep learning-based channel modeling approach

In this section, we present some learning-based channel-modeling algorithms. An important advantage of many machine-learning methods, especially artificial neural networks, is to automatically obtain the inherent features of input data and to find the mapping relationship between input and output. On the other hand, the purpose of wireless channel modeling is to accurately model MPCs in wireless channels, and on some level, it aims to find the mapping relationship between the channel parameters and the scenarios. Inspired by this, a number of papers have investigated channel-modeling using neural networks.

2.5.1 BP-based neural network for amplitude modeling

As early as 1997, [18] adaptively modeled nonlinear radio channels by using the odd and even BP algorithm for multilayer perceptron (MLP) neural networks. The BP algorithm is a classical algorithm for data training in neural networks; its flowchart is given in Figure 2.18.

The SV model [7] is adopted in [18] as a comparison, the simulation result is given in Figure 2.19.

Considering the advantages of neural network on regression problems, a multilayer neural network is used to find the mapping relationship between frequency and amplitude, where the architecture of the neural network is shown in Figure 2.20. For most of the neural networks, there are three parts in the framework: (i) input layer, (ii) hidden layer, and (iii) output layer. The measured data are used as training data, where the system input and output of the neural network are frequency and amplitude, respectively, and the sub-output of each layer is the sub-input of the next layer in the hidden layers. Through several iterations, the weight parameters of each layer can be obtained; thus, the mapping function between the system input and output can be modeled.

2.5.2 Development of neural-network-based channel modeling

Due to the good performance of finding mapping relationships, neural-network-based channel modeling has drawn a lot of attention. For a classical architecture of the neural network, the performance is related to the number of layers, and each layer is designed to classify basic elements. Theoretically, a neural network with more layers can thus achieve better performance. However, a neural network with too many layers leads to another problem: vanishing gradient. In the training process of a neural network, the difference between the system output and training data is assessed and fed back to the upper layer to adjust the weight parameters.

In a multilayer network, the feedback obviously has more influence on the layers closest to the system output and has a fairly limited influence on the layers at the front end. Nevertheless, the layers at the front end usually have a great effect on the final performance. As a result, the feedback in each iteration cannot be well transmitted to the front layers, as shown in Figure 2.21. Hence, the performance of the multilayer neural network can suffer. Besides, a multilayer neural network has high computational complexity, which was prohibitive given the limitations of the hardware at that time of that paper. Therefore, neural-network-based channel modeling gradually disappeared from public view.

To avoid the vanishing gradient, [52] proved that a three-layer MLP neural network can be multidimensionally approximated as an arbitrary function to any desired accuracy, which is important and useful for modeling a mapping function. Since there are only three layers in this network, the vanishing gradient does not impact the performance during the training process. In this case, a three-layer MLP neural network can be adopted for many channel-modeling applications without the limitation of vanishing gradients.

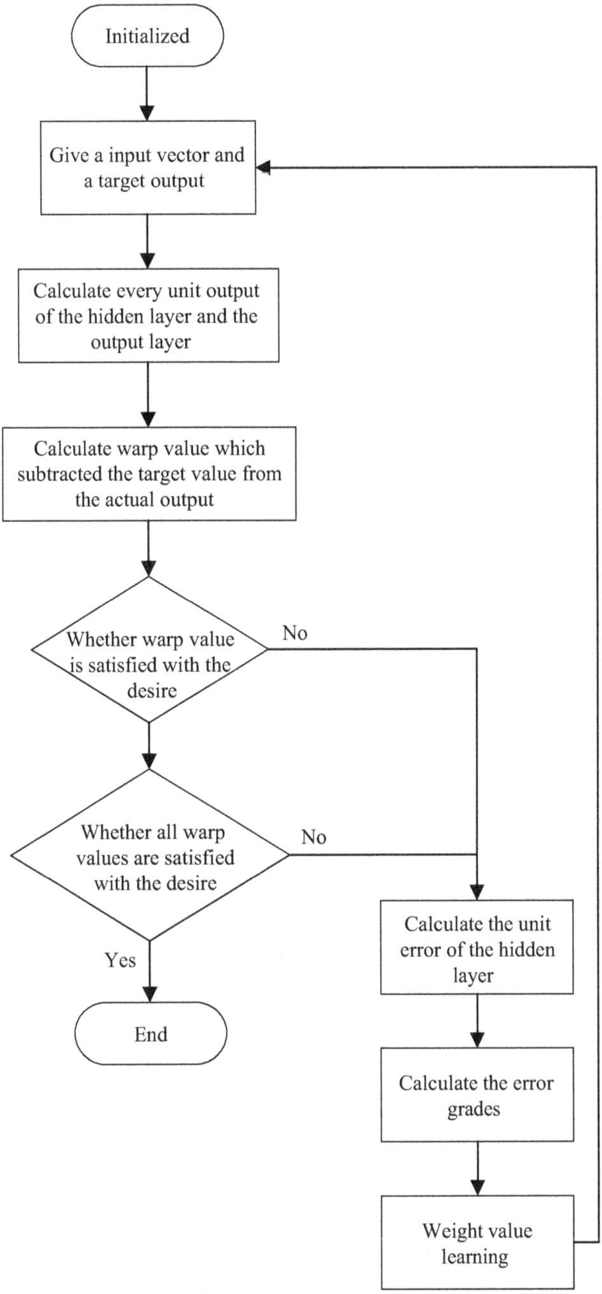

Figure 2.18 Flowchart of BP algorithm

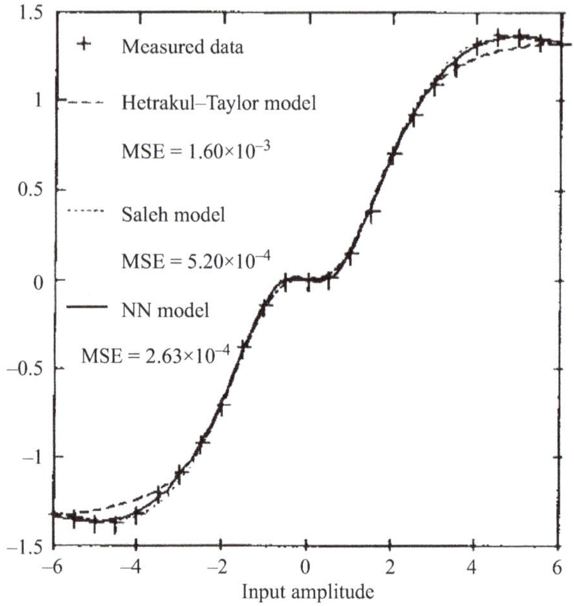

Figure 2.19 Simulation result of the comparison of SV model and ANN-based model in [18]

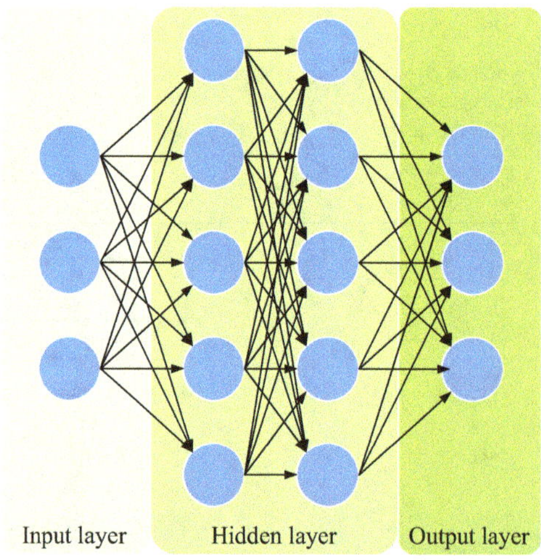

Figure 2.20 Illustrations of the common architecture of the neural network

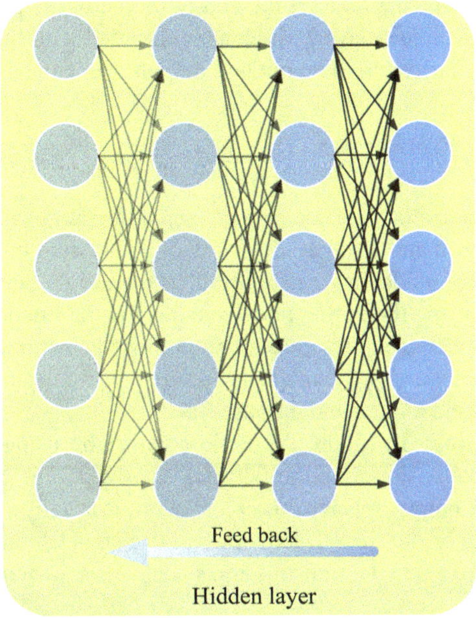

Figure 2.21 Demonstration of the vanishing gradient in training process of the neural network

In addition, there are some other novel frameworks for neural networks which can avoid the vanishing gradient, e.g., the restricted Boltzmann machines framework and the deep-learning method described in [53]. Hence, there are some propagation channel modeling studies, e.g., [54], where the amplitude frequency response of 11 paths is modeled by an MLP neural network.

2.5.3 RBF-based neural network for wireless channel modeling

Reference [19] introduces a RBF neural network for modeling a single nonfading path with additive white Gaussian noise (AWGN). The RBF neural network can approximate any arbitrary nonlinear function with any accuracy, just like an MLP network. However, it has a number of advantages: it only has one hidden layer, and the number of hidden layer nodes can be adaptively adjusted in the training stage, whereas the numbers of the hidden layer and the hidden layer nodes for MLP network is not easily determined. The main framework of the RBF-based neural network is expressed as follows.

1. **RBF neural network**: A RBF neural network is a three-layer feedforward network, which contains one input layer, one hidden layer, and one output layer. The input layer obtains the training data and transmits to the hidden layer. The hidden layer consists of a group of RBF, and the corresponding center vectors and width are the parameters of the RBF, where the Gaussian function is usually

adopted as the basis function. At the end of the network, the out layer receives the outputs of the hidden layer, which are combined with linear weighting. The mapping function between the input and output layer can be expressed as

$$y = f(x) = \sum_{i=1}^{m} \omega_i \phi(\|x - c_i\|, \sigma_i) = \sum_{i=1}^{m} \omega_i \exp\left| - \frac{\|x - c_i\|}{2\sigma_i^2} \right| \tag{2.45}$$

where the vector $x = (x_1, x_2, \ldots, x_m)$ represents the input data of the network, c_i and σ_i are the mean and standard deviation of a Gaussian function, respectively, m is the number of hidden layer neurons, ω_i is the weight between the link of the ith basis function and the output node, and $\| \bullet \|$ is the Euclidean norm.

The training process adjusts, through iterations, the parameters of the network including the center and width of each neuron in the hidden layer, and the weight vectors between the hidden and output layer.

2. **Channel modeling**: To model the radio channel by using a neural network, the mapping relationship/function between the input, i.e., transmit power and distance, and the output, receive power and delay, is usually a nonlinear function. Hence, the goal of the neural-network-based channel modeling is to use the network to approximate the transmission system, as shown in Figure 2.22.

In [19], the number of RBFs is set to the number of MPCs, to simulate the transmit signal with different time delays. The output layer gives the received signal. Besides, the width of the RBF network is obtained by

$$\sigma = \frac{d}{\sqrt{2M}} \tag{2.46}$$

where d is the maximum distance and M is the number of RBF nodes. In this case, once the nodes and width of RBF network are determined, the weights of the output layer can be obtained by solving linear equations.

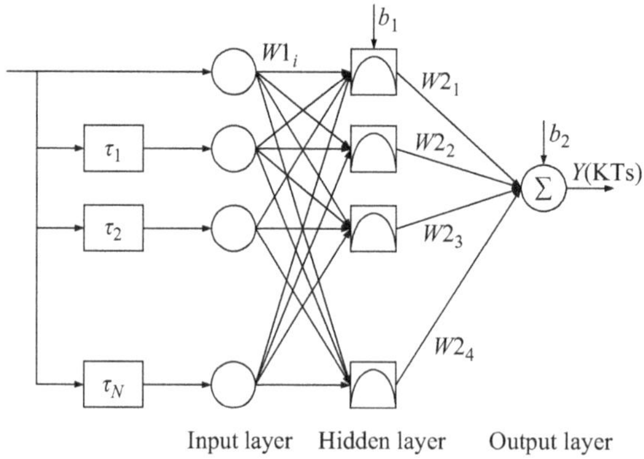

Figure 2.22 The wireless channel based on RBF neural network

Furthermore, the BP-based network model is compared with the RBF-based network model in [19]. A multipath channel with AWGN is simulated, where the results are given in Figure 2.23. From the simulation results, the RBF-based network model generally shows better accuracy than the BP-based network model.

Similarly, a neural network is also used in [55] to model the path loss in a mine environment, where the frequency and the distance are considered as the input data, and the received power is considered as output. The framework of the neural network is given in Figure 2.24, which contains two input nodes, 80 hidden nodes, and one output node. W_{1f} and W_{2f} are the weight parameters between the input layer and hidden layer, whereas w_{jk} and w_{Nk} are the weight parameters between the hidden and output layers.

2.5.4 Algorithm improvement based on physical interpretation

The high accuracy and flexibility make neural networks a powerful learning tool for channel modeling. Due to good learning ability and adaptability, neural networks are expected not only to model the channel of one single scenario but also to model the channels of multiple scenarios, as shown in Figure 2.25. Despite these high expectations, there are still many problems that need to be discussed and studied.

Based on past research, artificial intelligence has shown great power for the development of channel modeling, whether as a preprocessing tool or a learning tool. However, the difference between the two fields of artificial intelligence and channel modeling still needs to be considered. Although machine-learning approaches

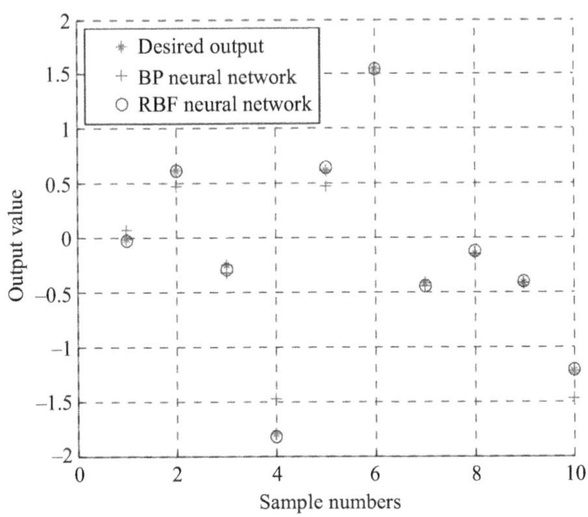

Figure 2.23 Simulation results of AWGN channel containing two pathways with doppler frequency shifts in [19]

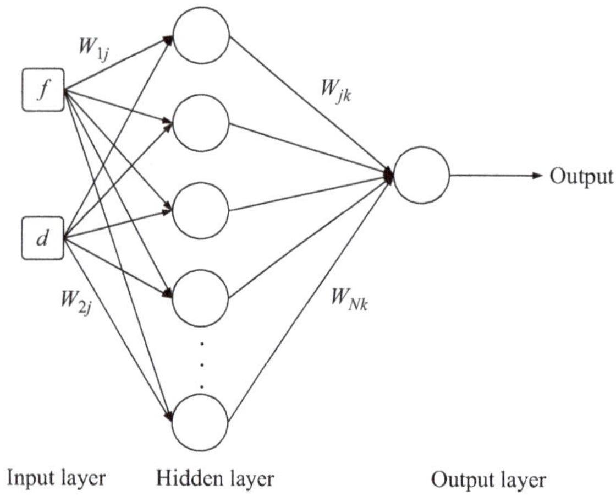

Figure 2.24 Neural network used in [55], which contains two input nodes, 80 hidden nodes, and one output node. The two input nodes correspond to frequency and distance, whereas the output node corresponds to the power of the received signal

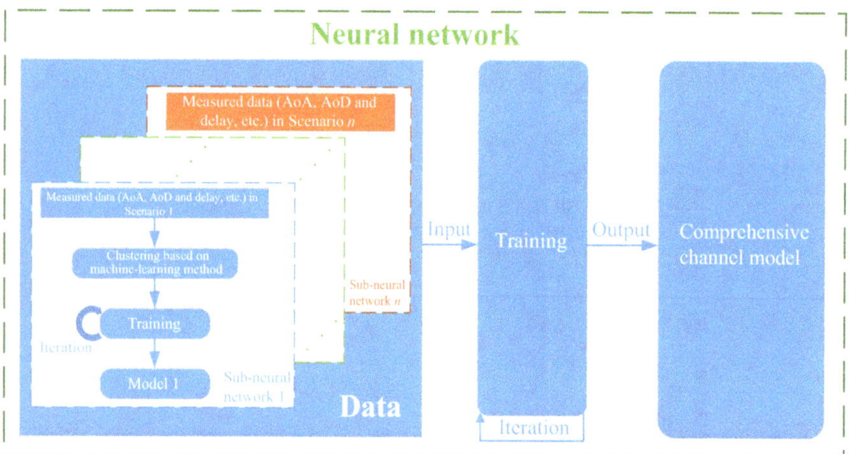

Figure 2.25 Framework of the learning-based channel modeling for multiple scenarios

generally have good performance at data processing, these approaches can be further improved by considering the physical characteristics of channel parameters. For example, *K*Means is a good conventional clustering algorithm for data processing, and the *K*PowerMeans in [13] is further developed by combining the physical interpretation of the power of MPCs with *K*Means, thus improving performance for clustering of MPCs than merely using the *K*Means. Moreover, the development of MCD is another example, where the physical characteristics of the MPCs are considered, and the MCD thus is a more accurate measure of the differences of MPCs than using the Euclidean distance for channel modeling. As for neural networks, the physical interpretation is also important to build an appropriate network, e.g., the description of the CIR needs to be considered while constructing the activation function for the neural network. In addition, the disadvantages of the adopted machine-learning methods cannot be neglected, e.g., the *K*Means is sensitive to initial parameters; this feature also appears in the *K*PowerMeans, e.g., the clustering result is sensitive to the assumed number of clusters and the position of cluster-centroids. Using the physical meaning of parameters of the approaches is a possible way to evade these disadvantages. Hence, the potential relationship between parameters of machine-learning techniques and physical variables of the radio channels needs to be further incorporated into the adopted algorithms to improve accuracy.

2.6 Conclusion

In this chapter, we presented some machine-learning-based channel modeling algorithms, including (i) propagation scenarios classification, (ii) machine-learning-based MPC clustering, (iii) automatic MPC tracking, and (iv) neural network-based channel modeling. The algorithms can be implemented to preprocess the measurement data, extract the characteristics of the MPCs, or model the channels by directly seeking the mapping relationships between the environments and received signals. The results in this chapter can provide references to other real-world measurement-based channel modeling.

References

[1] Fleury BH, Tschudin M, Heddergott R, *et al.* Channel Parameter Estimation in Mobile Radio Environments Using the SAGE Algorithm. IEEE Journal on Selected Areas in Communications. 1999;17(3):434–450.

[2] Vaughan RG, and Scott NL. Super-Resolution of Pulsed Multipath Channels for Delay Spread Characterization. IEEE Transactions on Communications. 1999;47(3):343–347.

[3] Richter A. Estimation of radio channel parameters: models and algorithms. Technischen Universität Ilmenau; 2005 December.

[4] Benedetto F, Giunta G, Toscano A, *et al.* Dynamic LOS/NLOS Statistical Discrimination of Wireless Mobile Channels. In: 2007 IEEE 65th Vehicular Technology Conference – VTC2007-Spring; 2007. p. 3071–3075.

[5] Guvenc I, Chong C, and Watanabe F. NLOS Identification and Mitigation for UWB Localization Systems. In: 2007 IEEE Wireless Communications and Networking Conference; 2007. p. 1571–1576.

[6] Zhou Z, Yang Z, Wu C, *et al.* WiFi-Based Indoor Line-of-Sight Identification. IEEE Transactions on Wireless Communications. 2015;14(11):6125–6136.

[7] Saleh AAM, and Valenzuela R. A Statistical Model for Indoor Multipath Propagation. IEEE Journal on Selected Areas in Communications. 1987;5(2):128–137.

[8] Molisch AF, Asplund H, Heddergott R, *et al.* The COST259 Directional Channel Model-Part I: Overview and Methodology. IEEE Transactions on Wireless Communications. 2006;5(12):3421–3433.

[9] Asplund H, Glazunov AA, Molisch AF, *et al.* The COST 259 Directional Channel Model-Part II: Macrocells. IEEE Transactions on Wireless Communications. 2006;5(12):3434–3450.

[10] Liu L, Oestges C, Poutanen J, *et al.* The COST 2100 MIMO Channel Model. IEEE Wireless Communications. 2012;19(6):92–99.

[11] RAN GT. Spatial channel model for multiple input multiple output (MIMO) simulations. Sophia Antipolis Valbonne, France: 3GPP, Tech. Rep.; 2008.

[12] Meinilä J, Kyösti P, Jämsä T, *et al.* WINNER II channel models. Radio Technologies and Concepts for IMT-Advanced, NOKIA; 2009. p. 39–92.

[13] Czink N, Cera P, Salo J, *et al.* A Framework for Automatic Clustering of Parametric MIMO Channel Data Including Path Powers. In: IEEE Vehicular Technology Conference; 2006. p. 1–5.

[14] Salmi J, Richter A, and Koivunen V. Detection and Tracking of MIMO Propagation Path Parameters Using State-Space Approach. IEEE Transactions on Signal Processing. 2009;57(4):1538–1550.

[15] Czink N, Tian R, Wyne S, *et al.* Tracking Time-Variant Cluster Parameters in MIMO Channel Measurements. In: 2007 Second International Conference on Communications and Networking in China; 2007. p. 1147–1151.

[16] Huang C, He R, Zhong Z, *et al.* A Novel Tracking-Based Multipath Component Clustering Algorithm. IEEE Antennas and Wireless Propagation Letters. 2017;16:2679–2683.

[17] Chen M, Challita U, Saad W, *et al.* Machine learning for wireless networks with artificial intelligence: A tutorial on neural networks. arXiv preprint arXiv:171002913. 2017.

[18] Ibukahla M, Sombria J, Castanie F, *et al.* Neural Networks for Modeling Nonlinear Memoryless Communication Channels. IEEE Transactions on Communications. 1997;45(7):768–771.

[19] Sha Y, Xu X, and Yao N. Wireless Channel Model Based on RBF Neural Network. In: 2008 Fourth International Conference on Natural Computation. vol. 2; 2008. p. 605–609.

[20] Bartlett MS. Smoothing Periodograms from Time Series with Continuous Spectra. Nature. 1948;161:343–347.

[21] He R, Ai B, Stuber GL, *et al.* Geometrical-Based Modeling for Millimeter-Wave MIMO Mobile-to-Mobile Channels. IEEE Transactions on Vehicular Technology. 2018;67(4):2848–2863.

[22] He R, Ai B, Molisch AF, *et al.* Clustering Enabled Wireless Channel Modeling Using Big Data Algorithms. IEEE Communications Magazine. 2018;56(5):177–183.

[23] Mota S, Perez-Fontan F, and Rocha A. Estimation of the Number of Clusters in Multipath Radio Channel Data Sets. IEEE Transactions on Antennas and Propagation. 2013;61(5):2879–2883.

[24] Mota S, Garcia MO, Rocha A, *et al.* Clustering of the Multipath Radio Channel Parameters. In: Proceedings of the 5th European Conference on Antennas and Propagation (EUCAP); 2011. p. 3232–3236.

[25] Schneider C, Bauer M, Narandzic M, *et al.* Clustering of MIMO Channel Parameters – Performance Comparison. In: VTC Spring 2009 – IEEE 69th Vehicular Technology Conference; 2009. p. 1–5.

[26] He R, Chen W, Ai B, *et al.* On the Clustering of Radio Channel Impulse Responses Using Sparsity-Based Methods. IEEE Transactions on Antennas and Propagation. 2016;64(6):2465–2474.

[27] He R, Chen W, Ai B, *et al.* A Sparsity-Based Clustering Framework for Radio Channel Impulse Responses. In: 2016 IEEE 83rd Vehicular Technology Conference (VTC Spring); 2016. p. 1–5.

[28] Candès JE, Wakin, MB, and Boyd SP. Enhancing Sparsity by Reweighted l_1 Minimization. Journal of Fourier Analysis and Applications. 2008;14(5):877–905. Available from: http://www.springerlink.com/content/wp246375t1037538/.

[29] He R, Li Q, Ai B, *et al.* A Kernel-Power-Density-Based Algorithm for Channel Multipath Components Clustering. IEEE Transactions on Wireless Communications. 2017;16(11):7138–7151.

[30] He R, Li Q, Ai B, *et al.* An Automatic Clustering Algorithm for Multipath Components Based on Kernel-Power-Density. In: 2017 IEEE Wireless Communications and Networking Conference (WCNC); 2017. p. 1–6.

[31] Spencer QH, Jeffs BD, Jensen MA, *et al.* Modeling the Statistical Time and Angle of Arrival Characteristics of an Indoor Multipath Channel. IEEE Journal on Selected Areas in Communications. 2000;18(3):347–360.

[32] Ester M, Kriegel HP, Sander J, *et al.* A Density-Based Algorithm for Discovering Clusters in Large Spatial Databases with Noise. In: Proceedings of the 2nd International Conference on Knowledge Discovery and Data Mining; 1996. p. 226–231. Available from: https://www.aaai.org/Papers/KDD/1996/KDD96-037.pdf.

[33] Samimi MK, and Rappaport TS. 3-D Millimeter-Wave Statistical Channel Model for 5G Wireless System Design. IEEE Transactions on Microwave Theory and Techniques. 2016;64(7):2207–2225.

[34] Sun S, MacCartney GR, and Rappaport TS. A Novel Millimeter-Wave Channel Simulator and Applications for 5G Wireless Communications. In: 2017 IEEE International Conference on Communications (ICC); 2017. p. 1–7.

[35] MacCartney GR, Rappaport TS, Samimi MK, *et al.* Millimeter-Wave Omnidirectional Path Loss Data for Small Cell 5G Channel Modeling. IEEE Access. 2015;3:1573–1580.

[36] Rappaport TS, MacCartney GR, Samimi MK, *et al.* Wideband Millimeter-Wave Propagation Measurements and Channel Models for Future Wireless Communication System Design. IEEE Transactions on Communications. 2015;63(9):3029–3056.

[37] Rappaport TS, Sun S, Mayzus R, *et al.* Millimeter Wave Mobile Communications for 5G Cellular: It Will Work!. IEEE Access. 2013;1:335–349.

[38] Sun S, MacCartney GR, Samimi MK, *et al.* Synthesizing Omnidirectional Antenna Patterns, Received Power and Path Loss from Directional Antennas for 5G Millimeter-Wave Communications. In: Global Communications Conference (GLOBECOM), 2015 IEEE. IEEE; 2015. p. 1–7.

[39] Huang C, He R, Zhong Z, *et al.* A Power-Angle Spectrum Based Clustering and Tracking Algorithm for Time-Varying Channels. IEEE Transactions on Vehicular Technology. 2019;68(1):291–305.

[40] Otsu N. A Threshold Selection Method from Gray-Level Histograms. IEEE Transactions on Systems, Man, and Cybernetics. 1979;9(1):62–66.

[41] Yacob A. Clustering of multipath parameters without predefining the number of clusters. 2015 Techn Univ, Masterarbeit TU Ilmenau; 2015.

[42] Schneider C, Ibraheam M, Hafner S, *et al.* On the Reliability of Multipath Cluster Estimation in Realistic Channel Data Sets. In: The 8th European Conference on Antennas and Propagation (EuCAP 2014); 2014. p. 449–453.

[43] Xie XL, and Beni G. A Validity Measure for Fuzzy Clustering. IEEE Transactions on Pattern Analysis & Machine Intelligence. 1991;(8):841–847.

[44] Sommerkorn G, Kaske M, Schneider C, *et al.* Full 3D MIMO Channel Sounding and Characterization in an Urban Macro Cell. In: 2014 XXXIth URSI General Assembly and Scientific Symposium (URSI GASS); 2014. p. 1–4.

[45] He R, Renaudin O, Kolmonen V, *et al.* A Dynamic Wideband Directional Channel Model for Vehicle-to-Vehicle Communications. IEEE Transactions on Industrial Electronics. 2015;62(12):7870–7882.

[46] He R, Renaudin O, Kolmonen V, *et al.* Characterization of Quasi-Stationarity Regions for Vehicle-to-Vehicle Radio Channels. IEEE Transactions on Antennas and Propagation. 2015;63(5):2237–2251.

[47] Czink N, Mecklenbrauker C, and d Galdo G. A Novel Automatic Cluster Tracking Algorithm. In: 2006 IEEE 17th International Symposium on Personal, Indoor and Mobile Radio Communications; 2006. p. 1–5.

[48] Karedal J, Tufvesson F, Czink N, *et al.* A Geometry-Based Stochastic MIMO Model for Vehicle-to-Vehicle Communications. IEEE Transactions on Wireless Communications. 2009;8(7):3646–3657.

[49] Yin X, Steinbock G, Kirkelund GE, *et al.* Tracking of Time-Variant Radio Propagation Paths Using Particle Filtering. In: 2008 IEEE International Conference on Communications; 2008. p. 920–924.

[50] Richter A, Enescu M, and Koivunen V. State-Space Approach to Propagation Path Parameter Estimation and Tracking. In: IEEE 6th Workshop on Signal Processing Advances in Wireless Communications, 2005; 2005. p. 510–514.

[51] Salmi J, Richter A, and Koivunen V. MIMO Propagation Parameter Tracking using EKF. In: 2006 IEEE Nonlinear Statistical Signal Processing Workshop; 2006. p. 69–72.

[52] Zhang QJ, Gupta KC, and Devabhaktuni VK. Artificial Neural Networks for RF and Microwave Design – From Theory to Practice. IEEE Transactions on Microwave Theory and Techniques. 2003;51(4):1339–1350.

[53] Hinton GE, Osindero S, and Teh YW. A Fast Learning Algorithm for Deep Belief Nets. Neural Computation. 2006;18(7):1527–1554.
[54] Ma Y-t, Liu K-h, and Guo Y-n. Artificial Neural Network Modeling Approach to Power-Line Communication Multi-Path Channel. In: 2008 International Conference on Neural Networks and Signal Processing; 2008. p. 229–232.
[55] Kalakh M, Kandil N, and Hakem N. Neural Networks Model of an UWB Channel Path Loss in a Mine Environment. In: 2012 IEEE 75th Vehicular Technology Conference (VTC Spring); 2012. p. 1–5.

Chapter 3

Channel prediction based on machine-learning algorithms

Xue Jiang[1] and Zhimeng Zhong[2]

In this chapter, the authors address the wireless channel prediction using state-of-the-art machine-learning techniques, which is important for wireless communication network planning and operation. Instead of the classic model-based methods, the authors provide a survey of recent advances in learning-based channel prediction algorithms. Some open problems in this field are then proposed.

3.1 Introduction

Modern wireless communication networks can be considered as large, evolving distributed databases full of context and information available from mobile devices, base stations, and environment. The wireless channel data in various scenarios including large-scale and small-scale parameters are one of the important and useful data that could be used for analyzing and making predictions.

A coverage map is often given as a set of radio measurements over discrete geographical coordinates and is typically obtained by drive tests. Accurate coverage maps are crucial for enabling efficient and proactive resource allocation. However, it is nearly impossible to obtain these maps completely from measurements. Thus, the coverage loss maps should be reconstructed with the available measurements. A reliable reconstruction of current and future coverage maps will enable future networks to better utilize the scarce wireless resources and to improve the quality-of-service experienced by the users. Reconstructing coverage maps is of particular importance in the context of (network-assisted) device-to-device (D2D) communication where no or only partial measurements are available for D2D channels [1].

As one of the hottest topics all over the world, machine-learning techniques have been applied in various research fields in recent years including reconstruction of coverage maps. These learning-based reconstruction approaches can be divided into two categories: batch algorithms and online algorithms. The former one mainly includes

[1]School of Electronic Information and Electrical Engineering, Shanghai Jiao Tong University, China
[2]Huawei Technologies Ltd., China

support vector machines (SVM) [2,3], artificial neural networks (ANN) [2,4], and matrix completion with singular value thresholding (SVT) [5]. Aside from that, Gaussian processes [6] and kriging-based techniques [7] have recently been successfully used for the estimation of radio maps. In [8], kriging-based techniques have been applied to track channel gain maps in a given geographical area. The proposed kriged Kalman filtering algorithm allows to capture both spatial and temporal correlations. These studies use batch schemes as well. In [9], an adaptive online reconstruction methodology is proposed with adaptive projected subgradient method (APSM) [10], which is the unique online coverage map reconstruction algorithm having been employed so far.

This chapter mainly generalizes and reviews the learning-based coverage maps reconstruction mentioned above. The rest of this survey is organized as follows. Section 3.2 introduces methodologies of obtaining measurements. Section 3.3 discusses the respective traits of batch algorithms and online algorithms. The corresponding approaches will be studied as well. Section 3.4 comprises of the techniques applied to label the measurements to get more accurate results. The final section draws the conclusion of the survey.

3.2 Channel measurements

Before discussing the learning-based algorithms, a set of radio measurements over discrete geographical coordinates should be obtained. In general, the methodology of achieving measurements can be divided into two types:

- **Conventional drive test:** Conventional drive test is a manual process. To collect network quality information, an operator often needs to send engineers directly to the concerning area and obtain radio measurements in a hand-operated manner. Typically, a measurement vehicle equipped with specially developed test terminals, measurement devices, and a global-positioning system receiver to obtain geographical location is used to check coverage outdoors [11]. With such measurement vehicle, engineers would perform test calls in the car and record measurement results along the drive route.
- **Minimization of drive test (MDT):** The main concept of MDT is to exploit commercial user equipment (UE) measurement capabilities and geographically spread nature for collecting radio measurements [12]. This methodology exploits information from the so-called crowdsourcing applications. With crowdsourcing, a user installs an application on an off-the-shelf smart phone and returns measurements to a database [13].

Conventional drive test is simple and stable. Nevertheless, this methodology consumes significant time and human efforts to obtain reliable data, and the cost will ascend vastly with the studied area getting larger. Thus, it would be more suitable to small-scale area. MDT is a relatively cost-efficient way to get the measurements, sacrificing for its stability. The application on users' smart phones would reduce their data budget and the battery lifetime of the mobile device. Wide fluctuation of the involved smart phones' function might also result in the systematic measurement errors.

3.3 Learning-based reconstruction algorithms

Over the past years, path-loss models in various types of networks have been proposed and analyzed. Despite the fact that path-loss modeling is useful in many applications, the deviation between the path loss, measured in a real propagation environment, and the one given by the model can be large [14]. For this reason, learning-based reconstruction has been extensively studied since it can be tailored to the specific environment under consideration and give more accurate results. As mentioned in the first section, learning-based reconstruction algorithms can be divided into two categories: batch and online algorithms (Table 3.1).

3.3.1 Batch algorithms

Batch algorithms assume the complete data to be available before performing the reconstruction algorithm. The performance of batch algorithm is excellent, and the cost is much cheaper than online algorithms, while storage is needed to store the samples. The batch algorithms, which have been employed in coverage loss reconstruction, includes SVM, ANN, and SVT.

3.3.1.1 Support vector machine

SVM was first introduced by Vapnik [16]. One of the main advantages of SVM over other classical machine-learning techniques (e.g., neural networks) is the absence of local minima in the optimization problems, the possibility of enforcing sparse solutions, and the capacity for controlling error margins in the prediction. SVMs were initially developed for classification tasks, but they have proven to be a powerful tool for regression problems (i.e., for function approximation problems), so they are natural candidates for the task of coverage map estimation [2]. In particular, in [2], an extended feature vector is used to train the SVM, and this feature vector includes environmental information about transmitters, receivers, buildings, and the transmit frequency, among others. To counter the curse of dimensionality owing to the high number of input features, the authors of [3] use a principal component analysis (PCA)

Table 3.1 Representative algorithms for channel map reconstruction

Type	Methods
Batch	SVM [2,3], ANN [2,4], matrix completion [5]
Online	APSM [10], multi-kernel [15]

based on dimensionality-reduction techniques. This operation is performed before applying the SVMs.

The estimate usually assumes the following form by SVM:

$$\tilde{y} = \sum_{j=1}^{m} \omega_j \phi_j(x) \tag{3.1}$$

where $\{\phi_j(x)\}_{j=1}^{m}$ is a set of m nonlinear basis functions. The loss function [17] used for determining the estimate is given by

$$L_\varepsilon(\tilde{y}, y) = \begin{cases} |\tilde{y} - y| - \varepsilon, & |\tilde{y} - y| > \varepsilon \\ 0, & \text{otherwise} \end{cases} \tag{3.2}$$

with ε being a small value. The problem can be formally stated as

$$\begin{aligned} \min \quad & \frac{1}{N}\sum_{i=1}^{N} L_\varepsilon(\tilde{y}^i, y^i) \\ \text{s.t.} \quad & \|\omega\| \le \alpha \end{aligned} \tag{3.3}$$

where $\omega \in \mathbb{R}^m$ and $\alpha \in \mathbb{R}_+$ is an arbitrarily chosen constant parameter. It is possible, by introducing some slack variables, to reformulate problem (3.3) as follows:

$$\begin{aligned} \min_{\xi^i, \bar{\xi}^i, \omega} \quad & \frac{1}{2}\|\omega\|^2 + C\sum_{i=1}^{N}\{\xi^i + \bar{\xi}^i\} \\ \text{s.t.}\ & y^i - \omega^T\phi(x^i) \le \varepsilon + \xi^i \quad i = 1, \ldots, N \\ & \omega^T\phi(x^i) - y^i \le \varepsilon + \bar{\xi}^i \quad i = 1, \ldots, N \\ & \xi^i, \bar{\xi}^i \ge 0 \quad i = 1, \ldots, N. \end{aligned} \tag{3.4}$$

Then the dual problem of (3.4) can be considered as

$$\begin{aligned} \max_{\alpha^i, \bar{\alpha}^i} \quad & Q(\alpha^i, \bar{\alpha}^i) = \sum_{i=1}^{N} y^i(\alpha^i - \bar{\alpha}^i) - \varepsilon\sum_{i=1}^{N}(\alpha^i + \bar{\alpha}^i) \\ & \qquad - \frac{1}{2}\sum_{i=1}^{N}\sum_{j=1}^{N}(\alpha^i - \bar{\alpha}^i)(\alpha^j - \bar{\alpha}^j)K(x^i, x^j) \\ \text{s.t.} \quad & \sum_{i=1}^{N} y^i(\alpha^i - \bar{\alpha}^i) = 0 \\ & 0 \le \alpha^i \le C \quad i = 1, \ldots, N \\ & 0 \le \bar{\alpha}^i \le C \quad i = 1, \ldots, N \end{aligned} \tag{3.5}$$

where ε and C are arbitrarily chosen constants, and $K(x^i, x^j)$ is the inner-product kernel:

$$K(x^i, x^j) = \phi(x^i)^T\phi(x^j) \tag{3.6}$$

defined in accordance with the Mercer's condition [16]. Once problem (3.4) is solved, $\alpha^i, \bar{\alpha}^i$ can be used to determine the approximating function:

$$f(x, \omega) = \sum_{i=1}^{N} (\alpha^i - \bar{\alpha}^i) K(x, x^i). \tag{3.7}$$

Data points for which $\alpha^i - \bar{\alpha}^i \neq 0$ are defined as support vectors. Parameters ε and C control in some way the machine complexity for whom control in nonlinear regression is a very tough task, which impacts directly on the performance of SVM.

3.3.1.2 Neural networks

Machine-learning algorithms have been proven to be effective tools for solving regression problems, and they have been effectively applied to the task of wave propagation prediction [2,4]. Channel prediction can be viewed as a regression problem related to wave propagation, where the input consists of information about transmitters, receivers, buildings, frequencies, among others, and the corresponding radio measurements represent the output to be calculated, so we can pose the radio estimation problem as that of finding a suitable input vector x to be used as the argument of a function f that best approximates (in some sense) the radio measurements. Stated in these general terms, we are in the typical setting of a machine-learning problem that can be naturally addressed with ANNs.

In more detail, ANNs are methods motivated by the way biological nervous systems, such as the human brain, process information. Their power lies in the fact that they learn representations of the input data that are suitable for the prediction of the output produced by possibly unseen inputs. In general, ANNs consist of several elementary-processing units called neurons, which are located in different layers and interconnected by a set of weighted edges. Neurons map their input information into an output information by means of nonlinear functions, from which a variety exists, each of them having its own estimation properties, but a principled means of choosing the functions remains an open research problem.

Very early attempts to path-loss prediction via ANNs have been made in [18] and [19]. These studies show that ANNs can give good estimates of path loss in rural environments by using discretized information about land cover and topography, the frequency of the radio waves, and the antenna height. These studies are built upon the well-known empirical Okumura–Hata's model for rural areas. In [18,19], the approximation of Okumura–Hata's model is carried out using a three-layer neural network with four input units corresponding to these four parameters. This approach shows a good predictive power, thus demonstrating the feasibility of neural networks for the task of path-loss prediction. In [20], the authors consider a semiempirical model for path-loss prediction. They use a semiempirical model field strength prediction combined with theoretical results from propagation loss algorithms and neural networks. They obtain good results for the case of dense urban areas and show that neural networks are efficient empirical methods, able to produce good models that integrate theoretical and experimental data. A similar approach is taken in [21], where neural networks are used to correct the biases generated by unknown environmental properties and algorithmic simplifications of path-loss estimations that are common in ray-tracing techniques. The considered neural networks show to substantially improve the results obtained by classic ray-tracing tools. In [22], radial basis function (RBF) neural networks are used instead of the classic multilayer perceptron (MLP). One of the major advantages of RBF neural networks is that they tend to learn much faster than MLP neural networks, because their learning process can be split into two stages for which relatively efficient algorithms exist. More specifically, a two-stage learning approach is taken, where the first stage is composed of an unsupervised clustering step via the rival penalized competitive learning approach. Then the centers of the radial basis function are adjusted, and, once fixed, the weights are then learned in a supervised fashion by using the celebrated recursive least squares algorithm.

In [23], a one-layer backpropagation ANN is proposed to gauge the performance of kriging-based coverage map estimation. A new distance measure that takes obstacles between two points into consideration is introduced, and it is defined as

$$d_{i,j} = \sqrt{(x_i - x_j)^2 + (y_i - y_j)^2 + 10^E} \tag{3.8}$$

where $E = (10c)^{(-1)} \sum_{r \in W_{i,j}} L_r$ with $W_{i,j}$ representing the set of obstacles between point i and j, L_r being the path loss of the respective obstacles, and c being the free space parameter. The first term, involving the square root, is simply the Euclidean distance between points i and j. The term 10^E expresses the path loss caused by obstacles. For an example, if one assumes that the path-loss factor of a wall between two points is 5 dB and the free space parameter c is 2 dB for the environment in which the wall resides, then the path loss between these two points due to the wall will equal to the free space path loss of $10^{5/(10\times2)}$. This increase of the path loss can be equivalently represented by an increase of the effective distance between the two points. This new measure for the distance improves the achievable estimation accuracy for prediction tools based on both kriging and ANNs.

A common problem that arises in learning tasks is that in general we have no or little prior knowledge of the relevance of the input data, and hence many candidate features are generally included in order to equip algorithms with enough degrees of freedom to represent the domain. Unfortunately, many of these features are irrelevant or redundant, and their presence does not improve the discrimination ability. Furthermore, many inputs and a limited number of training examples generally lead to the so-called *curse of dimensionality*, where the data is very sparse and provides a poor representation of the mapping. (Deep neural networks do not perform well with limited training data.) As a remedy to this problem, dimensionality-reduction techniques are applied to the data in practice, which transform the input into a reduced representation of features. Dimensionality-reduction techniques are usually divided into two classes, linear methods (e.g., independent component analysis) and nonlinear methods (e.g., nonlinear PCA). In [2], a two-step approach using learning machines and dimensionality-reduction techniques is proposed. SVMs and ANNs are used as the learning tools, and they are combined with two dimensionality-reduction techniques, namely, linear and nonlinear PCA. In more detail, in [2], the macrocellular path-loss model is defined as follows:

$$L(\text{dB}) = L_0 + \alpha_{\text{buildings}} = 32.4 + 20\log(d) + 20\log(f) + \alpha_{\text{buildings}} \tag{3.9}$$

where L_0 is the free space path loss in dB, d is the radio path, f is the radio frequency, and $\alpha_{\text{buildings}}$ is an attenuation term that depends on several parameters, such as height of base stations and receivers, the distance between consecutive buildings, the height of buildings. In [2], the function in (3.9) is learned by using a three-layer ANN, with the three parameters as input. The estimation using dimensionality-reduction techniques has shown to improve substantially the prediction power over methods that use the full dimensionality of the input. In addition, PCA-based prediction models provide better prediction performance than nonlinear PCA-based models, and ANNs-based models tend to perform slightly better than SVM-based predictors (in the scenarios considered in the above mentioned studies).

The applications of neural network discussed in this topic are considered as function approximation problems consisting of a nonlinear mapping from a set of input variables containing information about potential receiver onto a single output variable representing the predicted path loss. MLPs is applied to reconstruct the path loss in [24]. Figure 3.1 shows the configuration for an MLP with one hidden layer and output layer. The output of the neural network is described as

$$y = F_0 \sum_{j=0}^{M} w_{oj} \left(F_h \sum_{i=0}^{N} w_{ji} x_i \right) \tag{3.10}$$

where w_{oj} represents the synaptic weights from neuron j in the hidden layer to the single output neuron, x_i represents the ith element of the input vector, F_h and F_0 are the activation function of the neurons from the hidden and output layers, respectively, and w_{ji} are the connection weights between the neurons of the hidden layer and the inputs. The learning phase of the network proceeds by adaptively adjusting the free

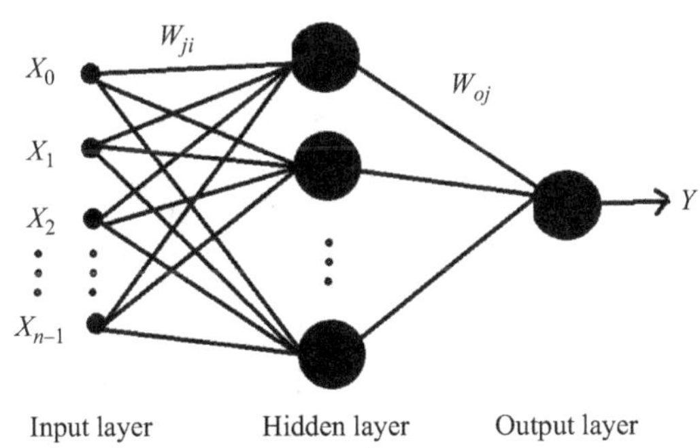

Figure 3.1 The configuration of the multilayer perceptron

parameters of the system based on the mean squares error described by (3.10), between predicted and measured path loss for a set of appropriately selected training examples:

$$E = \frac{1}{2}\sum_{i=1}^{m}(y_i - d_i)^2 \tag{3.11}$$

where y_i is the output value calculated by the network and d_i represents the expected output.

When the error between network output and the desired output is minimized, the learning process is terminated. Thus, the selection of the training data is critical to achieve good generalization properties [25,26]. In coverage map reconstruction, the neural networks are trained with the Levenberg–Marquardt algorithm, which provides faster convergence rate than the backpropagation algorithm with adaptive learning rates and momentum. The Levenberg–Marquardt rule for updating parameters is given by

$$\Delta W = \left(J^T J + \mu I\right)^{-1} J^T e \tag{3.12}$$

where e is an error vector, μ is a scalar parameter, W is a matrix of networks weights, and J is the Jacobian matrix of the partial derivations of the error components with respect to the weights.

An important problem that occurs during the neural network training is the over adaptation. That is, the network memorizes the train examples, and it does not learn to generalize the new situation. In order to avoid over adaptation and to achieve good generalization performances, the training set is separated in the actual training subset and the validation subset, typical 10%–20% of the full training set [26].

3.3.1.3 Matrix completion

In radio map reconstruction, if the sampling rate of the area of interest is high enough, classical signal-processing approaches can be used to reconstruct coverage maps. However, dense sampling can be very costly or impracticable, and in general only a subset of radio measurements of an area are available at a given time. By making assumptions on the spatial correlation properties of radio measurements, which are strongly related to structural properties of an area, and by fitting corresponding correlation models, statistical estimators such as kriging interpolation are able to produce precise estimates based on only few measurements. However, the price for this precision is the high computational complexity and questionable scalability. Nevertheless, the spatial correlation exploited by kriging approaches suggests that coverage maps contain redundant information, so, if represented by a matrix, radio maps can be assumed to be of low rank. This observation has led some authors to propose the framework of low-rank matrix completion for coverage map estimation, which is the topic of this section.

Matrix completion builds on the observation that a matrix that is of low rank or approximately low rank can be recovered by using just a subset of randomly observed data [27,28]. A major advantage of matrix completion is that it is able to recover a matrix by making no assumption about the process that generates the matrix, except that the resulting matrix is of low rank. In the context of radio map estimation, matrix completion has been successfully applied in [13,29,30]. Although in [31] standard matrix completion is used for radio map construction, the authors in [29] considered the non-consistency (continuity) observed in radio map constructed via matrix completion. More specifically, they add a smoothness constraint to the reconstruction problem. Building upon the approaches in [29,31], the authors in [13] use SVT techniques for the reconstruction. In order to increase the estimation quality, which generally degrades in areas with low spatial correlation, the query by committee (QbC) rationale is used in [13] to identify areas requiring more samples in order to obtain accurate predictions. An online algorithm for matrix completion is introduced by the same authors in a subsequent work [30], where they propose an alternating least squares (ALSs) algorithm as an alternative to the popular stochastic gradient descent approach, popularized in the context of the Netflix prize problem [27].

Matrix completion, i.e., low-rank matrix recovery with missing entries, has attracted much attention in recent years because it plays an important role in information retrieval and inference and has numerous applications in computer vision, data mining, signal processing, bioinformatics, and machine learning. For the following theoretical review on matrix completion, we denote by $P \in \mathbb{R}^{m \times n}$ the matrix to be recovered, which in the present case is a two-dimensional coverage map containing path-loss values. Without any assumptions, it is impossible to recover reliably P with a small number $d \ll mn$, of measurements. However, for the case that the rank of the matrix P is small enough compared to its dimensions, the matrix completion framework shows that full recovery of P is possible with high probability. More precisely, full recovery is feasible with high probability from $d \geq cn^{6/5} r \log(n)$ uniformly random measurements, with r being the matrix rank $r = \text{rank}(P)$ and $n > m$.

Due to the regular propagation of a radio wave in unobstructed environments, pass-loss maps exhibit spatial correlation and smooth patterns. Hence, they can be well approximated by low-rank matrices. For the specific case of coverage map estimation, first a matrix P representing the area of interest is defined, and this matrix contains measured values and the respective missing entries. This matrix is used to represent the physical space, where each cell corresponds to a physical position. The values of the matrix are either zero, for the case of a missing entry, or contain the measured path loss at the given cell. The problem of estimating missing entries using the matrix completion framework can be informally formulated as follows: compute a low-rank matrix A that has entries equal to the observation matrix P at the positions containing observed measurements.

Nuclear norm minimization-based methods

Let us denote by Ω the set of observed entries. Formally, the matrix completion problem is formulated as the following nonconvex optimization problem [27]:

$$\begin{aligned} &\min_{A} \ \text{rank}(A) \\ &\text{s.t. } A_{ij} = P_{ij},\ \forall i,j \in \Omega \end{aligned} \tag{3.13}$$

where P_{ij} and A_{ij} are the $\{i,j\}$th entry of P and A, respectively, $\{i,j\} \in \Omega$. Unfortunately, the problem of rank minimization is NP-hard. Therefore, existing approaches in the literature replace the intractable problem by a relaxed formulation that can be solved efficiently with convex optimization tools. (The relaxed problems are often analyzed to check the number of measurements required to recover the solution to the original NP-hard problem exactly, with high probability.) In particular, a common relaxation of the rank minimization problem is formulated as

$$\begin{aligned}&\min_{A} \|A\|_* \\ &\text{s.t. } A_{ij} = P_{ij},\ \forall i,j \in \Omega\end{aligned} \tag{3.14}$$

where $\|A\|_*$ denotes the nuclear norm of the matrix A which is defined as

$$\|A\|_* = \sum_{k=1}^{\min(m,n)} \sigma_k(A)$$

with $\sigma_k(\cdot)$ being the kth largest singular value of a matrix. Note that (3.14) can be converted into a semidefinite programming (SDP) and hence can be solved by interior-point methods. However, directly solving the SDP has a high complexity. Several algorithms faster than the SDP-based methods have been proposed to solve the nuclear norm minimization, such as SVT, fixed point continuation (FPC), and proximal gradient descent [5]. In radio map reconstruction, the authors of [13] opt for the SVT algorithm, which can be briefly described as follows. Starting from an initial zero matrix Y_0, the following steps take place at each iteration:

$$\begin{aligned}A_i &= \text{shrink}(Y_{i-1}, \tau) \\ Y_i &= Y_{i-1} + \mu \Pi_\Omega(P - X_i)\end{aligned} \tag{3.15}$$

with μ being a nonnegative step size. The operator $\Pi_\Omega(X)$ is the sampling operator associated with the set Ω. Entries not contained in the index set Ω are set to zero, the remaining entries are kept unchanged. The shrink operator $(\cdot, \tau)$ is the standard rank-reduction thresholding function, which sets singular values beneath a certain threshold $\tau > 0$ to be zero.

In [13], the authors also introduce a method to improve the path-loss reconstruction via matrix completion. The idea is to define a notion of "informative areas," which are regions in which samples are required to improve greatly the map reconstruction. The motivation for this approach is that, in coverage maps, there may exist nonsmooth transitions caused by abrupt attenuation of signals, which are common when radio waves impinge on obstacles such as large buildings, tunnels, metal constructions. Consequently, path loss in such areas exhibits low spatial correlation, which can lead to reconstruction artifacts that can only be mitigated by increasing the sampling rate in those regions. In order to identify those regions, which are mathematically represented by matrix entries, the authors of [13] resort to a family of active learning algorithms, and, in particular, they employ the QbC rationale. The general approach is to quantify the uncertainty of the prediction in each missing value in the matrix, so only measurements corresponding to the most uncertain entries are taken. In the QbC rationale, the missing matrix values are first estimated by means of many different algorithms, and only a subset of the available data is used. Assuming that the available data budget amounts to k measurements, first the coverage map is computed by only using $l < k$ of the available entries. Then, three different algorithms for matrix reconstruction are compared, and the top $K = k - l$ entries with the largest disagreement are chosen. New measurements for those K entries are then gathered, and a new coverage map is estimated by using the new samples. The three different reconstruction algorithms used in [13] are the SVT, the K-nearest neighbors, and the kernel APSM.

In a subsequent work [30], the authors from [13] derive an online algorithm based on the ALS method for matrix completion. They adopt the matrix factorization framework in which the low-rank matrix A is replaced by the low-rank product LR^T, with $L \in \mathbb{R}^{m \times \rho}$ and $R^T \in \mathbb{R}^{n \times \rho}$, and ρ is a prespecified overestimated of the rank of A. Based on this framework, the rank-minimization objective is replaced by the equivalent objective:

$$\begin{aligned} &\min_{L,R}\ \tfrac{1}{2}(\|L\|_F^2 + \|R\|_F^2) \\ &\text{s.t. } LR^T = A \\ &\quad A_{ij} = P_{ij},\ \forall i,j \in \Omega \end{aligned} \tag{3.16}$$

For noisy case, the objective function for matrix completion becomes in [30]:

$$\begin{aligned} &\min_{L,R}\ \|P - LR^T\|_F^2 + \gamma(\|L\|_F^2 + \|R\|_F^2) \\ &\text{s.t. } LR^T = A \\ &\quad A_{ij} = P_{ij},\ \forall i,j \in \Omega \end{aligned} \tag{3.17}$$

with γ being a regularization parameter that controls the trade-off between the closeness to data and the nuclear norm of the reconstructed matrix. The ALS method is a two-step iterative method in which the objective is consecutively minimized over one variable by holding the other constant. Hence, two quadratic programs have to be solved in each iteration step consecutively. This amounts to solving D least square problems to find the optimum solution of each row of L and R. This, however, amounts computing a $(\rho \times \rho)$ matrix inversion for each row, which might become prohibitive with increased number of samples. Therefore, the authors in [30] propose an approximation algorithm, in which the coefficients of the optimum row vector are computed one by one, which significantly reduces the computational complexity, especially for sparse datasets. In this mindset, the online version of the ALS is proposed in a way that, with new incoming data, only the respective coefficients are updated via this approximated update function.

In addition to the online reconstruction algorithm for matrix-completion-based coverage map reconstruction, the authors in [30] also derive a new adaptive sampling scheme, able to outperform the QbC rationale from their previous work. They assume that coverage maps are in general smooth. Therefore, for two neighboring matrix entries (i_1,j_1) and (i_2,j_2) that satisfy $|i_1 - i_2| \leq 1$ and $|j_1 - j_2| \leq 1$, the entry difference should be bounded by

$$|A_{i_1 j_1} - A_{i_2 j_2}| \leq \Delta$$

where Δ is a small positive number. Under this assumption, incomplete or erroneous reconstruction will very likely violate this condition. In order to detect eventual violations of the gradient bound, a two-dimensional edge-detector filter was proposed in [30] with the following kernel:

$$f = \frac{1}{9}\begin{bmatrix} -1 & -1 & -1 \\ -1 & 8 & -1 \\ -1 & -1 & -1 \end{bmatrix}.$$

The data smoothness condition on A implies that each entry of its filtered version $\bar{A}$ will be bounded by $|\bar{A}_{i,j}| \leq (8/9)\Delta$. With Y being the current estimate of A, the authors of that study propose to obtain measurements corresponding to entries for which $|\bar{Y}_{i,j}|$ is large. Since the filter is a bounded linear operator, if the reconstruction is reliable, then we should have

$$\|\bar{A} - \bar{Y}\| \leq M\|A - Y\| \leq \varepsilon$$

for ε small enough. By considering the triangular inequality, the following constraint can be imposed on each coefficient $|\bar{Y}_{i,j}|$:

$$|\bar{Y}_{i,j}| \leq |\bar{Y}_{i,j} - \bar{X}_{i,j}| \leq \varepsilon + \frac{8}{9}\Delta$$

which implies that coefficients of $\bar{Y}$ should be small. However, if the entries $\bar{Y}_{i,j}$ are large, typically larger than Δ, then $|\bar{Y}_{i,j} - \bar{X}_{i,j}|$ should also be large, and the matrix completion algorithm fails to reconstruct the coverage map correctly. As a consequence, we should obtain measurements in the region of the respective entry (i,j) in order to increase the accuracy of the estimation. The proposed online algorithm and adaptive sampling scheme are tested based on the image data. A 150×150 gray-scale image is considered. At first, only 8% of the entire dataset is considered for the reconstruction. In steps of $N = 20$ entries, which are selected based on the proposed adaptive sampling scheme and the QbC rationale, the reconstruction is refined. It is shown that 100 selected entries based on the adaptive sampling schemes resulted in better improvements than 1,000 randomly selected entries. The proposed adaptive sampling scheme outperformed the QbC approach. Further, it is also shown that in combination with the proposed online algorithm, the QbC approach, which is based on batch algorithms, is also outperformed in terms of computational complexity.

An alternative approach for producing smooth coverage maps is proposed in [29]. The low-rank minimization objective is here extended by adding a smoothness constraint term. The revised low-rank model is formulated as follows [29]:

$$\begin{aligned} &\min_{L,R} \tfrac{1}{2}(\|L\|_F^2 + \|R\|_F^2) + \lambda s(LR^T) \\ &\text{s.t. } LR^T = A \\ &\quad A_{ij} = P_{ij},\ \forall i,j \in \Omega \end{aligned} \tag{3.18}$$

where $s(A)$ is the smoothing term, and the regularization term λ is a weight that balances low rankness and smoothness. In [29], the smoothness constraint term is defined via the diversity of the row-wise and column-wise difference of $LR^T = A$, or, in mathematical terms:

$$s(LR^T) = \|D_x(LR^T)\|_F^2 + \|D_y(LR^T)\|_F^2$$

with the gradient operators $D_x(A)$ and $D_y(A)$ being defined as

$$\begin{aligned} D_x(i,j) &= A(i,j+1) - A(i,j) \\ D_y(i,j) &= A(i+1,j) - A(i,j) \end{aligned}$$

The smoothed matrix completion objective from is then stated as

$$\begin{aligned} &\min_{L,R} \tfrac{1}{2}(\|L\|_F^2 + \|R\|_F^2) + \lambda \left(\|D_x(LR^T)\|_F^2 + \|D_y(LR^T)\|_F^2\right) \\ &\text{s.t. } LR^T = A \\ &\quad A_{ij} = P_{ij},\ \forall i,j \in \Omega \end{aligned} \tag{3.19}$$

For the solution of the minimum problem in (3.19), an alternating iteration algorithm over L and R is adopted in [29]. At first, L and R are chosen at random, then L is fixed and R is optimized by a linear least square method. Then R is fixed and the cost function is optimized over L. This procedure is repeated until no progress is observed. In [29], the proposed smoothed low-rank reconstruction method is compared with interpolation methods such as radial basis interpolation and inverse distance weighting. The smoothed low-rank reconstruction method shows to achieve similar reconstruction properties with fewer samples compared to these methods.

Alternating projection methods

Note that these nuclear-norm-based algorithms require performing the full SVD of an $m \times n$ matrix. When m or n is large, computing the full SVD is time-consuming. Different from rank or nuclear norm minimization, a new strategy is adopted for matrix completion. The basic motivation of alternating projection algorithm (APA) is to find a matrix such that it has low rank and its entries over the sample set Ω are consistent with the available observations. Denote the known entries as P_Ω:

$$[P_\Omega]_{i,j} = \begin{cases} P_{i,j}, & \text{if } (i,j) \in \Omega \\ 0, & \text{otherwise} \end{cases}. \tag{3.20}$$

Then it can be formulated as the following feasibility problem [32]:

$$\begin{aligned} &\text{find } A \\ &\text{s.t. } \operatorname{rank}(A) \le r,\ A_\Omega = P_\Omega, \end{aligned} \tag{3.21}$$

where $r \ll \min(m, n)$ is the desired rank. It is obvious that (3.21) is only suitable for noise-free case. For noisy case, we use:

$$\begin{aligned} &\text{find } A \\ &\text{s.t. } \operatorname{rank}(A) \le r,\ \|A_\Omega - P_\Omega\|_F \le \varepsilon_2 \end{aligned} \tag{3.22}$$

to achieve robustness to the Gaussian noise. In the presence of outliers, we adopt:

$$\begin{aligned} &\text{find } A \\ &\text{s.t. } \operatorname{rank}(A) \le r,\ \|A_\Omega - P_\Omega\|_p \le \varepsilon_p \end{aligned} \tag{3.23}$$

where $\varepsilon_p > 0$ is a small tolerance parameter that controls the ℓ_p-norm of the fitting error and $\|\cdot\|_p$ denotes the element-wise ℓ_p-norm of a matrix, i.e.:

$$\|A_\Omega\|_p = \left(\sum_{(i,j)\in\Omega} \left|[A]_{i,j}\right|^p \right)^{1/p}. \tag{3.24}$$

Apparently, (3.23) reduces to (3.22) when $p = 2$. Also, (3.23) reduces to the noise-free case of (3.21) if $\varepsilon_p = 0$.

By defining the rank constraint set

$$\mathscr{S}_r := \{A|\text{rank}(A) \le r\} \tag{3.25}$$

and the fidelity constraint set

$$\mathscr{S}_p := \left\{A| \ \|A_\Omega - P_\Omega\|_p \le \varepsilon_p\right\}, \tag{3.26}$$

the matrix completion problem of (3.23) is formulated as finding a common point of the two sets, i.e.:

$$\text{find } X \in \mathscr{S}_r \cap \mathscr{S}_p. \tag{3.27}$$

For a given set $\mathscr{S}$, the projection of a point $Z \notin \mathscr{S}$ onto it, which is denoted as $\Pi_{\mathscr{S}}(Z)$, is defined as

$$\Pi_{\mathscr{S}}(Z) := \underset{X\in\mathscr{S}}{\arg\min} \ \|X - Z\|_F^2. \tag{3.28}$$

We adopt the strategy of alternating projection (AP) onto $\mathscr{S}_r$ and $\mathscr{S}_e$ to find a common point lying in the intersection of the two sets [32]. That is, we alternately project onto $\mathscr{S}_r$ and $\mathscr{S}_p$ in the kth iteration as

$$\begin{aligned} Y^k &= \Pi_{\mathscr{S}_r}(A^k) \\ A^{k+1} &= \Pi_{\mathscr{S}_p}(Y^k). \end{aligned} \tag{3.29}$$

The choice of $p = 1$ is quite robust to outliers. Other values of $p < 2$ may also be of interest. The case of $p < 1$ requires to compute the projection onto a nonconvex and nonsmooth ℓ_p-ball, which is difficult and hence not considered here. The $1 < p < 2$ involves the projection onto a convex ℓ_p-ball, which is not difficult to solve but requires an iterative procedure. Since the choice of $p = 1$ is more robust than $1 < p < 2$ and computationally simpler, we can use $p = 1$ for outlier-robust matrix completion.

By Eckart–Young theorem, the projection of $Z \notin \mathscr{S}_r$ onto $\mathscr{S}_r$ can be computed via truncated SVD of Z:

$$\Pi_{\mathscr{S}_r}(Z) = \sum_{i=1}^{r} \sigma_i u_i v_i^T \tag{3.30}$$

where $\{\sigma_i\}_{i=1}^r$, $\{u_i\}_{i=1}^r \in \mathbb{R}^m$, and $\{v_i\}_{i=1}^r \in \mathbb{R}^n$ are the r largest singular values and the corresponding left and right singular vectors of Z, respectively. Clearly, the AP does not need to perform the full SVD. Only truncated SVD is required. That is, we only calculate the r largest singular values and their corresponding singular vectors.

Without loss of generality, assuming $n \leq m$, the computational cost of full SVD is $\mathcal{O}(mn^2 + n^3)$, while that of truncated SVD is $\mathcal{O}(mnr)$. In practical applications, the rank r could be much smaller than the matrix dimension. Therefore, the computational cost of the AP is much lower than the nuclear norm minimization-based methods that need full SVD.

We then investigate computing the projection onto $\mathcal{S}_p$ for $p = 1$ and $p = 2$. Note that projection onto $\mathcal{S}_p$ only affects the entries indexed by Ω. Other entries $\{Z_{i,j}\}$ with $(i,j) \notin \Omega$ will remain unchanged through this projection. Define $p_\Omega \in \mathbb{R}^{|\Omega|}$, where $|\Omega|$ is the cardinality of Ω, as the vector that contains the observed entries of P, i.e., the nonzero entries of P_Ω. Also, $a_\Omega \in \mathbb{R}^{|\Omega|}$ is defined in a similar manner. Then the set $\mathcal{S}_p$ of (3.26) has the equivalent vector form:

$$\mathcal{B}_p := \left\{ a_\Omega \in \mathbb{R}^{|\Omega|} \middle| \, \|a_\Omega - p_\Omega\|_p \leq \varepsilon_p \right\} \tag{3.31}$$

which is an ℓ_p-ball with the observed vector p_Ω being the ball center. Now it is clear that the projection for matrices is converted into one for vectors of length $|\Omega|$. We consider the following three cases with different values of p and ε_p:

- For $\varepsilon_p = 0$, (3.31) reduces to the equality constraint of $a_\Omega = p_\Omega$. For any vector $z \in \mathbb{R}^{|\Omega|}$, the projection is calculated as $\Pi_{\mathcal{B}_p}(z) = p_\Omega$.
- For $p = 2$ and $\varepsilon_2 > 0$, $\mathcal{B}_2$ is the conventional ℓ_2-ball in the Euclidean space. For any vector $z \notin \mathcal{B}_2$, it is not difficult to derive the closed-form expression of the projection onto $\mathcal{B}_2$ as

  $$\Pi_{\mathcal{B}_2}(z) = p_\Omega + \frac{\varepsilon_2 (z - p_\Omega)}{\|z - p_\Omega\|_2}. \tag{3.32}$$

  With a proper value of ε_2 and $p = 2$, the robustness to Gaussian noise can be enhanced.
- For $p = 1$ and $\varepsilon_1 > 0$, $\mathcal{B}_1$ is an ℓ_1-ball. For any vector $z \notin \mathcal{B}_1$, the projection onto $\mathcal{B}_1$ is the solution of

  $$\min_a \frac{1}{2}\|a - z\|_2^2, \text{ s.t. } \|a - p_\Omega\|_1 \leq \varepsilon_1. \tag{3.33}$$

  Using the Lagrange multiplier method, we obtain the solution of (3.33):

  $$[\Pi_{\mathcal{B}_1}(\mathbf{z})]_i = \operatorname{sgn}([z - p_\Omega]_i) \max(|[z - p_\Omega]_i| - \lambda^\star, 0) \tag{3.34}$$

  where $i = 1, \ldots, |\Omega|$ and $\lambda^\star$ is the unique root of the nonlinear equation:

  $$\sum_{i=1}^{|\Omega|} \max(|[z - p_\Omega]_i| - \lambda, 0) = \varepsilon_1 \tag{3.35}$$

  in the interval $(0, \|z - p_\Omega\|_\infty)$ using the bisection method, where $\|\cdot\|_\infty$ is the ℓ_∞-norm of a vector. The computational complexity of projection onto ℓ_1-ball is $\mathcal{O}(|\Omega|)$, which is much lower than that of the projection onto $\mathcal{S}_r$.

The selection of ε_p is critical to the performance of APA. In the absence of noise, the optimum is $\varepsilon_p = 0$. For noisy case, ε_p is related to the noise level. Roughly speaking, larger noise requires a larger ε_p. If the probability of the noise is known *a priori*, we can estimate the probability distribution of the ℓ_p-norm of the noise. Then a proper value of ε_p can be determined according to the probability such that the true entries are located in the ℓ_p-ball. If the probability of the noise is unknown, one may resort to cross validation to determine a proper ε_p. Note that in the nuclear norm regularized problem:

$$\min_{A} \frac{1}{2}\|A_\Omega - P_\Omega\|_F^2 + \tau\|A\|_* \tag{3.36}$$

one also faces the issue of selecting the regularization parameter τ. Clearly, an advantage of the proposed formulation is that it is not difficult to determine ε_p from the *a priori* noise level but not easy for τ.

Remark: It should be pointed out that the APA is different from the iterative hard thresholding (IHT) and its variants [33,34] although they all use a rank-r projection. The IHT solves the rank constrained Frobenius norm minimization:

$$\min_{A} f(A) := \frac{1}{2}\|A_\Omega - P_\Omega\|_F^2, \quad \text{s.t. } \operatorname{rank}(A) \le r \tag{3.37}$$

using gradient projection with iteration step being:

$$A^{k+1} = \Pi_{\mathscr{S}_r}\left(A^k - \mu\nabla f(A^k)\right) \tag{3.38}$$

where $\mu > 0$ is the step size and ∇f is the gradient of f. Determining the step size with a line search scheme requires computing the projection $\Pi_{\mathscr{S}_r}(\cdot)$ for several times. Thus, the computational cost of the IHT is several times of the APA per iteration.

Convergence of the alternating projection for finding a common point of two sets was previously established for convex sets only [35]. Recently, the convergence of APA for nonconvex sets, which satisfies a regularity condition has been investigated [36,37]. Exploiting the fact that the rank constraint set of (3.25) satisfies the prox-regularity and according to Theorem 5.2 of [36], we can establish the convergence of the APA for matrix completion, as stated in the following proposition.

Proposition: The APA locally converges to a point in $\mathscr{S}_r \cap \mathscr{S}_p$ at a linear rate.

3.3.2 Online algorithms

The batch algorithms applied in coverage map reconstruction was widely studied as mentioned above. However, batch algorithms need a great number of storage and is poor in real-time performance. It is more practical that an algorithm is capable of updating the corresponding base station's current approximation of the unknown path loss function in its cell. Thus, coverage map reconstruction needs to be an online function itself. In [9], APSM-based [10] and multi-kernel learning techniques [38] are adopted for their capability of coping with large-scale problems where the huge number of measurements arrives to operators.

3.3.2.1 APSM-based algorithm

APSM is a recently developed tool for iteratively minimizing a sequence of convex cost functions [10]. And it can be easily combined with kernel-based tools from machine learning [10,39,40]. In particular, a variation of the APSM is proposed in this study.

In more detail, at each iteration, n, q sets are selected from the collection $\{S_1, \ldots, S_n\}$ with the approach described in [9]. The intersection of these sets is the set C_n and the index of the sets chosen from the collection is denoted by

$$I_{n,q} := \{i_{r_n}^{(n)}, i_{r_n-1}^{(n)}, \ldots, i_{r_n-q+1}^{(n)}\} \subseteq \{1, \ldots, n\}, \tag{3.39}$$

where $n \geq q$, and r_n is the size of dictionary. With this selection of sets, starting from $\hat{f}_0 = 0$, sequence $\{\hat{f}_n\}_{n \in N} \subset H$ by

$$\hat{f}_{n+1} := \hat{f}_n + \mu_n \sum_{j \in I_{n,q}} \omega_{j,n} P_{S_j}(\hat{f}_n) - \hat{f}_n, \tag{3.40}$$

where $\mu_n \in (0, 2M_n)$ is the step size, M_n is a scalar given by

$$M_n = \begin{cases} \dfrac{\sum_{j \in I_{n,q}} w_{j,n} \left\| P_{S_j}(f_n) - f_n \right\|^2}{\left\| \sum_{j \in I_{n,q}} w_{j,n} P_{S_j}(f_n) - f_n \right\|^2}, & \text{if } f_n \notin \cap_{j \in I_{n,q}} S_j, \\ 1, & \text{otherwise}, \end{cases} \tag{3.41}$$

and $\omega_{j,n} > 0$ are weights satisfying:

$$\sum_j \omega_{j,n} = 1 \tag{3.42}$$

The projection onto the hyperslab induced by measurement n is given by $P_{S_n}(f) = f + \beta_f \kappa(\tilde{x}_n, \cdot)$ where

$$\beta_f = \begin{cases} \dfrac{y - \langle f, \kappa(\tilde{x}_n, \cdot) \rangle - \varepsilon}{\kappa(\tilde{x}_n, \tilde{x}_n)}, & \text{if } \langle f, \kappa(\tilde{x}_n, \cdot) \rangle - y < -\varepsilon, \\ \dfrac{y - \langle f, \kappa(\tilde{x}_n, \cdot) \rangle + \varepsilon}{\kappa(\tilde{x}_n, \tilde{x}_n)}, & \text{if } \langle f, \kappa(\tilde{x}_n, \cdot) \rangle - y > \varepsilon, \\ 0, & \text{if } |\langle f, \kappa(\tilde{x}_n, \cdot) \rangle - y| \leq \varepsilon. \end{cases} \tag{3.43}$$

3.3.2.2 Multi-kernel algorithm

The choice of the kernel κ for a given estimation task is one of the main challenges for the application of kernel methods. To address this challenge in the path-loss estimation problem, we propose the application of the multi-kernel algorithm described in [38]. Briefly, this algorithm provides good estimates by selecting, automatically, both a reasonable kernel (the weighted sum of a few given kernels) and a sparse dictionary.

The APSM-based online algorithm demonstrated above has good performance in real-time and requires little storage. Nevertheless, its time complexity and accuracy are inferior to common batch algorithms. What should be noticed is that so far only the online algorithm mentioned in this section has been employed in coverage map reconstruction. So there is still a great deal of open questions in this topic.

3.4 Optimized sampling

Informative areas are areas from which we want to have samples, since such knowledge can improve the path-loss reconstruction. Note that some regions can be nonsmooth. This is the consequence of large buildings, obstacles, tunnels that can abruptly attenuate the propagating radio wave. Due to these, the path loss in such areas exhibits low spatial correlation and this can lead to poor reconstruction effects. Consequently, in this case, optimized samplings are required.

3.4.1 Active learning

Active learning is a special case of semi-supervised machine learning in which a learning algorithm is able to interactively query the user (or some other information source) to obtain the desired outputs at new data points. In active learning systems, an algorithm is able to choose its training data, as opposed to passively trained systems that must learn a behavior from a set of random observations.

3.4.1.1 Query by committee

A QbC training strategy for selecting new data points uses the disagreement between a committee of different algorithms to suggest new data points, which most rationally complement existing data, that is, they are the most informative data points.

In the application of [13], assume that the available budget corresponds to k measurements, coming from drive tests, the matrix is first completed using a number of $l < k$ observed entries. Subsequently, having access to a number of reconstructed matrices, finding the top $K := k - l$ entries with the largest "disagreement" according to a certain criterion and obtaining measurements from them. Finally, drive tests are performed to obtain the K samples indicated by the previous step and reconstruct the path-loss map exploiting the newly obtained information.

In general, one can employ any number of algorithms to reconstruct the matrix. These algorithms run in parallel using the same set of measurements as an input. After the estimation of the missing entries, the entries with the largest disagreement can be obtained according to the following simple rules. Supposing that three algorithms are employed as committees, the entries obtained by these algorithms are denoted by $a_{ij}(\xi), \xi = 1, 2, 3$. Then the disagreement equals to

$$d_{ij} = (a_{ij}^{(1)} - a_{ij}^{(2)})^2 + (a_{ij}^{(12)} - a_{ij}^{(3)})^2 + (a_{ij}^{(1)} - a_{ij}^{(3)})^2. \tag{3.44}$$

The K entries, which score the largest disagreement, are chosen, and we perform drive tests to obtain the path loss.

QbC algorithm is a simple one which can be implemented easily. And this algorithm enhances the accuracy of the reconstruction as illustrated in [13]. However, for that larger than two algorithms required to run parallel, it cannot be one of the most efficient algorithms. And the predicted results can be influenced greatly if the algorithms employed are not stable enough.

3.4.1.2 Side information

In [9], in order to predict the informative areas, the approach based on side information is applied to decide the weight of parallel projections. For instance, sets corresponding to measurements taken at pixels farther away from the route of the user of interest (UOI) should be given smaller weights than measurements of pixels that are close to the user's trajectory. The reason is that estimates should be accurate at the pixels the UOI is expected to visit because these are the pixels of interest to most applications (e.g., video caching based on channel conditions). Therefore, we assign large weights to measurements close to the UOI's route by proceeding as follows. Let $\chi_{\text{UOI}} \subset N^2$ be the set of pixels that belong to the path of the UOI. Then, for each weight $\omega_{i,n}$:

$$\omega_{i,n} = \frac{1}{d_{\min}(\tilde{x}_i, \chi_{\text{UOI}}) + \varepsilon_\omega} \tag{3.45}$$

where $d_{\min}(\tilde{x}_i, \chi_{\text{UOI}})$ denotes the minimum distance of measurement $\tilde{x}_i$ to the area of interest, and $\varepsilon_\omega > 0$ is a small regularization parameter. This distance can be obtained for each pixel $\tilde{x}_i$ by considering the distances of every pixel in χ_{UOI} to $\tilde{x}_i$ and by taking the minimum of these distances. Subsequently, the weights are normalized. Compared to an equal choice of the weights, the proposed method provides fast convergence to a given prediction quality for the UOI, but at the cost of degraded performance in other areas of the map.

3.4.2 Channel prediction results with path-loss measurements

We implement the matrix completion via the AP algorithm for channel prediction, whose performance is compared with the state-of-the-art method, i.e., SVT [5]. The experimental data is collected in Tengfei Industrial Park, and the scenario of data collection is illustrated in Figure 3.2. Specifically, the UE routes, including the

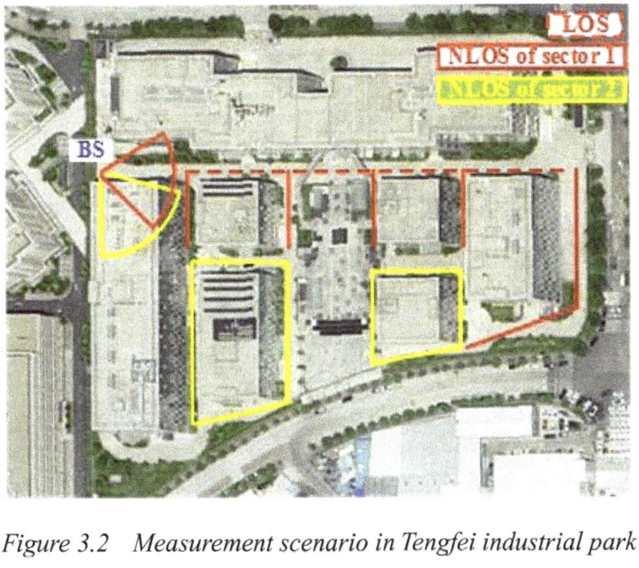

Figure 3.2 Measurement scenario in Tengfei industrial park

line-of-sight (LOS) and non-LOS routes, are in red lines and yellow lines, respectively. With the data points being sampled from these UE routes, our purpose is to predict the values of other points and hence to achieve the path-loss reconstruction.

We devise an experiment to evaluate the prediction performance of both AP and SVT. The experiment is based upon the fact that only the fractional sampling points are viewed as the known points. And the rest part of sampling points is considered to be unknown in advance. Hence, those points are supposed to be predicted. Assuming that Ω_1 is a subset, consisting of those predicted points, of the set Ω and $P_{i,j}, (i,j) \in \Omega_1$ denotes the true value of the (i,j) point. We compare the predicted value $\widehat{P}_{i,j}, (i,j)$ with its true value. If the condition

$$\left|\widehat{P}_{i,j}, (i,j) - P_{i,j}, (i,j)\right| \leq \delta$$

is satisfied, the prediction with respect to the (i,j) point is assumed to be successful, otherwise the prediction is failed. In our experiment, we set $\delta = 20$ and investigate the successful ratio of prediction with respect to the rank r. For each value of the matrix rank, 100 trials are carried out to calculate the average result. The proportion of known sampling points, which are randomly selected in each trial, is 85%. Hence, the rest of 15% sampling points is viewed as predicted points.

Note that AP's performance is affected by the parameter of the estimated rank r, while SVT's performance is determined by the parameter of the threshold of singular value. Hence, we evaluate the performance of each algorithm with respect to different parameters. In terms of the AP, Figure 3.3 plots the successful ratio of prediction versus the rank r. It is observed that using $r = 2$ yields the best performance. When $r > 2$, the successful ratio is successively decreased with the increase of the rank. This phenomenon shows that only when tight rank constraint with quite small r is adopted, the reasonable prediction can be obtained. With the SVT in contrast, Figure 3.4 plots its successful ratio of prediction versus the threshold of singular value. While an appropriate threshold of singular value yields the highest successful ratio, too small or too large threshold value will result in a decreased successful ratio.

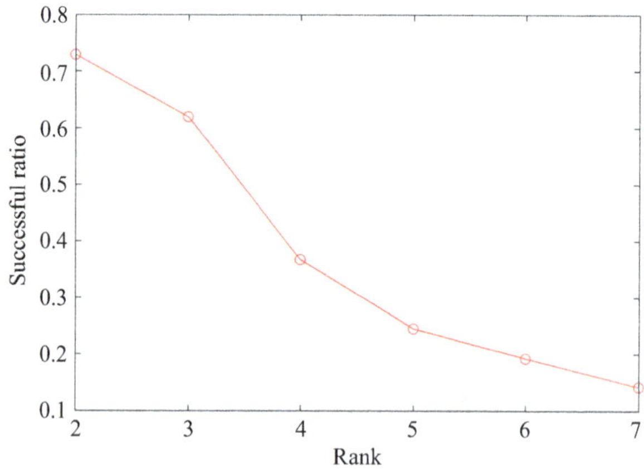

Figure 3.3 Successful ratio of prediction versus rank r with $\varepsilon_2 = 0$ via AP

Based upon the optimal parameters in two algorithms (rank $r = 2$ in the APA and threshold of singular value equal to 1.5×10^5 in the SVT), we compare the highest successful ratio that two algorithms can attain. Observe that the highest successful ratio of the APA is 72.9% and that of the SVT is 67.8%. We can see that the AP outperforms the SVT.

Then we evaluate the prediction errors of both algorithms by the root mean square error (RMSE) which is defined as

$$\text{RMSE} = 10 \log_{10} \sqrt{\text{E}\{\|\widehat{P}_{\Omega_1} - P_{\Omega_1}\|_F^2\}}. \tag{3.46}$$

Figure 3.5 plots the RMSE versus the rank r of AP, which demonstrates that the best performance can be achieved when $r = 2$. This RMSE result is consistent with the above successful ratio result. In those two aspects of evaluation, the choice of $r = 2$

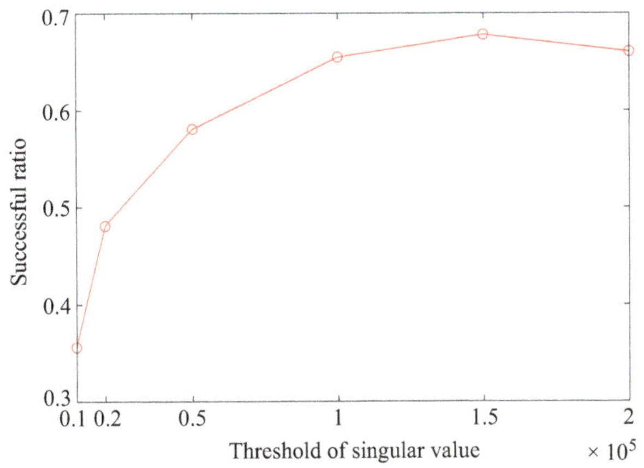

Figure 3.4 Successful ratio of prediction versus threshold of singular value via SVT

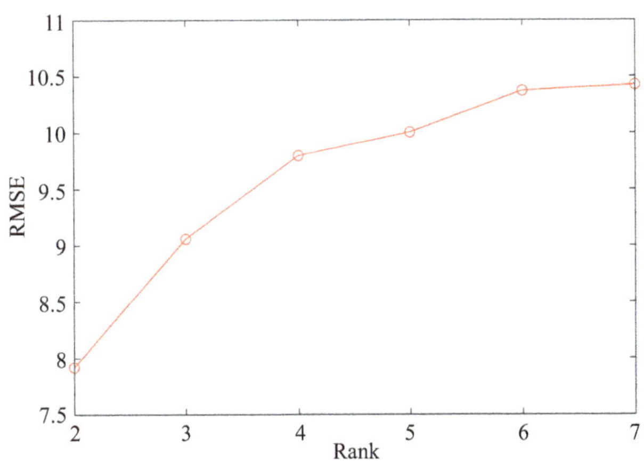

Figure 3.5 RMSE versus rank r with $\varepsilon_2 = 0$ via AP

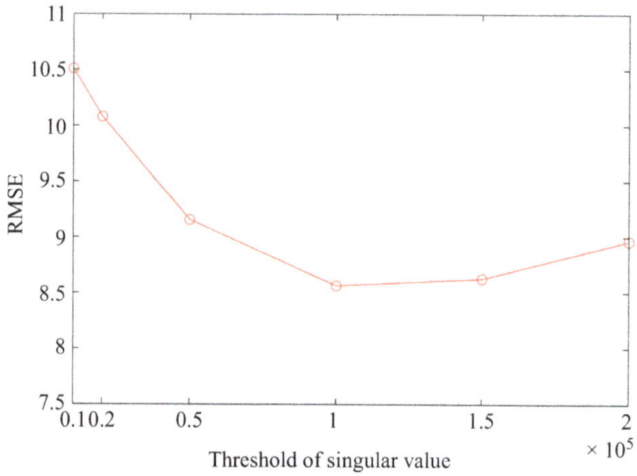

Figure 3.6 RMSE versus threshold of singular value via SVT

derives the best successful ratio and prediction error. Therefore, we can conclude that adopting $r = 2$ as the estimated rank for the AP yields the best prediction. In contrast, Figure 3.6 plots the RMSE versus the threshold of singular value for the SVT. While the SVT can attain the smallest RMSE value of 8.57 with threshold of singular value equal to 1×10^5, the AP can obtain the smallest RMSE value of 7.92 with rank $r = 2$. This comparison proves the better performance of the AP than the SVT.

3.5 Conclusion

Numerous wireless communication applications in future will greatly depend on accurate coverage loss map. Making sense of the coverage loss map reconstruction can be a precarious task for the uninitiated research. In this survey, methodologies of reconstruction are divided into three parts to discuss the following: approaches of achieving measurements, learning-based reconstruction algorithms, and optimized sampling of measurements. And different methodologies of each part are studied and analyzed, respectively. Mainly, two approaches can be applied to achieve measurements: conventional drive tests and MDTs. The former one is simple and stable, while the cost of it will rapidly ascend with the measured area getting larger. MDTs are a relatively cheap and efficient way without the consideration of absolute stability. Then learning-based reconstruction algorithms which can be categorized into two parts are discussed. Batch algorithms are usually efficient and cheap, but it requires huge amount of storage and performs in real-time calculating. Among batch algorithms, SVM is hard to implement for which resolving nonlinear regression should be complicated; SVT is easy to be implemented, but it is too simple to adapt to various scenarios. Over adaptation may occur in the training procedure of ANN, which manifests that it cannot be an adaptive algorithm. On the other hand, online algorithm has great performance in real-time and storage is unnecessary. The only online approach having been employed in path-loss reconstruction is APSM-based algorithm with multi-kernel techniques. It successfully updates the prediction of coverage map whenever new data comes in. Nevertheless, the time complexity and accuracy are inferior to common batch algorithms. The last part is optimized sampling. The main techniques applied in this part is active learning which is derived from machine learning. There are mainly two active learning algorithms in the procedure of optimized sampling: QbC and side information. QbC determines the informative areas through running various reconstruction algorithms in parallel. Its accuracy is considerable, but the stability and time complexity are poor. Side information is applied in the online algorithm mentioned above. Informative areas are selected through exerting different weights. This is a simple and efficient way, but whether it is robust should be studied more deeply. According to the generalization discussed above, learning-based algorithms employed in coverage loss map reconstruction has a great area to be explored, especially in online algorithm. How to revise an efficient and real-time learning-based algorithm may be a meaningful topic to study on.

References

[1] G. Fodor, E. Dahlman, G. Mildh, *et al.*, "Design aspects of network assisted device-to-device communications," *IEEE Communications Magazine*, vol. 50, no. 3, pp. 170–177, 2012.

[2] M. Piacentini and F. Rinaldi, "Path loss prediction in urban environment using learning machines and dimensionality reduction techniques," *Computational Management Science*, vol. 8, no. 4, pp. 371–385, 2011.

[3] R. Timoteo, D. Cunha, and G. Cavalcanti, "A proposal for path loss prediction in urban environments using support vector regression," in *The Tenth Advanced International Conference on Telecommunications (AICT)*, 2014, pp. 119–124.

[4] I. Popescu, I. Nafomita, P. Constantinou, A. Kanatas, and N. Moraitis, "Neural networks applications for the prediction of propagation path loss in urban environments," in *IEEE 53rd Vehicular Technology Conference (VTC)*, vol. 1, 2001, pp. 387–391.

[5] J.-F. Cai, E. J. Candès, and Z. Shen, "A singular value thresholding algorithm for matrix completion," *SIAM Journal on Optimization*, vol. 20, no. 4, pp. 1956–1982, 2010.

[6] R. Di Taranto, S. Muppirisetty, R. Raulefs, D. Slock, T. Svensson, and H. Wymeersch, "Location-aware communications for 5G networks: how location information can improve scalability, latency, and robustness of 5G," *IEEE Signal Processing Magazine*, vol. 31, no. 6, pp. 102–112, 2014.

[7] D. M. Gutierrez-Estevez, I. F. Akyildiz, and E. A. Fadel, "Spatial coverage cross-tier correlation analysis for heterogeneous cellular networks," *IEEE Transactions on Vehicular Technology*, vol. 63, no. 8, pp. 3917–3926, 2014.

[8] E. Dall'Anese, S.-J. Kim, and G. B. Giannakis, "Channel gain map tracking via distributed kriging," *IEEE Transactions on Vehicular Technology*, vol. 60, no. 3, pp. 1205–1211, 2011.

[9] M. Kasparick, R. L. Cavalcante, S. Valentin, S. Stańczak, and M. Yukawa, "Kernel-based adaptive online reconstruction of coverage maps with side information," *IEEE Transactions on Vehicular Technology*, vol. 65, no. 7, pp. 5461–5473, 2016.

[10] I. Yamada and N. Ogura, "Adaptive projected subgradient method for asymptotic minimization of sequence of nonnegative convex functions," *Numerical Functional Analysis and Optimization*, vol. 25, no. 7/8, pp. 593–617, 2005.

[11] M. Tomala, I. Keskitalo, G. Bodog, and C. Sartori, "Supporting function: minimisation of drive tests (MDT)," *LTE Self-Organising Networks (SON): Network Management Automation for Operational Efficiency*, John Wiley, Hoboken, NJ, pp. 267–310, 2011.

[12] W. A. Hapsari, A. Umesh, M. Iwamura, M. Tomala, B. Gyula, and B. Sebire, "Minimization of drive tests solution in 3GPP," *IEEE Communications Magazine*, vol. 50, no. 6, pp. 28–36, 2012.

[13] S. Chouvardas, S. Valentin, M. Draief, and M. Leconte, "A method to reconstruct coverage loss maps based on matrix completion and adaptive sampling," in *The 41st IEEE International Conference on Acoustics, Speech and Signal Processing (ICASSP)*. IEEE, 2016, pp. 6390–6394.

[14] C. Phillips, S. Raynel, J. Curtis, *et al.*, "The efficacy of path loss models for fixed rural wireless links," in *International Conference on Passive and Active Network Measurement*. Springer, 2011, pp. 42–51.

[15] M. Kasparick, R. L. G. Cavalcante, S. Valentin, S. Staczak, and M. Yukawa, "Kernel-based adaptive online reconstruction of coverage maps with side information," *IEEE Transactions on Vehicular Technology*, vol. 65, no. 7, pp. 5461–5473, 2016.

[16] V. N. Vapnik and V. Vapnik, *Statistical learning theory*. Wiley, New York, 1998, vol. 1.

[17] M. Kubat, "Neural networks: a comprehensive foundation by Simon Haykin, Macmillan, 1994, ISBN 0-02-352781-7,", *The Knowledge Engineering Review*, vol. 13, no. 4, pp. 409–412, 1999.

[18] K. E. Stocker and F. M. Landstorfer, "Empirical prediction of radiowave propagation by neural network simulator," *Electronics Letters*, vol. 28, no. 612, pp. 1177–1178, 1992.

[19] B. E. Gschwendtner and F. M. Landstorfer, "An application of neural networks to the prediction of terrestrial wave propagation," in *The 8th International Conference on Antennas and Propagation*, vol. 2, 1993, pp. 804–807.

[20] T. Balandier, A. Caminada, V. Lemoine, and F. Alexandre, "An application of neural networks to the prediction of terrestrial wave propagation," in *The 6th International Symposium on Personal, Indoor and Mobile Radio Communications*, vol. 1, 1995, pp. 120–124.

[21] O. Perrault, J. P. Rossi, and T. Balandier, "Field strength with a neural ray-tracing model," in *IEEE Global Telecommunications Conference (GLOBECOM)*, vol. 2, 1996, pp. 1167–1171.

[22] P.-R. Chang and W.-H. Yang, "Environment-adaptation mobile radio propagation prediction using radial basis function neural networks," *IEEE Transactions on Vehicular Technology*, vol. 46, no. 1, pp. 155–160, 1997.

[23] A. Konak, "Predicting coverage in wireless local area networks with obstacles using kriging and neural networks," *International Journal of Mobile Network Design and Innovation*, vol. 3, no. 4, pp. 224–230, 2011.

[24] I. Popescu, I. Nafomita, P. Constantinou, A. Kanatas, and N. Moraitis, "Neural networks applications for the prediction of propagation path loss in urban environments," in *Vehicular Technology Conference, 2001. VTC 2001 Spring. IEEE VTS 53rd*, vol. 1. IEEE, 2001, pp. 387–391.

[25] G. Wolfle and F. Landstorfer, "Field strength prediction in indoor environments with neural networks," in *Vehicular Technology Conference, 1997, IEEE 47th*, vol. 1. IEEE, 1997, pp. 82–86.

[26] S. Haykin and R. Lippmann, "Neural networks, a comprehensive foundation," *International Journal of Neural Systems*, vol. 5, no. 4, pp. 363–364, 1994.

[27] E. J. Candès and Y. Plan, "Matrix completion with noise," *Proceedings of the IEEE*, vol. 98, no. 6, pp. 925–936, 2010.

[28] M. A. Davenport and J. Romberg, "An overview of low-rank matrix recovery from incomplete observations," *IEEE Journal of Selected Topics in Signal Processing*, vol. 10, no. 4, pp. 608–622, 2016.

[29] Y. Hu, W. Zhou, Z. Wen, Y. Sun, and B. Yin, "Efficient radio map construction based on low-rank approximation for indoor positioning," *Mathematical Problems in Engineering*, vol. 2013, pp. 1–9, 2013.

[30] L. Claude, S. Chouvardas, and M. Draief, "An efficient online adaptive sampling strategy for matrix completion," in *The 42nd IEEE International Conference on Acoustics, Speech and Signal Processing (ICASSP)*. IEEE, 2017, pp. 3969–3973.

[31] S. Nikitaki, G. Tsagkatakis, and P. Tsakalides, "Efficient training for fingerprint based positioning using matrix completion," in *The 20th European Signal Processing Conference (EUSIPCO)*. IEEE, 2012, pp. 195–199.

[32] X. Jiang, Z. Zhong, X. Liu, and H. C. So, "Robust matrix completion via alternating projection," *IEEE Signal Processing Letters*, vol. 24, no. 5, pp. 579–583, 2017.

[33] P. Jain, R. Meka, and I. S. Dhillon, "Guaranteed rank minimization via singular value projection," in *Adv. Neural Inf. Process. Syst. (NIPS)*, 2010, pp. 937–945.

[34] J. Tanner and K. Wei, "Normalized iterative hard thresholding for matrix completion," *SIAM Journal on Scientific Computing*, vol. 35, no. 5, pp. S104–S125, 2013.

[35] L. Bregman, "The relaxation method of finding the common point of convex sets and its application to the solution of problems in convex programming," *USSR Computational Mathematics and Mathematical Physics*, vol. 7, no. 3, pp. 200–217, 1967.

[36] A. S. Lewis, D. R. Luke, and J. Malick, "Local linear convergence for alternating and averaged nonconvex projections," *Foundations of Computational Mathematics*, vol. 9, no. 4, pp. 485–513, 2009.

[37] D. R. Luke, "Prox-regularity of rank constraint sets and implications for algorithms," *Journal of Mathematical Imaging and Vision*, vol. 47, no. 3, pp. 231–328, 2013.

[38] M. Yukawa and R.-i. Ishii, "Online model selection and learning by multi-kernel adaptive filtering," in *Signal Processing Conference (EUSIPCO), 2013 Proceedings of the 21st European*. IEEE, 2013, pp. 1–5.
[39] K. Slavakis and S. Theodoridis, "Sliding window generalized kernel affine projection algorithm using projection mappings," *EURASIP Journal on Advances in Signal Processing*, vol. 2008, pp. 1–16, 2008.
[40] S. Theodoridis, K. Slavakis, and I. Yamada, "Adaptive learning in a world of projections," *IEEE Signal Processing Magazine*, vol. 28, no. 1, pp. 97–123, 2011.

Chapter 4
Machine-learning-based channel estimation

Yue Zhu[1], *Gongpu Wang*[1], *and Feifei Gao*[2]

Wireless communication has been a highly active research field [1]. Channel estimation technology plays a vital role in wireless communication systems [2]. Channel estimates are required by wireless nodes to perform essential tasks such as precoding, beamforming, and data detection. A wireless network would have good performance with well-designed channel estimates [3,4].

Recently, artificial intelligence (AI) has been a hot research topic which attracts worldwide attentions from both academic and industrial circles. AI, which aims to enable machines to mimic human intelligence, was first proposed and founded as an academic discipline in Dartmouth Conference in 1956 [5]. It covers a series of research areas, including natural language processing, pattern recognition, computer vision, machine learning (ML), robotics, and other fields as shown in Figure 4.1.

ML, a branch of AI, uses statistical techniques to develop algorithms that can enable computers to learn with data and make predictions or yield patterns. According to different learning styles, ML can be divided into supervised learning, unsupervised learning, semi-supervised learning, and reinforcement learning. Typical ML algorithms include support vector machine (SVM) [6], decision tree, expectation-maximization (EM) algorithm [7], artificial neural network (NN), ensemble learning, Bayesian model, and so on.

Currently, one of the most attractive branches of ML is deep learning proposed by Geoffrey Hinton in 2006 [8]. Deep learning is a class of ML algorithms that can use a cascade of multiple layers of nonlinear processing units for feature extraction and transformation. Its origin can be traced back to the McCulloch–Pitts (MP) model of neuron in the 1940s [9]. Nowadays, with the rapid development in data volume and also computer hardware and software facilities such as central processing unit, graphic processing unit, and TensorFlow library, deep learning demonstrates powerful abilities such as high recognition and prediction accuracy in various applications.

In short, ML is an important branch of AI, and deep learning is one key family among various ML algorithms. Figure 4.2 depicts a simplified relationship between AI, ML, and deep learning.

[1]School of Computer and Information Technology, Beijing Jiaotong University, China
[2]Tsinghua National Laboratory for Information Science and Technology, Tsinghua University, China

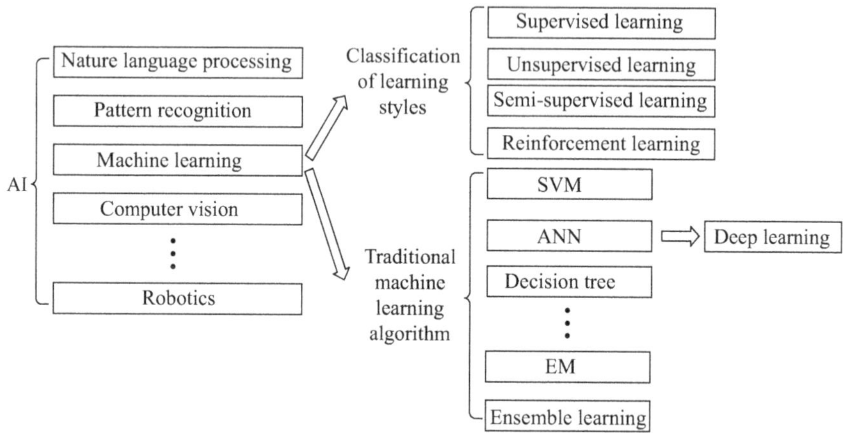

Figure 4.1 The research branches of AI and ML

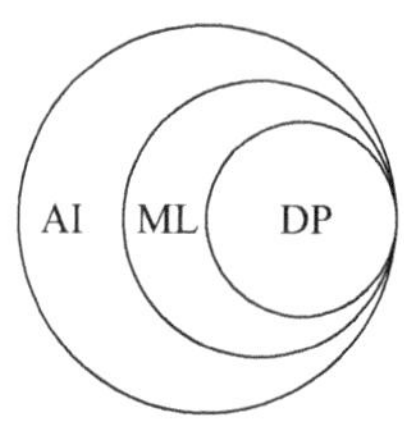

Figure 4.2 The relationship between AI, ML, and deep learning

In recent years, a range of ML algorithms have been exploited in wireless communication systems to address key issues. Reference [10] has proposed a Bayesian channel estimator with a substantial improvement over the conventional estimators in the presence of pilot contamination. Besides, a blind estimator based on EM algorithm [11] has been introduced which requires no training symbols and outperforms the existing training-aided estimators. Some deep-learning methods [12–14] have also been exploited to enhance channel estimation and detection performance of wireless communication systems. In addition, one new wireless communication architecture on the basis of an ML-aided autoencoder has been suggested in [15].

In this chapter, we first review the channel model for wireless communication systems and then describe two traditional channel estimation methods, and finally introduce two newly designed channel estimators based on deep learning and one EM-based channel estimator.

4.1 Channel model

The wireless channel is described by the response $h(t, \tau)$ at time t to an impulse transmitted at time $t - \tau$. The channel consists of several independent paths. For this multipath model, the general expression can be written as [16]:

$$h(\tau, t) = \sum_i a_i(t)\delta(\tau - \tau_i(t)), \tag{4.1}$$

where $a_i(t)$ is the attenuation and $\tau_i(t)$ is the delay from the transmitter to the receiver on the ith path. An example of a wireless channel with three paths is shown in Figure 4.3.

The general expression (4.1) is also known as a doubly selective channel since there are several paths and the attenuations and delays are functions of time. The following two special cases for $h(t, \tau)$ are widely used:

- Time-invariant frequency-selective channel: This channel occurs when the transmitter, receiver, and the environment are all stationary so that the attenuations $a_i(t)$ and propagation delays $\tau_i(t)$ do not depend on time t. However, the delays are significantly large compared to the symbol period.
- Time-varying (or time-selective) flat-fading channel: The delays $\tau_i(t)$ in this case are all approximately constant and small compared to the symbol period. This channel occurs when the transmitter or the receiver is mobile and when the symbol period of the transmitted signal significantly exceeds any of all the delays.

Since the symbol period T_s decreases when the data rate increases, the channel can be flat fading or frequency selective depending on the data rate. Moreover, the

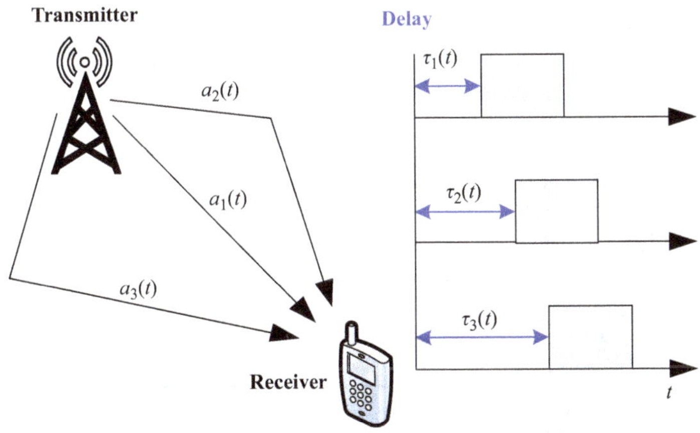

Figure 4.3 Wireless channel model

delay spread is another relevant parameter. Delay spread T_d is defined as the difference in propagation delay between the longest and shortest path:

$$T_d = \max_{i,j} |\tau_i(t) - \tau_j(t)|. \tag{4.2}$$

When T_s is much larger than T_d, the channel is **flat fading**. Otherwise, the channel is **frequency selective**. For example, the typical delay spread in a wireless channel in an urban area is 5 μs when the distance between transmitter and receiver is 1 km [1]. When the data rate is 1 kbps, the symbol period is 1 ms, and the channel is flat-fading since the delay is negligible compared to the symbol period. If the data rate increases to 1 Mbps, the symbol period T_s is 1 μs. Then the channel becomes frequency selective due to the non-negligible delays.

Furthermore, the mobility of transmitter or receiver will induce a shift in radio frequency, which is referred to as the Doppler shift D_s. Coherence time T_c, a parameter related to the Doppler shift, is defined as

$$T_c = \frac{1}{4D_s}. \tag{4.3}$$

If the coherence time T_c is comparable to the symbol period, the channel is **time-varying**. On the other hand, in **time-invariant** channels, the coherence time T_c is much larger than the symbol period (i.e., the channel remains constant). For example, if Doppler shift $D_s = 50$ Hz and the transmission data rate is 1 Mbps, then the coherence time $T_c = 2.5$ ms is much larger than one symbol duration 1 μs. In this case, the channel is time invariant.

The types of wireless channels are depicted in Table 4.1.

4.1.1 Channel input and output

In terms of the wireless channel $h(t, \tau)$, the relationship between input $s(t)$ and output $y(t)$ is given by

$$y(t) = \int_{-\infty}^{+\infty} h(t, \tau)s(t - \tau)d\tau + w(t), \tag{4.4}$$

where $w(t)$ is an additive white Gaussian complex noise signal. The receiver is required to recover data signal $s(t)$ from received signal $y(t)$; this process is called data detection.

For data detection, the receiver requires the knowledge of $h(t, \tau)$, which is referred to as channel state information (CSI). To help the receiver estimate CSI, special

Table 4.1 Different types of wireless channels

Types of channel	**Characteristic**
Time varying	$T_c \ll T_s$
Time invariant	$T_c \gg T_s$
Flat fading	$T_d \ll T_s$
Frequency selective	$T_d \gg T_s$

predefined symbols may be transmitted in addition to data symbols. These symbols are called pilot symbols or training symbols. Pilot symbols are utilized by the channel estimator at the receiver to obtain CSI.

In practice, channel estimation and data detection are done by using the discrete-time baseband signals. Define the samples $y(nT_s) = y(n)$ for $n = 0, 1, \ldots, N-1$. The discrete-time baseband model equivalent to (4.4) can then be obtained as

$$y(n) = \sum_{l=0}^{L} h(n,l)s(n-l) + w(n), \tag{4.5}$$

where $h(n,l)$ is the sampling version of $h(t,\tau)$, i.e., $h(n,l) = h(nT_s, lT_s)$, and $s(n-l)$ is the sampling version of $s(t)$, i.e., $s(n-l) = s((n-l)T_s)$, and $L+1$ is the number of multipaths and $w(n)$ is complex white Gaussian noise with mean zero and variance σ_w^2.

4.2 Channel estimation in point-to-point systems

4.2.1 Estimation of frequency-selective channels

For a frequency-selective time-invariant channel where $h(n,l)$ does not change with time index n, i.e., $h(n,l) = h(l)$, the model (4.5) can be simplified as

$$y(n) = \sum_{l=0}^{L} h(l)s(n-l) + w(n). \tag{4.6}$$

Define $\mathbf{y} = [y(0), y(1), \ldots, y(N-1)]^T$, $\mathbf{w} = [w(0), w(1), \ldots, w(N-1)]^T$, and $\mathbf{h} = [h(0), h(1), \ldots, h(L)]^T$, where N is the block length. We can further write (4.6) in the following vector form:

$$\mathbf{y} = \mathbf{S}\mathbf{h} + \mathbf{w}, \tag{4.7}$$

where $\mathbf{S}$ is a $N \times (L+1)$ circulant matrix with the first column $\mathbf{s} = [s(0), s(1), \ldots, s(N-1)]^T$. Note that the sequence $\mathbf{s}$ is the training sequence and depends on the choice of pilots and their values.

Two linear estimators are often utilized to obtain the estimate of $\mathbf{h}$ from the received signal $\mathbf{y}$. The first one is least square (LS). It treats $\mathbf{h}$ as deterministic constant and minimizes the square error. The LS estimate is [17]:

$$\hat{\mathbf{h}} = (\mathbf{S}^H\mathbf{S})^{-1}\mathbf{S}^H\mathbf{y}. \tag{4.8}$$

LS estimator can be derived as follows. The square error between the real value and the estimate value is

$$J(\mathbf{h}) = (\mathbf{y} - \mathbf{S}\mathbf{h})^H(\mathbf{y} - \mathbf{S}\mathbf{h}) = \mathbf{y}^H\mathbf{y} - 2\mathbf{y}^H\mathbf{S}\mathbf{h} + \mathbf{h}^H\mathbf{S}^H\mathbf{S}\mathbf{h} \tag{4.9}$$

To minimize the error, the gradient of $J(\mathbf{h})$ with respect to $\mathbf{h}$ is derived as

$$\frac{\partial J(\mathbf{h})}{\partial \mathbf{h}} = -2\mathbf{S}^H\mathbf{y} + 2\mathbf{S}^H\mathbf{S}\mathbf{h} \tag{4.10}$$

Setting the gradient to be zero, we can then obtain the LS estimate (4.8). For simplicity, denote $(\mathbf{S}^H\mathbf{S})^{-1}\mathbf{S}^H$ as $\mathbf{S}^\dagger$, the LS estimate can be rewritten as

$$\hat{\mathbf{h}} = \mathbf{S}^\dagger \mathbf{y}, \tag{4.11}$$

where $(\cdot)^\dagger$ represents the pseudo inverse. It can be readily checked that the minimum square error of LS is

$$J_{\min} = J(\mathbf{h}) = \mathbf{y}^T(\mathbf{I} - \mathbf{S}(\mathbf{X}^T\mathbf{S})^{-1}\mathbf{S}^T)\mathbf{y} \tag{4.12}$$

The second one is the linear minimum mean square error (LMMSE) estimator. It treats $\mathbf{h}$ as a random vector and minimizes the mean square error.

Define $\mathbf{R}_{yy} = \mathbb{E}(\mathbf{y}\mathbf{y}^H)$, $\mathbf{R}_{hh} = \mathbb{E}(\mathbf{h}\mathbf{h}^H)$, and $\mathbf{R}_{hy} = \mathbb{E}(\mathbf{h}\mathbf{y}^H)$, where $\mathbb{E}(\mathbf{x})$ is the expected value of a random variable $\mathbf{x}$. The LMMSE estimator can be expressed as

$$\hat{\mathbf{h}} = \mathbf{R}_{\mathbf{h}}\mathbf{S}^H(\mathbf{S}\mathbf{R}_{\mathbf{h}}\mathbf{S}^H + \sigma_w^2\mathbf{I})^{-1}\mathbf{y} \tag{4.13}$$

LMMSE estimator can be derived as follows. As a linear estimator, the estimate $\hat{\mathbf{h}}$ can be given as linear combination of the received signal $\mathbf{y}$:

$$\hat{\mathbf{h}} = \mathbf{A}\mathbf{y}. \tag{4.14}$$

LMMSE estimator aims to minimize the mean square error through choosing the linear combinator $\mathbf{A}$, i.e.:

$$\mathbf{A} = \arg\min_{\mathbf{A}} \mathbb{E}(\|\mathbf{h} - \hat{\mathbf{h}}\|^2) = \arg\min_{\mathbf{A}} \mathbb{E}(\|\mathbf{h} - \mathbf{A}\mathbf{y}\|^2). \tag{4.15}$$

The mean square error can be further obtained as

$$\mathbb{E}(\|\mathbf{h} - \mathbf{A}\mathbf{y}\|^2) = \mathbb{E}(tr\{(\mathbf{h} - \mathbf{A}(\mathbf{S}\mathbf{h} + \mathbf{e}))(\mathbf{h} - \mathbf{A}(\mathbf{S}\mathbf{h} + \mathbf{e}))^H\}) \tag{4.16}$$

$$= tr\{\mathbf{R}_{\mathbf{h}}\} - tr\{\mathbf{R}_{\mathbf{h}}\mathbf{S}^{\mathbf{H}}\mathbf{A}^{\mathbf{H}}\} - tr\{\mathbf{A}\mathbf{S}\mathbf{R}_{\mathbf{h}}\} + tr\{\mathbf{W}(\mathbf{S}\mathbf{R}_{\mathbf{h}}\mathbf{S}^H + \sigma_w^2\mathbf{I})\mathbf{A}^H\} \tag{4.17}$$

where $tr\{\mathbf{A}\}$ denotes the trace operation of the matrix $\mathbf{A}$ and σ_w^2 is the noise variance.

Setting the derivative of MSE to the parameter $\mathbf{A}$ as zero, we can derive:

$$\mathbf{A} = \mathbf{R}_{\mathbf{h}}\mathbf{S}^H(\mathbf{S}\mathbf{R}_{\mathbf{h}}\mathbf{S}^H + \sigma_w^2\mathbf{I})^{-1}. \tag{4.18}$$

Substituting (4.18) into (4.14) will generate (4.13).

The LS estimator is simpler compared with the LMMSE estimator. But the LMMSE estimator outperforms the LS estimator because it exploits the statistics of $\mathbf{h}$.

4.3 Deep-learning-based channel estimation

4.3.1 History of deep learning

Deep learning, suggested by Geoffrey Hinton in 2006, is rooted from NN. The earliest idea about NN originated from the MP neuron model [9] proposed by Warren McCulloch and Walter Pitts in 1943. Interestingly, we can find that there exists three up-and-down tides of NN development during the past 70 years.

The first tide, as illustrated in Figure 4.4, took place from 1940s to 1960s [18]. The MP neuron model, created in 1943, laid the foundation for the research of NN. Then in 1958, Frank Rosenblatt created the first machine referred to as perceptron [19] which exhibited the ability of simple image recognition. The perception aroused huge interests and large investments during its first decade. However, in 1969, Marvin Minsky discovered that perceptrons were incapable of realizing the exclusive OR function. He also pointed out that the computers, due to the limited computing ability at that time, cannot effectively complete the large amount of computation work required by large-scale NN [20], such as adjusting the weights. The two key factors leaded to the first recession in the development of NN.

The second wave started from the 1980s and ended at the 1990s. In 1986, David Rumelhart, Geoffrey Hinton, and Ronald J. Williams successfully utilized the back propagation (BP) algorithm [21] and effectively solved the nonlinear problems for NN with multiple layers. From then on, BP algorithms gained much popularization, which resulted in the second upsurge of NN. Unfortunately, in the early 1990s, it was pointed out that there existed three unsolved challenges for BP algorithms. The first is that the optimization method obtains the local optimal value, instead of global, when training the multilayer NN. The second is the vanishing gradient problem that the neuron weights closed to the inputs have little changes. The third is the over-fitting problem caused by the contradiction between training ability and prediction results. In addition, the data sets for training the NN and the computing capability at the time also cannot fully support the requirements from the multilayer NN. Besides, SVM [6] attracted much attentions and became one hot research topic. These factors led to a second winter in the NN development.

The third wave emerged in 2006 when Geoffrey Hinton proposed deep belief networks [8] to solve the problem of gradient disappearance through pretraining and supervised fine tuning. The term deep learning became popular ever since then. Later the success of ImageNet in 2012 provided abundant pictures for training sets and

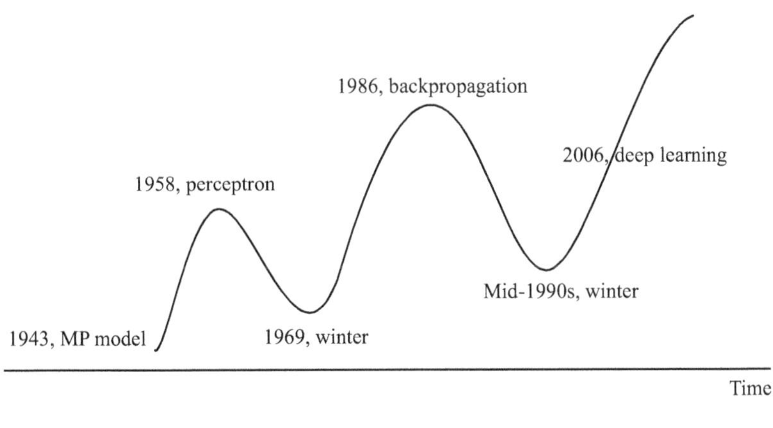

Figure 4.4 Development trend of deep learning

set a good example for deep-learning research. So far, the third wave is still gaining momentums.

4.3.2 Deep-learning-based channel estimator for orthogonal frequency division multiplexing (OFDM) systems

Figure 4.5 illustrates the functional diagram and the basic elements of a digital orthogonal frequency division multiplexing (OFDM) communication system. At the transmitter, the source bits $X(k)$ follows the operations of modulation, inverse discrete Fourier transform (IDFT), and adding cyclic prefix (CP), respectively.

Denote the multipath fading channels as $h(0), h(1), \ldots, h(L-1)$. The signal arrived at the receiver is

$$y(n) = x(n) \otimes h(n) + w(n), \tag{4.19}$$

where $x(n)$ and $w(n)$ indicate the transmitted signal and the noise, respectively, and $\otimes$ represents the circular convolution. After removing CP and performing DFT operation, the received signals can be obtained as

$$Y(k) = X(k)H(k) + W(k) \tag{4.20}$$

where $Y(k)$, $X(k)$, $H(K)$, and $W(k)$ are the DFT of $y(n)$, $x(n)$, $h(n)$, and $w(n)$ respectively. Finally, the source information is recovered from $Y(k)$ through frequency domain equalization and demodulation. Generally, the traditional OFDM receiver first estimates the CSI $H(k)$ using the pilot and then detects the source signal with the channel estimates $\hat{H}(k)$.

Different from the classical design shown in Figure 4.5, a deep-learning-based transceiver is proposed in [12] which can estimate CSI implicitly and recover the signal directly. This approach considers the whole receiver as a black box, takes the received signal as input of a deep NN (DNN), and outputs the recovered source bits after calculation and transformation in the hidden layers of the NN.

The OFDM transmission frame structure is shown in Figure 4.6. One OFDM frame consists of two OFDM blocks: one for pilot symbols and the other for data symbols. Assume that the channel parameters remain unchanged in each frame and may vary between frames.

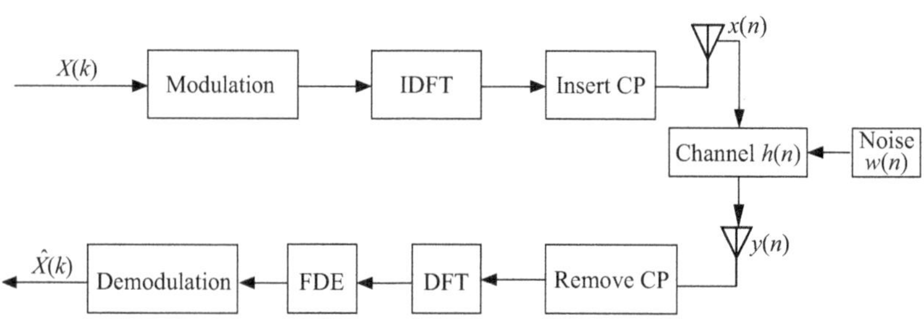

Figure 4.5 Basis elements of an OFDM system

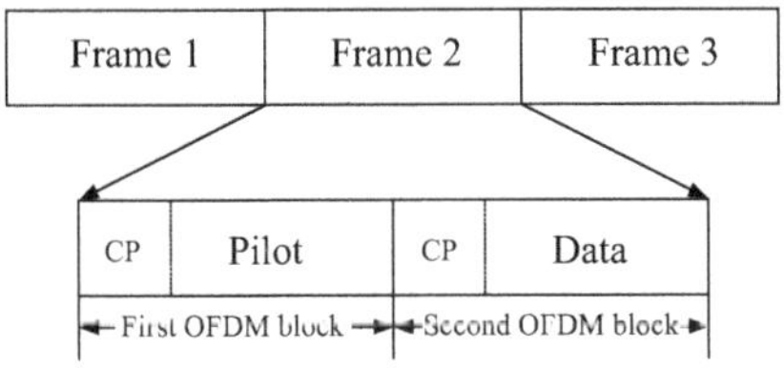

Figure 4.6 OFDM frame structure for deep-learning-based transceiver

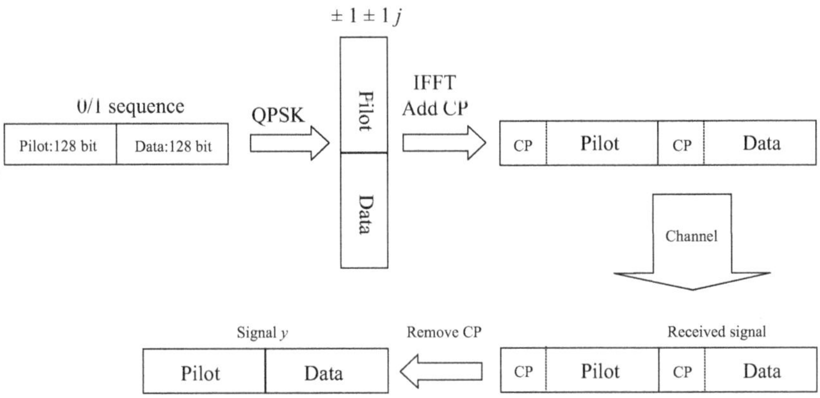

Figure 4.7 The data-generation process

The data-generation process is depicted in Figure 4.7. Suppose the OFDM system has 64 subcarriers and the length of CP is 16. In each frame, the first block contains fixed pilot symbols, and the second data block consists of 128 random binary bits.

After QPSK modulation, IDFT, and CP insertion, the whole frame data is convolved with the channel vector. The channel vector is randomly selected from the generated channel parameter sets based on the WINNER model [22]. The maximum multipath delay is set as 16.

At the receiver side, the received signals including noise and interference in one frame will be collected as the input of DNN after removing CP. The DNN model aims to learn the wireless channel parameters and recover the source signals.

As illustrated in Figure 4.8, the architecture of the DNN model has five layers: input layer, three hidden layers, and output layer. The real and imaginary parts of the signal are treated separately. Therefore, the number of neurons in the first layer is 256. The number of neurons in the three hidden layers are 500, 250, and 120, respectively. The active function of the hidden layers are rectified linear unit (ReLU) function and that of the last payer is Sigmoid function. Every 16 bits of the transmitted data are detected for one model which indicates the dimension of the output layer is 16. For example, the model in Figure 4.8 aims to predict the 16th–31st data bits in the second block. Since the data block contains 128 binary bits, eight DNN models are needed to recover the whole transmitted data part as indicated in Figure 4.9.

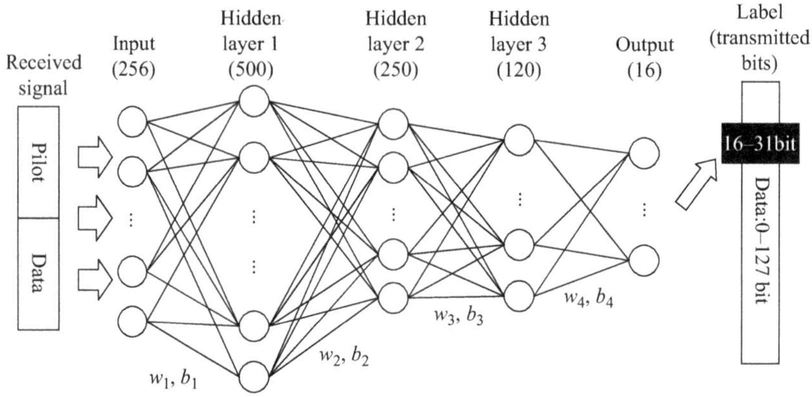

Figure 4.8 Architecture of one DNN model

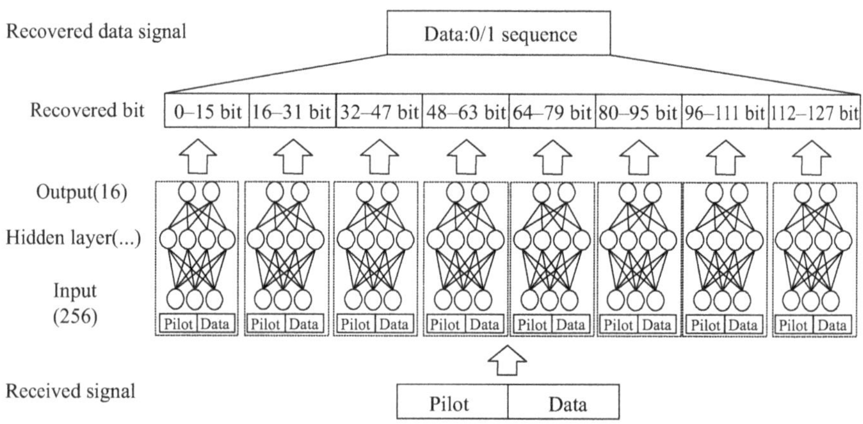

Figure 4.9 Eight DNN models

The objective function for optimization is L_2 loss function and optimal parameters is obtained with the root mean square prop (RMSProp)[1] optimizer algorithm, where Python[2] environment and TensorFlow[3] architecture are utilized. Table 4.2 lists some key parameters for training the DNN.

Figure 4.10 illustrates the bit error rate performance of the DNN method and traditional estimators: LS and LMMSE. It can be seen that the LS method performs worst and that the DNN method has the same performance as the LMMSE at low SNR. When the SNR is over 15 dB, the LMMSE slightly outperforms the DNN method.

[1]RMSProp is a stochastic gradient descent method with adapted learning rates.
[2]Python is a high-level programming language.
[3]TensorFlow is an open-source software library for dataflow programming.

Table 4.2 Key parameters for training the DNN

Parameters	Value
Epoch	60
Batch size	2,000
Batch	400
Learning rate	0.001
Test set	10,000

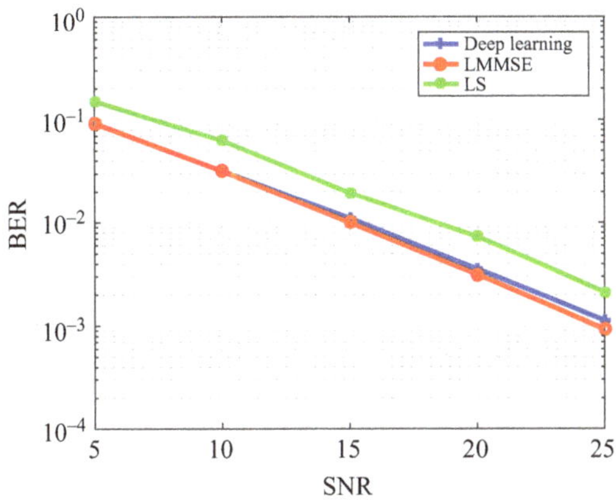

Figure 4.10 DNN, LMMSE, and LS performance

4.3.3 Deep learning for massive MIMO CSI feedback

Massive multiple-input and multiple-output (MIMO) wireless communication systems have attracted enormous attentions from both academy and industry. For massive MIMO with frequency division duplex mode, the user equipment (UE) estimates the downlink channel information and returns it to the base station (BS) via the feedback link. The main challenge of this CSI feedback mechanism is the large overhead. Existing systems usually utilize compressed sensing (CS)-based methods to obtain the sparse vector and then restructure the matrix as the estimate. These methods require that the channel should be sparse in some bases.

Different from the CS method, a feedback mechanism based on DNN is proposed in [13]. The suggested deep-learning-based CSI network (CsiNet) can sense and recover the channel matrix. The workflow of the CsiNet is shown in Figure 4.11.

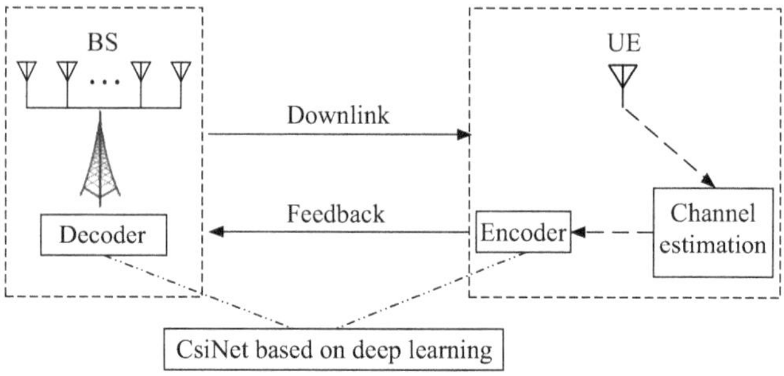

Figure 4.11 Deep-learning-based feedback mechanism

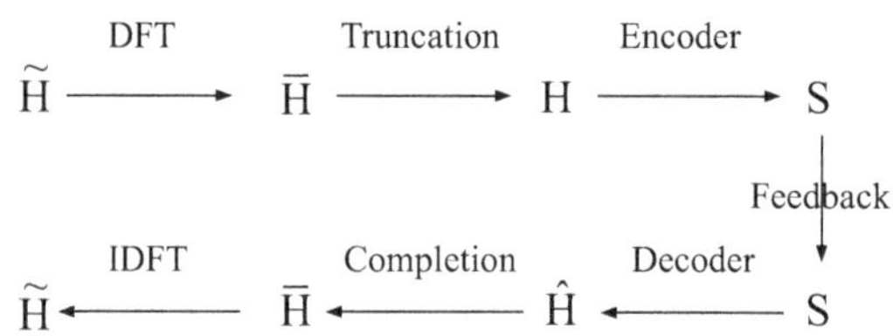

Figure 4.12 CSI feedback approach

Suppose the BS has N_t transmit antennas and UE has one, and the OFDM system has $\widetilde{N}_c$ carriers. Denote the estimated downlink channel matrix as $\widetilde{\mathbf{H}} \in C^{N_t \times \widetilde{N}_c}$. Once the UE estimates the channel $\widetilde{\mathbf{H}}$, it will apply the following DFT and then obtain:

$$\bar{\mathbf{H}} = \mathbf{F}_d \widetilde{\mathbf{H}} \mathbf{F}_a^H \tag{4.21}$$

where $\mathbf{F}_d$ and $\mathbf{F}_a$ are $\widetilde{N}_c \times \widetilde{N}_c$ and $N_t \times N_t$ DFT matrices, respectively. Next the UE selects the first N_c rows of $\mathbf{H}$ since the CSI are mainly included in these rows. Let $\mathbf{H}$ represent the truncated matrix, i.e., $\mathbf{H} = \bar{\mathbf{H}}(:, 1 : N_c)$. Clearly, the matrix $\mathbf{H}$ contains $N = N_c \times N_t$ elements, which indicates that the number of the feedback parameters is cut down to N.

Based on the deep-learning method, the CsiNet designs an encoder to convert $\mathbf{H}$ to a vector $\mathbf{s}$ that only has M elements. Next the UE sends the codeword $\mathbf{s}$ to the BS. The BS aims to reconstruct the original channel matrix $\mathbf{H}$ with the code word $\mathbf{s}$. The compression ratio is $\gamma = M/N$. Then the decoder in CsiNet can recover $\mathbf{s}$ to $\hat{\mathbf{H}}$. After completing $\hat{\mathbf{H}}$ to $\bar{\mathbf{H}}$, IDFT is used to obtain the final channel matrix. In summary, the CSI feedback approach is shown as Figure 4.12.

The CsiNet is an autoencoder model based on convolution NN. Figure 4.13 shows the architecture of the CsiNet which mainly consists of an encoder (Figure 4.14) and a decoder (Figure 4.15).

The detailed structure of the encoder is shown in Figure 4.14. It contains two layers. The first layer is a convolutional layer and the second is a reshape layer. The real and imaginary parts of the truncated matrix $\mathbf{H}$ with dimensions 8×8 is the input of the convolutional layer. This convolutional layer uses a 3×3 kernel to generate two feature maps, which is the output of the layer, i.e., two matrices with dimensions 8×8. Then the output feature maps are reshaped into a 128×1 vector. The vector enters

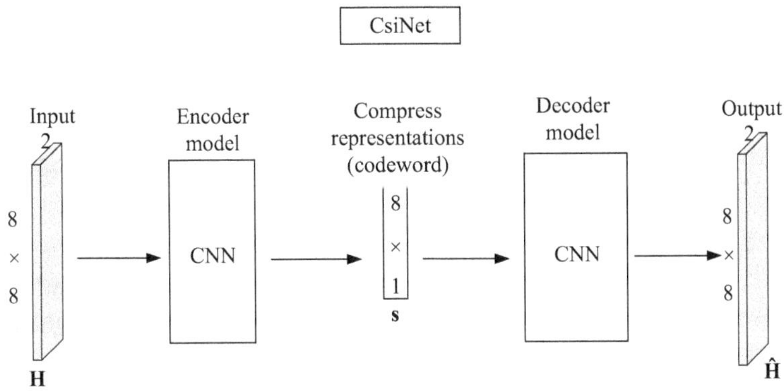

Figure 4.13 Architecture of CsiNet

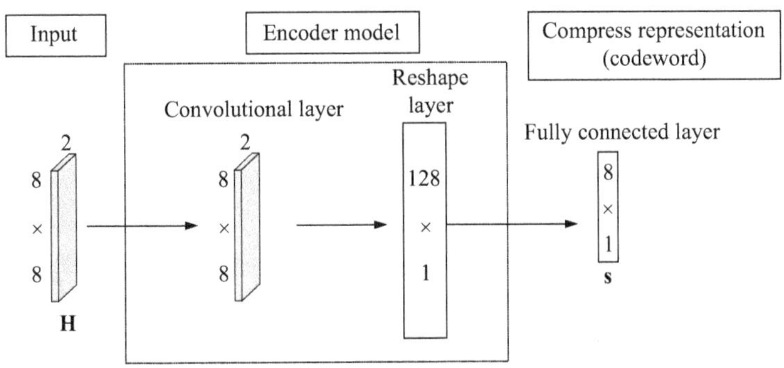

Figure 4.14 Structure of the encoder

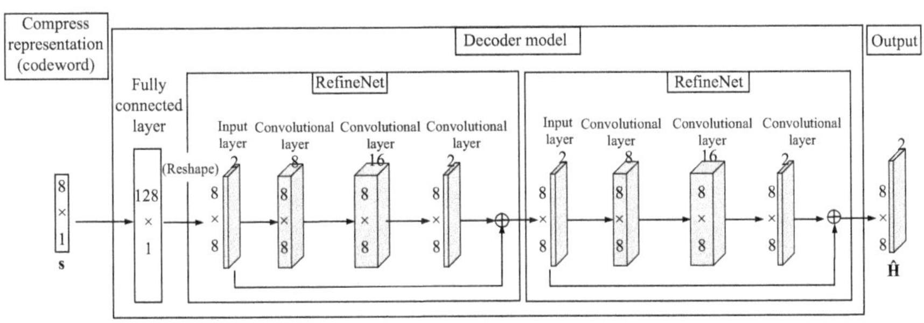

Figure 4.15 Structure of the decoder

into a fully connected layer and the output of the connected layer is the compressed codeword **s** with eight elements.

The goal of the decoder is to recover the codeword to the matrix **H**. The detailed structure of the decoder is shown in Figure 4.15. The decoder is comprised of three main parts: a fully connected layer (also referred to as dense layer) and two RefineNet. The first fully connected layer transforms the codeword **s** into a 128×1 vector. And then the vector is reshaped into two 8×8 matrices which is considered as the initial estimate of $\mathbf{H}$. Next, two RefineNets are designed to refine the estimates.

Each RefineNet has four layers: one input layer and three convolutional layers. The input layer has two feature maps, and the three convolutional layers have eight, sixteen, and two feature maps, respectively. All the feature maps are of the same size as the input channel matrix size 8×8.

It is worth noting that in each RefineNet, there is a direct data flow from the input layer to the end of last convolutional layer so as to avoid the gradient disappearance.

Each layer of the CsiNet executes normalization and employs a ReLU function to activate the neurons. After two RefineNet units, the refined channel estimates will be delivered to the final convolutional layer and the sigmoid function is also exploited to activate the neurons.

The end of the second RefineNet in the decoder will output two matrices with size 8×8, i.e., the real and imaginary parts of $\hat{\mathbf{H}}$, which is the recovery of $\mathbf{H}$ at the BS.

MSE is chosen as the loss function for optimization, and the optimal parameters are obtained through ADAM algorithm. Simulation experiments are carried out in Python environment with TensorFlow and Keras[4] architecture. The key parameters needed to train the network are listed in Table 4.3.

Here, we provide one example of $\mathbf{H}$. The UE obtain channel estimate and transform it into $\mathbf{H}$. The real and image parts of $\mathbf{H}$ are as shown in Tables 4.4 and 4.5, respectively.

The UE inputs $\mathbf{H}$ to CsiNet. The encoder of the CsiNet then generates a 8×1 codeword:

$$\begin{aligned}\mathbf{s} = [&-0.17767, -0.035453, -0.094305, -0.072261,\\ &-0.34441, -0.34731, 0.14061, 0.089002]\end{aligned} \tag{4.22}$$

Table 4.3 The parameters for training CsiNet

Parameters	**Value**
Training set	100,000
Validation set	3,000
Test set	2,000
Epoch	1,000
Batch size	200
Batch	50
Learning rate	0.01

[4]Keras is an open-source neural network library which contains numerous implementations of commonly used neural network building blocks.

Table 4.4 Real parts of the channel matrix $\mathbf{H}$

h(0)	*h*(1)	*h*(2)	*h*(3)	*h*(4)	*h*(5)	*h*(6)	*h*(7)
0.49238	0.48270	0.57059	0.48917	0.50353	0.50847	0.46970	0.46497
0.49181	0.47943	0.53719	0.51973	0.50382	0.51015	0.47846	0.50859
0.49114	0.47463	0.52339	0.51216	0.50416	0.51259	0.48357	0.50004
0.49035	0.46690	0.51880	0.50956	0.50456	0.51650	0.48655	0.49873
0.48941	0.45239	0.51662	0.50631	0.50503	0.52370	0.48838	0.49633
0.48828	0.41570	0.51161	0.50951	0.50561	0.54108	0.48948	0.49984
0.48687	0.16565	0.51290	0.50482	0.50634	0.62954	0.48995	0.49621
0.48507	0.81859	0.50914	0.50794	0.50726	0.33268	0.49038	0.49959

Table 4.5 Image parts of the channel matrix $\mathbf{H}$

h(0)	*h*(1)	*h*(2)	*h*(3)	*h*(4)	*h*(5)	*h*(6)	*h*(7)
0.50008	0.50085	0.49364	0.44990	0.49925	0.49954	0.48900	0.34167
0.50012	0.50115	0.49413	0.51919	0.49924	0.49975	0.49007	0.57858
0.50016	0.50161	0.49481	0.50429	0.49924	0.50010	0.49050	0.55735
0.50022	0.50237	0.49508	0.50391	0.49924	0.50074	0.49014	0.50956
0.50029	0.50383	0.49503	0.50028	0.49926	0.50203	0.48907	0.50658
0.50037	0.50749	0.49465	0.50244	0.49929	0.50539	0.48700	0.50755
0.50049	0.52458	0.49375	0.49802	0.49933	0.52117	0.48301	0.50263
0.50064	0.44605	0.49326	0.49966	0.49941	0.46087	0.48067	0.50367

The decoder can utilize this codeword $\mathbf{s}$ to reconstruct the channel matrix $\hat{\mathbf{H}}$. Define the distance between $\mathbf{H}$ and $\hat{\mathbf{H}}$ as $d = \|\mathbf{H} - \hat{\mathbf{H}}\|_2^2$. In this case, we obtain $d = 3.98 \times 10^{-4}$. The compression ratio is $\gamma = ((8 \times 8 \times 2)/8) = 1/16$.

4.4 EM-based channel estimator

4.4.1 Basic principles of EM algorithm

EM algorithm [7] is an iterative method to obtain maximum likelihood estimates of parameters in statistical models which depend on unobserved latent or hidden variables. Each iteration of the EM algorithm consists of two steps: calculating the expectation (E step) and performing the maximization (M step).

Next, we will introduce the principles of the EM algorithm in detail. And for clarity and ease of understanding EM algorithm, the notations used throughout this section are illustrated in Table 4.6.

In probabilistic models, there may exist observable variables and latent variables. For example, in the wireless communication systems, the received signals can be given as

$$y(i) = hx(i) + w(i), \qquad i = 1, 2, \ldots, N \tag{4.23}$$

Table 4.6 Parameters description

Notations	Description	Corresponding item in the model (4.23)
y	Observed variable	$[y(1), y(2), \ldots, y(N)]^T$
z	Latent variable (hidden variable)	$[x(1), x(2), \ldots, x(N)]^T$
θ	Parameter to be estimated	h
$\theta^{(j)}$	The *j*th iterative estimate of parameter θ	$h^{(j)}$
$L(\theta)$	The log-likelihood function about parameter θ	$L(h)$
$P(\mathbf{y};\theta)$	The probability with the parameter θ	$\prod_{i=1}^{N} P(y(i);h)$
$P(\mathbf{y},\mathbf{z};\theta)$	The joint probability for variables **y**, **z** with the parameter θ	$P(y(i), x_k; h)$
$P(\mathbf{z}\|\mathbf{y};\theta^{(j)})$	The conditional probability given **y** and parameter	$P(x_k\|y(i);h^{(j)})$
$LB(\theta,\theta^{(j)})$	The lower bound of log likelihood function	$LB(h, h^{(j)})$
$Q(\theta,\theta^{(j)})$	The expected value of the log likelihood function of θ given the current estimates of the parameter θ	$Q(h, h^{(j)})$
$Q(\theta^{(j)},\theta^{(j)})$	The value of the log likelihood function when θ is equal to $\theta^{(j)}$	$Q(h^{(j)}, h^{(j)})$

where h is the flat-fading channel to be estimated, and $x(i)$ is the unknown modulated BPSK signal, i.e., $x(i) \in \{+1, -1\}$. In this statistical model (4.23) that aims to estimate h with unknown BPSK signals $x(i)$, the received signals $y(i)$ are the observable variables and the transmitted signals $x(i)$ can be considered as latent variables.

Denote **y**, **z**, and θ as the observed data, the latent variable, and the parameter to be estimated, respectively. For the model (4.23), we can have $\mathbf{y} = [y(1), y(2), \ldots, y(N)]^T$, $\mathbf{z} = [x(1), x(2), \ldots, x(N)]^T$, and $\theta = h$.

If the variable **z** can be available, the parameter θ can be estimated by maximum likelihood approach or Bayesian estimation. Maximum likelihood estimator solves the log-likelihood function:

$$L(\theta) = \ln P(\mathbf{y};\theta) \tag{4.24}$$

where **z** is a parameter in the probability density function $P(\mathbf{y};\theta)$. Clearly, there is only one unknown parameter θ in the log-likelihood function $L(\theta|\mathbf{y})$, and therefore the estimate of θ can be obtained through maximizing[5] $L(\theta)$:

$$\hat{\theta} = \max_{\theta} L(\theta). \tag{4.25}$$

However, if the variable **z** are unknown, we cannot find the estimate $\hat{\theta}$ from (4.25) since the expression of $L(\theta)$ contains unknown parameter **z**. To address this problem, EM algorithm was proposed in [7] in 1977. EM algorithm estimates the parameter θ

[5]One dimensional search or setting the derivative as zero can obtain the optimal value of θ.

iteratively. Denote the jth estimate of θ as $\theta^{(j)}$. The basic principle of EM algorithm is as follows.

Noting that the relationship between the marginal probability density function $P(\mathbf{y})$ and the joint density function $P(\mathbf{y}, \mathbf{z})$ is

$$P(\mathbf{y}) = \sum_{\mathbf{z}} P(\mathbf{y}, \mathbf{z}). \tag{4.26}$$

Hence, we can rewrite the likelihood function $L(\theta)$ as

$$L(\theta) = \ln P(\mathbf{y}; \theta) = \ln \left\{ \sum_{\mathbf{z}} P(\mathbf{y}, \mathbf{z}; \theta) \right\} \tag{4.27}$$

$$= \ln \left\{ \sum_{\mathbf{z}} P(\mathbf{z}; \theta) P(\mathbf{y}|\mathbf{z}; \theta) \right\}, \tag{4.28}$$

where the Bayesian equation $P(\mathbf{y}, \mathbf{z}) = P(\mathbf{z})P(\mathbf{y}|\mathbf{z})$ is utilized in the last step in (4.28).

Equation (4.28) is often intractable since it contains not only the logarithm operation of the summation of multiple items and also the unknown parameter $\mathbf{z}$ in the function $P(\mathbf{y}|\mathbf{z}; \theta)$.

To address this problem of the unknown parameter $\mathbf{z}$, EM algorithm rewrites the likelihood function $L(\theta)$ as

$$L(\theta) = \ln \left(\sum_{\mathbf{z}} P(\mathbf{z}|\mathbf{y}; \theta^{(j)}) \frac{P(\mathbf{z}; \theta) P(\mathbf{y}|\mathbf{z}; \theta)}{P(\mathbf{z}|\mathbf{y}; \theta^{(j)})} \right) \tag{4.29}$$

where $P(\mathbf{z}|\mathbf{y}; \theta^{(j)})$ is the probability distribution function of the latent variable $\mathbf{z}$ given $\mathbf{y}$. Since the distribution $P(\mathbf{z}|\mathbf{y}; \theta^{(j)})$ can be readily obtained, it is possible to generate a likelihood function that only contains one unknown parameter θ.

Using Jensen's inequality [23]:

$$f(\mathbb{E}[\mathbf{x}]) \geq \mathbb{E}[f(\mathbf{x})], \tag{4.30}$$

where $\mathbf{x}$ is a random variable, $f(\mathbf{x})$ is a concave function, and $\mathbb{E}[\mathbf{x}]$ is the expected value of $\mathbf{x}$, we can have

$$\ln(\mathbb{E}[\mathbf{x}]) \geq \mathbb{E}[\ln(\mathbf{x})]. \tag{4.31}$$

Thus, we can find

$$L(\theta) \geq \mathbb{E}\left[\ln \left(\frac{P(\mathbf{z}; \theta) P(\mathbf{y}|\mathbf{z}; \theta)}{P(\mathbf{z}|\mathbf{y}; \theta^{(j)})} \right) \right]$$

$$= \sum_{\mathbf{z}} P(\mathbf{z}|\mathbf{y}; \theta^{(j)}) \ln \frac{P(\mathbf{z}; \theta) P(\mathbf{y}|\mathbf{z}; \theta)}{P(\mathbf{z}|\mathbf{y}; \theta^{(j)})} = LB(\theta, \theta^{(j)}), \tag{4.32}$$

where $LB(\theta, \theta^{(j)})$ is defined as the lower bound of the likelihood function $L(\theta)$.

We can further simplify $LB(\theta, \theta^{(j)})$ as

$$LB(\theta, \theta^{(j)}) = \sum_{\mathbf{z}} P(\mathbf{z}|\mathbf{y}; \theta^{(j)}) \ln \frac{P(\mathbf{y}, \mathbf{z}; \theta)}{P(\mathbf{z}|\mathbf{y}; \theta^{(j)})}. \tag{4.33}$$

It is worth noting that there is only one unknown parameter θ in the above expression (4.33) of $LB(\theta, \theta^{(j)})$. Therefore, we can find the $(j+1)$th iterative estimate $\theta^{(j+1)}$ through:

$$\theta^{(j+1)} = \arg\max_{\theta} LB(\theta, \theta^{(j)}) \tag{4.34}$$

$$= \arg\max_{\theta} \sum_{\mathbf{z}} P(\mathbf{z}|\mathbf{y}; \theta^{(j)})(\ln P(\mathbf{y}, \mathbf{z}; \theta) - \ln P(\mathbf{z}|\mathbf{y}; \theta^{(j)})) \tag{4.35}$$

which is the focus of the M step.

Since the item $P(\mathbf{z}|\mathbf{y}; \theta^{(j)})$ does not contain the variable θ, it can be negligible in the maximization optimization operation (4.35). Therefore, we can further simplify (4.35) as

$$\theta^{(j+1)} = \arg\max_{\theta} \underbrace{\sum_{\mathbf{z}} P(\mathbf{z}|\mathbf{y}; \theta^{(j)}) \ln P(\mathbf{y}, \mathbf{z}; \theta)}_{Q(\theta, \theta^{(j)})} \tag{4.36}$$

where the function $Q(\theta, \theta^{(j)})$ is defined as the corresponding item.

Interestingly, we can find the function $Q(\theta, \theta^{(j)})$ can be written as

$$Q(\theta, \theta^{(j)}) = \sum_{\mathbf{z}} P(\mathbf{z}|\mathbf{y}; \theta^{(j)}) \ln P(\mathbf{y}, \mathbf{z}; \theta) \tag{4.37}$$

$$= \mathbb{E}_{\mathbf{z}|\mathbf{y};\theta^{(j)}}[\ln P(\mathbf{y}, \mathbf{z}; \theta)], \tag{4.38}$$

which indicates that the function $Q(\theta, \theta^{(j)})$ is the expected value of the log likelihood function $\ln P(\mathbf{y}, \mathbf{z}; \theta)$ with respect to the current conditional distribution $P(\mathbf{z}|\mathbf{y}; \theta^{(j)})$ given the observed data $\mathbf{y}$ and the current estimate $\theta^{(j)}$. This is the reason for the name of the E step.

Next the M step is to find the parameter θ which can maximize the function $Q(\theta, \theta^{(j)})$ found on the E step, and we set the optimal value of θ as $\theta^{(j+1)}$. That is,

$$\theta^{(j+1)} = \arg\max_{\theta} Q(\theta, \theta^{(j)}). \tag{4.39}$$

Till now, the jth iteration ends. The estimate $\theta^{(j+1)}$ is then used in the next round of iteration (the E step and the M step).

The termination condition of the iterative process is

$$\|\theta^{(j+1)} - \theta^{(j)}\| < \varepsilon, \tag{4.40}$$

or

$$\|Q(\theta^{(j+1)}, \theta^{(j)}) - Q(\theta^{(j)}, \theta^{(j)})\| < \varepsilon, \tag{4.41}$$

where ε is a predefined positive constant.

4.4.2 An example of channel estimation with EM algorithm

In this section, we will use the EM method to estimate the channel h in the signal transmission model (4.23) without training symbols which indicates that the BPSK signals

$x(i)$ are unknown to the receiver. We assume that the BPSK signals are equiprobable, i.e., $P(x(i) = +1) = P(x(i) = -1) = 1/2, i = 1, 2, \ldots, N$.

Suppose $x_1 = +1$ and $x_2 = -1$, and clearly the BPSK signals $x(i)$ can be either x_1 or x_2. The conditional probability density function of received signal $y(i)$ given $x(i) = x_k, k = 1, 2$ can be expressed as

$$P(y(i)|x_k; h) = \frac{1}{\sqrt{2\pi}\sigma_w} \exp\left\{\frac{(y(i) - hx_k)^2}{-2\sigma_w^2}\right\}, \tag{4.42}$$

And the joint probability density function of x_k and $y(i)$ is

$$\begin{aligned} P(y(i), x_k; h) &= P(y(i)|x_k; h)P(x_k) \\ &= \frac{1}{2\sqrt{2\pi}\sigma_w} \exp\left\{\frac{(y(i) - hx_k)^2}{-2\sigma_w^2}\right\}. \end{aligned} \tag{4.43}$$

The core of the EM algorithm is to iteratively calculate the function $Q(h, h^{(j)})$ where h is the channel to be estimated and $h^{(j)}$ denoted the jth estimate of h. According to (4.38), we need the joint distribution $P(y(i), x_k; h)$ and also the conditional distribution $P(x_k|y(i); h^{(j)})$ to derive the function $Q(h, h^{(j)})$. Since the joint distribution $P(y(i), x_k; h)$ is given in (4.43), our next focus is to calculate the conditional probability $P(x_k|y(i); h^{(j)})$.

The conditional probability $P(x_k|y(i); h^{(j)})$ can be derived as

$$\begin{aligned} P(x_k|y(i); h^{(j)}) &= \frac{P(x_k, y(i); h^{(j)})}{P(y(i); h^{(j)})} \\ &= \frac{P(y(i)|x_k; h^{(j)})P(x_k)}{\sum_{m=1}^{2} P(y(i)|x_m; h^{(j)})P(x_m)} \\ &= \frac{(1/(2\sqrt{2\pi}\sigma_w)) \exp\{(y(i) - h^{(j)}x_k)^2/-2\sigma_w^2\}}{\sum_{m=1}^{2} (1/(2\sqrt{2\pi}\sigma_w)) \exp\{(y(i) - h^{(j)}x_m)^2/-2\sigma_w^2\}} \\ &= \frac{\exp\{(y(i) - h^{(j)}x_k)^2/-2\sigma_w^2\}}{\sum_{m=1}^{2} \exp\{(y(i) - h^{(j)}x_m)^2/-2\sigma_w^2\}} \end{aligned} \tag{4.44}$$

Subsequently, in the E step, the expectation of $\ln P(y(i), x_k; h)$ with respect to the current conditional distribution $P(x_k|y(i); h^{(j)})$ given $y(i)$ can be found as

$$Q(h, h^{(j)}) = \sum_{i=1}^{N} \sum_{k=1}^{2} P(x_k|y(i); h^{(j)}) \ln P(y(i), x_k; h) \tag{4.45}$$

Substituting (4.43) and (4.44) into (4.45), we can further obtain

$$Q(h,h^{(j)}) = \sum_{i=1}^{N}\sum_{k=1}^{2} P(x_k|y(i);h^{(j)})\left(-\ln 2\sqrt{2\pi}\sigma_w - \frac{(y(i)-hx_k)^2}{-2\sigma_w^2}\right)$$
$$= \sum_{i=1}^{N}\sum_{k=1}^{2} \frac{\exp\{(y(i)-h^{(j)}x_k)^2/-2\sigma_w^2\}}{\sum_{m=1}^{2}\exp\{(y(i)-h^{(j)}x_m)^2/-2\sigma_w^2\}}\left\{-\ln 2\sqrt{2\pi}\sigma_w - \frac{(y(i)-hx_k)^2}{-2\sigma_w^2}\right\} \tag{4.46}$$

It is worth noting that the expression (4.46) of $Q(h,h^{(j)})$ only contains one unknown parameter h. Therefore, the $(j+1)$th estimate of the channel h can be calculated through setting the derivative of (4.46) with respect to h as zero. Accordingly, it can be readily obtained that

$$h^{(j+1)} = \frac{\sum_{i=1}^{N}\sum_{k=1}^{2} P(x_k|y(i);h^{(j)})y(i)x_k}{\sum_{i=1}^{N}\sum_{k=1}^{2} P(x_k|y(i);h^{(j)})x_k^2}$$
$$= \frac{\sum_{i=1}^{N}\sum_{k=1}^{2}\left(\left(\exp\{(y(i)-h^{(j)}x_k)^2/-2\sigma_w^2\}y(i)x_k\right)/\left(\sum_{m=1}^{2}\exp\{(y(i)-h^{(j)}x_m)^2/-2\sigma_w^2\}\right)\right)}{\sum_{i=1}^{N}\sum_{k=1}^{2}\left(\left(\exp\{(y(i)-h^{(j)}x_k)^2/-2\sigma_w^2\}x_k^2\right)/\left(\sum_{m=1}^{2}\exp\{(y(i)-h^{(j)}x_m)^2/-2\sigma_w^2\}\right)\right)} \tag{4.47}$$

In conclusion, the EM algorithm preset a value for θ and then calculates $h^{(j+1)}$ iteratively according to (4.47) until the convergence condition (4.40) or (4.41) is satisfied.

In the following part, we provide simulation results to corroborate the proposed EM-based channel estimator. Both real and complex Gaussian channels are simulated. For comparison, Cramér–Rao Lower Bound (CRLB) is also derived. CRLB determines a lower bound for the variance of any unbiased estimator. First, since N observations are used in the estimation, the probability density function of $\mathbf{y}$ is

$$P(\mathbf{y};h) = \prod_{i}^{N} \frac{1}{\sqrt{2\pi}\sigma_w}\exp\left\{\frac{(y(i)-hx_k)^2}{-2\sigma_w^2}\right\}, \tag{4.48}$$

and its logarithm likelihood function is

$$\ln P(\mathbf{y};h) = -\frac{N}{2}\ln 2\pi\sigma_w^2 - \frac{1}{2\sigma_w^2}\sum_{i}^{N}(y(i)-hx(i))^2. \tag{4.49}$$

The first derivative of $\ln P(\mathbf{y};h)$ with respect to h can be derived as

$$\frac{\partial \ln P(\mathbf{y};h)}{\partial h} = \frac{1}{\sigma_w^2}\sum_{i}^{N}(y(i)-hx(i))x(i). \tag{4.50}$$

Thus, the CRLB can be expressed as [17]

$$\text{var}(h) \geq \frac{1}{-E\left[(\partial^2 \ln p(\mathbf{y};h))/\partial^2 h\right]} = \frac{\sigma_w^2}{E[\sum_i^N x_i^2]} = \frac{\sigma_w^2}{NP_x} = \text{CRLB} \tag{4.51}$$

where P_x is the average transmission power of the signals $x(n)$.

Figures 4.16 and 4.17 depict the MSEs of the EM estimator versus SNR for the following two cases separately: when the channel h is generated from $N(0, 1)$, i.e., real channel; and when it is generated from $CN(0, 1)$, i.e., Rayleigh channel. The observation length is set as $N = 6$. For comparison with the EM estimator, the MSE curves of LS method are also plotted when the length of pilot is $N/2$ and N,

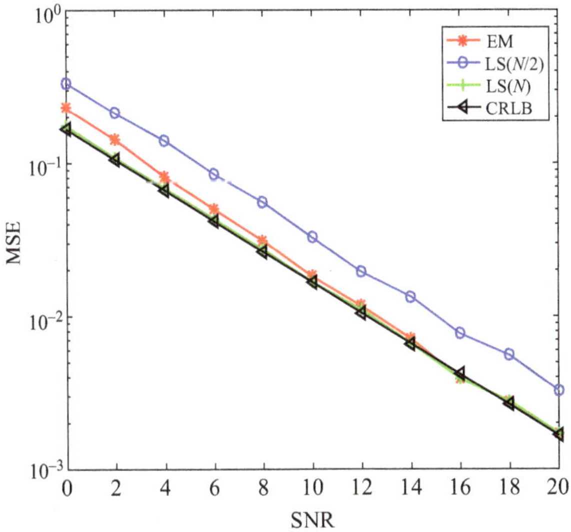

Figure 4.16 Real channel estimation MSEs versus SNR

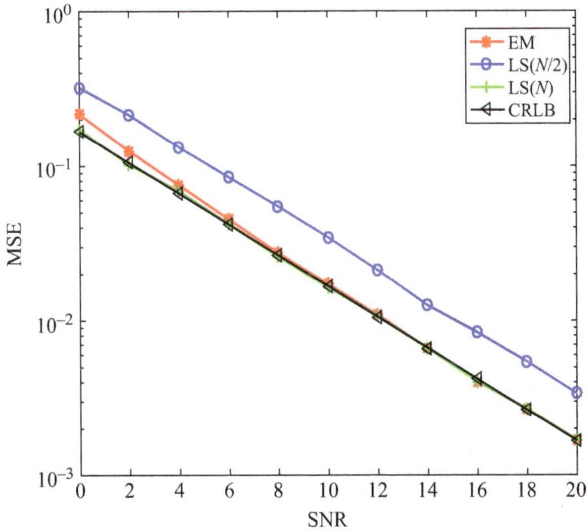

Figure 4.17 Complex channel estimation MSEs versus SNR

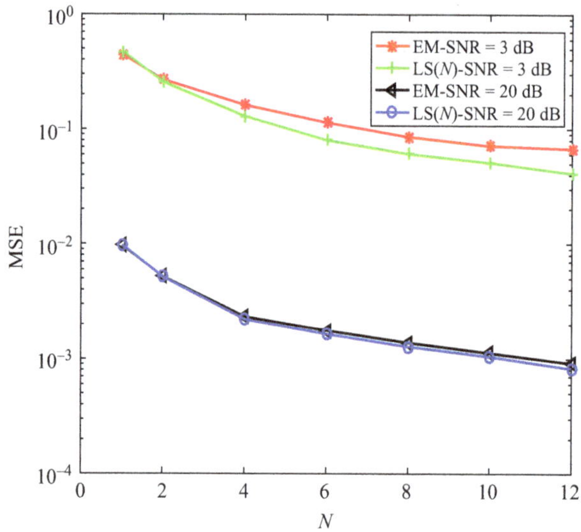

Figure 4.18 Channel estimation MSEs versus N

respectively. The CRLBs are also illustrated as benchmarks. It can be seen from Figures 4.16 and 4.17 that the EM-based blind channel estimator performs well and approaches CRLB at high SNR. It can also be found that the EM-based blind channel estimator with no pilots exhibits almost the same performance with the LS estimator with N pilots and fully outperforms the LS estimator with $N/2$ pilots.

Figure 4.18 demonstrates the MSEs of LS method and EM algorithm versus the length of signal N when SNR = 3 and 20 dB, respectively. As we expected, the MSEs of the two estimator witness a downward trend when the length N increases. It can be also found that the EM algorithm without pilot has almost consistent performance with the LS method with N pilots when SNR is 20 dB.

4.5 Conclusion and open problems

In this chapter, we introduce two traditional, one EM-based, and two deep-learning-based channel estimators. Exploiting ML algorithms to design channel estimators is a new research area that involves many open problems [24]. For example, developing new estimators in the relay or cooperative scenarios is worthy of further studies. Besides, design of new deep-learning-based channel estimators for wireless communications on high-speed railways is another interesting challenge since the railway tracks are fixed and a large amount of historical data along the railways can be exploited [25] through various deep-learning approaches to enhance estimation performance.

References

[1] Tse D., and Viswanath P. *Fundamentals of Wireless Communication*. Newyork: Cambridge University Press; 2005. p. 1.

[2] Cavers J.K. An analysis of pilot symbol assisted modulation for Rayleigh fading channels. *IEEE Transactions on Vehicular Technology*. 1991; 40(4):686–693.

[3] Wang G., Gao F., and Tellambura C. Joint frequency offset and channel estimation methods for two-way relay networks. *GLOBECOM 2009–2009 IEEE Global Telecommunications Conference*. Honolulu, HI; 2009. pp. 1–5.

[4] Wang G., Gao F., Chen W., *et al.* Channel estimation and training design for two-way relay networks in time-selective fading environments. *IEEE Transactions on Wireless Communications*. 2011; 10(8):2681–2691.

[5] Russell S.J., and Norvig P. *Artificial Intelligence: A Modern Approach* (3rd ed.). Upper Saddle River, NJ: Prentice Hall; 2010.

[6] Cortes C., Vapnik V. Support-vector networks. *Machine Learning*. 1995; 20(3):273–297.

[7] Dempster A.P. Maximum likelihood from incomplete data via the EM algorithm. *Journal of Royal Statistical Society B*. 1977; 39(1):1–38.

[8] Hinton G.E., Osindero S., and Teh Y.-W. A fast learning algorithm for deep belief nets. *Neural Computation*. 2006; 18(7):1527–1554.

[9] McCulloch W.S., and Pitts W. A logical calculus of the ideas immanent in nervous activity. *The Bulletin of Mathematical Biophysics*. 1943; 5(4):115–133.

[10] Wen C.K., Jin S., Wong K.K., *et al.* Channel estimation for massive MIMO using Gaussian-mixture Bayesian learning. *IEEE Transactions on Wireless Communications*. 2015; 14(3):1356–1368.

[11] Wang X., Wang G., Fan R., *et al.* Channel estimation with expectation maximization and historical information based basis expansion model for wireless communication systems on high speed railways. *IEEE Access*. 2018; 6:72–80.

[12] Ye H., Li G.Y., and Juang B.H.F. Power of deep learning for channel estimation and signal detection in OFDM Systems. *IEEE Wireless Communications Letters*. 2017; 7(1):114–117.

[13] Wen C., Shih W.T., and Jin S. Deep learning for massive MIMO CSI feedback. *IEEE Wireless Communications Letters*. 2018; 7(5):748–751.

[14] Samuel N., Diskin T., and Wiesel A. Deep MIMO detection. *2017 IEEE 18th International Workshop on Signal Processing Advances in Wireless Communications (SPAWC)*. Sapporo: Japan; 2017. pp. 1–5.

[15] Dorner S., Cammerer S., Hoydis J., *et al.* Deep learning based communication over the air. *IEEE Journal of Selected Topics in Signal Processing*. 2018; 12(1):132–143.

[16] Jakes W.C. *Microwave Mobile Communications*. New York: Wiley; 1974.

[17] Steven M.K. *Fundamentals of Statistical Signal Processing: Estimation Theory*. Upper Saddle River, NJ: PTR Prentice Hall; 1993.

[18] Goodfellow I., Bengio Y., Courville A. *Deep Learning*. Cambridge: MIT Press; 2016.

[19] Rosenblatt F. The perception: a probabilistic model for information storage and organization in the brain. *Psychological Review*. 1958; 65(6):386–408.
[20] Minsky M.L., and Papert S. *Perceptron*. Cambridge: MIT Press; 1969.
[21] Rumelhart D.E. Learning representations by back-propagating errors. *Nature*. 1986; 323:533–536.
[22] He R., Zhong Z., Ai B., *et al.* Measurements and analysis of propagation channels in high-speed railway viaducts. *IEEE Transactions on Wireless Communication*. 2013; 12(2):794–805.
[23] Boyd S., and Vandenberghe L. *Convex Optimization*. Cambridge: Cambridge University Press; 2004.
[24] Wang T., Wen C., Wang H., *et al.* Deep learning for wireless physical layer: opportunities and challenges. *China Communications*. 2017; 4(11):92–111.
[25] Wang G., Liu Q., He R., *et al.* Acquisition of channel state information in heterogeneous cloud radio access networks: challenges and research directions. *IEEE Wireless Communications*. 2015; 22(3):100–107.

Chapter 5
Signal identification in cognitive radios using machine learning

Jingwen Zhang[1] *and Fanggang Wang*[1]

As an intelligent radio, cognitive radio (CR) allows the CR users to access and share the licensed spectrum. Being a typical noncooperative system, the applications of signal identification in CRs have emerged. This chapter introduces several signal identification techniques, which are implemented based on the machine-learning theory.

The background of signal identification techniques in CRs and the motivation of using machine learning to solve signal identification problems are introduced in Section 5.1. A typical signal-identification system contains two parts, namely, the modulation classifier and specific emitter identifier, which are respectively discussed in Sections 5.2 and 5.3. Conclusions are drawn in Section 5.3.5.

5.1 Signal identification in cognitive radios

CR was first proposed by Joseph Mitola III in 1999 [1], with its original definition as a software-defined radio platform, which can be fully configured and dynamically adapt the communication parameters to make the best use of the wireless channels. In 2005, Simon Haykin further developed the concept of CR to spectrum sharing [2], where the CR users are allowed to share the spectrum with the licensed users and hence mitigates the scarcity problem of limited spectrum. With the advent of the upcoming fifth generation (5G) cellular-communication systems, the challenges faced by 5G are neglectable. In particular, the explosion of mobile data traffic, user demand, and new applications contradicts with the limited licensed spectrum. The existing cellular network is built based on the legacy command-and-control regulation, which in large part limits the ability of potential users to access the spectrum. In such a case, CR provides a promising solution to address the above bottleneck faced by 5G cellular system.

In general, CR network is a noncooperative system. Within the framework, the CR users and the licensed users work in separate and independent networks. For a CR user, it has little *a prior* knowledge of the parameters used by the licensed

[1]State Key Laboratory of Rail Traffic Control and Safety, Beijing Jiaotong University, China

user and other CR users and barely knows the identities (whether they are legal or evil) of other users in the network. Hence, signal identification plays a key role in CRs in order to successfully proceed the received signals and to guarantee the safety and fairness of the networks. In this chapter, two signal identification techniques are introduced, modulation classification and specific emitter identification (SEI). Figure 5.1 illustrates the diagram of a typical signal-identification system. The signal identification system is capable of solving two typical problems that the CR networks are confronted with: one is that the CR users have little information of the parameters used by the licensed users and/or other CR users; the other is that with the ability that allows any unlicensed users to access the spectrum, a certain user lacks the identity information of other users in the network. Modulation classification can be adopted to solve the unknown parameter problem and has vital applications in CRs. For the case when the licensed and cognitive users share the same frequency band for transmission and reception, the received signal at the cognitive receiver is the superposition of signals from the licensed transmitter and its own transmitter, which implies that the signal from the licensed user can be treated as an interference with higher transmission power. By applying the modulation classification techniques, the CR receiver can blindly recognize the modulation format adopted by the licensed signal and is capable of demodulating, reconstructing, and canceling the interference caused by the licensed user, which is the basis to proceed its own signal. Furthermore, to solve the problem that the CRs is exposed to high possibility of being attacked or harmed by the evil users, SEI offers a way to determine the user's identity and guarantees the safety and fairness of the CR networks.

The task of signal identification is to blindly learn from the signal and the environment to make the classification decision, behind which is the idea of the artificial intelligence. As an approach to implement the artificial intelligence, machine learning has been introduced in signal identification for the designing of the identification algorithms. With little knowledge of the transmitted signal and the transmission

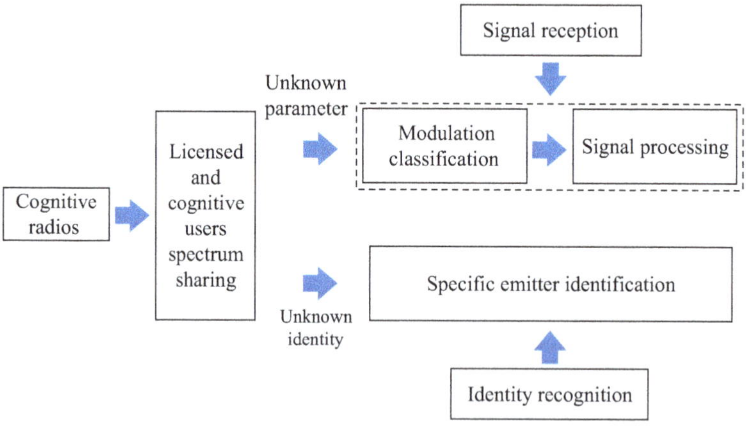

Figure 5.1 The diagram of a typical signal identification system

environments, blind signal identification remains a challenging task for conventional methods. Machine learning provides a novel and promising solution for the signal-identification problem. In general, a signal identifier/classifier consists of two parts, one is the feature extraction subsystems to extract distinguishable features and the other one is the classifier, which implements the classification task and makes decisions by training and learning from the extracted features. In the remaining sections, details of two signal identification techniques, i.e., the modulation classification and SEI algorithms, based on machine learning is described.

5.2 Modulation classification via machine learning

The automatic modulation classification, hereinafter referred to as modulation classification, is a technique adopted at an intelligent receiver to automatically determine the unknown modulation format used by a detected signal of interest. As an indispensable process between signal detection and demodulation, it finds its applications in military communications, along with CRs and adaptive systems.

Depending on the theories that used, typical modulation classification techniques can be categorized into two classes, i.e., the decision-theoretic and the pattern-recognition algorithms. The decision-theoretic algorithms are based on the likelihood theory, where the modulation-classification problem is formulated as a multiple-hypothesis test [3,4]. On the other hand, the pattern-recognition algorithms are based on the pattern-recognition and machine-learning theory, where certain classification features are extracted from the received signal and then inputted into the classifiers to decide the modulation format [5,6].

More recently, the modulation classification investigation has focused on a more challenging task of implementing the classification problem in the realistic environment, e.g., in the presence of complicated channel conditions and with no knowledge of many transmission parameters. The expectation–maximization (EM) algorithm is a commonly known algorithm in machine learning, which can be widely adopted to clustering and dimension reduction. For a probabilistic model with unobserved latent variables, the EM algorithm provides a feasible way to obtain the maximum likelihood estimates (MLEs) of the unknown parameters.

In this section, we introduce two modulation classifiers, one is for determining the typical constellation-based modulation formats (e.g., quadrature amplitude modulation (QAM) and phase-shift keying (PSK)), which are widely adopted in the mobile communication systems long term evolution (LTE) and new radio (NR), and the other is for classifying continuous phase modulation (CPM) types, which have critical applications in satellite-communication systems. Both of the two classification problems are considered in unknown fading channels, where the EM-based algorithms are proposed to obtain the MLEs of the unknown channel parameters and further determine the modulation formats. Note that in Section 5.2.2, the exact algorithm used in the classification problem is the Baum–Welch (BW) algorithm. However, it is clarified that the idea behind the BW algorithm is using the EM algorithm to find the MLEs of the unknowns in a hidden Markov model (HMM).

5.2.1 Modulation classification in multipath fading channels via expectation–maximization

In this section, we investigate the classification problem of classifying QAM/PSK modulations in the presence of unknown multipath channels. QAM/PSK is the most commonly used modulation formats in the existing LTE systems and the upcoming NR systems, of which the classification problem has been thoroughly investigated for decades. However, the classification problem in real-world scenarios, such as in the unknown multipath fading channels, is still challenging and needs further study. A hybrid maximum likelihood (also known as hybrid likelihood ratio test) based classifier is proposed to solve this problem. Specifically, the likelihood function is computed by averaging over the unknown transmitted constellation points and then maximizing over the unknown channel coefficients. Solutions to this problem cannot be obtained in a computationally efficient way, where the EM algorithm is developed to compute the MLEs of the unknowns tractably.

5.2.1.1 Problem statement

We consider a typical centralized cooperation system with one transmitter, K receivers, and one fusion center; the fusion center collects data from the K receivers to enhance the classification performance.[1] The transmit signal undergoes a multipath channel with L resolvable paths. Then, the received signal at the kth receiver is written as

$$y_k(t) = \sum_n \sum_{l=0}^{L-1} a_k(l) e^{j\phi_k(l)} x_s(n) g(t - nT - lT) + w_k(t), \quad 0 \leq t \leq T_0 \tag{5.1}$$

where T and T_0 are the symbol and observation intervals, respectively, with $T_0 \gg T$, $g(\cdot)$ is the real-valued pulse shape, $j = \sqrt{-1}$, $a_k(l) > 0$ and $\phi_k(l) \in [0, 2\pi)$ are the unknown amplitude and phase of the lth path at the kth receiver, respectively, $w_k(t)$ is the complex-valued zero-mean white Gaussian noise process with noise power σ_k^2, and $x_s(n)$ is the nth transmit constellation symbol drawn from an unknown modulation format s. The modulation format s belongs to a modulation candidate set $\{1, \ldots, \mathcal{S}\}$, which is known at the receivers. The task of the modulation classification problem is to determine the correct modulation format to which the transmit signal belongs based on the received signal.

The maximum likelihood algorithm is adopted as the classifier, which is optimal when each modulation candidate is equally probable. Let $\mathcal{H}_s$ denote the hypothesis that transmit symbols are drawn from the modulation format s, the likelihood function under the hypothesis $\mathcal{H}_s$ is given by

$$p_s(\boldsymbol{y}|\boldsymbol{\theta}) = \sum_{\boldsymbol{x}_s} p_s\left(\boldsymbol{y}|\boldsymbol{x}_s, \boldsymbol{\theta}\right) p\left(\boldsymbol{x}_s|\boldsymbol{\theta}\right) \tag{5.2}$$

[1] Multiple receivers are considered to obtain the diversity gain, while the proposed algorithm is applicable to the case with only one receiver.

where $\boldsymbol{y}^{\mathrm{T}} = [\boldsymbol{y}_1^{\mathrm{T}}, \ldots, \boldsymbol{y}_K^{\mathrm{T}}]$, with $\boldsymbol{y}_k^{\mathrm{T}}$ as the vector representation of $y_k(t)$, $\boldsymbol{\theta}^{\mathrm{T}} = [\boldsymbol{\theta}_1^{\mathrm{T}}, \ldots, \boldsymbol{\theta}_K^{\mathrm{T}}]$, with $\boldsymbol{\theta}_k = [a_k(0), \phi_k(0), \ldots, a_k(L-1), \phi_k(L-1)]^{\mathrm{T}}$ as the vector of unknown parameters, and $\boldsymbol{x}_s = [x_s(1), \ldots, x_s(N)]^{\mathrm{T}}$ is the vector representation of the transmit symbol, with $N = (T_0/T)$, and $(\cdot)^{\mathrm{T}}$ as the transpose of a vector/matrix. We assume that the multiple receivers are spatially divided; therefore, the received signals at multiple receivers are independent. Then, $p_s(\boldsymbol{y}|\boldsymbol{x}_s, \boldsymbol{\theta})$ is expressed as

$$p_s(\boldsymbol{y}|\boldsymbol{x}_s, \boldsymbol{\theta}) = \prod_{k=1}^{K} p_s(\boldsymbol{y}_k|\boldsymbol{x}_s, \boldsymbol{\theta}_k) \propto \exp\left\{ -\sum_{k=1}^{K} \int_0^{T_0} \frac{|y_k(t) - f_k(x_s(n), t)|^2 \mathrm{d}t}{\sigma_k^2} \right\} \tag{5.3}$$

where

$$f_k(x_s(n), t) = \sum_{n=1}^{N} \sum_{l=0}^{L-1} a_k(l) e^{j\phi_k(l)} x_s(n) g(t - nT - lT).$$

Define $x_{s,i}(n)$ as the nth transmit symbol that maps to the ith constellation point under the hypothesis $\mathcal{H}_s$ and assume that each constellation symbol has equal prior probability, i.e., $p(x_{s,i}(n)) = (1/M)$, with M as the modulation order of modulation format s. The log-likelihood function $L_s(\boldsymbol{\theta})$ is then obtained by

$$L_s(\boldsymbol{\theta}) = \ln p_s(\boldsymbol{y}|\boldsymbol{\theta}) = \ln\left(\sum_{i=1}^{M} \frac{1}{M} \exp\left\{ -\sum_{k=1}^{K} \int_0^{T_0} \frac{|y_k(t) - f_k(x_{s,i}(n), t)|^2 dt}{\sigma_k^2} \right\} \right). \tag{5.4}$$

The classifier makes the final decision on the modulation format by

$$\hat{s} = \arg\max_s L_s(\boldsymbol{\theta}_s^{(\star)}) \tag{5.5}$$

where $\boldsymbol{\theta}_s^{(\star)}$ is the MLE of the unknown parameters under the hypothesis $\mathcal{H}_s$, which can be obtained by

$$\boldsymbol{\theta}_s^{(\star)} = \arg\max_{\boldsymbol{\theta}} L_s(\boldsymbol{\theta}). \tag{5.6}$$

It should be noted that (5.26) is a high-dimensional non-convex optimization problem with no closed-form solutions. Essentially, the computation of the MLEs suffers from high-computational complexity, which is impractical in applications of modulation classification.

5.2.1.2 Modulation classification via EM

In this section, an EM-based algorithm is proposed to solve the problem in (5.26) in a tractable way. The expectation step (E-step) and maximization step (M-step) under the hypothesis $\mathcal{H}_s$ can be mathematically formulated as [7]:

$$\text{E-step:} \quad J(\boldsymbol{\theta}|\boldsymbol{\theta}_s^{(r)}) = \mathbb{E}_{\boldsymbol{z}|\boldsymbol{y}, \boldsymbol{\theta}_s^{(r)}} [\ln p(\boldsymbol{z}|\boldsymbol{\theta})] \tag{5.7}$$

$$\text{M-step:} \quad \boldsymbol{\theta}_s^{(r+1)} = \arg\max_{\boldsymbol{\theta}} J(\boldsymbol{\theta}|\boldsymbol{\theta}_s^{(r)}) \tag{5.8}$$

where z is the complete data, which cannot be directly observed at the receivers. Instead, the complete data is related to the observations, i.e., the received signals, by $\boldsymbol{y} = \mathcal{K}(\boldsymbol{z})$, where $\mathcal{K}(\cdot)$ is a deterministic and non-invertible transformation. The non-invertible property of $\mathcal{K}(\cdot)$ implies that there exits more than one possible definitions of the complete data to generate the same observations. It should be noted that these choices have great impact on the complexity and convergence result of the EM algorithm, bad choices of the complete data make the algorithm invalid.

In our problem, the received signal that undergoes multipath channels is equivalent to a superposition of signals from different independent paths; therefore, the complete data can be defined as

$$z_{kl}(t) = \sum_{n} a_k(l)e^{j\phi_k(l)}x_s(n)g(t - nT - lT) + w_{kl}(t) \tag{5.9}$$

where $w_{kl}(t)$ is the lth noise component, which is obtained by arbitrarily decomposing the total noise $w_k(t)$ into L independent and identically distributed components, i.e., $\sum_{l=0}^{L-1} w_{kl}(t) = w_k(t)$. Assume that $w_{kl}(t)$ follows the complex-valued zero-mean Gaussian process with power σ_{kl}^2. The noise power σ_{kl}^2 is defined as $\sigma_{kl}^2 = \beta_{kl}\sigma_k^2$, where β_{kl} is a positive real-valued random noise decomposition factor following $\sum_{l=0}^{L-1} \beta_{kl} = 1$ [8]. Hence, we can rewrite the transmission model in (5.1) as

$$y_k(t) = \mathcal{K}(\boldsymbol{z}_k) = \mathbf{1}_L \boldsymbol{z}_k^{\mathrm{T}}$$

where $\mathbf{1}_L$ is a $L \times 1$ vector with all entries equal to 1, and $\boldsymbol{z}_k$ is the vector representation of $z_{kl}(t)$. Since the multiple receivers are assumed to be independent, we reduce the E-step in (5.27) to

$$\begin{aligned} J(\boldsymbol{\theta}|\boldsymbol{\theta}_s^{(r)}) &= \sum_{k=1}^{K} J_k(\boldsymbol{\theta}|\boldsymbol{\theta}_{s,k}^{(r)}) \\ &= \sum_{k=1}^{K} \mathbb{E}_{\boldsymbol{z}_k|\boldsymbol{y}_k,\boldsymbol{\theta}_{s,k}^{(r)}} [\ln p(\boldsymbol{z}_k|\boldsymbol{\theta}_k)] \end{aligned} \tag{5.10}$$

which indicates that $J(\boldsymbol{\theta}|\boldsymbol{\theta}_s^{(r)})$ can be computed locally at each receiver.

In order to derive the E-step and M-step, the posterior expectations of the unknown transmit symbols should be estimated first. Define $\rho_{s,i}^{(r)}(n)$ as the posterior probability of the nth transmit symbol mapping to the ith constellation point under the hypothesis $\mathcal{H}_s$ at iteration r, which is given by

$$\begin{aligned} &\rho_{s,i}^{(r)}(n) \\ &= p_s\left(x_s(n) = X_{s,i}|\boldsymbol{y}, \boldsymbol{\theta}_s^{(r)}\right) \\ &= \frac{p_s\left(\boldsymbol{y}|x_s(n) = X_{s,i}, \boldsymbol{\theta}_s^{(r)}\right) p(x_s(n) = X_{s,i}|\boldsymbol{\theta}_s^{(r)})}{\sum_{j=1}^{M} p_s\left(\boldsymbol{y}|x_s(n) = X_{s,j}, \boldsymbol{\theta}_s^{(r)}\right) p(x_s(n) = X_{s,j}|\boldsymbol{\theta}_s^{(r)})} \\ &= \frac{\exp\left\{-\sum_{k=1}^{K}\left(\left|y_k(n) - \sum_{l=0}^{L-1} a_{s,k}^{(r)}(l)e^{j\phi_{s,k}^{(r)}(l)}x_s^{(r)}(n-l)\right|^2/\sigma_k^2\right)\right\}}{\sum_{j=1}^{M}\exp\left\{-\sum_{k=1}^{K}\left(\left|y_k(n) - \sum_{l=0}^{L-1} a_{s,k}^{(r)}(l)e^{j\phi_{s,k}^{(r)}(l)}x_s^{(r)}(n-l)\right|^2/\sigma_k^2\right)\right\}} \end{aligned} \tag{5.11}$$

where the last equality is computed with the assumption that each symbol has equal prior probability, i.e., $p\left(x_s(n) = X_{s,i}|\boldsymbol{\theta}_s^{(r)}\right) = (1/M)$, $y_k(n)$ is the nth received symbol at discrete time nT, $X_{s,i}$ is the ith constellation point for modulation format s. By obtaining the posterior probability $\rho_{s,i}^{(r)}(n)$, we can compute $x_s^{(r)}(n)$ as

$$x_s^{(r)}(n) = \sum_{i=1}^{M} \rho_{s,i}^{(r)}(n) X_{s,i}. \tag{5.12}$$

We define $\bar{z}_{kl}(t) = \sum_n a_k(l) e^{j\phi_k(l)} x_s(n) g(t - nT - lT)$. By computing (5.12), $x_s(n)$ turns to a deterministic symbol. Furthermore, since flat fading is assumed for each path, the channel amplitude $a_k(l)$ and phase $\phi_k(l)$ are treated as unknown deterministic parameters. Thus, we have that $\bar{z}_{kl}(t)$ is an unknown deterministic signal. Note that $w_{kl}(t)$ is a zero-mean white Gaussian noise process, $\ln p(\mathbf{z}_k|\boldsymbol{\theta}_k)$ is then given by [9]:

$$\ln p(\mathbf{z}_k|\boldsymbol{\theta}_k) = C_1 - \sum_{l=0}^{L-1} \int_0^{T_0} \frac{1}{\sigma_{kl}^2} |z_{kl}(t) - \bar{z}_{kl}(t)|^2 \, \mathrm{d}t \tag{5.13}$$

where C_1 is a term irrelevant to the unknowns. Taking the conditional expectation of (5.13) given $\boldsymbol{y}_k$ and $\boldsymbol{\theta}_{s,k}^{(r)}$, $J_k(\boldsymbol{\theta}|\boldsymbol{\theta}_{s,k}^{(r)})$ is obtained by [10]:

$$J_k(\boldsymbol{\theta}|\boldsymbol{\theta}_{s,k}^{(r)}) = C_2 - \sum_{l=0}^{L-1} \int_0^{T_0} \frac{1}{\sigma_{kl}^2} \left|\hat{z}_{s,kl}^{(r)}(t) - \bar{z}_{s,kl}^{(r)}(t)\right|^2 \mathrm{d}t \tag{5.14}$$

where C_2 is a term independent of the unknowns, and

$$\hat{z}_{s,kl}^{(r)}(t) = \bar{z}_{s,kl}^{(r)}(t) + \beta_{kl} \left(y_k(t) - \sum_{l=0}^{L-1} \bar{z}_{s,kl}^{(r)}(t) \right). \tag{5.15}$$

It is noted from (5.14) that the maximization of $J_k(\boldsymbol{\theta}|\boldsymbol{\theta}_{s,k}^{(r)})$ with respect to $\boldsymbol{\theta}$ is equivalent to the minimization of each of the L summations. Hence, the E-step and M-step in (5.10) and (5.28) are respectively simplified as

E-step: for $l = 0, \ldots, L-1$ compute

$$\hat{z}_{s,kl}^{(r)}(t) = \bar{z}_{s,kl}^{(r)}(t) + \beta_{kl} \left(y_k(t) - \sum_{l=0}^{L-1} \bar{z}_{s,kl}^{(r)}(t) \right) \tag{5.16}$$

M-step: for $l = 0, \ldots, L-1$ compute:

$$\boldsymbol{\theta}_{s,k}^{(r+1)}(l) = \arg\min_{\boldsymbol{\theta}_k(l)} \int_0^{T_0} \frac{\left(\hat{z}_{s,kl}^{(r)}(t) - \bar{z}_{s,kl}^{(r)}(t)\right)^2 \mathrm{d}t}{\sigma_{kl}^2}. \tag{5.17}$$

By taking the derivative of (5.17) with respect to $a_{s,k}(l)$ and setting it to zero, we can obtain that

$$a_{s,k}^{(r+1)}(l) = \frac{1}{E^{(r)}} \sum_{n=1}^{N} \Re \left\{ x_s^{(r)}(n)^* e^{-j\phi_{s,k}^{(r)}(l)} \int_0^{T_0} \hat{z}_{s,kl}^{(r)}(t) g^*(t - nT - lT) \mathrm{d}t \right\} \tag{5.18}$$

where $E^{(r)} = E_g \sum_{n=1}^{N} |x_s^{(r)}(n)|^2$, with $E_g = \int_{-\infty}^{\infty} g^2(t)\mathrm{d}t$ as the pulse energy, $\Re\{\cdot\}$ represents the real component of a complex variable, and $(\cdot)^*$ denotes the conjugation of a variable. Apparently, the second derivative of (5.17) with respect to $a_{s,k}(l)$ is a negative definite matrix, which implies that (5.18) is the optimal estimate of $a_{s,k}(l)$. By substituting (5.18) into (5.17), with the assumption that $E^{(r)}$ is independent of $\phi_{s,k}^{(r)}(l)$, the M-step in (5.17) is rewritten as

M-step: for $l = 0, \ldots, L-1$ compute

$$\phi_{s,k}^{(r+1)}(l) = \tan^{-1} \frac{\Im\left\{\boldsymbol{x}_{s,l}^{(r)\mathrm{H}} \hat{\boldsymbol{z}}_{s,kl}^{(r)}\right\}}{\Re\left\{\boldsymbol{x}_{s,l}^{(r)\mathrm{H}} \hat{\boldsymbol{z}}_{s,kl}^{(r)}\right\}} \tag{5.19}$$

$$a_{s,k}^{(r+1)}(l) = \frac{1}{E^{(r)}} \sum_{n=1}^{N} \Re\left\{x_s^{(r)}(n)^* e^{-j\phi_{s,k}^{(r+1)}(l)} \times \int_0^{T_0} \hat{z}_{s,kl}^{(r)}(t) g^*(t - nT - lT)\mathrm{d}t\right\} \tag{5.20}$$

where $\boldsymbol{x}_{s,l}^{(r)} = \left[\mathbf{0}_l^{\mathrm{T}}, x_s^{(r)}(1), \ldots, x_s^{(r)}(N-l)\right]^{\mathrm{T}}$, with $\mathbf{0}_l$ as a $l \times 1$ vector with all elements equal zero, $\hat{\boldsymbol{z}}_{s,kl}^{(r)} = \left[\hat{z}_{s,kl}^{(r)}(1), \ldots, \hat{z}_{s,kl}^{(r)}(N)\right]^{\mathrm{T}}$, $\Im\{\cdot\}$ represents the imaginary component of a complex variable, and $(\cdot)^{\mathrm{H}}$ denotes the conjugate transpose of a vector/matrix.

It should be noted from (5.16), (5.19), and (5.20) that by employing the EM algorithm and properly designing the complete data, the multivariate optimization problem in (5.26) is successfully decomposed into L separate ones, where only one unknown parameter is optimized at each step, solving the original high-dimensional and non-convex problem in a tractable way.

Fourth-order moment-based initialization

The most prominent problem for the EM algorithm is how to set proper initialization points of the unknowns, from which the EM algorithm takes iterative steps to converge to some stationary points. Since the EM algorithm has no guarantee of the convergence to the global maxima, poor initializations enhance its probability to converge to the local maxima. In general, the most commonly adopted initialization schemes for the EM algorithm include the simulated annealing (SA) [11] and random restart. However, since our problem considers multipath channels and multiple users, where a $(2 \times K \times L)$-dimensional initial value should be selected, it is computationally expensive for the SA and random restart algorithms to find proper initials.

In this section, we employ a simple though effective method to find the initial values of the unknown fadings. A modified version of the fourth-order moment-based estimator proposed in [12] is applied to roughly estimate the multipath channels, which are then used as the initialization points of the EM algorithm. The estimator is expressed as

$$\hat{h}_k(l) = \frac{m_4^{y_k}(p,p,p,l)}{m_4^{y_k}(p,p,p,p)} \tag{5.21}$$

where $m_4^{y_k}(\tau_1, \tau_2, \tau_3, \tau_4) = \mathbb{E}\{y_k(n+\tau_1)y_k(n+\tau_2)y_k(n+\tau_3)y_k(n+\tau_4)\}$ is the fourth-order moment of $y_k(n)$, and $h_k(p)$ denotes the coefficient of the dominant path of the channel between the transmitter and kth receiver. Without loss of generality, the dominant path is assumed to be the leading path, i.e., $p = 0$.

The overall modulation classification algorithm is summarized as follows:

EM-based modulation classifier

1. Set the stopping threshold Λ and the maximum number of iterations I;
2. FOR $s = 1, \ldots, S$;
3. Set $r = 0$;
4. Initialize the unknown parameters $\boldsymbol{\theta}_{s,k}^{(0)}(l)$ using the fourth-order moment-based estimator;
5. For $n = 1, \ldots, N$, compute $x_s^{(r)}(n)$ according to (5.12);
6. Compute the likelihood function $L_s(\boldsymbol{\theta}_s^{(r)})$ according to (5.4);
7. Set $r = r + 1$;
8. Perform over (5.16), (5.19) and (5.20) to estimate $\phi_{s,k}^{(r+1)}(l)$ and $a_{s,k}^{(r+1)}(l)$;
9. Compute the likelihood function $L_s(\boldsymbol{\theta}_s^{(r+1)})$ with the new estimates;
10. If $\left|\left(L_s(\boldsymbol{\theta}_s^{(r+1)}) - L_s(\boldsymbol{\theta}_s^{(r)})\right)/L_s(\boldsymbol{\theta}_s^{(r)})\right| > \Lambda$ or $r \leq I$, go to Step 5; otherwise, set $\boldsymbol{\theta}_s^{(*)} = \boldsymbol{\theta}_s^{(r+1)}$, and continue;
11. ENDFOR
12. Final decision is made by $\hat{s} = \arg\max_s L_s(\boldsymbol{\theta}_s^{(*)})$.

5.2.1.3 Numerical results

In this section, various numerical experiments are provided to examine the classification performance of the proposed algorithm, using the probability of correct classification P_c as a performance metric. The number of paths is set to $L = 6$, where without loss of generality, the coefficient of the leading path $h_k(0) = a_k(0)e^{j\phi_k(0)}$ is set to 1, and the remaining channel coefficients follow zero-mean complex Gaussian distribution with parameter $\varsigma_k^2 = 0.1$. The number of receivers is set to $K = 3$, and we assume that the noise power at all receivers is the same. The number of samples per receiver is set to $N = 500$. For the EM algorithm, we set the stopping threshold $\Lambda = 10^{-3}$ and the maximum number of iterations $I = 100$.

Since the convergence of the EM algorithm is highly dependent on the initializations of the unknowns, we first evaluate the impact of the initial values of the unknowns on the classification performance of the proposed algorithm, which are formulated as the true values plus bias. Let $\phi_k(l)$ and $a_k(l)$ denote the true values, respectively, and $\delta_{\phi_k(l)}$ and $\delta_{a_k(l)}$ denote the maximum errors for each unknown parameter. We randomly take the initial values of the unknown channel phase and amplitude within $[\phi_k(l) - \delta_{\phi_k(l)}, \phi_k(l) + \delta_{\phi_k(l)}]$ and $[0, a_k(l) + \delta_{a_k(l)}]$, respectively [13].

Figure 5.2(a) and (b) shows the classification performance of the proposed algorithm for QAM and PSK, respectively, with curves parameterized by different initialization points of the unknowns. For QAM modulations, the candidate set is {4, 16, 64}-QAM, and for PSK modulations, it is {QPSK, 8-PSK, 16-PSK}.

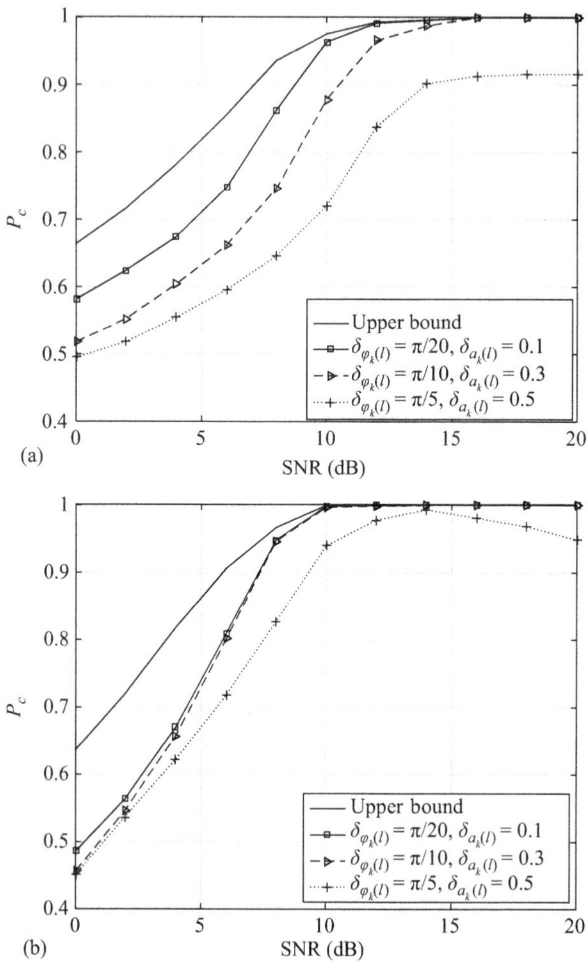

Figure 5.2 (a) Impact of the initial values of unknowns on the proposed algorithm for QAM and (b) impact of the initial values of unknowns on the proposed algorithm for PSK

Three sets of maximum errors are examined, namely, $(\delta_{\phi_k(l)} = \pi/20, \delta_{a_k(l)} = 0.1)$, $(\delta_{\phi_k(l)} = \pi/10, \delta_{a_k(l)} = 0.3)$, and $(\delta_{\phi_k(l)} = \pi/5, \delta_{a_k(l)} = 0.5)$. Moreover, the classification performance is compared to the performance upper bound, which is obtained by using the Cramér–Rao lower bounds of the estimates of the unknowns as the variances. It is apparent that the classification performance decreases with the increase of the maximum errors. To be specific, for both QAM and PSK modulations, the classification performance is not sensitive to the initial values for the first two sets with smaller biases, especially for the PSK modulation, where the classification performance is more robust against smaller initialization errors, while for the third set with larger bias, the classification performance degrades, especially in the low signal to noise ratio (SNR) region. In our problem, we consider a complicated case with multiple receivers in the presence of multipath channels; therefore, the likelihood function contains large amounts of local extrema. Then, the EM algorithm easier converges to local extrema when the initial values are far away from the true values. In addition, we can see from Figure 5.2(a) and (b) that, in the high SNR region, the classification performance with smaller maximum errors is close to the upper bounds. It indicates that with proper initialization points, the proposed algorithm can provide promising performance.

Next, we consider the classification performance of the proposed algorithm using the fourth-order moment-based initialization scheme, as shown in Figure 5.3(a) and (b) for QAM and PSK modulations, respectively. Figure 5.3(a) depicts that the classification performance of the proposed algorithm for QAM modulations using the fourth-order moment-based initialization method attains $P_c \geq 0.8$ for SNR $>$ 10 dB. When compared to that taking the true values plus bias as the initial values, we can see that the classification performance of the proposed algorithm is comparable in the SNR region ranges from 6 to 10 dB. The results show that the fourth-order moment-based method is feasible in the moderate SNR region. In addition, we also compare the classification performance of the proposed algorithm with the cumulant-based methods in [14,15]. The number of samples per receiver for the cumulant-based methods is set to $N_c = 2,000$. Note that the difference of cumulant values between higher order modulation formats (e.g., 16-QAM and 64-QAM) is small; therefore, the classification performance of the cumulant-based approaches is limited, which saturates in the high SNR region. It is apparent that the classification performance of the proposed algorithm outperforms that of the cumulant-based ones. Meanwhile, it indicates that the propose algorithm is more sample efficient than the cumulant-based ones. Similar results can be seen from Figure 5.3(b) when classifying PSK modulations. The advantage in the probability of correct classification of the proposed algorithm is obvious when compared to that of cumulant-based methods.

On the other hand, however, we can see from Figure 5.3(a) and (b) that the classification performance of the proposed algorithm decreases in the high SNR region. The possible reason is that the likelihood function in the low SNR region is dominated by the noise, which contains less local extrema and is not sensitive to the initialization errors. In contrast, the likelihood function in the high SNR region is dominated by the signal, which contains more local extrema. In such a case, the convergence result is more likely to be trapped at the local extrema, even when the initial values are slightly far away from the true values. In addition, when comparing Figure 5.3(a) with (b), it is noted that using the moment-based estimator, the PSK modulations are more sensitive to initialization errors in the high SNR region.

To intuitively demonstrate the impact of the noise decomposition factor β_{kl} on the classification performance, we evaluate the probability of correct classification versus the SNR in Figure 5.4, with curves parameterized by different choices of β_{kl}. The lines illustrate the classification performance with fixed $\beta_{kl} = 1/L$, and the markers show that with random β_{kl}. As we can see, different choices of the noise decomposition factor β_{kl} does not affect the classification performance of the proposed algorithm.

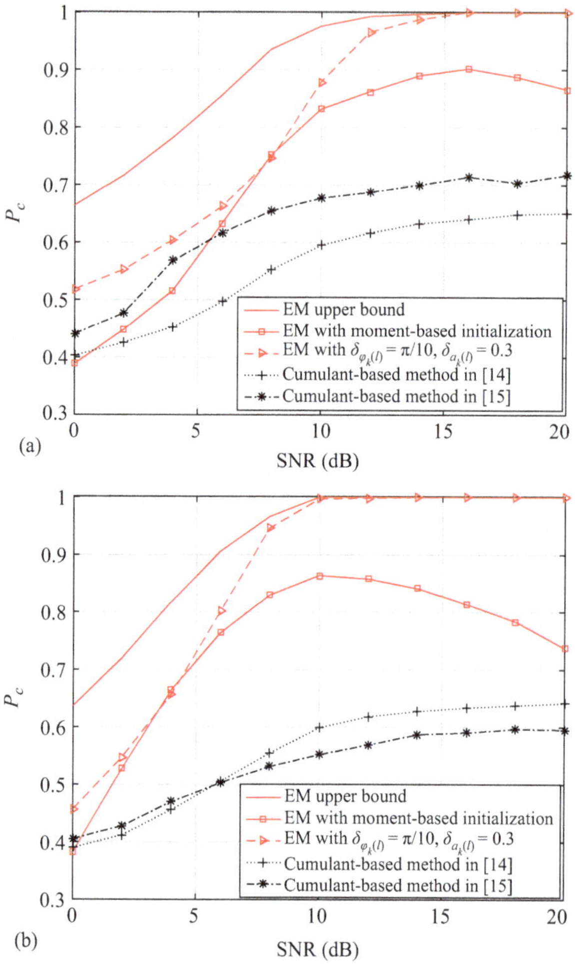

Figure 5.3 (a) The classification performance of the proposed algorithm for QAM with the fourth-order moment-based initialization scheme and (b) the classification performance of the proposed algorithm for PSK with the fourth-order moment-based initialization scheme

5.2.2 Continuous phase modulation classification in fading channels via Baum–Welch algorithm

In contrast to constellation-based modulation formats (e.g., QAM and PSK) in the previous section, CPM yields high spectral and power efficiency, which are widely adopted in wireless communications, such as the satellite communications. Most of the existing literature that classifies CPM signals is merely investigated in additive white Gaussian noise (AWGN) channels [16] [17], while the effective algorithms in the presence of the unknown fading channels have not been adequately studied yet.

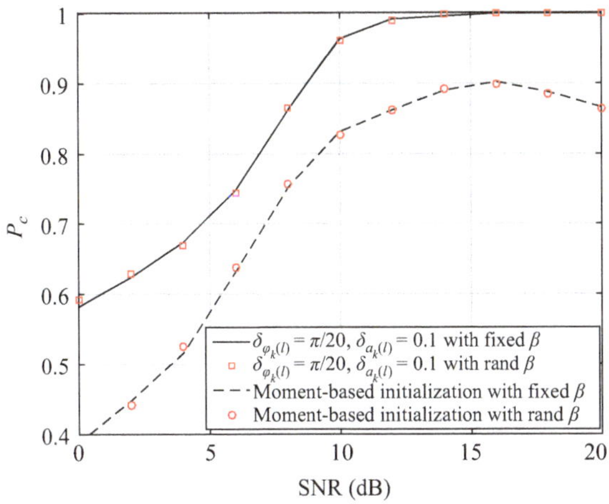

Figure 5.4 The classification performance of the proposed algorithm for QAM with curves parameterized by different choices of β_{kl}

In this section, we consider a classification problem of CPM signals under unknown fading channels, where the CPM signal is formulated as an HMM corresponding to the memorable property of the continuous phases. A likelihood-based classifier is proposed using the BW algorithm, which obtains the MLEs of the unknown parameters of the HMM based on the EM algorithm.

5.2.2.1 Problem statement

Consider a typical centralized cooperation system with one transmitter and K receivers, and a fusion center is introduced to fuse data from the K receivers to enhance the classification performance. At the kth receiver, the discrete-time transmission model is given by

$$y_{k,n} = g_k x_n + w_{k,n}, \quad n = 1, \dots, N, \quad k = 1, \dots, K \tag{5.22}$$

where g_k is the unknown complex channel fading coefficient from the transmitter to the kth receiver, $w_{k,n}$ is circularly symmetric complex Gaussian with the distribution $\mathcal{CN}(0, \sigma_k^2)$, and x_n is the transmit CPM signal. The continuous-time complex CPM signal is expressed as

$$x(t) = \sqrt{\frac{2E}{T}} e^{j\Theta(t;I)} \tag{5.23}$$

where E is the energy per symbol, T is the symbol duration, and $\Theta(t; I)$ is the time-varying phase. For $t \in [nT, (n+1)T]$, the time-varying phase is represented as

$$\Theta(t; I) = \pi h \sum_{l=-\infty}^{n-L} I_l + 2\pi h \sum_{l=n-L+1}^{n} I_l q(t - lT) \tag{5.24}$$

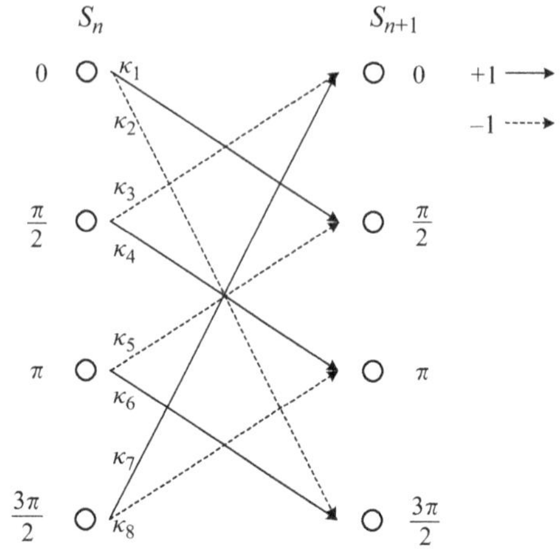

Figure 5.5 A trellis of state transition for CPM with parameter $\{M = 2, h = (1/2), L = 1\}$. In this example, the number of states is $Q_0 = 4$, and the information data is drawn from $\{+1, -1\}$. The transition from state S_n to S_{n+1}, denoted as $\kappa_q, q = 1, \ldots, 8$, is represented by solid line, if +1 is transmitted, and dotted line, if −1 is transmitted

where h is the modulation index, I_l is the lth information symbol drawn from the set $\{\pm 1, \ldots, \pm(M-1)\}$, with M as the symbol level, $q(t)$ is the integral of the pulse shape $u(t)$, i.e., $q(t) = \int_0^t u(\tau)d\tau,\ t \leq LT$, and L is the pulse length. From (5.24), we can see that a CPM format is determined by a set of parameters, denoted as $\{M, h, L, u(t)\}$. Basically, by setting different values of these parameters, infinite number of CPM formats can be generated. Let $S_n = \{\theta_n, I_{n-1}, \ldots, I_{n-L+1}\}$ be the state of the CPM signal at $t = nT$, where $\theta_n = \pi h \sum_{l=-\infty}^{n-L} I_l$. The modulation index h is a rational number, which can be represented by $h = (h_1/h_2)$, with h_1 and h_2 as coprime numbers. Then, we define h_0 as the number of states of θ_n, which is given by

$$h_0 = \begin{cases} h_2, & \text{if } h_1 \text{ is even} \\ 2h_2, & \text{if } h_1 \text{ is odd.} \end{cases}$$

Hence, we can obtain that the number of states of S_n is $Q_0 = h_0 M^{L-1}$. A trellis of state transition for CPM with parameter $\{M = 2, h = (1/2), L = 1\}$ is shown in Figure 5.5.

Let $\mathcal{S}$ be the set of CPM candidates, which is known at the receivers. The classification task is to identify the correct CPM format $s \in \{1, \ldots, \mathcal{S}\}$ based on the received signals. Let $\boldsymbol{y} = [\boldsymbol{y}_1, \ldots, \boldsymbol{y}_K]$, where $\boldsymbol{y}_k$ is the received signal at the kth receiver, $\boldsymbol{g} = \{g_k\}_{k=1}^K$, and $\boldsymbol{x} = \{x_n\}_{n=1}^N$, which are the observations, unknown parameters, and hidden variables of the HMM, respectively. We formulate this problem as a multiple composite hypothesis testing problem, and the likelihood-based classifier is adopted to solve it. For hypothesis $\mathcal{H}_s$, meaning that the transmit signal uses the CPM format s, the log-likelihood function $\ln p_s(\boldsymbol{y}|\boldsymbol{g})$ is computed. The classifier makes the final decision on the modulation format by

$$\hat{s} = \arg\max_{s \in S} \ln p_s(\boldsymbol{y}|\boldsymbol{g}_s^{(\dagger)}) \tag{5.25}$$

where $\boldsymbol{g}_s^{(\dagger)}$ is the MLEs of the unknown parameters $\boldsymbol{g}$ under hypothesis $\mathcal{H}_s$, which can be computed by

$$\boldsymbol{g}_s^{(\dagger)} = \arg\max_{\boldsymbol{g}} \ln p_s(\boldsymbol{y}|\boldsymbol{g}). \tag{5.26}$$

Unlike the case for constellation-based modulation formats, where the likelihood function $p_s(\boldsymbol{y}|\boldsymbol{g})$ is obtained by averaging over all the unknown constellation symbols $\mathcal{A}$, i.e., $p_s(\boldsymbol{y}|\boldsymbol{g}) = \sum_{\boldsymbol{x} \in \mathcal{A}} p_s(\boldsymbol{y}|\boldsymbol{x}, \boldsymbol{g}) p_s(\boldsymbol{x}|\boldsymbol{g})$, since the CPM signal is time correlated, its likelihood function cannot be calculated in such a way.

5.2.2.2 Classification of CPM via BW

In this section, the CPM signal is first formulated as an HMM regarding to its memorable property, a likelihood-based classifier is then proposed based on the BW algorithm, which utilizes the EM algorithm to calculate the MLEs of the unknown parameters in the HMM [18].

HMM description for CPM signals

According to the phase memory of the CPM signal, it can be developed as an HMM. We parameterize the HMM by $\lambda = (\boldsymbol{A}, \boldsymbol{B}, \boldsymbol{\pi})$, which are defined as follows

1. $\boldsymbol{A}$ denotes the state probability matrix, the element $\alpha_{ij} = \Pr\{S_{n+1} = j | S_n = i\}$ is expressed as

$$\alpha_{ij} = \begin{cases} \frac{1}{M}, & \text{if } i \to j \text{ is permissible} \\ 0, & \text{otherwise.} \end{cases}$$

2. $\boldsymbol{B}$ represents the conditional probability density function vector, the element $\beta_i(y_{k,n}) = p(y_{k,n}|S_n = i, g_k)$ is given by

$$\beta_i(y_{k,n}) = \frac{1}{\pi \sigma_k^2} \exp\left\{ -\frac{|y_{k,n} - g_k x(S_n = i)|^2}{\sigma_k^2} \right\}$$

with $x(S_n = i)$ as the transmit CPM signal at $t = nT$ corresponding to the state S_n.
3. $\boldsymbol{\pi}$ is the initial state probability vector, the element is defined as

$$\pi_i = \Pr\{S_1 = i\} = \frac{1}{Q_0}.$$

BW-based modulation classifier

The BW algorithm provides a way to compute the MLEs of the unknowns in the HMM by using the EM algorithm. To be specific, under the hypothesis $\mathcal{H}_s$, the E-step and M-step at iteration r are written as

$$\text{E-step:} \quad J(\boldsymbol{g}_s^{(r)}, \boldsymbol{g}) = \sum_{\boldsymbol{x}} p_s(\boldsymbol{y}, \boldsymbol{x}|\boldsymbol{g}_s^{(r)}) \log p_s(\boldsymbol{y}, \boldsymbol{x}|\boldsymbol{g}) \tag{5.27}$$

$$\text{M-step:} \quad \boldsymbol{g}_s^{(r+1)} = \arg\max_{\boldsymbol{g}} J(\boldsymbol{g}_s^{(r)}, \boldsymbol{g}). \tag{5.28}$$

As shown in Figure 5.5, we denote κ_q as the transition from state S_n to S_{n+1}, where κ_q, $q = 1, \ldots, Q$ is drawn from the information sequence, with $Q = Q_0 M$. Let $\ell(x_n, \kappa_q)$ denote the transmit signal at $t = nT$ that corresponding to κ_q. To simplify the notifications, denote $\boldsymbol{x}_{1:n-1}$ and $\boldsymbol{x}_{n+1:N}$ as the transmit symbol sequences $\{x_1, \ldots, x_{n-1}\}$ and $\{x_{n+1}, \ldots, x_N\}$, respectively; we then can rewrite (5.27) as

$$\begin{aligned}
J(\boldsymbol{g}_s^{(r)}, \boldsymbol{g}) &= \sum_{\boldsymbol{x}} p_s(\boldsymbol{y}, \boldsymbol{x}|\boldsymbol{g}_s^{(r)}) \left(\log p_s(\boldsymbol{x}|\boldsymbol{g}) + \log p_s(\boldsymbol{y}|\boldsymbol{x}, \boldsymbol{g})\right) \\
&= \Lambda_1 + \sum_{k=1}^{K} \sum_{n=1}^{N} \log p_s(y_{k,n}|x_n, g_k) \sum_{\boldsymbol{x}} p_s(\boldsymbol{y}, \boldsymbol{x}|\boldsymbol{g}_s^{(r)}) \\
&= \Lambda_1 + \sum_{k=1}^{K} \sum_{n=1}^{N} \sum_{q=1}^{Q} \log p_s(y_{k,n}|\ell(x_n, \kappa_q), g_k) \\
&\quad \times \sum_{\boldsymbol{x} \backslash x_n} p_s(\boldsymbol{y}, \boldsymbol{x}_{1:n-1}, \ell(x_n, \kappa_q), \boldsymbol{x}_{n+1:N}|\boldsymbol{g}_s^{(r)}) \\
&= \Lambda_2 - \sum_{k=1}^{K} \sum_{n=1}^{N} \sum_{q=1}^{Q} p_s(\boldsymbol{y}, \ell(x_n, \kappa_q)|\boldsymbol{g}_s^{(r)}) \\
&\quad \times \frac{1}{\sigma_k^2} |y_{k,n} - g_k x(S_n = z(q, Q_0))|^2
\end{aligned} \tag{5.29}$$

where $z(q, Q_0)$ defines the remainder of (q/Q_0).

Forward–backward algorithm: Define $\eta_s(n, q) = p_s(\boldsymbol{y}, \ell(x_n, \kappa_q)|\boldsymbol{g}_s^{(r)})$. Note that $\ell(x_n, \kappa_q)$ is equivalent to the event $\{S_n = i, S_{n+1} = j\}$, we can then derive $\eta_s(n, q)$ as

$$\begin{aligned}
\eta_s(n, q) &= \prod_{k=1}^{K} p_s(\boldsymbol{y}_k, S_n = i, S_{n+1} = j|g_k) \\
&= \prod_{k=1}^{K} p_s(\boldsymbol{y}_k|S_n = i, S_{n+1} = j, g_k) p_s(S_n = i|g_k) \\
&\quad \times p_s(S_{n+1} = j|S_n = i, g_k) \\
&= \prod_{k=1}^{K} \upsilon_{k,n}(i) \omega_{k,n+1}(j) p_s(S_{n+1} = j|S_n = i, g_k)
\end{aligned} \tag{5.30}$$

where $\upsilon_{k,n}(i) = p_s(\mathbf{y}_{k,1:n}|S_n = i, g_k)p_s(S_n = i|g_k)$ and $\omega_{k,n+1}(j) = p_s(\mathbf{y}^N_{k,n+1}|S_{n+1} = j, g_k)$ are the forward and backward variables, respectively, which can be inductively obtained by performing the forward–backward procedure as follows:

- Compute forward variable $\upsilon_{k,n}(i)$
  Initialize: $\upsilon_{k,1}(i) = \pi_i$
  Induction: $\upsilon_{k,n}(i) = \sum_{j=1}^{Q} \upsilon_{k,n-1}(j)\alpha_{ji}\beta_i(y_{k,n})$
- Compute backward variable $\omega_{k,n+1}(j)$
  Initialize: $\omega_{k,N+1}(j) = 1$
  Induction: $\omega_{k,n}(j) = \sum_{i=1}^{Q} \omega_{k,n+1}(i)\alpha_{ij}\beta_i(y_{k,n})$.

Finally, by taking the derivative of (5.29) with respect to $\mathbf{g}$ and setting it to zero, the unknown channel fading can be estimated by

$$g_{s,k}^{(r+1)} = \frac{\sum_{n=1}^{N}\sum_{q=1}^{Q}\eta_s(n,q)y_{k,n}x(S_n = z(q,Q_0))^*}{\sum_{n=1}^{N}\sum_{q=1}^{Q}\eta_s(n,q)\|x(S_n = z(q,Q_0))\|}. \tag{5.31}$$

The proposed BW-based modulation classifier is summarized in the following box:

BW-based modulation classifier for CPM

1. Set stopping threshold ε and maximum number of iterations N_m;
2. FOR $s = 1, \ldots, |S|$;
3. Set $r = 0$;
4. Initialize the unknown parameters $g_{s,k}^{(0)}$;
5. For $n = 1, \ldots, N$, $q = 1, \ldots, Q$, perform the forward–backward procedure to compute $\eta_s(n,q)$ in (5.30);
6. Compute $J(\mathbf{g}_s^{(r)}, \mathbf{g})$ according to (5.29);
7. Set $r = r + 1$;
8. Compute $g_{s,k}^{(r+1)}$ according to (5.31);
9. Compute $J(\mathbf{g}_s^{(r+1)}, \mathbf{g})$ using the new estimates;
10. If $\left|(J(\mathbf{g}_s^{(r+1)}, \mathbf{g}) - J(\mathbf{g}_s^{(r)}, \mathbf{g}))/J(\mathbf{g}_s^{(r)}, \mathbf{g})\right| > \varepsilon$ or $r \leq N_m$, go to Step 5; otherwise, set $\mathbf{g}_s^{(\mathrm{opt})} = \mathbf{g}_s^{(r+1)}$, and continue;
11. ENDFOR
12. Final decision is made by $\hat{s} = \arg\max_s J(\mathbf{g}_s^{(\mathrm{opt})}, \mathbf{g})$.

5.2.2.3 Numerical results

In this section, the classification performance of the proposed algorithm is examined through various numerical simulations. The probability of correct classification P_c is adopted as the measure metric. To comprehensively evaluate the proposed algorithm, two experiments with different CPM candidate sets are considered and is compared to the conventional approximate entropy (ApEn)-based method in [17]:

Experiment 1: Denote the CPM parameters as $\{M, h, L, u(t)\}$, four CPM candidates are considered, namely, {2, 1/2, 1, rectangular pulse shape}, {4, 1/2, 2, raised cosine pulse shape}, {2, 3/4, 3, Gaussian pulse shape}, and {4, 3/4, 3, Gaussian pulse shape}. For the Gaussian pulse shape, the bandwidth-time product is set to $B = 0.3$.

Experiment 2: Consider the case where various values of M, h, and L are set, with $M = \{2, 4\}$, $h = \{1/2, 3/4\}$, $L = \{1, 2, 3\}$. The pulse shapes for $L = 1, 2, 3$ are set to rectangular, raised cosine, and Gaussian, respectively. In such a case, 12 different CPM candidates are considered.

For the proposed algorithm, we assume that the number of symbols per receiver is $N = 100$, the stopping threshold is $\varepsilon = 10^{-3}$, and maximum iterations is $N_m = 100$. For the ApEn-based algorithm, the simulation parameters are set according to [17]. Without loss of generality, we assume that the noise power at all receivers is the same.

Comparison with approximate entropy-based approach

We first consider the scenario with one receiver in the presence of AWGN and fading channels. Figure 5.6 evaluates the classification performance of the proposed algorithm when considering experiment 1 and is compared to that of the ApEn-based algorithm in [17] as well. From Figure 5.6, we can see that in AWGN channels, the proposed algorithm attains an acceptable classification performance, i.e., a classification probability of 0.8, when SNR = 5 dB, and it achieves an error-free classification performance at SNR = 10 dB.

Furthermore, we consider the classification performance in the presence of fading channels. We first initiate the unknown fading channel coefficients with the true values with bias for the proposed algorithm [13]. Let a_k and ϕ_k denote the true values of the magnitude and phase of the fading channel, respectively, and let Δa_k and $\Delta\phi_k$ denote the maximum errors of the magnitude and phase, respectively. The initial values of the unknown magnitude and phase are arbitrarily chosen within $[0, a_k + \Delta a_k]$

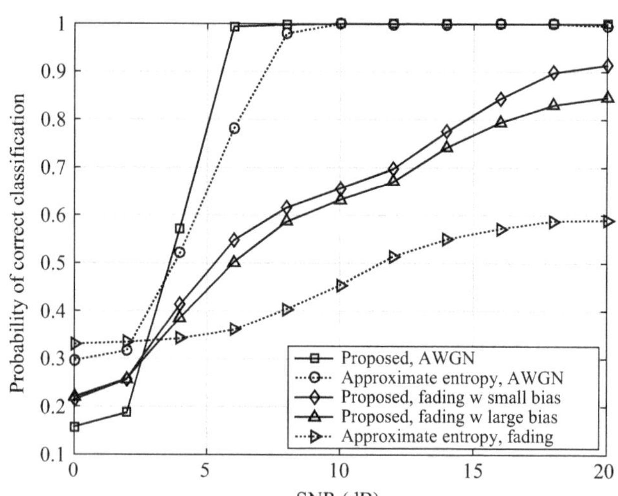

Figure 5.6 The classification performance of the proposed algorithm under AWGN and fading channels

and $[\phi_k - \Delta\phi_k, \phi_k + \Delta\phi_k]$, respectively. Two sets of initials of the unknowns are evaluated, whose maximum errors are set to $(\Delta a_k, \Delta\phi_k) = (0.1, \pi/20)$ and $(0.3, \pi/10)$, respectively. It is noted that the classification performance of the proposed algorithm outcomes that of the ApEn-based algorithm in the fading channels. In particular, the proposed algorithm provides classification probability of 0.8 for SNR > 15 dB, while the probability of correct classification of the ApEn-based algorithm saturates around 0.6 in the high SNR region, which is invalid for the classification in fading channels.

Impact of initialization of unknowns

Note that the BW algorithm uses the EM algorithm to estimate the unknowns; therefore, its estimation accuracy highly relies on the initial values of the unknowns. In Figure 5.7, we examine the impact of the initializations of the unknowns on the classification performance of the proposed algorithm. Both experiments 1 and 2 are evaluated. We consider multiple receivers to enhance the classification performance, and the number of receivers is set to $K = 3$. In such a case, the unknown fadings are first estimated at each receiver independently, the estimations are then forwarded to a fusion center to make the final decision. The initial values of the unknown parameters are set as the true values with bias, as previously described. It is seen from Figure 5.7 that, with smaller bias, the classification performance of the proposed algorithm provides promising classification performance. For experiment 1, the proposed classifier achieves $P_c > 80\%$ when SNR > 10 dB, and for experiment 2, $P_c > 80\%$ is obtained when SNR = 14 dB. Apparently, the cooperative classification further enhances the classification performance when compared to that with a single receiver. Furthermore, with large bias, it is noticed that the classification performance of the proposed algorithm degrades in the high SNR region. This phenomenon occurs when using the EM algorithm [19]. The main reason is that the estimation results of the EM are not

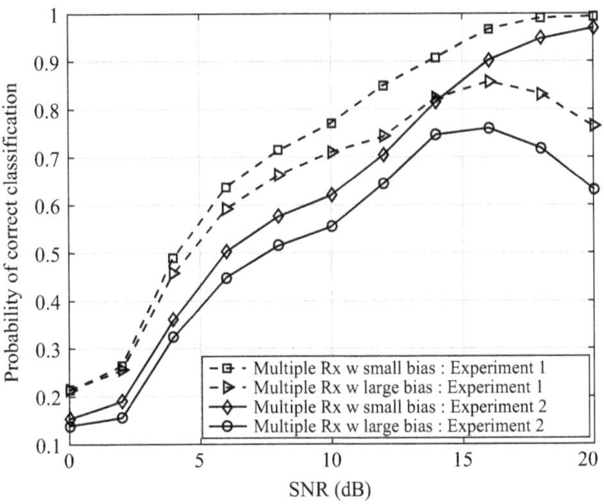

Figure 5.7 The impact of different initial values of the unknowns on the classification performance under fading channels

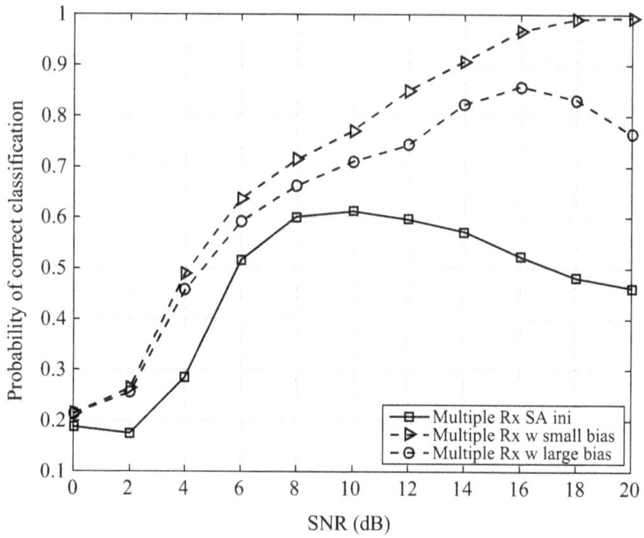

Figure 5.8 The classification performance under fading channels with simulated annealing initialization method

guaranteed to converge at the global maxima and are sensitive to the initialization points of the unknowns. In the high SNR region, the likelihood function is dominated by the signal and it has more local maxima than that in the low SNR region. Thus, with large bias, the proposed scheme is more likely to converge at a local maxima, which causes the degradation of the classification performance.

Performance with simulated annealing initialization
Next, we evaluate the classification performance of the proposed algorithm with the SA initialization method, as illustrated in Figure 5.8. The parameters of the SA method are set as in [13]. Experiment 1 is considered. Figure 5.8 shows that, using the SA method to generate the initial values of the unknowns, the classification performance of the proposed algorithm monotonically increases in the low-to-moderate SNR region (0–10 dB). It implies that the SA scheme can provide appropriate initializations for the proposed algorithm. Apparently, a gap is noticed between the classification performance with the SA scheme and that with the true values of the unknowns plus bias. However, note that how to determine proper initials could be an interesting topic for future research, which is out of the scope of this chapter.

5.3 Specific emitter identification via machine learning

SEI, motivated by the identification and tracking task of unique emitters of interest in military communications, is a technique to discriminate individual emitters by extracting the identification feature from the received signal and comparing it with a categorized feature set [20]. As a key technology in military communications, the SEI has been intensively investigated over the past few decades. More recently, with the advent of CRs and adaptive techniques, it has become increasingly important for commercial applications.

In general, the SEI technique is based on the machine-learning theory, where a simplified SEI system includes three main parts, namely, signal preprocessing, feature extraction, and identification classifier. In the feature-extraction subsystem, identification features are extracted from the received signal, and then the features are input to the identification classifier to determine the class of the emitters. Hence, the most important and challenging part of SEI based on machine learning is to design proper and robust identification features, which have strong separation capability among different emitters and are robust against the real-world scenarios.

In this section, we introduce three SEI algorithms, which extract identification features via adaptive time–frequency analysis and implement classification task using the support vector machine (SVM). We investigate the SEI approaches under realistic scenarios. The three SEI algorithms are applicable to both single-hop and relaying systems and are robust against several nonideal channel conditions, such as the non-Gaussian and fading channels.

5.3.1 System model

5.3.1.1 Single-hop scenario

A time-division multiple-access communication system is considered with K emitters/transmitters and one receiver, as shown in Figure 5.9(a). The number of the emitters is assumed to be *a priori* known at the receiver, which can be achieved by algorithms for estimating the number of transmitters. Since the target of the SEI is to distinguish different emitters by extracting the unique feature carried by the transmit signals, a model that describes the difference between emitters is first introduced below.

For an emitter, a power amplifier is the main component, whose nonlinear system response characteristic is one of the principle sources of the specific feature, as known as the fingerprint of an emitter. We use the Taylor series to describe the nonlinear system [21,22], and let L_s denote the order for the Taylor series. For emitter k, the system response function of the power amplifier is then defined as

$$\Phi^{[k]}(x(t)) = \sum_{l=1}^{L_s} \alpha_l^{[k]} (x(t))^l \tag{5.32}$$

where $x(t) = s(t)e^{j2\pi nfT}$ is the input signal at the power amplifier—with $s(t)$ as the baseband-modulated signal, f as the carrier frequency, and T as the sampling period—$\{\alpha_l^{[k]}\}$ denotes the coefficients of the Taylor series, and $\Phi^{[k]}(x(t))$ denotes the output signal at the power amplifier of the kth emitter, i.e., the transmit signal of the kth emitter. Apparently, for emitters with the same order L_s, their different coefficients represent the specific fingerprints, which are carried through the transmit signals $\Phi^{[k]}(x(t))$.

At the receiver, the received signal is given by

$$r(t) = H_{sd}^{[k]}\Phi^{[k]}(x(t)) + w(t), \quad k = 1, \ldots, K \tag{5.33}$$

where $H_{sd}^{[k]}$ is the unknown channel fading coefficient between the kth emitter and the receiver, and $w(t)$ is the additive noise. By substituting (5.32) into (5.33), we can rewrite the received signal as

$$r(t) = H_{sd}^{[k]}\sum_{l=1}^{L_s}\alpha_l^{[k]}(x(t))^l + w(t). \tag{5.34}$$

5.3.1.2 Relaying scenario

Next, we expand the single-hop scenario to the multi-hop one. Note that two-hop communication systems are the most commonly adopted, such as the satellite communications; therefore, we focus on the two-hop communication system with an amplify-and-forward relay, as shown in Figure 5.9(b). The received signal at the relay is expressed as

$$y(t) = H_{sr}^{[k]}\Phi^{[k]}(x(t)) + \eta(t), \quad k = 1, \ldots, K \tag{5.35}$$

where $H_{sr}^{[k]}$ is the unknown channel fading from the kth emitter to the relay, and $\eta(t)$ is the additive noise.

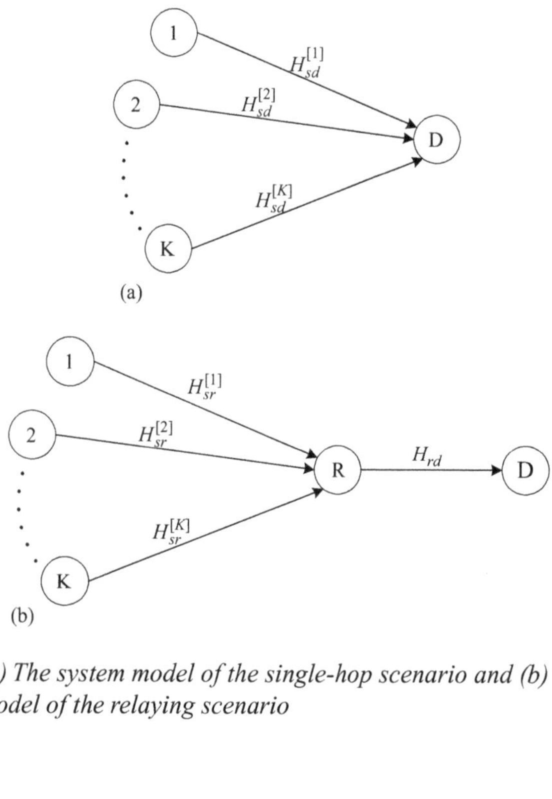

Figure 5.9 (a) The system model of the single-hop scenario and (b) the system model of the relaying scenario

Then, the received signal at the receiver, which is forwarded by the relay, is written as

$$\begin{aligned} r(t) &= H_{rd}\Psi\left(y(t)\right) + \upsilon(t) \\ &= H_{rd}\Psi\left(H_{sr}^{[k]}\Phi^{[k]}\left(x(t)\right) + \eta(t)\right) + \upsilon(t) \end{aligned} \tag{5.36}$$

where $\Psi(\cdot)$ denotes the system response characteristic of the power amplifier of the relay, H_{rd} is the unknown channel fading coefficient from the relay to the receiver, and $\upsilon(t)$ is the additive noise. Similarly, we use the Taylor series to define $\Psi(\cdot)$, which is given by

$$\Psi(y(t)) = \sum_{m=1}^{L_r} \beta_m \left(y(t)\right)^m \tag{5.37}$$

where L_r denotes the order of Taylor series for the power amplifier of the relay, and $\{\beta_m\}$ represent the fingerprint of the relay. Hence, the received signal is further expressed as

$$r(t) = H_{rd}\sum_{m=1}^{L_r} \beta_m \left(y(t)\right)^m + \upsilon(t) \tag{5.38}$$

$$= H_{rd}\sum_{m=1}^{L_r} \beta_m \left(H_{sr}^{[k]}\sum_{l=1}^{L_s} \alpha_l^{[k]} \left(x(t)\right)^l + \eta(t)\right)^m + \upsilon(t). \tag{5.39}$$

It is obvious that the features carried by the received signal are the combinations of the fingerprint of both the emitter and the relay, meaning that the fingerprint of the emitter is contaminated by that of the relay, which causes negative effect on SEI.

5.3.2 *Feature extraction*

In this section, three feature-extraction algorithms are introduced, namely, the entropy and first- and second-order moments-based algorithm (EM^2), the correlation-based (CB) algorithm, and the Fisher discriminant ratio (FDR)-based algorithm. All the three algorithms extract identification features from the Hilbert spectrum, which is obtained by employing the Hilbert–Huang transform (HHT).

5.3.2.1 Hilbert–Huang transform

The HHT is a powerful and adaptive tool to analyze nonlinear and nonstationary signals [23]. The principle of the HHT being adaptive relies on the empirical mode decomposition (EMD), which is capable of decomposing any signals into a finite number of intrinsic mode functions (IMFs). By performing the Hilbert transform on the IMFs, we can obtain a time–frequency–energy distribution of the signal, also known as the Hilbert spectrum of the signal.

Empirical mode decomposition

The target of the EMD is to obtain the instantaneous frequency with physical meaning, therefore, the IMF should satisfy two conditions [23]: (1) the number of extrema and the number of zero-crossings should either be equal, or the difference is one at most; (2) at any point, the sum of the upper and lower envelopes, respectively, defined by the local maxima and minima, should be zero.

Let $z(t)$ be the original signal, the EMD uses an iteration process to decompose the original signal into the IMFs, which is described as follows [23]:

1. First, identify all local maxima and minima, then employing the cubic spline fitting to obtain the upper and lower envelopes of the signal;
2. Compute the mean of the upper and lower envelopes, denoted by $\mu_{10}(t)$. Subtract $\mu_{10}(t)$ from $z(t)$ to obtain the first component $z_{10}(t)$, i.e., $z_{10}(t) = z(t) - \mu_{10}(t)$;
3. Basically, since the original signal is complicated, the first component does not satisfy the IMF conditions. Thus, steps 1 and 2 are repeated p times until $z_{1p}(t)$ becomes an IMF:

$$z_{1p}(t) = z_{1(p-1)}(t) - \mu_{1p}(t), \quad p = 1, 2, \ldots, \tag{5.40}$$

where $\mu_{1p}(t)$ is the mean of the upper and lower envelopes of $z_{1(p-1)}(t)$. We define that

$$\xi = \sum_{t=0}^{T_s} \frac{|z_{1(p-1)}(t) - z_{1p}(t)|^2}{z_{1(p-1)}^2(t)} \tag{5.41}$$

where T_s is the length of the signal. Then, the stopping criterion of this shifting process is when $\xi < \varepsilon$. Note that an empirical value of ε is set between 0.2 and 0.3.
4. Denote $c_1(t) = z_{1p}(t)$ as the first IMF. Subtract it from the $z(t)$ to obtain the residual, which is

$$d_1(t) = z(t) - c_1(t). \tag{5.42}$$

5. Consider the residual as a new signal, repeat steps 1 to 4 on all residuals $d_q(t)$, $q = 1, \ldots, Q$, to extract the remaining IMFs, i.e.:

$$\begin{aligned} d_2(t) &= d_1(t) - c_2(t), \\ &\cdots \\ d_Q(t) &= d_{Q-1}(t) - c_Q(t) \end{aligned} \tag{5.43}$$

where Q is the number of IMFs. The stopping criterion of the iteration procedure is when $d_Q(t) < \varepsilon$, or it becomes a monotonic function without any oscillation. From (5.42) and (5.43), we can rewrite $z(t)$ as

$$z(t) = \sum_{q=1}^{Q} c_q(t) + d_Q(t). \tag{5.44}$$

Hilbert spectrum analysis

We omit the residual $d_Q(t)$ in (5.44), and implement the Hilbert transform on the extracted IMFs, the original signal is then expressed as

$$z(t) = \Re\left(\sum_{q=1}^{Q} a_q(t) \exp\left(j \int \omega_q(t) dt\right)\right) \tag{5.45}$$

where $\Re(\cdot)$ denotes the real component of a complex variable, $j = \sqrt{-1}$, $a_q(t) = \sqrt{c_q^2(t) + \hat{c}_q^2(t)}$ and $\omega_q(t) = (d\theta_q(t)/dt)$ are the instantaneous amplitude and frequency of the IMF $c_q(t)$, respectively, in which $\hat{c}_q(t) = \frac{1}{\pi}\int_{-\infty}^{\infty}(c_q(\tau)/(t-\tau))d\tau$ is the Hilbert transform, and $\theta_q(t) = \arctan(\hat{c}_q(t)/c_q(t))$ is the phase function.

With the instantaneous amplitude and frequency of the IMFs, we can obtain the Hilbert spectrum of the original signal, denoted by $\mathcal{H}(\omega, t)$. In this section, we use the squared value of the instantaneous amplitude, therefore, the Hilbert spectrum provides the time–frequency–energy distribution.

5.3.2.2 Entropy and first- and second-order moments-based algorithm

The EM^2 algorithm extracts features by measuring the uniformity of the Hilbert spectrum. The feature contains three-dimensional data, which are the energy entropy, the first-order and second-order moments, respectively.

Energy entropy

We use the definition of information entropy to define the energy entropy of the Hilbert spectrum. First, the Hilbert spectrum is divided into several time–frequency slots. Denote $\mathcal{H}_{ij}(\omega, t)$, $i = 1, \ldots, G_t$, $j = 1, \ldots, G_\omega$ as the (i,j)th time–frequency slot, where G_t is the number of time slots with resolution Δt and G_ω is the number of frequency slots with resolution $\Delta\omega$. By using the expression of the information entropy [24], the energy entropy of the Hilbert spectrum is defined as

$$I = -\sum_{i=1}^{G_t}\sum_{j=1}^{G_\omega} p_{ij}\log p_{ij} \tag{5.46}$$

where $p_{ij} = E_{ij}/E$ is the proportion of the energy of each time–frequency slot, with E as the total energy of the Hilbert spectrum, and E_{ij} as the energy of the (i,j)th time–frequency slot, given by

$$E_{ij} = \int_{(i-1)\Delta t}^{i\Delta t}\int_{(j-1)\Delta\omega}^{j\Delta\omega} \mathcal{H}_{ij}(\omega, t)d\omega dt. \tag{5.47}$$

First- and second-order moments

The first- and second-order moments adopt the concept of color moments in image processing, which measure the color distribution of an image. To compute the first- and second-order moments, we first map the Hilbert spectrum into a gray scale image, where the Hilbert spectrum elements are described by shades of gray with intensity information, which is

$$B_{m,n} = \left\lfloor (2^\zeta - 1) \times \frac{\mathcal{H}_{m,n}}{\max_{\{m,n\}} \mathcal{H}_{m,n}} \right\rfloor \tag{5.48}$$

where $B_{m,n}$ is the (m, n)th value of the gray scale image matrix $\boldsymbol{B}$, $\mathcal{H}_{m,n}$ is the (m, n)th element of the Hilbert spectrum matrix $\boldsymbol{\mathcal{H}}$,[2] and $\lfloor\cdot\rfloor$ is the floor function, which equals

[2]As a two-dimensional spectrum, we represent the Hilbert spectrum through a matrix, referred to as the Hilbert spectrum matrix, where the indices of the columns and rows correspond to the sampling point and instantaneous frequency, respectively, and the elements of the matrix are combinations of the instantaneous energy of the IMFs.

the nearest lower integer value. Taking ς-bit gray scale as an example, the largest value of the Hilbert spectrum is converted to the gray scale $(2^{\varsigma} - 1)$, while other values are linearly scaled.

The first- and second-order moments of the gray scale image are, respectively, defined as

$$\mu = \frac{1}{N_H} \sum_{m=1}^{M} \sum_{n=1}^{N} B_{m,n} \tag{5.49}$$

$$\varsigma = \left(\frac{1}{N_H} \sum_{m=1}^{M} \sum_{n=1}^{N} (B_{m,n} - \mu)^2 \right)^{1/2} \tag{5.50}$$

where $N_H = M \times N$ is the total number of pixels (elements) of the gray scale image matrix. Note that the first-order moment interprets the average intensity of the gray scale image, and the second-order moment describes the standard deviation of the shades of gray.

5.3.2.3 Correlation-based algorithm

It is observed that the shape of the time–frequency–energy distribution of signals from the same emitter are similar, while that from different emitters are diverse; therefore, the correlation coefficients between Hilbert spectra can be extracted as the identification features.

Let $\mathcal{H}_i$ and $\mathcal{H}_j$, $i, j = 1, \ldots, N_R$ represent the Hilbert spectrum matrices of the ith and jth training sequence, respectively, where N_R is the total number of training sequences over K classes. The correlation coefficient between $\mathcal{H}_i$ and $\mathcal{H}_j$ is expressed as

$$\rho^{(i,j)} = \frac{\sum_m \sum_n \left(\mathcal{H}_{i,m,n} - \mathbb{E}(\mathcal{H}_i)\right) \left(\mathcal{H}_{j,m,n} - \mathbb{E}(\mathcal{H}_j)\right)}{\sqrt{\left(\sum_m \sum_n \left(\mathcal{H}_{i,m,n} - \mathbb{E}(\mathcal{H}_i)\right)^2\right) \left(\sum_m \sum_n \left(\mathcal{H}_{j,m,n} - \mathbb{E}(\mathcal{H}_j)\right)^2\right)}} \tag{5.51}$$

where $\mathcal{H}_{i,m,n}$ ($\mathcal{H}_{j,m,n}$) denotes the (m, n)th element of the Hilbert spectrum matrix $\mathcal{H}_i$ ($\mathcal{H}_j$), and $\mathbb{E}(\cdot)$ is the mean of the elements. Equation (5.51) depicts the linear dependence between $\mathcal{H}_i$ and $\mathbf{H}_j$; larger $\rho^{(i,j)}$ implies that $\mathcal{H}_i$ and $\mathcal{H}_j$ are more likely from the same emitter; otherwise, $\rho^{(i,j)}$ close to zero indicates that $\mathcal{H}_i$ and $\mathcal{H}_j$ are from diverse emitters.

5.3.2.4 Fisher's discriminant ratio-based algorithm

It should be noted that large number of elements in a Hilbert spectrum are featureless, meaning that they have little discrimination. In contrast to the EM^2 and CB algorithms that exploits all the elements of a Hilbert spectrum, the FDR algorithm selects the elements of a Hilbert spectrum that provides well separation between two classes.

Let (k_1, k_2) be a possible combination of two classes arbitrarily selected from K classes, $k_1 \neq k_2$. Note that the total number of possible combinations (k_1, k_2) for all K

classes is $C = K(K-1)/2$. For (k_1, k_2), we define the FDR at time–frequency spot (ω, t) as

$$\mathcal{F}^{(k_1,k_2)}(\omega,t) = \frac{\left(\mathbb{E}_i\left(\mathcal{H}_i^{[k_1]}(\omega,t)\right) - \mathbb{E}_i\left(\mathcal{H}_i^{[k_2]}(\omega,t)\right)\right)^2}{\sum_{k=k_1,k_2} \mathbb{D}_i\left(\mathcal{H}_i^{[k]}(\omega,t)\right)} \tag{5.52}$$

where $\mathcal{H}_i^{[k]}(\omega,t)$, $i = 1, \ldots, \bar{N}_0$ is the Hilbert spectrum of the ith training sequence of the kth class at (ω, t), with $\bar{N}_0$ as the number of training sequences for each class, and $\mathbb{E}_i\left(\mathcal{H}_i^{[k]}(\omega,t)\right)$ and $\mathbb{D}_i\left(\mathcal{H}_i^{[k]}(\omega,t)\right)$ denote the mean and variance of the training sequences of class k at (ω, t), respectively. From (5.52), we can see that the FDR $\mathcal{F}^{(k_1,k_2)}(\omega,t)$ measures the separability of the time–frequency spot (ω, t) between classes k_1 and k_2. It indicates that the time–frequency spot (ω, t) with larger FDR provides larger separation between the mean of two classes and smaller within-class variance, which shows stronger discrimination.

For each combination (k_1, k_2), we define $\Omega = \{\mathcal{F}_1^{(k_1,k_2)}(\omega,t), \ldots, \mathcal{F}_{N_H}^{(k_1,k_2)}(\omega,t)\}$ as the original FDR sequence. Sort Ω in descending order and denote the new FDR sequence as $\tilde{\Omega} = \{\tilde{\mathcal{F}}_1^{(k_1,k_2)}(\omega,t), \ldots, \tilde{\mathcal{F}}_{N_H}^{(k_1,k_2)}(\omega,t)\}$, i.e., $\tilde{\mathcal{F}}_1^{(k_1,k_2)}(\omega,t) \geq \cdots \geq \tilde{\mathcal{F}}_{N_H}^{(k_1,k_2)}(\omega,t)$. Let $\{(\tilde{\omega}_1, \tilde{t}_1), \ldots, (\tilde{\omega}_{N_H}, \tilde{t}_{N_H})\}$ be the time–frequency slots which correspond to the rearranged FDR sequence $\tilde{\Omega}$. Then, we select the time–frequency spots that correspond to the S largest FDR as optimal time–frequency spots, denoted as $\mathcal{Z}^{(c)} = \{(\tilde{\omega}_s^{(c)}, \tilde{t}_s^{(c)}),\ s = 1, \ldots, S,\ c = 1, \ldots, C\}$. The total set of optimal time–frequency spots is defined as the union of $\mathcal{Z}^{(c)}$, i.e., $\mathcal{Z} = \bigcup_{c=1}^{C} \mathcal{Z}^{(c)}$. For the same $(\tilde{\omega}_s, \tilde{t}_s)$ between different combinations (k_1, k_2), only one is retained in order to avoid duplication, i.e., $\mathcal{Z} = \{(\tilde{\omega}_1, \tilde{t}_1), \ldots, (\tilde{\omega}_D, \tilde{t}_D)\}$, where D is the number of optimal time–frequency spots without duplication, with $D \leq S \times ((K(K-1))/2)$.

5.3.3 Identification procedure via SVM

By using the feature extraction algorithms proposed in the previous section to extract the identification features, the SVM is then adopted to implement the identification procedure. SVM is a widely used supervised learning classifier in machine learning, which is originally designed for two-class classification problems. Input a set of labeled training samples, the SVM outputs an optimal hyperplane, which can classify new samples.

Linear SVM: Suppose that $\{\boldsymbol{v}_i, \iota_i\}$ is the training set, with $\boldsymbol{v}_i$ as the training vector and $\iota_i \in \{1, -1\}$ as the class label. We first consider the simplest linear case, where the labeled data can be separated by a hyperplane $\chi(\boldsymbol{v}) = 0$. The decision function $\chi(\boldsymbol{v})$ is expressed as [25]:

$$\chi(\boldsymbol{v}) = \boldsymbol{w}^{\mathrm{T}}\boldsymbol{v} + b \tag{5.53}$$

where $\boldsymbol{w}$ is the normal vector to the hyperplane and $(b/\|\boldsymbol{w}\|)$ determines the perpendicular offset of the hyperplane from the origin, with $\|\cdot\|$ as the Euclidean norm.

Given a set of training data, labeled as positive and negative ones, we define the closest distances from the positive and negative points to the hyperplane as $m_{d,+}$ and $m_{d,-}$, respectively. Then, the optimization task of the hyperplane is to make the margin, $m_d = m_{d,+} + m_{d,-}$, the largest. To simplify the derivation, two hyperplanes that bound the margin are defined as

$$\boldsymbol{w}^{\mathrm{T}}\boldsymbol{v} + b = 1 \tag{5.54}$$

$$\boldsymbol{w}^{\mathrm{T}}\boldsymbol{v} + b = -1. \tag{5.55}$$

Note that the distance between the two hyperplanes is $m_d = (2/\|\boldsymbol{w}\|)$, the original problem of maximizing m_d can be converted to a constrained minimization problem, which is [25,26]:

$$\min \quad \frac{1}{2}\|\boldsymbol{w}\|^2 \tag{5.56}$$

$$\text{s.t.} \quad \iota_i\left(\boldsymbol{w}^{\mathrm{T}}\boldsymbol{v}_i + b\right) \geq 1, \quad i = 1, \ldots, \bar{N} \tag{5.57}$$

where $\bar{N}$ is the number of training examples. By introducing nonnegative Langrage multipliers $\lambda_i \geq 0$, (5.56) is transformed to a dual quadratic programming optimization problem, given by [25,26]:

$$\max_{\boldsymbol{\lambda}} \quad \sum_{i=1}^{\bar{N}} \lambda_i - \frac{1}{2}\sum_{i,j=1}^{\bar{N}} \lambda_i\lambda_j\iota_i\iota_j\langle \boldsymbol{v}_i, \boldsymbol{v}_j\rangle \tag{5.58}$$

$$s.t. \quad \lambda_i \geq 0, \quad i = 1, \ldots, \bar{N} \tag{5.59}$$

$$\sum_{i=1}^{\bar{N}} \lambda_i\iota_i = 0 \tag{5.60}$$

where $\boldsymbol{w}$ and b are represented by $\boldsymbol{w} = \sum_{i=1}^{\bar{N}} \lambda_i\iota_i\boldsymbol{v}_i$ and $b = -(1/2)(\max_{i:\iota_1=-1} \boldsymbol{w}^{\mathrm{T}}\boldsymbol{v}_i + \min_{i:\iota_1=1} \boldsymbol{w}^{\mathrm{T}}\boldsymbol{v}_i)$, respectively.

The decision function is obtained by solving the optimization problem in (5.58):

$$\chi(\boldsymbol{v}) = \sum_{i=1}^{\bar{N}} \lambda_i\iota_i\langle \boldsymbol{v}_i, \boldsymbol{v}\rangle + b \tag{5.61}$$

where $\langle\cdot,\cdot\rangle$ denotes the inner product. The decision criterion of the correct classification is $\iota_l\chi(\mathbf{u}_l) > 0$, i.e., the testing example $\boldsymbol{u}_l$ that satisfies $\chi(\boldsymbol{u}_l) > 0$ is labeled as 1; otherwise, it is labeled as -1.

Nonlinear SVM: For the case where $\boldsymbol{v}_i$ cannot be simply distinguished by the linear classifier, a nonlinear mapping function ϕ is utilized to map $\boldsymbol{v}_i$ to a high-dimensional space $\mathfrak{F}$, in which the categorization can be done by a hyperplane. Similarly, the decision function is expressed as [25]:

$$\chi(\boldsymbol{v}) = \sum_{i=1}^{\bar{N}} \lambda_i\iota_i\langle \phi(\boldsymbol{v}_i), \phi(\boldsymbol{v})\rangle + b. \tag{5.62}$$

In such a case, a kernel function $\kappa(\boldsymbol{v}_i, \boldsymbol{v})$ is defined to avoid the computation of the inner product $\langle \phi(\boldsymbol{v}_i), \phi(\boldsymbol{v}) \rangle$, which is generally intractable in high-dimensional spaces [27].[3] Using $\boldsymbol{w} = \sum_{i=1}^{\bar{N}} \lambda_i \iota_i \boldsymbol{v}_i$ and the kernel function, we rewrite (5.62) as

$$\chi(\boldsymbol{v}) = \sum_{i=1}^{\bar{N}} \lambda_i \iota_i \kappa(\boldsymbol{v}_i, \boldsymbol{v}) + b. \tag{5.63}$$

The decision rule is the same as that for the linear classifier.

Multi-class SVM: Next, we consider the case of multiple classes. The multi-class classification problem is solved by reducing it to several binary classification problems. Commonly adopted methods include one-versus-one [28], one-versus-all [28] and binary tree architecture [29] techniques. In this section, we employ the one-versus-one technique for the multi-class problem, by which the classification is solved using a max-win voting mechanism, and the decision rule is to choose the class with the highest number of votes.

The training and identification procedures of the three proposed algorithms using the SVM are summarized as follows:

Training and identification procedures of the EM^2 algorithm

Training procedure: Let $\mathcal{H}_i$, $i = 1, \ldots, \bar{N}$, denote the Hilbert spectrum matrix of the training sequence i, with $\bar{N}$ as the total number of training sequences over all K classes.

1. From (5.46), we compute the energy entropy of $\mathcal{H}_i$, denoted as $\bar{I}_i$.
2. Map $\mathcal{H}_i$ to $\boldsymbol{B}_i$ using (5.48), with $\boldsymbol{B}_i$ as the Hilbert gray scale image matrix of the training sequence i. From (5.49) and (5.50), we compute the first- and second-order moments, denoted as $(\bar{\mu}_i, \bar{\varsigma}_i)^{\mathrm{T}}$, respectively.
3. Generate the three-dimensional training vector, i.e., $\boldsymbol{v}_i = \left(\bar{I}_i, \bar{\mu}_i, \bar{\varsigma}_i\right)^{\mathrm{T}}$.
4. Let $\{\boldsymbol{v}_i, \iota_i\}$ be the training set, where $\iota_i \in \{1, \ldots, K\}$ is the label of each class. Use the labeled training set to train the optimal hyperplane $\chi(\boldsymbol{v})$.

Identification procedure: Let $\mathcal{H}_l$, $l = 1, \ldots, N$, denote the Hilbert spectrum matrix of the lth test sequence of an unknown class, where N is the number of test sequences.

1. Use (5.46)–(5.50) to compute the energy entropy I_l, the first- and second-order moments $(\mu_l, \varsigma_l)^{\mathrm{T}}$ of $\mathcal{H}_l$.
2. Generate the test vector, denoted as $\boldsymbol{u}_l = (I_l, \mu_l, \varsigma_l)^{\mathrm{T}}$.
3. The identification task is implemented by employing the SVM classifier defined in the training procedure. For $K = 2$, $\boldsymbol{u}_l$ which satisfies $\chi(\boldsymbol{u}_l) > 0$ is labeled as class 2; otherwise, it is labeled as class 1. For $K > 2$, the one-versus-one technique is applied, where the decision depends on the max-win voting mechanism, i.e., the class with the highest number of votes is considered as the identification result.

[3] Typical kernel functions include the Gaussian radial-base function (RBF), $\kappa(\boldsymbol{x}, \boldsymbol{y}) = e^{-\|\boldsymbol{x}-\boldsymbol{y}\|^2/2\gamma^2}$, and the polynomial kernel, $\kappa(\boldsymbol{x}, \boldsymbol{y}) = (\boldsymbol{x}, \boldsymbol{y})^d$, with d as the sum of the exponents in each term.

Training and identification procedures of the CB algorithm

Training procedure: Let $\mathcal{H}_i$ and $\mathcal{H}_j$ denote the Hilbert spectrum matrices of the training sequences i and j, respectively, where $i, j = 1, \ldots, \bar{N}$, with $\bar{N}$ as the total number of the training sequences over all K classes.

1. From (5.51), we calculate the correlation coefficient $\bar{\rho}^{(i,j)}$ between $\mathcal{H}_i$ and $\mathcal{H}_j$. For the training sequence i, the training vector is $\bar{N}$-dimensional, denoted as $\bar{\boldsymbol{\rho}}_i = \left[\bar{\rho}^{(i,1)}, \ldots, \bar{\rho}^{(i,\bar{N})}\right]^{\mathrm{T}}$.
2. Let $\left\{\bar{\boldsymbol{\rho}}_i, \iota_i\right\}$ be the set of training data with $\iota_i \in \{1, \ldots, K\}$ as the label of each class. Then, input the labeled training set to the SVM classifier to optimize the decision hyperplane $\chi(\bar{\boldsymbol{\rho}})$.

Identification procedure: Let $\mathcal{H}_l,\ l = 1, \ldots, N$, be the Hilbert spectrum matrix of the test sequence of an unknown class, where N is the number of test sequences.

1. For the test sequence l, the correlation coefficient $\rho^{(l,i)}$ between $\mathcal{H}_l$ and $\mathcal{H}_i$ is calculated from (5.51), the $\bar{N}$-dimensional test vector is denoted as $\boldsymbol{\rho}_l = [\rho^{(l,1)}, \ldots, \rho^{(l,\bar{N})}]^{\mathrm{T}}$.
2. Classify the test sequence by employing the SVM classifier. For $K = 2$, $\boldsymbol{\rho}_l$ which satisfies $\chi(\boldsymbol{\rho}_l) > 0$ is labeled as class 2; otherwise, it is labeled as class 1. For $K > 2$, the one-versus-one technique is applied, where the decision depends on the max-win voting mechanism, i.e., the class with the highest number of votes is considered as the identification result.

Training and identification procedures of the FDR algorithm

Training procedure: Let $\mathcal{H}_i(\omega, t)$ denote the Hilbert spectrum of the training sequence i at time–frequency spot (ω, t), where $i = 1, \ldots, \bar{N}$, with $\bar{N}$ as the total number of the training sequences over all K classes.

1. For the combination (k_1, k_2), we compute the original FDR sequence $\Omega = \{\mathcal{F}_1^{(k_1,k_2)}(\omega, t), \ldots, \mathcal{F}_{NH}^{(k_1,k_2)}(\omega, t)\}$ from (5.52);
2. Obtain the FDR sequence (descending order of elements) $\tilde{\Omega} = \{\tilde{\mathcal{F}}_1^{(k_1,k_2)}(\omega, t), \ldots, \tilde{\mathcal{F}}_{NH}^{(k_1,k_2)}(\omega, t)\}$;
3. Select the time–frequency spots that correspond with the S largest FDR in $\tilde{\Omega}$ to form the optimal time–frequency spots set $\mathcal{Z}^{(c)} = \{(\tilde{\omega}_s^{(c)}, \tilde{t}_s^{(c)}), s = 1, \ldots, S, c = 1, \ldots, C\}$;
4. Repeat steps 1–3 over all $C = K(K-1)/2$ possible combinations (k_1, k_2) to obtain the total set of optimal time–frequency spots, $\mathcal{Z} = \bigcup_{c=1}^{C} \mathcal{Z}^{(c)}$. For the same $(\tilde{\omega}_s, \tilde{t}_s)$ between different combinations (k_1, k_2), only one of them is retained in order to avoid duplication;
5. For the training sequence i, the Hilbert spectrum elements corresponding to $(\tilde{\omega}_1, \tilde{t}_1), \ldots, (\tilde{\omega}_D, \tilde{t}_D)$ are extracted to form a D-dimensional training vector, expressed as $\mathbf{v}_i = \left[\mathcal{H}_i(\tilde{\omega}_1, \tilde{t}_1), \ldots, \mathcal{H}_i(\tilde{\omega}_D, \tilde{t}_D)\right]^{\mathrm{T}}$;

6. Let $\{\boldsymbol{v}_i, \iota_i\}$ be the set of training data with $\iota_i \in \{1, \ldots, K\}$ as the label of each class. Then, the data is input into the SVM classifier for training, i.e., to obtain the optimal $\boldsymbol{w}$ and b of the decision hyperplane $\chi(\boldsymbol{v})$.

Identification procedure: Let $\mathcal{H}_l(\omega, t)$, $l = 1, \ldots, N$, denote the Hilbert spectrum of test sequence l at time–frequency spot (ω, t) of a unknown class, where N is the number of test sequences.

1. For the test sequence l, extract the elements corresponding to the D optimal time–frequency spots as the test vector, i.e., $\boldsymbol{u}_l = [\mathcal{H}_l(\tilde{\omega}_1, \tilde{t}_1), \ldots, \mathcal{H}_l(\tilde{\omega}_D, \tilde{t}_D)]^{\mathrm{T}}$;
2. Utilize the SVM classifier to identify the test sequence. For $K = 2$, $\boldsymbol{u}_l$ which satisfies $\chi(\boldsymbol{u}_l) > 0$ is labeled as class 2; otherwise, it is labeled as class 1. For $K > 2$, one-versus-one technique is applied, where the decision depends on the max-win voting mechanism, i.e., the class with the highest number of votes is considered as the identification results.

5.3.4 Numerical results

In this section, we provide various numerical experiments based on the probability of correct identification P_c, to evaluate the identification performance of the proposed algorithms, and compare it with that of some conventional approaches. The identification performance of the proposed algorithms are evaluated under three scenarios, namely, the AWGN channels, the flat-fading channels, and the non-Gaussian noise channels. It is noted that distinguishing the identity of different emitters is a difficult task even in the presence of the AWGN channels, the results in AWGN channels provides the baseline of the proposed algorithms, which show that they can obtain better identification performance than the conventional methods. Furthermore, we examine the identification performance of the proposed algorithms in the presence of flat-fading channels and non-Gaussian noise channels to illustrate that the proposed algorithms are robust against more realistic and complicated scenarios.

For the power amplifiers of both the emitters and the relay, we assume that the order of the Taylor polynomial is $L_s = L_r = 3$. In the simulations, we consider three cases, i.e., the number of emitters are set to $K = 2$, 3 and 5. For the power amplifier of the emitters, denote the coefficients of the Taylor polynomial as $\boldsymbol{\alpha}^{[k]} = \left(\alpha_1^{[k]}, \ldots, \alpha_{L_s}^{[k]}\right)$, and each coefficient is set to $\boldsymbol{\alpha}^{[1]} = (1, 0.5, 0.3)^{\mathrm{T}}$, $\boldsymbol{\alpha}^{[2]} = (1, 0.08, 0.6)^{\mathrm{T}}$, $\boldsymbol{\alpha}^{[3]} = (1, 0.01, 0.01)^{\mathrm{T}}$, $\boldsymbol{\alpha}^{[4]} = (1, 0.01, 0.4)^{\mathrm{T}}$ and $\boldsymbol{\alpha}^{[5]} = (1, 0.6, 0.08)^{\mathrm{T}}$, respectively. To be specific, for the case that $K = 2$, the coefficient matrix is $\mathcal{A}_2^{\mathrm{T}} = \left(\boldsymbol{\alpha}^{[1]}; \boldsymbol{\alpha}^{[2]}\right)$; for the case that $K = 3$, it is $\mathcal{A}_3 = \left(\boldsymbol{\alpha}^{[1]}; \boldsymbol{\alpha}^{[2]}; \boldsymbol{\alpha}^{[3]}\right)$; and for $K = 5$, it is $\mathcal{A}_5 = \left(\boldsymbol{\alpha}^{[1]}; \boldsymbol{\alpha}^{[2]}; \boldsymbol{\alpha}^{[3]}; \boldsymbol{\alpha}^{[4]}; \boldsymbol{\alpha}^{[5]}\right)$. For the coefficient matrix of the power amplifier Taylor polynomial model of the relay, we set it to $\mathcal{B}^{\mathrm{T}} = (1, 0.1, 0.1)$. The SVM classifier is implemented by using the LIBSVM toolbox, in which we adopt the Gaussian RBF, $\kappa(\boldsymbol{x}, \boldsymbol{y}) = e^{-\|\boldsymbol{x}-\boldsymbol{y}\|^2/2\gamma^2}$, as the kernel function with parameter $\gamma = 0.1$. For each class, we set the number of training and test sequences as $\bar{N}_0 = N_0 = 50$.

- Algorithms performance in AWGN channel

In the AWGN channels, we first evaluate the identification performance of the EM^2 algorithm for $K = 2$ and $K = 3$ in the single-hop and relaying scenarios, shown as Figure 5.10. For $K = 2$, the acceptable identification performance ($P_c \geq 0.8$) is attained for SNR $\geq$ 0 dB and SNR > 2 dB under the single-hop and relaying scenarios, respectively. For $K = 3$, P_c achieves 0.8 at SNR = 6 dB in the single-hop scenario and SNR = 8 dB in the relaying scenario. When compared to the conventional algorithm in [30], the advantage of the EM^2 algorithm is apparent. In particular, P_c enhances more than 30% at SNR = 8 dB for both $K = 2$ and $K = 3$ in both single-hop and relaying scenarios. In addition, it is seen that in high SNR region, the identification accuracy of the EM^2 algorithm in the single-hop and relaying scenarios is similar, while the conventional algorithm presents a gap of 10% between these cases. It implies that the EM^2 algorithm can effectively combat the negative effect caused by the relay on the identification of emitters at high SNR.

We then consider the identification performance of the EM^2 algorithm versus the number of training and test samples. In Figure 5.11, P_c is plotted as a function of the number of training sequences for each class $\bar{N}_0$, with curves parameterized by the number of test sequences for each class N_0. We take the case that $K = 2$ and $K = 3$ in the relaying scenario at SNR = 14 dB as an example. It is noted that the number of training and test samples barely influences the identification performance.

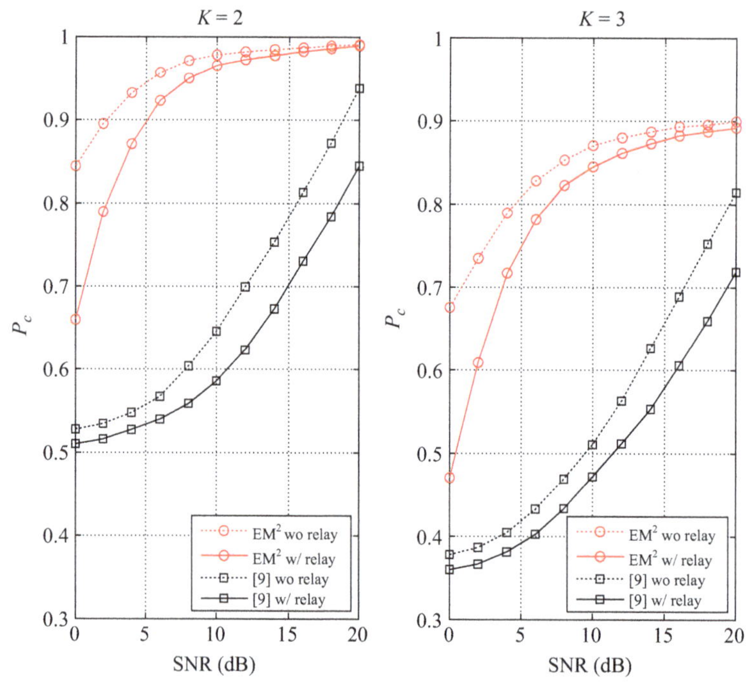

Figure 5.10 The identification performance of the EM^2 algorithm in AWGN channel

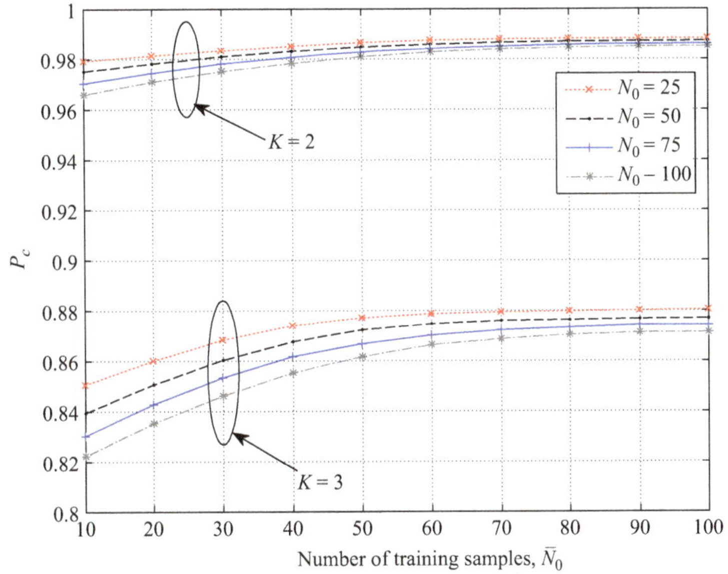

Figure 5.11 The identification performance of the EM^2 algorithm in AWGN channel. The number of test sequences for each class is $N_0 = 25, 50, 75, 100$ *and SNR* $= 14$ *dB*

Next, we depict the identification performance of the CB algorithm in Figure 5.12. where the identification performance for $K = 2$ and $K = 3$ in the single-hop and relaying scenarios is illustrated. When $K = 2$, the CB algorithm reaches $P_c \geq 0.8$ for SNR ≥ 0 dB in both scenarios. When $K = 3$, the identification performance attains $P_c = 0.8$ at SNR $= 4$ dB in the single-hop scenario and at SNR $= 7$ dB in the relaying scenario. Moreover, we compare the identification performance of the CB algorithm with that of the EM^2 algorithm, and the results show that the CB algorithm provides better identification performance for $K = 2$ and $K = 3$ in both single-hop and relaying scenarios.

In Figure 5.13, the identification performance of the FDR algorithm is illustrated, along with that of the EM^2 and CB algorithms when classifying five emitters for single-hop and relaying scenarios. Apparently, the FDR attains a reliable performance when $K = 5$ and outperforms that of the EM^2 and CB algorithms. Specifically, the FDR algorithm obtains $P_c \geq 0.8$ at SNR $= 5$ dB in the single-hop scenario and at SNR $= 7$ dB in the relaying scenario. The reason is that the FDR algorithm extracts elements with strong separability as identification features, which can provide better identification performance.

Table 5.1 summarizes the identification performance of the three proposed algorithms, along with that of the algorithm in [30], for SNR = 4, 12, and 20 dB in AWGN channels. It can be seen that when $K = 2$ and $K = 3$, all the proposed algorithms attain good identification performance in both single-hop and relaying scenarios.

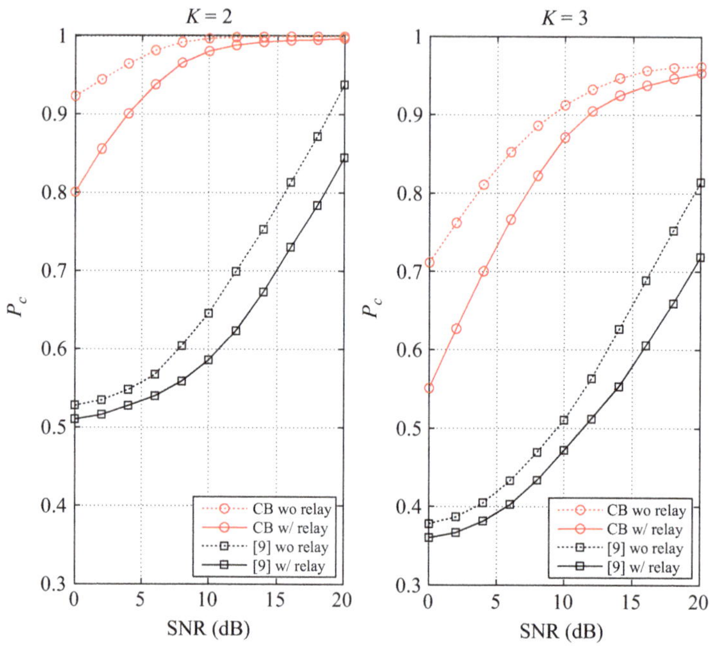

Figure 5.12 The identification performance of the CB algorithm in AWGN channel

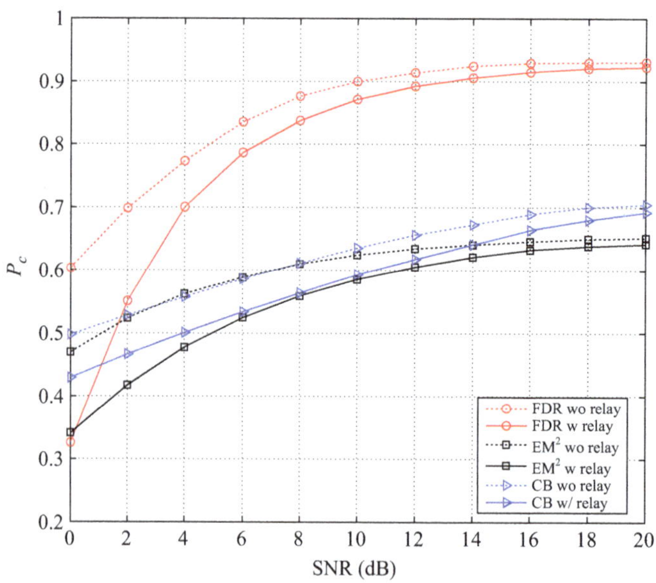

Figure 5.13 The identification performance of the FDR algorithm in AWGN channel

Table 5.1 Identification performance P_c for SNR = 4,12,20 dB in AWGN channels

		4 dB				12 dB				20 dB			
		EM^2	CB	FDR	[30]	EM^2	CB	FDR	[30]	EM^2	CB	FDR	[30]
$K = 2$	Single-hop	0.93	0.97	0.99	0.55	0.98	0.99	1.00	0.70	0.99	1.00	1.00	0.94
	Relaying	0.87	0.90	0.94	0.53	0.97	0.99	0.99	0.62	0.99	0.99	0.99	0.85
$K = 3$	Single-hop	0.79	0.81	0.97	0.41	0.88	0.93	0.99	0.56	0.90	0.96	0.99	0.81
	Relaying	0.72	0.70	0.89	0.38	0.86	0.91	0.98	0.51	0.89	0.95	0.99	0.72
$K = 5$	Single-hop	0.56	0.56	0.77	0.25	0.63	0.66	0.92	0.28	0.65	0.71	0.93	0.33
	Relaying	0.48	0.50	0.70	0.25	0.61	0.62	0.89	0.26	0.64	0.69	0.92	0.30

As expected, the FDR algorithm obtains the best identification performance, then followed by the CB algorithm and finally the EM^2 algorithm. The FDR algorithm can effectively identify emitters when $K = 5$ since it extracts features with strong separability. In addition, the identification performance of the proposed algorithms outperforms that of the algorithm in [30], especially in the relaying scenario.

- Algorithms performance comparison in non-Gaussian noise channel

Next, the identification performance of the proposed algorithms is evaluated under the non-Gaussian noise channels and compared to that in [30]. In the simulations, we assume that the non-Gaussian noise follows the Middleton Class A model, where the probability distribution function (pdf) is expressed as [31,32]:

$$f_{ClassA}(x) = e^{-A} \sum_{m=0}^{\infty} \frac{A^m}{m!\sqrt{(2\pi\sigma_m^2)}} e^{-(x^2/2\sigma_m^2)} \tag{5.64}$$

where A is the impulse index, and $\sigma_m^2 = (((m/A) + \Gamma)/(1 + \Gamma))$ is the noise variance, with Γ as the ratio of the intensity of the independent Gaussian component and the intensity of the impulsive non-Gaussian component. We set $A = 0.1$ and $\Gamma = 0.05$, respectively. In addition, we assume that the number of terms in the Class A pdf M is finite, i.e., $m \in [0, M - 1]$, and set $M = 500$ [31]. No fading is considered.

Table 5.2 summarizes the identification accuracy of the proposed algorithms and the algorithm in [30], at SNR = 4, 12 and 20 dB in the presence of the non-Gaussian noise channels. Comparing the results with that in the AWGN channels, it is noticed that for $K = 2$ and $K = 3$ in the single-hop scenario, the proposed algorithms have little degradation in the identification performance, and in the relaying scenario, the proposed algorithms effectively combat the negative effect of the non-Gaussian noise in the high SNR region. The results indicate that the proposed algorithms are applicable to non-Gaussian noise channels. Furthermore, it is obvious that all proposed algorithms outperform that of the conventional method in [30], which performs poorly especially in the relaying scenario.

- Algorithms performance comparison in flat-fading channel

Table 5.2 *Identification performance P_c for SNR = 4, 12, 20 dB in non-Gaussian noise channels*

		4 dB				12 dB				20 dB			
		EM²	**CB**	**FDR**	**[30]**	**EM²**	**CB**	**FDR**	**[30]**	**EM²**	**CB**	**FDR**	**[30]**
$K = 2$	Single-hop	0.90	0.91	0.98	0.55	0.98	0.99	0.99	0.70	0.99	0.99	0.99	0.94
	Relaying	0.82	0.83	0.86	0.53	0.96	0.97	0.99	0.62	0.98	0.99	0.99	0.85
$K = 3$	Single-hop	0.74	0.75	0.94	0.40	0.86	0.91	0.99	0.56	0.89	0.95	0.99	0.81
	Relaying	0.60	0.62	0.81	0.38	0.85	0.87	0.97	0.51	0.88	0.94	0.99	0.72
$K = 5$	Single-hop	0.53	0.50	0.73	0.25	0.61	0.62	0.90	0.28	0.64	0.69	0.93	0.33
	Relaying	0.41	0.43	0.63	0.25	0.57	0.57	0.88	0.26	0.62	0.67	0.91	0.30

Table 5.3 *Identification performance P_c for SNR = 4, 12, 20 dB in fading channels*

		4 dB				12 dB				20 dB			
		EM²	**CB**	**FDR**	**[30]**	**EM²**	**CB**	**FDR**	**[30]**	**EM²**	**CB**	**FDR**	**[30]**
$K = 2$	Single-hop	0.81	0.90	0.97	0.51	0.96	0.98	0.99	0.56	0.99	0.99	0.99	0.68
	Relaying	0.63	0.72	0.82	0.50	0.90	0.90	0.99	0.53	0.98	0.98	0.99	0.59
$K = 3$	Single-hop	0.65	0.73	0.91	0.36	0.86	0.89	0.98	0.40	0.90	0.93	0.99	0.54
	Relaying	0.63	0.50	0.77	0.35	0.80	0.76	0.96	0.39	0.98	0.92	0.98	0.45
$K = 5$	Single-hop	0.50	0.49	0.71	0.21	0.60	0.61	0.89	0.23	0.62	0.66	0.92	0.29
	Relaying	0.35	0.40	0.58	0.20	0.55	0.55	0.86	0.21	0.60	0.64	0.91	0.25

Next, the identification performance in the presence of flat-fading channels is examined. We assume that the channel-fading coefficients are unknown at the receiver.

Table 5.3 summarizes the identification performance of all the proposed algorithms and compared to that in [9], for SNR = 4, 12 and 20 dB in fading channels for both single-hop and relaying scenarios. As expected, the fading degrades the performance, especially for the CB algorithm in the relaying scenario. The main reason lies in that the fading channel significantly corrupts the similarity between Hilbert spectra, which leads to a severe identification performance loss for the CB algorithm. In addition, it is noted that the FDR algorithm provides the most promising identification performance among all the three proposed algorithms, which has a relatively reduced performance degradation in the fading channels.

5.3.5 Conclusions

This chapter discusses two main signal identification issues in CRs, that is the modulation classification and the SEI. New challenges to the signal identification techniques have raised considering the real-world environments. More advanced and intelligent theory is required to solve the blind recognition tasks. The machine-learning theory-based algorithms are introduced to solve the modulation classification and SEI problems. Numerical results demonstrate that the proposed algorithms provide promising identification performance.

References

[1] Mitola J, and Maguire GQJ. Cognitive radio: making software radios more personal. IEEE Personal Communications Magazine. 1999;6(4):13–18.
[2] Haykin S. Cognitive radio: brain-empowered wireless communications. IEEE Journal on Selected Areas of Communications. 2005;23(2):201–220.
[3] Huang C, and Polydoros A. Likelihood methods for MPSK modulation classification. IEEE Transactions on Communications. 1995;43(2/3/4):1493–1504.
[4] Kebrya A, Kim I, Kim D, *et al.* Likelihood-based modulation classification for multiple-antenna receiver. IEEE Transactions on Communications. 2013;61(9):3816–3829.
[5] Swami A, and Sadler B. Hierarchical digital modulation classification using cumulants. IEEE Transactions on Communications. 2000;48(3):416–429.
[6] Wang F, Dobre OA, and Zhang J. Fold-based Kolmogorov–Smirnov modulation classifier. IEEE Signal Processing Letters. 2017;23(7):1003–1007.
[7] Dempster AP, Laird NM, and Rubin DB. Maximum likelihood from incomplete data via the EM algorithm. Journal of the Royal Statistical Society: Series B. 1977;39(1):1–38.
[8] Xie Y, and Georghiades C. Two EM-type channel estimation algorithms for OFDM with transmitter diversity. IEEE Transactions on Communications. 2003;51(1):106–115.
[9] Trees HL. Detection Estimation and Modulation Theory, Part I – Detection, Estimation, and Filtering Theory. New York, NY: John Wiley and Sons; 1968.
[10] Gelb A. Applied Optimal Estimation. Cambridge, MA: The MIT Press; 1974.
[11] Lavielle M, and Moulines E. A simulated annealing version of the EM algorithm for non-Gaussian deconvolution. Statistics and Computing. 1997;7(4):229–236.
[12] Orlic VD, and Dukic ML. Multipath channel estimation algorithm for automatic modulation classification using sixth-order cumulants. IET Electronics Letters. 2010;46(19):1348–1349.
[13] Ozdemir O, Wimalajeewa T, Dulek B, *et al.* Asynchronous linear modulation classification with multiple sensors via generalized EM algorithm. IEEE Transactions on Wireless Communications. 2015;14(11):6389–6400.
[14] Markovic GB, and Dukic ML. Cooperative modulation classification with data fusion for multipath fading channels. IET Electronics Letters. 2013;49(23): 1494–1496.
[15] Zhang Y, Ansari N, and Su W. Optimal decision fusion based automatic modulation classification by using wireless sensor networks in multipath fading

channel. In: IEEE Global Communications Conference (GLOBECOM). Houston, TX, USA; 2011. p. 1–5.

[16] Bianchi P, Loubaton P, and Sirven F. Non data-aided estimation of the modulation index of continuous phase modulations. IEEE Transactions on Signal Processing. 2004;52(10):2847–2861.

[17] Pawar SU, and Doherty JF. Modulation recognition in continuous phase modulation using approximate entropy. IEEE Transactions on Information Forensics and Security. 2011;6(3):843–852.

[18] Rabiner RL. A tutorial on hidden Markov models and selected applications in speech recognition. Proceedings of the IEEE. 1989;77(2):257–286.

[19] Zhang J, Cabric D, Wang F, *et al.* Cooperative modulation classification for multipath fading channels via expectation-maximization. IEEE Transactions on Wireless Communications. 2017;16(10):6698–6711.

[20] Talbot KI, Duley PR, and Hyatt MH. Specific emitter identification and verification. Technology Review Journal. 2003;Spring/Summer:113–133.

[21] Carrol TL. A non-linear dynamics method for signal identification. Chaos: An Interdisciplinary Journal of Nonlinear Science. 2007;17(2):1–7.

[22] Couch LW. Digital and Analog Communication System. Upper Saddle River, NJ: Prentice Hall; 1996.

[23] Huang NE, Shen Z, Long SR, *et al.* The empirical mode decomposition and the Hilbert spectrum for nonlinear and non-stationary time series analysis. Proceedings of the Royal Society of London A: Mathematical, Physical and Engineering Sciences. 1998;454(1971):903–995.

[24] Shannon C. A mathematical theory of communication. Bell System Technical Journal. 1948;27(3):379–423.

[25] Cristianini N, and Taylor J. An Introduction to Support Vector Machines. Cambridge: Cambridge University Press; 2000.

[26] Burges CJC. A tutorial on support vector machine for pattern recognition. Data Mining and Knowledge Discovery. 1998;2(2):121–167.

[27] Muller KR, Mika S, Ratsch G, *et al.* An introduction to kernel-based learning algorithms. IEEE Transactions on Neural Networks. 2001;12(2):181–201.

[28] Hsu C, and Lin C. A comparison of methods for multiclass support vector machines. IEEE Transactions on Neural Networks. 2002;13(2):415–425.

[29] Madzarov G, and Gjorgjevikj D. Multi-class classification using support vector machines in decision tree architecture. In: IEEE EUROCON. St. Petersburg, Russia; 2009. p. 288–295.

[30] Xu S. On the identification technique of individual transmitter based on signalprints [dissertation]. Huazhong University of Science and Technology; 2007.

[31] Srivatsa A. RFI/impulsive noise toolbox 1.2 for Matlab 2009.

[32] Middleton D. Non-Gaussian noise models in signal processing for telecommunications: new methods and results for Class A and Class B noise models. IEEE Transactions on Information Theory. 1999;45(4):1129–1149.

Chapter 6
Compressive sensing for wireless sensor networks

Wei Chen[1]

Over the past two decades, the rapid development of technologies in sensing, computing and communication has made it possible to employ wireless sensor networks (WSNs) to continuously monitor physical phenomena in a variety of applications, for example, air-quality monitoring, wildlife tracking, biomedical monitoring and disaster detection. Since the development of these technologies will continue to reduce the size and the cost of sensors in the next few decades, it is believed that WSNs will be involved more and more in our daily lives increasing the impact on the way we live our lives.

A WSN can be defined as a network of sensor nodes, which can sense the physical phenomena in a monitored field and transmit the collected information to a central information-processing station, namely, the fusion center (FC), through wireless links. A wireless sensor node is composed of three basic elements, i.e., a sensing unit, a computation unit and a wireless communication unit, although the node's physical size and shape may differ in various applications. The rapid development of WSNs with various types of sensors has resulted in a dramatic increase in the amount of data that has to be transmitted, stored and processed. As number and resolution of the sensors grow, the main constraints in the development of WSNs are limited battery power, limited memory, limited computational capability, limited wireless bandwidth, the cost and the physical size of the wireless sensor node. While the sensor node is the performance bottleneck, the FC (or any back-end processor) usually has a comparatively high-computational capability and power. The asymmetrical structure of WSNs motivates us to exploit compressive-sensing (CS)-related techniques and to incorporate those techniques into a WSN system for data acquisition.

CS, also called compressed sensing or sub-Nyquist sampling, was initially proposed by Candès, Romberg and Tao in [1] and Donoho in [2], who derived some important theoretical results on the minimum number of random samples needed to reconstruct a signal. By taking advantage of the sparse characteristic of the natural physical signals of interest, CS makes it possible to recover sparse signals from far fewer samples that is predicted by the Nyquist–Shannon sampling theorem.

[1]State Key Laboratory of Rail Traffic Control and Safety, Beijing Jiaotong University, China

The typical CS measurement of signals can be seen as a randomly weighted linear combination of samples in some basis different from the basis where the signal is sparse. Since the number of CS measurements is smaller than the number of elements in a discrete signal, the task of CS reconstruction is to solve an underdetermined matrix equation with a constraint on the sparsity of the signal in some known basis. In contrast to the conventional data-acquisition structure, i.e., doing compression at the sensor node and decompression at the FC, the CS process trade-off an increase in the computational complexity of post-processing at the FC against the convenience of a smaller quantity of data acquisition and lower demands on the computational capability of the sensor node.

As indicated, the rapid development of WSNs with various types of sensors has resulted in a dramatic increase in the amount of data that needs to be processed, transported and stored. CS is a promising technique to deal with the flood of data, as it enables the design of new kinds of data-acquisition systems that combine sensing, compression and data processing in one operation by using the so-called compressive sensors, instead of conventional sensors. In addition, CS is a nonadaptive compression method, and the characteristics of the signals are exploited at the recovery stage. Therefore, compressive sensors do not need the intra-sensor and inter-sensor correlation information to perform compression, which facilitates distributed compression in WSNs.

This chapter introduces the fundamental concepts that are important in the study of CS. We present the mathematical model of CS where the use of sparse signal representation is emphasized. We describe three conditions, i.e., the null space property (NSP), the restricted isometry property (RIP) and mutual coherence, that are used to evaluate the quality of sensing matrices and to demonstrate the feasibility of reconstruction. We briefly review some widely used numerical algorithms for sparse recovery, which are classified into two categories, i.e., convex optimization algorithms and greedy algorithms. Finally, we illustrate various examples where the CS principle has been applied to deal with various problems occurring in WSNs.

6.1 Sparse signal representation

6.1.1 Signal representation

Most naturally occurring signals, that people are interested in monitoring, have a very high degree of redundancy of information. Therefore, by removing redundancy, signals can be transformed to some compressed version that is convenient for storage and transportation.

Transform compression reduces the redundancy of an n dimensional signal $\mathbf{f} \in \mathbb{R}^n$ by representing it with a sparse or nearly sparse representation $\mathbf{x} \in \mathbb{R}^n$ in some basis $\boldsymbol{\Psi} \in \mathbb{R}^{n \times n}$, i.e.:

$$\mathbf{f} = \boldsymbol{\Psi}\mathbf{x} \tag{6.1}$$

Here, sparse means that all elements in vector $\mathbf{x}$ are zeros except for a small number of them. Thus, we say the signal $\mathbf{f}$ is s-sparse if its sparse representation $\mathbf{x}$

has only s nonzero elements. Most naturally occurring signals are not exactly but nearly sparse under a given transform basis, which means the values of the elements in $\mathbf{x}$, when sorted, decay rapidly to zero, or follow power-law distributions, i.e., the ith element of the sorted representation $\grave{\mathbf{x}}$ satisfies:

$$|\grave{x}_i| \leq c \cdot i^{-p} \tag{6.2}$$

for each $1 \leq i \leq n$, where c denotes a constant and $p \geq 1$.

Transforming signals into a sparse domain has been widely used in data reduction. For example, audio signals are compressed by projecting them into the frequency domain, and images are compressed by projecting them into the wavelet domain and curvelet domain. Furthermore, sometimes it is easier to manipulate or process the information content of signals in the projected domain than in the original domain where signals are observed. For example, by expressing audio signals in the frequency domain, one can acquire the dominant information more accurately than by expressing them as the amplitude levels over time. In this case, people are more interested in the signal representation in the transformed domain rather than the signal itself in the observed domain.

6.1.2 Representation error

By preserving a small number of the largest components of the representation, the original signals are compressed subject to a tolerable distortion. If the signals are exactly sparse, then perfect reconstruction of the signals is possible. However, as most signals of interest are nearly sparse in the transformed domain, discarding the small elements of the representations will result in a selective loss of the least significant information. Therefore, the transform compression is typically lossy.

A representation error $\mathbf{x} - \mathbf{x}_s$ occurs if $\mathbf{x}_s$ is used to recover the signal. Assume $\mathbf{x}$ follows power-law distribution as in (6.2), then the squared representation error is given by

$$\begin{aligned} \|\mathbf{x} - \mathbf{x}_s\|_2^2 &\leq \sum_{i=s+1}^{n} \left(c \cdot i^{-p}\right)^2 \approx c^2 \int_{s+1}^{n} z^{-2p} dz \\ &= \frac{c^2}{2p-1}\left((s+1)^{1-2p} - n^{1-2p}\right) \\ &< \frac{c_0}{(s+1)^{2p-1}} \end{aligned} \tag{6.3}$$

where $c_0 = c^2/(2p-1)$ is a constant and $p \geq 1$. Obviously, the more elements that are kept in the representation, the smaller error the compressed signal has. It is noted that the squared sparse representation error decays in the order of $1/(s^{2p-1})$.

Lossy compression is most commonly used to compress multimedia signals. For example, the image in Figure 6.1 with 256×256 pixel can be represented by coefficients of the discrete wavelet transform. By sorting the coefficients in order of their magnitude, we note that only a small portion of the coefficients have a large amplitude. We then choose to neglect all the coefficients whose amplitudes are

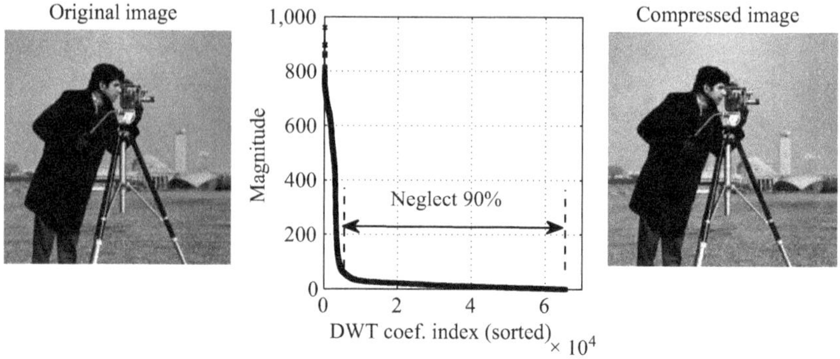

Figure 6.1 The original cameraman image vs. the compressed version

smaller than 10 and show the compressed image in the right-hand image in Figure 6.1. The perceptual loss in quality of the compressed image compared to the original is imperceptible.

6.2 CS and signal recovery

6.2.1 CS model

Given a signal $\mathbf{f} \in \mathbb{R}^n$, we consider a measurement system that acquires m ($m \leq n$) linear measurements by projecting the signal with a sensing matrix $\mathbf{\Phi} \in \mathbb{R}^{m \times n}$. This sensing system can be presented as

$$\mathbf{y} = \mathbf{\Phi}\mathbf{f}, \tag{6.4}$$

where $\mathbf{y} \in \mathbb{R}^m$ denotes the measurement vector.

The standard CS framework assumes that the sensing matrices are randomized and nonadaptive, which means each measurement is derived independently to the previously acquired measurements. In some settings, it is interesting to design fixed and adaptive sensing matrices which can lead to improved performance. More details about the design of sensing matrices are given in [3–7]. For now, we will concentrate on the standard CS framework.

Remembering that the signal $\mathbf{f}$ can be represented by an s-sparse vector $\mathbf{x}$ as expressed in (6.1), the sensing system can be rewritten as

$$\mathbf{y} = \mathbf{\Phi}\mathbf{\Psi}\mathbf{x} = \mathbf{A}\mathbf{x}, \tag{6.5}$$

where $\mathbf{A} = \mathbf{\Phi}\mathbf{\Psi}$ denotes an equivalent sensing matrix. The simplified model with the equivalent sensing matrix $\mathbf{A}$ will be frequently used in this dissertation unless we need to specify the basis, not only to simplify nomenclature but also because many important results are given in the product form of $\mathbf{\Phi}$ and $\mathbf{\Psi}$. More generally, measurements are considered to be contaminated by some noise term $\mathbf{n} \in \mathbb{R}^m$

owing to the sampling noise or the quantization process. Then the CS model can be described as

$$\mathbf{y} = \mathbf{A}\mathbf{x} + \mathbf{n}. \tag{6.6}$$

In generally, it is not possible to solve (6.6) even if the noise term is equal to zero, as there are infinite number of solutions satisfying (6.6). However, a suitable sparsity constraint may rule out all the solutions except for the one that is expected. Therefore, the most natural strategy to recover the sparse representation from the measurements uses ℓ_0 minimization, which can be written as

$$\begin{aligned} \min_{\mathbf{x}} \quad & \|\mathbf{x}\|_0 \\ \text{s.t.} \quad & \mathbf{A}\mathbf{x} = \mathbf{y}. \end{aligned} \tag{6.7}$$

The solution of (6.7) is the most sparse vector satisfying (6.5). However, (6.7) is a combinatorial optimization problem and thus computationally intractable.

Consequently, as a convex relaxation of ℓ_0 minimization, ℓ_1 minimization is used instead to solve the sparse signal representation, which leads to a linear program and thus straight forward to solve [8]. Therefore, the optimization problem becomes:

$$\begin{aligned} \min_{\mathbf{x}} \quad & \|\mathbf{x}\|_1 \\ \text{s.t.} \quad & \mathbf{A}\mathbf{x} = \mathbf{y}. \end{aligned} \tag{6.8}$$

This program is also known as basis pursuit (BP).

In the presence of noise, the equality constraint in (6.8) can never be satisfied. Instead, the optimization problem (6.8) can be relaxed by using the BP de-noising (BPDN) [9], which is

$$\begin{aligned} \min_{\mathbf{x}} \quad & \|\mathbf{x}\|_1 \\ \text{s.t.} \quad & \|\mathbf{A}\mathbf{x} - \mathbf{y}\|_2^2 \le \varepsilon, \end{aligned} \tag{6.9}$$

where ε is an estimate of the noise level. It has been demonstrated that only $m = \mathcal{O}(s \log(n/s))$ measurements [10] are required for robust reconstruction in the CS framework.

This standard CS framework only exploits the sparse characteristics of the signal to reduce the dimensionality required for sensing the signal. A recent growing trend relates to the use of more complex signal models that go beyond the simple sparsity model to further enhance the performance of CS. For example, Baraniuk *et al.* [11] have introduced a model-based CS, where more realistic signal models such as wavelet trees or block sparsity are leveraged in order to reduce the number of measurements required for reconstruction. In particular, it has been shown that robust signal recovery is possible with $m = \mathcal{O}(s)$ measurements in model-based CS [11]. Ji *et al.* [12] introduced Bayesian CS, where a signal statistical model instead is exploited to reduce the number of measurements for reconstruction. In [13,14], reconstruction methods have been proposed for manifold-based CS, where the signal is assumed to belong to a manifold. Other works that consider various sparsity models that go beyond that of simple sparsity in order to improve the performance of traditional CS include [15–18].

6.2.2 Conditions for the equivalent sensing matrix

Another theoretical question in CS is what conditions should the equivalent sensing matrix $\mathbf{A}$ satisfy in order to preserve the information in the sparse representation $\mathbf{x}$. In this subsection, three different conditions for the matrix $\mathbf{A}$ are presented, i.e., the NSP, the RIP and mutual coherence. The information in the sparse representation $\mathbf{x}$ is recoverable by CS if any property can be satisfied.

6.2.2.1 Null space property

For any pair of distinct s-sparse vectors, $\mathbf{x}$ and $\mathbf{x}'$, a proper equivalent sensing matrix $\mathbf{A}$ must have $\mathbf{Ax} \neq \mathbf{Ax}'$. Otherwise, it is impossible to differentiate each other from the measurements $\mathbf{y}$ in conjunction with the s-sparse constraint. Since $\mathbf{x} - \mathbf{x}'$ could be any $2s$-sparse vector and $\mathbf{A}(\mathbf{x} - \mathbf{x}') \neq \mathbf{0}$, we can deduce that there exists a unique s-sparse vector $\mathbf{x}$ satisfying $\mathbf{Ax} = \mathbf{y}$ if and only if the null space of matrix $\mathbf{A}$ does not contain any $2s$-sparse vector.

This condition corresponds to the ℓ_0 norm constraint, which is used in (6.7). However, as mentioned previously, it is painful to solve (6.7) directly. Therefore, it is desirable to evaluate the quality of matrix $\mathbf{A}$ corresponding to the ℓ_1 norm operation which is computationally tractable. Based on this consideration, we now give the definition of the NSP as in [19].

Definition 6.1. *A matrix $\mathbf{A} \in \mathbb{R}^{m \times n}$ satisfies the NSP in ℓ_1 of order s if and only if the following inequality:*

$$\|\mathbf{x}_{\mathscr{J}}\|_1 < \|\mathbf{x}_{\mathscr{J}^c}\|_1 \tag{6.10}$$

holds for all $\mathbf{x} \in Null(\mathbf{A})$ and $\mathbf{x} \neq \mathbf{0}$ with $\|\mathscr{J}\|_0 = s$.

The NSP highlights that vectors in the null space of the equivalent sensing matrix $\mathbf{A}$ should not concentrate on a small number of elements. Based on the definition of the NSP, the following theorem [19] guarantees the success of ℓ_1 minimization with the equivalent sensing matrix $\mathbf{A}$ satisfying the NSP condition.

Theorem 6.1. *Let $\mathbf{A} \in \mathbb{R}^{m \times n}$. Then every s-sparse vector $\mathbf{x} \in \mathbb{R}^n$ is the unique solution of the ℓ_1 minimization problem in (6.8) with $\mathbf{y} = \mathbf{Ax}$ if and only if $\mathbf{A}$ satisfies the NSP in ℓ_1 of order s.*

This theorem claims that the NSP is both necessary and sufficient for successful sparse recovery by ℓ_1 minimization. However, it does not consider the presence of noise as in (6.9). Furthermore, it is very difficult to evaluate the NSP condition for a given matrix, since it includes calculation of the null space and testing all vectors in this space.

6.2.2.2 Restricted isometry property

A stronger condition, named RIP, is introduced by Candès and Tao in [20].

Definition 6.2. *A matrix* $\mathbf{A} \in \mathbb{R}^{m\times n}$ *satisfies the RIP of order s with a restricted isometry constant (RIC)* $\delta_s \in (0, 1)$ *being the smallest number such that*

$$(1 - \delta_s)\|\mathbf{x}\|_2^2 \leq \|\mathbf{A}\mathbf{x}\|_2^2 \leq (1 + \delta_s)\|\mathbf{x}\|_2^2 \tag{6.11}$$

holds for all $\mathbf{x}$ *with* $\|\mathbf{x}\|_0 \leq s$.

The RIP quantifies the notion that the energy of sparse vectors should not be scaled too much when projected by the equivalent sensing matrix $\mathbf{A}$. It has been established in [21] that the RIP provides a sufficient condition for exact or near exact recovery of a sparse signal via ℓ_1 minimization.

Theorem 6.2. *Let* $\mathbf{A} \in \mathbb{R}^{m\times n}$. *Then the solution* $\mathbf{x}^*$ *of (6.9) obeys:*

$$\|\mathbf{x}^* - \mathbf{x}\|_2 \leq c_1 s^{-1/2}\|\mathbf{x} - \mathbf{x}_s\|_{\ell_1} + c_2\varepsilon, \tag{6.12}$$

where $c_1 = (2+(2\sqrt{2}-2)\delta_{2s})/(1-(\sqrt{2}+1)\delta_{2s})$, $c_2 = (4\sqrt{1+\delta_{2s}})/(1-(\sqrt{2}+1)\delta_{2s})$, *and* δ_{2s} *is the RIC of matrix* $\mathbf{A}$.

This theorem claims that with a reduced number of measurements, the reconstructed vector $\mathbf{x}^*$ is a good approximation to the original signal representation $\mathbf{x}$. In addition, for the noiseless case, any sparse representation $\mathbf{x}$ with support size no larger than s, can be exactly recovered by ℓ_1 minimization if the RIC satisfies $\delta_{2s} < \sqrt{2} - 1$. Improved bounds based on the RIP are derived in [22–24].

For any arbitrary matrix, computing the RIP by going through all possible sparse signals is exhaustive. Baraniuk *et al.* prove in [10] that any random matrix whose entries are independent identical distributed (i.i.d.) realizations of certain zero-mean random variables with variance $1/m$, e.g., Gaussian distribution and Bernoulli distribution,[1] satisfies the RIP with a very high possibility when the number of samples $m = \mathcal{O}(s\log(n/s))$.

Note that the RIP is a sufficient condition for successful reconstruction, but it is too strict. In practice, signals with sparse representations can be reconstructed very well even though the sensing matrices do not satisfy the RIP.

6.2.2.3 Mutual coherence

Another way to evaluate a sensing matrix, which is not as computationally intractable as the NSP and the RIP, is via the mutual coherence of the matrix [26], which is given by

$$\mu = \max_{1\leq i,j\leq n, i\neq j} |\mathbf{A}_i^T\mathbf{A}_j|. \tag{6.13}$$

Small mutual coherence means that any pair of vectors in matrix $\mathbf{A}$ has a low coherence, which eases the difficulty in discriminating components from the measurement vector $\mathbf{y}$.

[1] In most of the experiments we have conducted, we use random matrices with elements drawn from i.i.d. Gaussian distributions, since it is the typical setting found in the literature and its performance is no worse than one with elements drawn from i.i.d. Bernoulli distributions [25].

Donoho, Elad and Temlyakov demonstrated in [26] that every $\mathbf{x}$ is the unique sparsest solution of (6.7) if $\mu < 1/(2s - 1)$, and the error of the solution (6.8) is bounded if $\mu < 1/(4s - 1)$. According to the relationship between the RIC and the mutual coherence, i.e., $\delta_s \leq (s - 1)\mu$ [19], it is clear that if a matrix possesses a small mutual coherence, it also satisfies the RIP condition. It means that the mutual coherence condition is a stronger condition than the RIP. However, the mutual coherence is still very attractive for sensing matrix design owing to its convenience in evaluation.

6.2.3 Numerical algorithms for sparse recovery

By applying CS, the number of samples required is reduced and the compression operation is simpler than that for traditional compression methods. However, the convenience of the compression operation leads to the increased complexity of the decoding operation, i.e., the decoder requires sophisticated algorithms to recover the original signal. Many approaches and their variants have been proposed in the literature. Most of those algorithms can be classified into two categories—convex optimization algorithms and greedy algorithms. Here, we briefly review some of these convex optimization algorithms and greedy pursuit algorithms, and refer the interested readers to the literature for other classes of algorithms including Bayesian approaches [12,27–30] and nonconvex optimization approaches [31].

6.2.3.1 Convex optimization algorithms

Replacing the ℓ_0 norm with the ℓ_1 norm as in (6.9) is one approach to solve a convex optimization problem that is computationally tractable. The reason for selecting the ℓ_1 norm is that the ℓ_1 norm is the closest convex function to the ℓ_0 norm, which is illustrated in Figure 6.2. It is clear that any ℓ_p norm with $0 < p < 1$ is not convex,

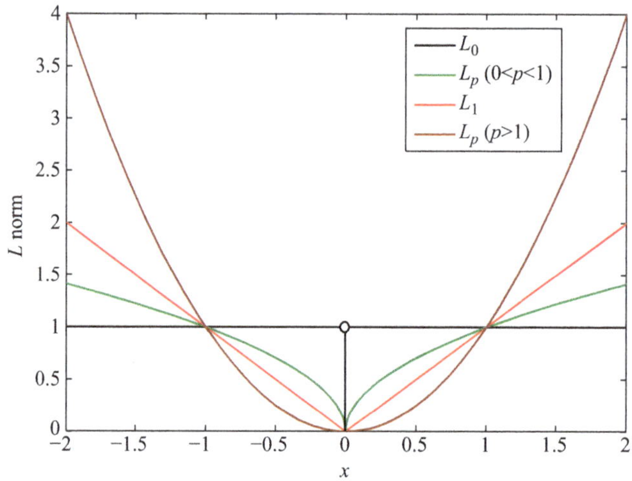

Figure 6.2 Norm functions

and the curve of the ℓ_1 norm is closer to the curve of the ℓ_0 norm than any other ℓ_p norms with $p > 1$.

Some equivalent formulations to (6.9) exist. For example, the least absolute shrinkage and selection operator (LASSO) [32] instead minimizes the energy of detection error with an ℓ_1 constraint:

$$\begin{aligned} \min_{\mathbf{x}} \quad & \|\mathbf{A}\mathbf{x} - \mathbf{y}\|_2^2 \\ \text{s.t.} \quad & \|\mathbf{x}\|_1 \leq \eta, \end{aligned} \tag{6.14}$$

where $\eta \geq 0$. Both BPDN and LASSO can be written as an unconstrained optimization problem with some $\tau \geq 0$ for any $\eta \geq 0$ in (6.14) and $\varepsilon \geq 0$ (6.9):

$$\min_{\mathbf{x}} \quad \frac{1}{2}\|\mathbf{A}\mathbf{x} - \mathbf{y}\|_2^2 + \tau\|\mathbf{x}\|_1. \tag{6.15}$$

Note that the value of τ is an unknown coefficient to make these problems equivalent. How to choose τ is discussed in [33].

There are several methods, such as the steepest descent and the conjugate gradient, to search for the global optimal solution for these convex-relaxed problems. Interior-point (IP) methods, developed in the 1980s to solve convex optimization, are used in [9,34] for sparse reconstruction. Figueiredo, Nowak and Wright propose a gradient projection approach with one level of iteration [35], while the IP approaches in [9,34] have two iteration levels, and ℓ_1-magic [9,34] has three iteration levels. Other algorithms proposed to solve (6.15) include the homotopy method [36,37], the iterative soft thresholding algorithm [38] and the approximately message passing algorithm [39].

Generally, algorithms in this category have better performances than greedy algorithms in terms of the number of measurements required for successful reconstruction. However, their high computing complexity makes them unsuitable for applications where high-dimensional signals are required to be reconstructed within a short time.

6.2.3.2 Greedy pursuit algorithms

While we have seen the success of convex optimization algorithms for sparse recovery, greedy pursuit algorithms also attract much attention because of their simple forms and ease of implementation. The main idea of greedy pursuit algorithms is to iteratively compute the support of the sparse representation and so estimate the representation corresponding to the identified support until some stopping criterion is satisfied.

One main branch of those algorithms is known as matching pursuit (MP) [40]. The outline of MP can be described as follows:

- **Step 0**: Initialize the signal $\mathbf{x} = \mathbf{0}$, the residual $\mathbf{r} = \mathbf{y}$ and the index set $\mathscr{I} = \emptyset$.
- **Step 1**: Find the largest coordinate i of $\mathbf{A}^T\mathbf{r}$.
- **Step 2**: Add the coordinate to the index set $\mathscr{I} = \mathscr{I} \bigcup \{i\}$, and calculate the product $u = \mathbf{r}^T\mathbf{A}_i$.
- **Step 3**: Update the residual $\mathbf{r} = \mathbf{r} - u\mathbf{A}_i$ and the signal $x_i = x_i + u$.
- **Step 4**: Return to step 1 until the stopping criterion is satisfied.

The disadvantage of MP is its poor performance although asymptotic convergence is guaranteed. A modified version called orthogonal MPs (OMPs) is proposed in [41]. OMP converges faster than MP by ensuring full backward orthogonality of the error. Tropp and Gilbert claimed in [42] that OMP can reliably reconstruct a signal with s nonzero entries out of n coordinates using $\mathcal{O}(s \ln n)$ random linear measurements of that signal. Comparing with BP, OMP requires more measurements for reconstruction. Another shortcoming of OMP is that only one component of the support of the signal is selected in each iteration. The number of iterations would be large for recovering a signal having a lot of nonzero components. To overcome this defect, Donoho *et al.* propose an alternative greedy approach, called stagewise OMP (StOMP) [43]. Instead of choosing one component in each iteration, StOMP selects all of the components whose values are above a specified threshold. Obviously, StOMP is faster than OMP, and in some cases, it even outperforms BP. However, a suitable threshold for StOMP is difficult to acquire in practice, which significantly effects the reconstruction performance. Many other algorithms based on OMP have been proposed in the literature such as regularized OMP [44], compressive sampling MP [45] and subspace MP [46], all of which require prior knowledge of the number of nonzero components.

6.3 Optimized sensing matrix design for CS

The sensing matrices used in CS play a key role for successful reconstruction in underdetermined sparse-recovery problems. A number of conditions, such as the NSP, the RIP and mutual coherence, have been put forth in order to study the quality of the sensing matrices and recovery algorithms. These conditions are mainly used to address the worst case performance of sparse recovery [25,47,48]. However, the actual reconstruction performance in practice is often much better than the worst case performance, so that this viewpoint can be too conservative. In addition, the worst case performance is a less typical indicator of quality in signal-processing applications than the expected-case performance. This motivates us to investigate the design of sensing matrices with adequate expected-case performance. Furthermore, a recent growing trend relates to the use of more complex signal models that go beyond the simple sparsity model to further enhance the performance of CS. The use of additional signal knowledge also enables one to replace the conventional random sensing matrices by optimized ones in order to further enhance CS performance (e.g., see [3–7,49–53]).

6.3.1 Elad's method

The mutual coherence denotes the maximum coherence between any pair of columns in the equivalent sensing matrix $\mathbf{A}$. In [49], Elad considered a different coherence indicator, called t-averaged mutual coherence, which reflects the average coherence between columns. The t-averaged mutual coherence is defined as the average of all normalized absolute inner products between different columns in the equivalent sensing matrix that are not smaller than a positive number t.

Definition 6.3. *For a matrix* $\mathbf{A} \in \mathbb{R}^{m \times n}$*, its t-averaged mutual coherence is*

$$\mu_t(\mathbf{A}) = \frac{\sum_{1 \le i,j \le n \text{ and } i \ne j} \mathbf{1}(\left|\mathbf{A}_i^T \mathbf{A}_j\right| \ge t) \cdot \left|\mathbf{A}_i^T \mathbf{A}_j\right|}{\sum_{1 \le i,j \le n \text{ and } i \ne j} \mathbf{1}(\left|\mathbf{A}_i^T \mathbf{A}_j\right| \ge t)}, \tag{6.16}$$

where $t \ge 0$*, and the function* $\mathbf{1}(\bullet)$ *is equal to 1 if its input expression is true, otherwise it is equal to 0.*

If $t = 0$, the t-averaged mutual coherence is the average of all coherence between columns. If $t = \mu$, then the t-averaged mutual coherence μ_t is exactly equal to the mutual coherence μ. Elad claimed that the equivalent sensing matrix $\mathbf{A}$ will have a better performance if one can reduce the coherence of columns. Iteratively reducing the mutual coherence by adjusting the related pair of columns is not an efficient approach to do this since the coherence of all column pairs is not improved except for the worst pair in each iteration. The t-averaged mutual coherence includes the contribution of a batch of column pairs with high coherence. Thus, one can improve the coherence of many column pairs by reducing the t-averaged mutual coherence. Elad proposes an iterative algorithm to minimize $\mu_t(\mathbf{A}) = \mu_t(\mathbf{\Phi\Psi})$ with respect to the sensing matrix $\mathbf{\Phi}$, assuming the basis $\mathbf{\Psi}$ and the parameter t are fixed and known.

In each iteration, the Gram matrix $\mathbf{G} = \mathbf{A}^T\mathbf{A}$ is computed, and the values above t are forced to reduce by multiplication with γ $(0 < \gamma < 1)$, which can be expressed as

$$\hat{G}_{i,j} = \begin{cases} \gamma G_{i,j} & |G_{i,j}| \ge t \\ \gamma t \cdot \mathrm{sign}(G_{i,j}) & t > |G_{i,j}| \ge \gamma t \\ G_{i,j} & \gamma t > |G_{i,j}| \end{cases}, \tag{6.17}$$

where sign($\bullet$) denotes the sign function. The shrunk Gram matrix $\hat{\mathbf{G}}$ becomes full rank in the general case due to the operation in (6.17). To fix this, the Gram matrix $\hat{\mathbf{G}}$ is forced to be of rank m by applying the singular value decomposition (SVD) and setting all the singular values to be zero except for the m largest ones. Then one can build the square root of $\hat{\mathbf{G}}$, i.e., $\tilde{\mathbf{A}}^T\tilde{\mathbf{A}} = \hat{\mathbf{G}}$, where the square root $\tilde{\mathbf{A}}$ is of size $m \times n$. The last step in each iteration is to find a sensing matrix $\mathbf{\Phi}$ that makes $\mathbf{\Phi\Psi}$ closest to $\tilde{\mathbf{A}}$ by minimizing $\|\tilde{\mathbf{A}} - \mathbf{\Phi\Psi}\|_F^2$.

The outline of Elad's algorithm is given as follows:

- **Step 0**: Generate an arbitrary random matrix $\mathbf{\Phi}$.
- **Step 1**: Generate a matrix $\mathbf{A}$ by normalizing the columns of $\mathbf{\Phi\Psi}$.
- **Step 2**: Compute the Gram matrix $\mathbf{G} = \mathbf{A}^T\mathbf{A}$.
- **Step 3**: Update the Gram matrix $\hat{\mathbf{G}}$ by (6.17).
- **Step 4**: Apply SVD and set all the singular values of $\hat{\mathbf{G}}$ to be zero except for the m largest ones.
- **Step 5**: Build the square root $m \times n$ matrix $\tilde{\mathbf{A}}$ by $\tilde{\mathbf{A}}^T\tilde{\mathbf{A}} = \hat{\mathbf{G}}$.
- **Step 6**: Update the sensing matrix $\mathbf{\Phi}$ by minimizing $\left\|\tilde{\mathbf{A}} - \mathbf{\Phi\Psi}\right\|_F^2$.
- **Step 7**: Return to step 1 if some halting condition is not satisfied.

Elad's method aims to minimize the large absolute values of the off-diagonal elements in the Gram matrix and thus reduces the t-averaged mutual coherence. This method updates a number of columns at the same time in each iteration. Therefore, it converges to a good matrix design faster than directly working on and updating the mutual coherence iteratively. Empirical knowledge is required to determine the value of t and γ, which affects the matrix quality and the convergence rate, respectively.

6.3.2 Duarte-Carvajalino and Sapiro's method

In [50], Duarte-Carvajalino and Sapiro propose an algorithm to iteratively optimize both the sensing matrix and the basis simultaneously. For any given basis, they propose an m-step algorithm to optimize the sensing matrix. Their aim is to make the Gram matrix of the equivalent sensing matrix as close as possible to an identity matrix, which can be described as

$$\mathbf{G} = \mathbf{A}^T\mathbf{A} = \mathbf{\Psi}^T\mathbf{\Phi}^T\mathbf{\Phi}\mathbf{\Psi} \approx \mathbf{I}_n. \tag{6.18}$$

By multiplying both sides of (6.18) with $\mathbf{\Psi}$ on the left and $\mathbf{\Psi}^T$ on the right, we have

$$\mathbf{\Psi}\mathbf{\Psi}^T\mathbf{\Phi}^T\mathbf{\Phi}\mathbf{\Psi}\mathbf{\Psi}^T \approx \mathbf{\Psi}\mathbf{\Psi}^T. \tag{6.19}$$

Now $\mathbf{\Psi}\mathbf{\Psi}^T$ can be decomposed to $\mathbf{V}\mathbf{\Lambda}\mathbf{V}^T$ by eigen-decomposition, where $\mathbf{V} \in \mathbb{R}^{n\times n}$ is an orthonormal matrix and $\mathbf{\Lambda} \in \mathbb{R}^{n\times n}$ is a diagonal matrix. Thus, (6.19) can be rewritten as

$$\mathbf{V}\mathbf{\Lambda}\mathbf{V}^T\mathbf{\Phi}^T\mathbf{\Phi}\mathbf{V}\mathbf{\Lambda}\mathbf{V}^T \approx \mathbf{V}\mathbf{\Lambda}\mathbf{V}^T, \tag{6.20}$$

which is equivalent to

$$\mathbf{\Lambda}\mathbf{V}^T\mathbf{\Phi}^T\mathbf{\Phi}\mathbf{V}\mathbf{\Lambda} \approx \mathbf{\Lambda}. \tag{6.21}$$

After defining a matrix $\mathbf{\Gamma} = \mathbf{\Phi}\mathbf{V}$, they formulated the following optimization problem:

$$\min_{\mathbf{\Gamma}} \quad \|\mathbf{\Lambda} - \mathbf{\Lambda}\mathbf{\Gamma}^T\mathbf{\Gamma}\mathbf{\Lambda}\|_F^2, \tag{6.22}$$

and then let the sensing matrix $\mathbf{\Phi} = \mathbf{\Gamma}\mathbf{V}^T$.

Duarte-Carvajalino and Sapiro select (6.18) as their optimal design without giving a theoretical analysis, although the approach appears reasonable and the superiority of the design is witnessed in their experimental results. As a closed form for its solution cannot be determined, they propose an algorithm that requires m iterations to determine a solution to (6.22).

The outline of their algorithm is given as follows:

- **Step 0**: Generate an arbitrary random matrix $\mathbf{\Phi}$.
- **Step 1**: Apply eigen-decomposition $\mathbf{\Psi}\mathbf{\Psi}^T = \mathbf{V}\mathbf{\Lambda}\mathbf{V}^T$.
- **Step 2**: Let $\mathbf{\Gamma} = \mathbf{\Phi}V$, $\mathbf{Z} = \mathbf{\Gamma}\mathbf{\Lambda}$ and $i = 1$.
- **Step 3**: Find the largest eigenvalue τ and corresponding eigenvector $\mathbf{u}$ of $\|\mathbf{\Lambda} - \mathbf{\Lambda}\mathbf{\Gamma}^T\mathbf{\Gamma}\mathbf{\Lambda} + \mathbf{Z}_i\mathbf{Z}_i^T\|_F^2$.
- **Step 4**: Let $\mathbf{Z}_i = \sqrt{\tau}\mathbf{u}$ and $i = i + 1$.
- **Step 5**: Update $\mathbf{\Gamma}$ according to $\mathbf{\Gamma} = \mathbf{Z}\mathbf{\Lambda}^{-1}$.
- **Step 6**: Return to step 2 unless $i = m + 1$.
- **Step 7**: Compute the sensing matrix $\mathbf{\Phi} = \mathbf{\Gamma}\mathbf{V}^T$.

6.3.3 Xu et al.'s method

Xu *et al.* considered the equiangular tight frame as their target design and proposed an iterative algorithm to make the sensing matrix close to that design [51]. Tight frames will be introduced in the next section, but for now, we only need to mention that the equiangular tight frame is a class of matrix with some good properties, for example, if the coherence value between any two columns of the equivalent sensing matrix $\mathbf{A}$ is equal, then $\mathbf{A}$ is an equiangular tight frame.

Although the equiangular tight frame does not exist for any arbitrary selection of dimensions and finding the equiangular tight frame for given dimensions is difficult, the achievable lower bound of the Gram matrix of the equiangular tight frame, i.e., $\mathbf{G} = \mathbf{A}^T\mathbf{A}$, has been derived [54]:

$$|G_{i,j}| \geq \sqrt{\frac{n-m}{m(n-1)}} \qquad \text{for } i \neq j. \tag{6.23}$$

Being aware of the difficulty in generating equiangular tight frames, Xu *et al.* propose an optimization approach that iteratively makes the Gram matrix of the equivalent sensing matrix close to the lower bound in (6.23). In each iteration, a new Gram matrix is calculated by

$$\hat{G}_{i,j} = \begin{cases} 1 & i = j \\ G_{i,j} & |G_{i,j}| < \sqrt{\dfrac{n-m}{m(n-1)}} \\ \sqrt{\dfrac{n-m}{m(n-1)}} \cdot \text{sign}(G_{i,j}) & |G_{i,j}| \geq \sqrt{\dfrac{n-m}{m(n-1)}} \end{cases}. \tag{6.24}$$

Then they update the Gram matrix by

$$\mathbf{G} = \alpha\mathbf{G}_{\text{prev}} + (1-\alpha)\hat{\mathbf{G}}, \tag{6.25}$$

where $\mathbf{G}_{\text{prev}}$ denotes the Gram matrix in the previous iteration, and $0 < \alpha < 1$ denotes the forgetting parameter. Then they update the sensing matrix $\mathbf{\Phi}$ as the matrix with the minimum distance to the Gram matrix, given by

$$\min_{\mathbf{\Phi}} \quad \|\mathbf{\Psi}^T\mathbf{\Phi}^T\mathbf{\Phi}\mathbf{\Psi} - \mathbf{G}\|_F^2, \tag{6.26}$$

which can be solved using QR factorization with eigen-decomposition. As the Gram matrix is forced to the bound of the equiangular tight frame in each iteration, they expect the equivalent sensing matrix in turn to be close to an equiangular tight frame.

The outline of their algorithm is given as follows:

- **Step 0**: Generate an arbitrary random matrix $\mathbf{\Phi}$.
- **Step 1**: Compute the Gram matrix $\mathbf{G} = \mathbf{\Psi}^T\mathbf{\Phi}^T\mathbf{\Phi}\mathbf{\Psi}$.
- **Step 2**: Update the Gram matrix by (6.24).
- **Step 3**: Update the Gram matrix by (6.25).
- **Step 4**: Apply SVD and set all the singular values of $\mathbf{G}$ to be zero except for the m largest ones.

- **Step 5**: Build the square root $m \times n$ matrix $\tilde{\mathbf{A}}$ by $\tilde{\mathbf{A}}^T\tilde{\mathbf{A}} = \mathbf{G}$.
- **Step 6**: Update the sensing matrix $\mathbf{\Phi}$ by minimizing $\left\|\tilde{\mathbf{A}} - \mathbf{\Phi}\mathbf{\Psi}\right\|_F^2$.
- **Step 7**: Return to step 1 if some halting condition is not satisfied.

Xu *et al.* make the equiangular tight frame lower bound as the target of their design, as the equiangular tight frame has minimum mutual coherence [51]. However, the lower bound can never be achieved if an equiangular tight frame for dimensions $m \times n$ does not exist. Although the design target is based on knowledge of the bound, an improved performance has been shown for arbitrary dimensions.

6.3.4 Chen et al.'s method

While previous work considers iterative methods, Chen *et al.* proposed noniterative methods that use tight frame principles for CS sensing matrix design. For finite-dimensional real spaces, a frame can be seen as a matrix $\mathbf{\Phi} \in \mathbb{R}^{m\times n}$ such that for any vector $\mathbf{z} \in \mathbb{R}^m$:

$$a\|\mathbf{z}\|_2^2 \leq \|\mathbf{\Phi}^T\mathbf{z}\|_2^2 \leq b\|\mathbf{z}\|_2^2, \tag{6.27}$$

where $a > 0$ and $b > 0$ are called frame bounds. Tight frames are a class of frames with equal frame bounds, i.e., $a = b$. Any tight frame can be scaled with the frame bound equal to 1 by multiplying $1/\sqrt{a}$. A tight frame is called a Parseval tight frame if its frame bound is equal 1, i.e.:

$$\|\mathbf{\Phi}^T\mathbf{z}\|_2^2 = \|\mathbf{z}\|_2^2, \tag{6.28}$$

for any $\mathbf{z}$. A tight frame represents a matrix whose coherence matrix is as close as possible to an orthonormal matrix corresponding to the Frobenius norm. Tight frames have been widely used in many applications such as denoising, code division multiple access systems and multi-antenna code design. Equal-norm tight frames require one more condition than a general tight frame, i.e., the ℓ_2 norm of all columns are equal. If all the columns of a tight frame are equal to 1, it is called a unit-norm tight frame.

In [4], Chen *et al.* proposed the use of unit-norm tight frames, which has been justified from optimization considerations and has been shown to lead to MSE performance gains when used in conjunction with standard sparse recovery algorithms. Finding a matrix, which has minimum mutual coherence and columns with ℓ_2 norm equal to 1, is equivalent to finding n points on the sphere in $\mathbb{R}^m$ so that the points are as orthogonal to each other as possible. However, the equiangular tight frame does not exist for any arbitrary selection of dimensions and finding the equiangular tight frame for any given dimension is in general very difficult. An alternative approach is to find the equilibrium of the points on the sphere under some applied "force." Equilibrium in this case means that in such a state, the points will return to their original positions if slightly disturbed.

In [3], Chen *et al.* proposed a closed-form sensing matrix design, given an over-complete dictionary $\mathbf{\Psi}$. The proposed sensing matrix design is given by

$$\hat{\mathbf{\Phi}} = \mathbf{U}_{\hat{\Phi}}\mathbf{\Lambda}_{\hat{\Phi}}\mathbf{\Pi}^T\mathbf{U}_{\Psi}^T, \tag{6.29}$$

where $\mathbf{U}_{\hat{\Phi}}$ is an arbitrary orthonormal matrix, $\mathbf{\Lambda}_{\Phi} = \left[\text{Diag}\left((1/\lambda_m^{\Psi}), \ldots, (1/\lambda_1^{\Psi})\right) \; \mathbf{O}_{m\times(n-m)}\right]$ and $\mathbf{\Pi} = \mathbf{J}_n$. It uncovers the key operations performed by this optimal sensing matrix design. In particular, this sensing matrix design (i) exposes the modes (singular values) of the dictionary; (ii) passes the m strongest modes and filters out the $n - m$ weakest modes and (iii) weighs the strongest modes. This is also accomplished by taking the matrix of right singular vectors of the sensing matrix to correspond to the matrix of left singular vectors of the dictionary and taking the strongest modes of the dictionary. It leads immediately to the sensing matrix design, which is consistent with the sensing cost constraint $\|\hat{\mathbf{\Phi}}\|_F^2 = n$, as follows:

$$\mathbf{\Phi} = \frac{\sqrt{n}\hat{\mathbf{\Phi}}}{\|\hat{\mathbf{\Phi}}\|_F} = \frac{\sqrt{n}\mathbf{\Lambda}_{\hat{\Phi}}\mathbf{J}_n\mathbf{U}_{\Psi}^T}{\|\mathbf{\Lambda}_{\hat{\Phi}}\|_F}. \tag{6.30}$$

The sensing matrix design can be generated as follows:

- **Step 0**: For a given $n \times k$ dictionary $\mathbf{\Psi}$, perform SVD, i.e., $\mathbf{\Psi} = \mathbf{U}_{\Psi}\mathbf{\Lambda}_{\Psi}\mathbf{V}_{\Psi}^T$ $(\lambda_1^{\Psi} \geq \cdots \geq \lambda_n^{\Psi})$.
- **Step 1**: Build an $m \times n$ matrix $\mathbf{\Lambda}_{\Phi}$ whose diagonal entries are $\lambda_i^{\Phi} = (1/\lambda_i^{\Psi})$ $(i = 1, \ldots, m)$ and all the other entries are zeros.
- **Step 2**: Build the sensing matrix $\mathbf{\Phi} = \mathbf{U}_{\hat{\Phi}}\mathbf{\Lambda}_{\Phi}\mathbf{U}_{\Psi}^T$, where $\mathbf{U}_{\hat{\Phi}}$ is an arbitrary orthonormal matrix.
- **Step 3**: Normalize the sensing matrix energy using $\mathbf{\Phi} = (\sqrt{n}\mathbf{\Phi}/\|\mathbf{\Phi}\|_F)$.

In [5], Ding, Chen and Wassell proposed sensing matrix designs for tensor CS. Previous study concerning CS implicitly assume the sampling costs for all samples are equal and suggest random sampling as an appropriate approach to achieve good reconstruction accuracy. However, this assumption does not hold for applications such as WSNs which have significant variability in sampling cost owing to the different physical conditions at particular sensors. To exploit this sampling cost nonuniformity, Chen *et al.* proposed cost-aware sensing matrix designs that minimize the sampling cost with constraints on the regularized mutual coherence of the equivalent sensing matrix [6,7].

6.4 CS-based WSNs

The CS principle can be applied to various practical applications where the sensing matrices represent different systems. In this section, we only illustrate five examples where the CS principle is used to deal with different problems occurring in WSNs, although CS can address a much wider range of applications.

6.4.1 Robust data transmission

Data loss in WSNs is inevitable due to the wireless fading channel. Various error-detection and error-correction schemes have been proposed to fight against channel fading in the physical layer. In addition, to guarantee reliable data transmission, retransmission schemes are often applied in the application layer. Considering the poor computing capability of wireless sensor devices and the bandwidth overhead, many error protection and retransmission schemes are not suitable for WSNs. In addition, for small and battery-operated sensor devices, sophisticated source coding cannot be afforded in some cases. However, as naturally occurring signals are often compressible, CS can be viewed as a compression process. What is more interesting is that these CS data with redundant measurements are robust against data loss, i.e., the original signal can be recovered without retransmission even though some data are missing.

As shown in Figure 6.3, the conventional sequence of sampling, source coding and channel coding is replaced by one CS procedure. Each CS measurement contains some information about the whole signal owing to the mixture effect of the sensing matrix. Thus, any lost measurement will not cause an inevitable information loss. With some redundant measurements, the CS system can combat data loss and successfully recover the original signal.

This CS-coding scheme has a low-encoding cost especially if random sampling is used. All the measurements are acquired in the same way, and thus the number of redundant measurements can be specified according to fading severity of the wireless channel. In addition, one can still use physical layer channel coding on the CS measurements. In this case, CS can be seen as a coding strategy that is applied at the application layer where the signal characteristics are exploited, and also can be seen as a replacement of traditional sampling and source-coding procedures. If channel coding fails, the receiver is still able to recover the original signal in the application layer.

In [56], Davenport *et al.* demonstrate theoretically that each CS measurement carries roughly the same amount of signal information if random matrices are used. Therefore, by slightly increasing the number of measurements, the system is robust to the loss of a small number of arbitrary measurements. Charbiwala *et al.* show that this CS coding approach is efficient for dealing with data loss, and cheaper than several other approaches including Reed–Solomon encoding in terms of energy consumption using a MicaZ sensor platform [57].

Note that the fountain codes [58]—in particular random linear fountain codes—and network coding [59] can also be used to combat data loss by transmitting mixed

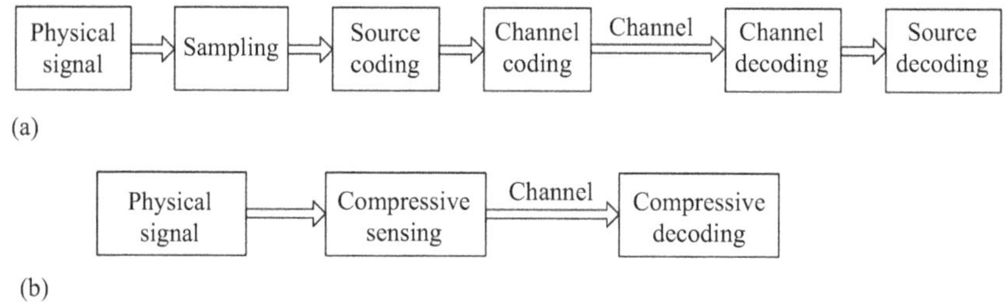

Figure 6.3 Conventional transmission approach vs. CS approach: (a) the conventional sequence of source and channel coding and (b) joint source and channel coding using CS

symbols, which are "equally important." The number of received symbols for both approaches should be no smaller than the original number of symbols for decoding, which is not necessary in the CS-based approach owing to the use of the sparse signal characteristic.

6.4.2 Compressive data gathering

Typical WSNs consist of a large number of sensors distributed in the field to collect information of interest for geographical monitoring, industrial monitoring, security and climate monitoring. In these WSNs, signals sensed in the physical field usually have high spatial correlations. Generally, it is difficult to compress the signals at the sensor nodes due to their distributed structure. However, by exploiting the CS principle, the signals can be gathered and transmitted in an efficient way.

6.4.2.1 WSNs with single hop communications

The proposed architecture in [60] for efficient estimation of sensor field data considers single hop communications between n sensor nodes and an FC as shown in Figure 6.4. Each sensor has a monitored parameter f_i $(i = 1, \ldots, n)$ to report. The conventional method to transmit the total number of parameters, n, requires n time slots by allocating one time slot to each sensor node, while the new strategy only needs m $(m < n)$ time slots to transmit all the information.

In the new strategy, each sensor generates a random sequence $\mathbf{\Phi}_i \in \mathbb{R}^m$ $(i = 1, \ldots, n)$ by using its network address as the seed of a pseudorandom number generator. Each sensor sequentially transmits the product of the random sequence and the sensed parameter f_i in m time slots, while the transmission power is reduced to $1/m$ of its default value in each time slot. All the sensors transmit in an analog fashion. The received signal at the FC in m time slots can be written as

$$\mathbf{y} = \mathbf{H} \circ \mathbf{\Phi}\mathbf{f} + \mathbf{n}, \tag{6.31}$$

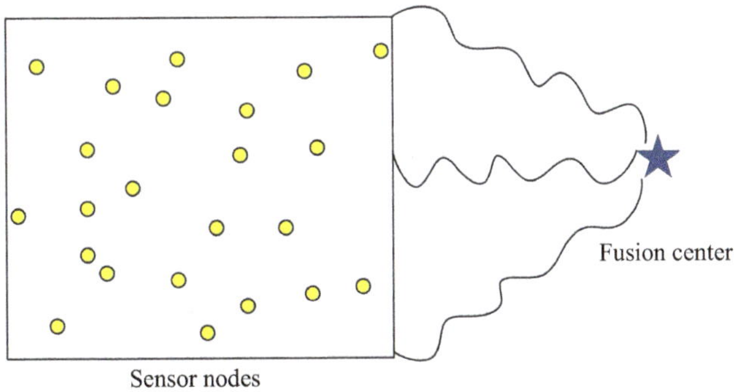

Figure 6.4 A WSN with single hop communication

where $\mathbf{\Phi} \in \mathbb{R}^{m\times n}$ denotes the random matrix corresponding to n different random sequences, $\mathbf{H} \in \mathbb{R}^{m\times n}$ denotes the channel path gain matrix, and $\mathbf{n}$ denotes the noise term. It is assumed that the channel path gain $h_{i,j}$ can be calculated using:

$$h_{j,i} = \frac{1}{d_i^{\alpha/2}}, \tag{6.32}$$

where d_i is the distance between the ith sensor and the FC, and α is the propagation loss factor. It is also assumed that the FC knows the network address of each sensor node and its distance to each sensor node.

Authors in [61] propose a method exploiting both the intra-sensor and inter-sensor correlation to reduce the number of samples required for reconstruction of the original signals. In [62], a sampling rate indicator feedback scheme is proposed to enable the sensor to adjust its sampling rate to maintain an acceptable reconstruction performance while minimizing the number of samples. In [6], authors propose a cost-aware activity-scheduling approach that minimizes the sampling cost with constraints on the regularized mutual coherence of the equivalent sensing matrix.

As the sensed field signal has high spatial correlation, it has a sparse representation in some basis. In [63], different 2-D transformations which sparsify the spatial signals are discussed with real data. In [64], the authors use principal component analysis to find transformations that sparsify the signal. Then, by exploiting the CS principle, the FC can recover the original signal $\mathbf{f}$ from the received signal $\mathbf{y}$.

Note that multiterminal source coding can also be used for joint decoding multiple correlated sources, where statistical correlation models are considered typically. However, the CS-based approach relies on a sparse transform, e.g., wavelet transform, which is appropriate for a specific class of signals.

6.4.2.2 WSNs with multi-hop communications

Sensor readings in some WSNs are transmitted to the FC through multi-hop routing. In [65], Luo *et al.* propose a compressive data gathering (CDG) scheme to reduce the communication cost of WSNs using multi-hop transmissions. A chain-type WSN, with n sensor nodes as shown in Figure 6.5(a), requires $\mathcal{O}(n^2)$ total message transmissions in the network and $\mathcal{O}(n)$ maximum message transmissions for any single sensor node. On the other hand, in the CDG scheme shown in Figure 6.5(b), the jth sensor node transmits m ($m < n$) messages that are the sum of the received message vector $\sum_{i=1}^{j-1} \boldsymbol{\phi}_i f_i$ and its own message vector $\boldsymbol{\phi}_j f_j$ generated by multiplying its monitored element f_j with a spreading code $\boldsymbol{\phi}_j \in \mathbb{R}^m$. By exploiting the CS principle, the CDG scheme only requires $\mathcal{O}(mn)$ total message transmissions in the network and $\mathcal{O}(m)$ maximum message transmissions for any single node.

This CDG scheme can also be applied to a tree-type topology [65]. Although the transmission cost for the tree-type topology is different to the chain-type topology, the superiority of the CDG scheme over the baseline scheme remains [65].

6.4.3 *Sparse events detection*

Another important usage of WSNs is to detect anomalies. For example, WSNs can be used to detect poisonous gas or liquid. Not only an alarm concerning the leakage but also the leaking positions and the volumes of the leaks are reported to the monitoring system. Other anomalies, such as abnormal temperature, humidity and so on, can also be detected by WSNs. All of these anomaly detection problems can be analyzed using the same model [66–69].

As shown in Figure 6.6, the n grid intersection points denote sources to be monitored, the m yellow nodes denote sensors, and the s red hexagons denote anomalies. The monitored phenomena is modeled as a vector $\mathbf{x} \in \mathbb{R}^n$ where x_i denotes the value at the ith monitored position. The normal situation is represented by $x_i = 0$ and $x_i \neq 0$ represents the anomaly. The measurements of the sensors are denoted by a vector $\mathbf{y} \in \mathbb{R}^m$ where y_j represents the jth sensor's measurement. The relationship between the events $\mathbf{x}$ and measurements $\mathbf{y}$ can be written as

$$\mathbf{y} = \mathbf{A}\mathbf{x} + \mathbf{n}, \tag{6.33}$$

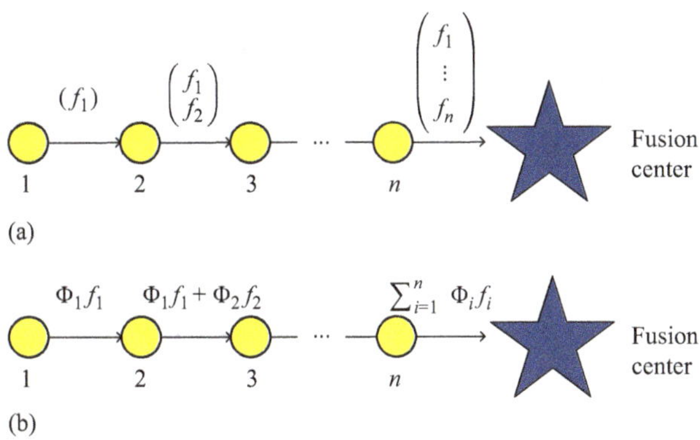

Figure 6.5 Chain-type WSNs: (a) baseline data gather and (b) compressive data gather

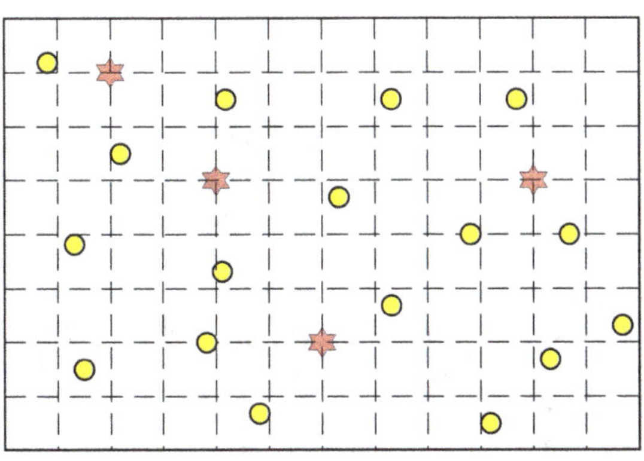

Figure 6.6 A WSN for sparse event detection

where $\mathbf{n} \in \mathbb{R}^m$ is the noise vector. The channel response matrix is denoted by $\mathbf{A} \in \mathbb{R}^{m\times n}$. The influence of the ith source to the jth sensor can be calculated as

$$A_{i,j} = \frac{|g_{i,j}|}{d_{i,j}^{\alpha/2}}, \tag{6.34}$$

where $d_{i,j}$ is the distance from the ith source to the jth sensor, α is the propagation loss factor and $g_{i,j}$ is the Rayleigh fading factor.

Assume that the total number of sources n is large while the number of anomalies s is relatively very small, i.e., $(s \ll n)$. The number of sensors m follows $s < m \ll n$. Therefore, to solve $\mathbf{x}$ from a reduced number of measurements, $\mathbf{y}$ turns out to be a typical sparse recovery problem. Consequently, WSNs are able to accurately detect multiple events at high spatial resolution by using measurements from a small number of sensor devices.

Various research have been performed using this framework. Meng, Li and Han formulate the problem for sparse event detection in WSNs as a CS problem in [66], where they assume the events are binary in nature, i.e., either $x_i = 1$ or $x_i = 0$. Ling and Tian develop a decentralized algorithm for monitoring localized phenomena in WSNs using a similar CS model [67]. Zhang *et al.* rigorously justify the validity of the formulation of sparse event detection by ensuring that the equivalent sensing matrix satisfies the RIP [68]. Liu *et al.* further exploit the temporal correlation of the events to improve the detection accuracy [69].

6.4.4 Reduced-dimension multiple access

In WSNs, medium access control (MAC) plays an important role in data transmission, where n sensor nodes share a wireless channel with m $(m < n)$ degree of freedom. Uncoordinated channel access from different sensor nodes will lead to packet collisions and retransmissions, which reduces both the efficiency of bandwidth usage and the lifetime of the sensor nodes. The simplest MAC protocols in WSNs are designed to avoid simultaneous transmissions. For example, the 802.11 protocol protects a communication link by disabling all other nodes using request-to-send and clear-to-send messages. Simultaneous transmissions can be realized by using multiuser communication techniques in the physical layer, which will result in an improved network throughput performance. One of these techniques is multiuser detection, where the received signal consists of a noisy version of the superposition of a number of transmitted waveforms, and the receiver has to detect the symbols of all users simultaneously. However, multiuser detectors in general have high complexities, as the number of correlators used at the receiver's front-end is equal to the number of users in the system.

In WSNs, the number of active sensor nodes s at any time is much smaller than the total number of sensor nodes n, which can be exploited to reduce the dimension in multiuser detection. We assume that the duration of the period of one time frame is less than the coherence time of both the monitored signal and the wireless channel, i.e., both the environment and the channel remain static in one time frame. Each time frame can be divided into m time slots, in which sensor nodes can transmit one symbol.

Each sensor node has a unique signature sequence $\boldsymbol{\phi}_i \in \mathbb{R}^m$, which is multiplied by its transmitted symbol. The received signal $\mathbf{y} \in \mathbb{R}^m$ in one time frame at the receiver is given by

$$\mathbf{y} = \sum_{i=1}^{n} \boldsymbol{\phi}_i h_i f_i + \mathbf{n}, \tag{6.35}$$

where h_i and f_i denote the channel gain and transmitted symbol corresponding to the ith sensor node, and $\mathbf{n} \in \mathbb{R}^m$ denotes a white Gaussian noise vector. We assume that both the channel gains h_i $(i = 1, \ldots, n)$ and sensor signature sequences $\boldsymbol{\phi}_i$ $(i = 1, \ldots, n)$ are known at the receiver. It is in general impossible to solve $\mathbf{f} \in \mathbb{R}^n$ with m received measurements. However, as there are very few active sensor nodes in each time frame, the transmitted symbols can be reconstructed exploiting CS reconstruction algorithms.

This reduced-dimension MAC design has been proposed in WSN applications to save channel resource and power consumption [70,71]. Various linear and nonlinear detectors are given and analyzed by Xie, Eldar and Goldsmith [72]. In [73], in addition to the sparsity of active sensor nodes, the authors exploit additional correlations that exist naturally in the signals to further improve the performance in terms of power efficiency.

6.4.5 Localization

Accurate localization is very important in many applications including indoor location-based services for mobile users, equipment monitoring in WSNs and radio frequency identification-based tracking. In the outdoor environment, the global positioning system (GPS) works very well for localization purpose. However, this solution for the outdoor environment is not suitable for an indoor environment. For one thing, it is difficult to detect the signal from the GPS satellites in most buildings due to the penetration loss of the signal. For the other thing, the precision of civilian GPS is about 10 m [74], while indoor location-based services usually requires a much higher accuracy than GPS provides.

Using trilateration, the position of a device in a 2-D space can be determined by the distances from the device to three reference positions. The precision of localization can be improved by using an increased number of distance measurements, which are corrupted by noises in real application. One localization technique considered for the indoor environment uses the received signal strength (RSS) as a distance proxy where the distance corresponding to a particular RSS value can be looked up from a radio map on the server. However, the RSS metric in combination with the trilateration is unreliable owing to the complex nature of indoor radio propagation [75]. Another approach is to compare the online RSS readings with off-line observations of different reference points, which is stored in a database. The estimated position of a device is a grid point in the radio map. However, owing to the dynamic and unpredictable nature of indoor radio propagation, accurate localization requires a large number of RSS measurements. CS can be used to accurately localize a target with a small number of RSS measurements, where the sparsity level of the signal representation is equal to 1.

The efficiency of this CS-based localization system is demonstrated in [75]. Further improved localization systems and multiple target localization systems based on the CS principle are proposed in [75–77].

6.5 Summary

This chapter reviews the fundamental concepts of CS and sparse recovery. Particularly, it has been shown that compressively sensed signals can be successfully recovered if the sensing matrices satisfy any of the three given conditions. The focus of this chapter is on the applications in WSNs, and five cases in WSNs are presented where the CS principle has been used to solve different problems. There are many new emerging directions and many challenges that have to be tackled. For example, it would be interesting to study better signal models beyond sparsity, computational efficient algorithms, compressive information processing, data-driven approaches, multidimensional data and so on.

References

[1] Candès EJ, Romberg JK, and Tao T. Stable Signal Recovery from Incomplete and Inaccurate Measurements. Communications on Pure and Applied Mathematics. 2006;59(8):1207–1223.
[2] Donoho DL. Compressed Sensing. IEEE Transactions on Information Theory. 2006;52(4):1289–1306.
[3] Chen W, Rodrigues MRD, and Wassell IJ. Projection Design for Statistical Compressive Sensing: A Tight Frame Based Approach. IEEE Transactions on Signal Processing. 2013;61(8):2016–2029.
[4] Chen W, Rodrigues MRD, and Wassell IJ. On the Use of Unit-Norm Tight Frames to Improve the Average MSE Performance in Compressive Sensing Applications. IEEE Signal Processing Letters. 2012;19(1):8–11.
[5] Ding X, Chen W, and Wassell IJ. Joint Sensing Matrix and Sparsifying Dictionary Optimization for Tensor Compressive Sensing. IEEE Transactions on Signal Processing. 2017;65(14):3632–3646.
[6] Chen W, and Wassell IJ. Cost-Aware Activity Scheduling for Compressive Sleeping Wireless Sensor Networks. IEEE Transactions on Signal Processing. 2016;64(9):2314–2323.
[7] Chen W, and Wassell IJ. Optimized Node Selection for Compressive Sleeping Wireless Sensor Networks. IEEE Transactions on Vehicular Technology. 2016;65(2):827–836.
[8] Baraniuk RG. Compressive Sensing. IEEE Signal Processing Magazine. 2007;24(4):118–121.
[9] Chen SS, Donoho DL, and Saunders MA. Atomic Decomposition by Basis Pursuit. SIAM Review. 2001;43(1):129–159.

[10] Baraniuk R, Davenport M, DeVore R, *et al.* A Simple Proof of the Restricted Isometry Property for Random Matrices. Constructive Approximation. 2008; 28(3):253–263.

[11] Baraniuk RG, Cevher V, Duarte MF, *et al.* Model-Based Compressive Sensing. IEEE Transactions on Information Theory. 2010;56(4):1982–2001.

[12] Ji S, Xue Y, and Carin L. Bayesian Compressive Sensing. IEEE Transactions on Signal Processing. 2008;56(6):2346–2356.

[13] Hegde C, and Baraniuk RG. Signal Recovery on Incoherent Manifolds. IEEE Transactions on Information Theory. 2012;58(12):7204–7214.

[14] Chen M, Silva J, Paisley J, *et al.* Compressive Sensing on Manifolds Using a Nonparametric Mixture of Factor Analyzers: Algorithm and Performance Bounds. IEEE Transactions on Signal Processing. 2010;58(12): 6140–6155.

[15] Ding X, He L, and Carin L. Bayesian Robust Principal Component Analysis. IEEE Transactions on Image Processing. 2011;20(12):3419–3430.

[16] Fu C, Ji X, and Dai Q. Adaptive Compressed Sensing Recovery Utilizing the Property of Signal's Autocorrelations. IEEE Transactions on Image Processing. 2012;21(5):2369–2378.

[17] Zhang Z, and Rao BD. Sparse Signal Recovery With Temporally Correlated Source Vectors Using Sparse Bayesian Learning. IEEE Journal of Selected Topics in Signal Processing. 2011;5(5):912–926.

[18] Peleg T, Eldar YC, and Elad M. Exploiting Statistical Dependencies in Sparse Representations for Signal Recovery. IEEE Transactions on Signal Processing. 2012;60(5):2286–2303.

[19] Rauhut H. Compressive Sensing and Structured Random Matrices. Theoretical Foundations and Numerical Methods for Sparse Recovery. 2010;9:1–92.

[20] Candès E, and Tao T. Decoding by Linear Programming. IEEE Transactions on Information Theory. 2005;51(12):4203–4215.

[21] Candès EJ. The Restricted Isometry Property and Its Implications for Compressed Sensing. Comptes Rendus-Mathématique. 2008;346(9–10): 589–592.

[22] Foucart S, and Lai MJ. Sparsest Solutions of Underdetermined Linear Systems via ℓ_q-Minimization for $0 < q \leq 1$. Applied and Computational Harmonic Analysis. 2009;26(3):395–407.

[23] Davies ME, and Gribonval R. Restricted Isometry Constants Where ℓ_p Sparse Recovery Can Fail for $0 < p \leq 1$. IEEE Transactions on Information Theory. 2009;55(5):2203–2214.

[24] Cai TT, Wang L, and Xu G. New Bounds for Restricted Isometry Constants. IEEE Transactions on Information Theory. 2010;56(9):4388–4394.

[25] Yu L, Barbot JP, Zheng G, *et al.* Compressive Sensing With Chaotic Sequence. IEEE Signal Processing Letters. 2010;17(8):731–734.

[26] Donoho DL, Elad M, and Temlyakov VN. Stable Recovery of Sparse Overcomplete Representations in the Presence of Noise. IEEE Transactions on Information Theory. 2006;52(1):6–18.

[27] Baron D, Sarvotham S, and Baraniuk RG. Bayesian Compressive Sensing Via Belief Propagation. IEEE Transactions on Signal Processing. 2010;58(1): 269–280.

[28] Chen W. Simultaneous Sparse Bayesian Learning With Partially Shared Supports. IEEE Signal Processing Letters. 2017;24(11):1641–1645.

[29] Chen W, Wipf D, Wang Y, *et al.* Simultaneous Bayesian Sparse Approximation With Structured Sparse Models. IEEE Transactions on Signal Processing. 2016;64(23):6145–6159.

[30] Chen W, and Wassell IJ. A Decentralized Bayesian Algorithm For Distributed Compressive Sensing in Networked Sensing Systems. IEEE Transactions on Wireless Communications. 2016;15(2):1282–1292.

[31] Chartrand R. Exact Reconstruction of Sparse Signals via Nonconvex Minimization. IEEE Signal Processing Letters. 2007;14(10):707–710.

[32] Tibshirani R. Regression Shrinkage and Selection via the Lasso. Journal of the Royal Statistical Society Series B (Methodological). 1996;58(1): 267–288.

[33] Eldar YC. Generalized SURE for Exponential Families: Applications to Regularization. IEEE Transactions on Signal Processing. 2009;57(2): 471–481.

[34] Kim SJ, Koh K, Lustig M, *et al.* An Interior-Point Method for Large-Scale l1-Regularized Least Squares. IEEE Journal of Selected Topics in Signal Processing. 2007;1(4):606–617.

[35] Figueiredo MAT, Nowak RD, and Wright SJ. Gradient Projection for Sparse Reconstruction: Application to Compressed Sensing and Other Inverse Problems. IEEE Journal of Selected Topics in Signal Processing. 2007;1(4):586–597.

[36] Donoho DL, and Tsaig Y. Fast Solution of ℓ_1 -Norm Minimization Problems When the Solution May Be Sparse. IEEE Transactions on Information Theory. 2008;54(11):4789–4812.

[37] Garrigues PJ, and Ghaoui L. An Homotopy Algorithm for the Lasso with Online Observations. In: Neural Information Processing Systems (NIPS). vol. 21; 2008.

[38] Bredies K, and Lorenz D. Linear Convergence of Iterative Soft-Thresholding. Journal of Fourier Analysis and Applications. 2008;14:813–837.

[39] Donoho DL, Maleki A, and Montanari A. Message-Passing Algorithms for Compressed Sensing. Proceedings of the National Academy of Sciences. 2009;106(45):18914.

[40] Mallat SG, and Zhang Z. Matching Pursuits with Time-Frequency Dictionaries. IEEE Transactions on Signal Processing. 1993;41(12):3397–3415.

[41] Pati YC, Rezaiifar R, and Krishnaprasad PS. Orthogonal Matching Pursuit: Recursive Function Approximation with Applications to Wavelet Decomposition. In: Signals, Systems and Computers, 1993. 1993 Conference Record of The Twenty-Seventh Asilomar Conference on. vol. 1; 1993. p. 40–44.

[42] Tropp JA, and Gilbert AC. Signal Recovery From Random Measurements Via Orthogonal Matching Pursuit. IEEE Transactions on Information Theory. 2007;53(12):4655–4666.

[43] Donoho DL, Tsaig Y, Drori I, *et al.* Sparse Solution of Underdetermined Systems of Linear Equations by Stagewise Orthogonal Matching Pursuit. IEEE Transactions on Information Theory. 2012;58(2):1094–1121.

[44] Needell D, and Vershynin R. Uniform Uncertainty Principle and Signal Recovery via Regularized Orthogonal Matching Pursuit. Foundations of Computational Mathematics. 2009;9(3):317–334.

[45] Needell D, and Tropp J. CoSaMP: Iterative Signal Recovery from Incomplete and Inaccurate Samples. Applied and Computational Harmonic Analysis. 2009;26(3):301–321.

[46] Dai W, and Milenkovic O. Subspace Pursuit for Compressive Sensing Signal Reconstruction. IEEE Transactions on Information Theory. 2009;55(5): 2230–2249.

[47] Candès E, and Tao T. The Dantzig Selector: Statistical Estimation When p Is Much Larger Than n. The Annals of Statistics. 2007;35(6): 2313–2351.

[48] Sarvotham S, and Baraniuk RG. Deterministic Bounds for Restricted Isometry of Compressed Sensing Matrices. IEEE Transactions on Information Theory. submitted for publication.

[49] Elad M. Optimized Projections for Compressed Sensing. IEEE Transactions on Signal Processing. 2007;55(12):5695–5702.

[50] Duarte-Carvajalino JM, and Sapiro G. Learning to Sense Sparse Signals: Simultaneous Sensing Matrix and Sparsifying Dictionary Optimization. IEEE Transactions on Image Processing. 2009;18(7):1395–1408.

[51] Xu J, Pi Y, and Cao Z. Optimized Projection Matrix for Compressive Sensing. EURASIP Journal on Advances in Signal Processing. 2010;43:1–8.

[52] Zelnik-Manor L, Rosenblum K, and Eldar YC. Sensing Matrix Optimization for Block-Sparse Decoding. IEEE Transactions on Signal Processing. 2011; 59(9):4300–4312.

[53] Carson WR, Chen M, Rodrigues MRD, Calderbank R, and Carin L. Communications-Inspired Projection Design with Application to Compressive Sensing. SIAM Journal on Imaging Sciences. 2012;5(4):1185–1212.

[54] Strohmer T, and Jr RWH. Grassmannian Frames with Applications to Coding and Communication. Applied and Computational Harmonic Analysis. 2003; 14(3):257–275.

[55] Sardy S, Bruce AG, and Tseng P. Block Coordinate Relaxation Methods for Nonparametric Wavelet Denoising. Journal of Computational and Graphical Statistics. 2000;9(2):361–379.

[56] Davenport MA, Laska JN, Boufounos PT, *et al.* A Simple Proof that Random Matrices are Democratic. Rice University ECE Technical Report TREE 0906; 2009 Nov.

[57] Charbiwala Z, Chakraborty S, Zahedi S, *et al.* Compressive Oversampling for Robust Data Transmission in Sensor Networks. In: INFOCOM, 2010 Proceedings IEEE; 2010. p. 1–9.

[58] MacKay DJC. Fountain Codes. IEE Proceedings Communications. 2005;152(6):1062–1068.

[59] Fragouli C, Le Boudec JY, and Widmer J. Network Coding: An Instant Primer. ACM SIGCOMM Computer Communication Review. 2006;36(1):63–68.
[60] Bajwa W, Haupt J, Sayeed A, *et al.* Joint Source-Channel Communication for Distributed Estimation in Sensor Networks. IEEE Transactions on Information Theory. 2007;53(10):3629–3653.
[61] Chen W, Rodrigues MRD, and Wassell IJ. A Frechet Mean Approach for Compressive Sensing Date Acquisition and Reconstruction in Wireless Sensor Networks. IEEE Transactions on Wireless Communications. 2012;11(10): 3598–3606.
[62] Chen W. Energy-Efficient Signal Acquisition in Wireless Sensor Networks: A Compressive Sensing Framework. IET Wireless Sensor Systems. 2012;2:1–8.
[63] Quer G, Masiero R, Munaretto D, *et al.* On the Interplay between Routing and Signal Representation for Compressive Sensing in Wireless Sensor Networks. In: Information Theory and Applications Workshop, 2009; 2009. p. 206–215.
[64] Masiero R, Quer G, Munaretto D, *et al.* Data Acquisition through Joint Compressive Sensing and Principal Component Analysis. In: Global Telecommunications Conference, 2009. GLOBECOM 2009. IEEE; 2009. p. 1–6.
[65] Luo C, Wu F, Sun J, *et al.* Efficient Measurement Generation and Pervasive Sparsity for Compressive Data Gathering. IEEE Transactions on Wireless Communications. 2010;9(12):3728–3738.
[66] Meng J, Li H, and Han Z. Sparse Event Detection in Wireless Sensor Networks Using Compressive Sensing. In: Information Sciences and Systems, 2009. CISS 2009. 43rd Annual Conference on; 2009. p. 181–185.
[67] Ling Q, and Tian Z. Decentralized Sparse Signal Recovery for Compressive Sleeping Wireless Sensor Networks. IEEE Transactions on Signal Processing. 2010;58(7):3816–3827.
[68] Zhang B, Cheng X, Zhang N, *et al.* Sparse Target Counting and Localization in Sensor Networks based on Compressive Sensing. In: INFOCOM, 2011 Proceedings IEEE; 2011. p. 2255–2263.
[69] Liu Y, Zhu X, Ma C, *et al.* Multiple Event Detection in Wireless Sensor Networks Using Compressed Sensing. In: Telecommunications (ICT), 2011 18th International Conference on; 2011. p. 27–32.
[70] Fletcher AK, Rangan S, and Goyal VK. On-off Random Access Channels: A Compressed Sensing Framework. IEEE Transactions on Information Theory. submitted for publication.
[71] Fazel F, Fazel M, and Stojanovic M. Random Access Compressed Sensing for Energy-Efficient Underwater Sensor Networks. IEEE Journal on Selected Areas in Communications. 2011;29(8):1660–1670.
[72] Xie Y, Eldar YC, and Goldsmith A. Reduced-Dimension Multiuser Detection. IEEE Transactions on Information Theory. 2013;59(6):3858–3874.
[73] Xue T, Dong X, and Shi Y. A Covert Timing Channel via Algorithmic Complexity Attacks: Design and Analysis. In: Communications (ICC), 2012 IEEE International Conference on; 2012.

[74] Panzieri S, Pascucci F, and Ulivi G. An Outdoor Navigation System Using GPS and Inertial Platform. IEEE/ASME Transactions on Mechatronics. 2002;7(2): 134–142.

[75] Feng C, Au WSA, Valaee S, *et al.* Compressive Sensing Based Positioning Using RSS of WLAN Access Points. In: INFOCOM, 2010 Proceedings IEEE; 2010.

[76] Feng C, Valaee S, Au WSA, *et al.* Localization of Wireless Sensors via Nuclear Norm for Rank Minimization. In: Global Telecommunications Conference (GLOBECOM 2010), 2010 IEEE; 2010.

[77] Abid MA. 3D Compressive Sensing for Nodes Localization in WNs Based On RSS. In: Communications (ICC), 2012 IEEE International Conference on; 2012.

Chapter 7

Reinforcement learning-based channel sharing in wireless vehicular networks

Andreas Pressas[1]*, Zhengguo Sheng*[1]*, and Falah Ali*[1]

In this chapter, the authors study the enhancement of the proposed IEEE 802.11p medium access control (MAC) layer for vehicular use by applying reinforcement learning (RL). The purpose of this adaptive channel access control technique is enabling more reliable, high-throughput data exchanges among moving vehicles for cooperative awareness purposes. Some technical background for vehicular networks is presented, as well as some relevant existing solutions tackling similar channel sharing problems. Finally, some new findings from combining the IEEE 802.11p MAC with RL-based adaptation and insight of the various challenges appearing when applying such mechanisms in a wireless vehicular network are presented.

7.1 Introduction

Vehicle-to-vehicle (V2V) technology aims to enable safer and more sophisticated transportation starting with minor, inexpensive additions of communication equipment on conventional vehicles and moving towards network-assisted fully autonomous driving. It will be a fundamental component of the intelligent transportation services and the Internet of Things (IoT). This technology allows for the formation of vehicular *ad hoc* networks (VANETs), a new type of network which allows the exchange of kinematic data among vehicles for the primary purpose of safer and more efficient driving as well as efficient traffic management and other third-party services. VANETs can help minimize road accidents and randomness in driving with on-time alerts as well as enhance the whole travelling experience with new infotainment systems, which allow acquiring navigation maps and other information from peers.

The V2V radio technology is based on the IEEE 802.11a stack, adjusted for low overhead operations in the dedicated short-range communications (DSRCs) spectrum (30 MHz in the 5.9 GHz band for Europe). It is being standardized as IEEE 802.11p [1]. The adjustments that have been made are mainly for enabling exchanges

[1]Department of Engineering and Design, University of Sussex, UK

without belonging in a basic service set (SS) (BSS). Consequently, communication via IEEE 802.11p is not managed by a central access point (AP) as in typical wireless LANs (WLANs). This allows faster, *ad hoc* communication, necessary for mission critical applications.

Enabling such applications for various traffic scenarios to prevent safety issues is a challenging topic to deal with in VANETs; since the bandwidth is limited, the topology is highly mobile, and there is a lack of central coordination. Additionally, a significant amount of data would have to be exchanged via DSRC links in dense, urban scenarios. Every vehicle has in-car sensor and controller networks, collecting kinematic, engine, safety, environmental and other information and passing some of them to on-board units (OBUs) to be exchanged via DSRC links. In this chapter, we look into the various subsystems for wireless vehicular networks and also suggest a new RL-based protocol, first presented in [2] to efficiently share the DSRC control channel (CCH) for safety communications among multiple vehicle stations. We begin from how the data is generated, proceed to presenting the DSRC stack for transmitting the vehicle's information via wireless links, and then go on to present a self-learning MAC protocol that is able to improve performance when network traffic becomes too heavy for the baseline DSRC MAC (i.e. in urban scenarios, city centres).

7.1.1 Motivation

VANETs are the first large-scale network to operate primarily on broadcast transmissions, since the data exchanges are often relevant for vehicles within an immediate geographical region of interest (ROI) of the host vehicle. This allows the transmission of broadcast packets (packets not addressed to a specific MAC address), so that they can be received from every vehicle within range without the overhead of authentication and association with an AP. Broadcasting has always been controversial for the IEEE 802.11 family of protocols [3] since they treat unicast and broadcast frames differently. Radio signals are likely to overlap with others in a geographical area, and two or more stations will attempt to transmit using the same channel leading to contention. Broadcast transmissions are inherently unreliable and more prone to contention since the MAC specification in IEEE 802.11 does not request explicit acknowledgements (ACKs packets) on receipt of broadcast packets to avoid the ACK storm phenomenon, which appears when all successful receivers attempt to send back an ACK simultaneously and consequently congest the channel. This has not changed in the IEEE 802.11p amendment.

A MAC protocol is part of the data link layer (L2) of the Open Systems Interconnection model (OSI model) and defines the rules of how the various network stations share access to the channel. The *de facto* MAC layer used in IEEE 802.11-based networks is called carrier sense multiple access (CSMA) with collision avoidance (CA) (CSMA/CA) protocol. It is a simple decentralized contention-based access scheme which has been extensively tested in WLANs and mobile *ad hoc* networks (MANETs). The IEEE 802.11p stack also employs the classic CSMA/CA MAC. Although the proposed stack works fine for sparse VANETs with few nodes, it quickly shows its inability to accommodate increased network traffic because of the lack of ACKs.

The lack of ACKs not only makes transmissions unreliable but also does not provide any feedback mechanism for the CSMA/CA backoff mechanism. So it cannot adapt and resolve contention among stations when the network is congested.

The DSRC operation requires that L1 and L2 must be built in a way that they can handle a large number of contenting nodes in the communication zone, on the order of 50–100. The system should not collapse from saturation even if this number is exceeded. Useful data for transportation purposes can be technical (i.e. vehicular, proximity sensors, radars), crowd-sourced (i.e. maps, environment, traffic, parking) or personal (i.e. Voice over Internet Protocol (VoIP), Internet radio, routes). We believe that a significant part of this data will be exchanged through V2V links, making system scalability a critical issue to address. There is a need for an efficient MAC protocol for V2V communication purposes that adapts to the VANET's density and transmitted data rate, since such network conditions are not known *a priori*.

7.1.2 Chapter organization

The chapter is organized as follows: Section 7.2 refers to the architecture and various components of vehicular networks, from in-vehicle to V2V and vehicle-to-anything (V2X). Section 7.3 introduces the DSRC networking stack for V2V and Section 7.4 gets into the intrinsics of the DSRC channel access control protocol. Section 7.5 is an overview of the congestion problem in such networks and methods of resolution. Section 7.6 is an introduction to learning-based networking protocols, Markov decision processes (MDPs) and *Q*-learning. Section 7.7 describes the operation of a novel-proposed *Q*-learning-based channel access control protocol for DSRC. Finally, Sections 7.8 and 7.9 introduce a simulation environment for wireless vehicular networks and present the protocol performance evaluation regarding packet delivery and achieved delay.

7.2 Connected vehicles architecture

The various electronic systems incorporated on a modern-vehicle-enable connected vehicles, since these provide the data input (sensors), actuators as well as the local processing and communication capabilities.

7.2.1 Electronic control units

Electronic control, rather than purely mechanical control, governs every function in modern vehicles from activating various lights to adaptive cruise control. Electronic control units (ECUs) are embedded systems that collect data from the vehicle's sensors and perform real-time calculations on these. Then they drive various electronic systems/actuators accordingly so which maximum driving efficiency can be achieved at all times. Each unit works independently, running its own firmware, but cooperation among ECUs can be done if needed for more complex tasks.

7.2.2 Automotive sensors

Vehicle Sensors are components that enable autonomous vehicles applications as well as advanced driver assistance/safety systems. One type of sensor that is commonly deployed for such application is **radar sensors**, which use radio waves to detect nearby objects and determine their relative position and velocity [4]. In actual applications, usually an array of radar sensors is placed on the vehicle. **Cameras** are another option for object detection and are used as sensing inputs for safety and driver-assistance applications. Radars, as mentioned, have the ability to detect distance and velocity, but cameras have a better angle of detection. These sensors are often used together to provide more accurate-detection capabilities. Radars are usually employed as primary sensors, with cameras extending the area of detection on the sides, for enhanced, reliable object detection.

7.2.3 Intra-vehicle communications

The various electronic subsystems incorporated on a vehicle have to communicate with each other so that sensor readings, control signals and other data can be exchanged to perform complex tasks. This is a challenging research task, considering the progressive advancements in the car electronics density. A typical car would feature around 8–10 ECUs in the early 1990s, around 50 ECUs at 2000, and today it is common to have around 100 ECUs exchanging up to 2,500 signals between them [5]. One can understand the complexities involved in designing a network of ECUs that dense, and the need for advancements in intra-vehicle networks. Table 7.1 presents the most common in-vehicle communication protocols.

7.2.4 Vehicular ad hoc networks

The technological advances of the past few years in the field of communications, regarding both software and hardware, are enablers of new types of networks targeted for previously unexplored environments. The VANET is a type of wireless networks that has received a lot of interest from researchers, standardization bodies and developers the past few years, since it has the potential to improve road safety, enhance traffic and travel efficiency as well as make transportation more convenient and comfortable for both drivers and passengers [6]. It is envisioned to be a fundamental building block of intelligent transport services (ITS), the smart city as well as the IoT. VANETs are self-organized networks composed of mobile and stationary nodes connected with

Table 7.1 Network protocols used in automotive networks

Protocol	Max. bit-rate	Medium	Protocol
CAN	1 Mbps	Shielded twisted pair	CSMA/CD
LIN	20 kbps	Single wire	Serial
FlexRay	2×10 Mbps	Twisted pair/fibre optic	TDMA
MOST	150 Mbps	Fibre optic	TDMA
Ethernet	1 Gbps	Coaxial/twisted pair/fibre optic	CSMA/CD
HomePlug AV	>150 Mbps	Vehicular power lines	CSMA/CA

wireless links [7]. They are a subclass of MANETs, but in this case, the mobile stations are embedded on vehicles and the stationary nodes are roadside units (RSUs). There are more differences from classic MANETs, since VANETs are limited to road topology while moving, meaning that potentially we could predict the future positions of the vehicles to be used for, i.e. better routing and traffic management. Additionally, since vehicles do not have the energy restrictions of typical MANET nodes, they can feature significant computational, communication and sensing capabilities [8]. Because of these capabilities and opportunities, many applications are envisioned for deployment on VANETs, ranging from simple exchange of status or safety messages between vehicles to large-scale traffic management, Internet Service provisioning and other infotainment applications.

7.2.5 Network domains

The VANET system architecture comprises three domains: the in-vehicle, the *ad hoc* and the infrastructure domain, as seen in [9]. The **in-vehicle** domain (whose components are described in Sections 7.2.1 and 7.2.3) is composed of an on-board communication unit (OBU) and multiple control units. The connections between them are usually wired, utilizing the protocols in Section 7.2.3, and sometimes wireless. The ***ad hoc* domain** is composed of vehicles equipped with such OBUs and RSUs. The OBUs can be seen as the mobile nodes of a wireless *ad hoc* network, and likewise RSUs are static nodes. Additionally, RSUs can be connected to the Internet via gateways, as well as communicate with each other directly or via multi-hop. There are two types of **infrastructure domain** access, RSUs and hot spots (HSs). These provide OBUs access to the Internet. In the absence of RSUs and HSs, OBUs could also use cellular radio networks (GSM, GPRS, LTE) for the same purpose. The various networking domains and their respective components can be seen in Figure 7.1.

7.2.6 Types of communication

In-vehicle communication refers to a car's various electronic controllers communicating within the in-vehicle domain. The in-vehicle communication system can detect the vehicle's performance regarding the internal systems (electrical and mechanical) as well as driver's fatigue and drowsiness [10], which is critical for driver and public safety.

In the *ad hoc* domain, **V2V** communication can provide a data-exchange platform for the drivers to share information and warning messages, so as to expand driver assistance and prevent road accidents. **Vehicle-to-road infrastructure (V2I)** communication enabled by VANETs allows real-time traffic updates for drivers, a sophisticated and efficient traffic light system as well, as could provide environmental sensing and monitoring.

Vehicle-to-broadband cloud communication means that vehicles may communicate via wireless broadband mechanisms such as 3G/4G (infrastructure domain). As the broadband cloud includes more traffic information and monitoring data as well as infotainment, this type of communication will be useful for active driver assistance and vehicle tracking as well as other infotainment services [11]. Figure 7.2 presents the various applications envisioned for the different types of communications.

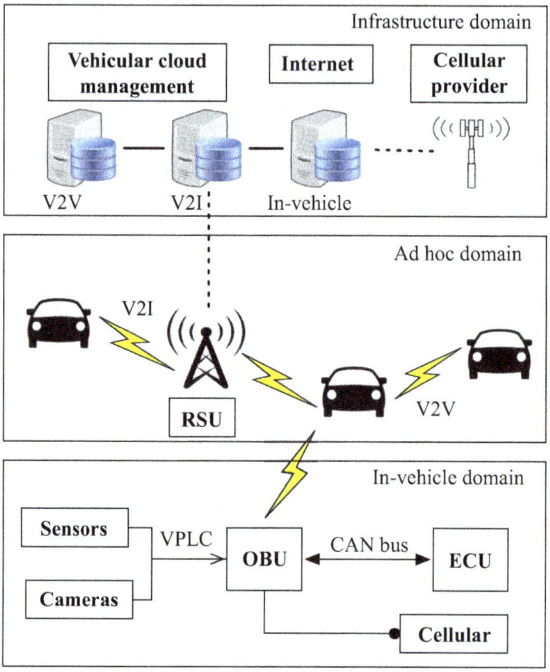

Figure 7.1 Internet of Vehicles network domains

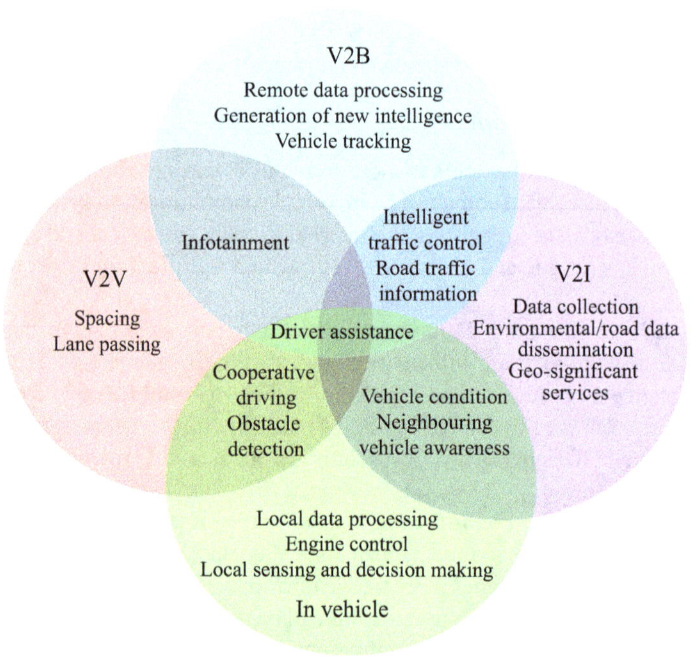

Figure 7.2 Key functionality for every type of automotive communication

7.3 Dedicated short range communication

The primary functionality of VANETs is advanced active road safety via V2V and V2I communication. A vehicular safety communication network is *ad hoc*, highly mobile with a large number of contending nodes. The safety messages are very short as it is their useful lifetime relevance and must be received with high probability [12]. The key-enabling technology, specifying Layer 1 and 2 (L1, L2) of the protocol stack used in V2X (*ad hoc* domain), is DSRC. The DSRC radio technology is essentially IEEE 802.11a adjusted for low overhead operations in the DSRC spectrum (70 MHz in the 5.9 GHz band for North America). It is being standardized as IEEE 802.11p [1].

Work in [13] shows that IEEE 802.11p exhibits lower latency and higher delivery ratio than LTE in scenarios fewer than 50 vehicles. More specifically, for smaller network densities, the standard allows end-to-end delays less than 100 ms and throughput of 10 kbps which satisfies the requirements set by active road safety applications and few of the lightweight cooperative traffic awareness applications. However, as the number of vehicles increases, the standard is unable to accommodate the increased network traffic and support performance requirements for more demanding applications.

7.3.1 IEEE 802.11p

In the architecture of classic IEEE 802.11 networks, there are three modes of operation:

- A BSS, which includes an AP node that behaves as the controller/master station (STA).
- The (independent BSS) IBSS, which is formed by STAs without infrastructure (AP/s). Networks formed like this are called *ad-hoc* networks.
- The (Extended Service Set), which is the union of two or more BSSs connected by a distribution system [14].

The most suitable architecture for a VANET would be the IBSS. An STA(node) within an IBSS acts as the AP and periodically broadcasts the SSID and other information. The rest of the nodes receive these packets and synchronize their time and frequency accordingly. Communication can only be established as long as the STAs belong in the same SS.

The IEEE 802.11p amendment defines a mode called "Outside the context of BSS" in its L2, that enables exchanging data without the need for the station to belong in a BSS, and thus, without the overhead required for these association and authentication procedures with an AP before exchanging data.

DSRC defines seven licenced channels, as seen in Figure 7.3, each of 10 MHz bandwidth, six service channels (SCHs) and one CCH. All safety messages, whether transmitted by vehicles or RSUs, are to be sent in the CCH, which has to be regularly monitored by all vehicles. The CCH could be also used by RSUs to inform approaching vehicles of their services, then use the SCH to exchange data with interested vehicles.

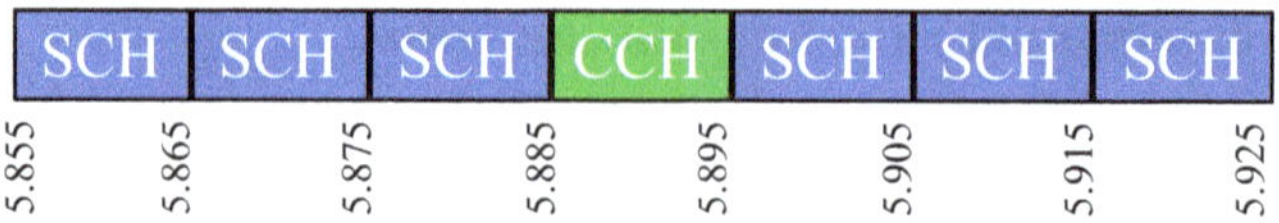

Figure 7.3 The channels available for 802.11p

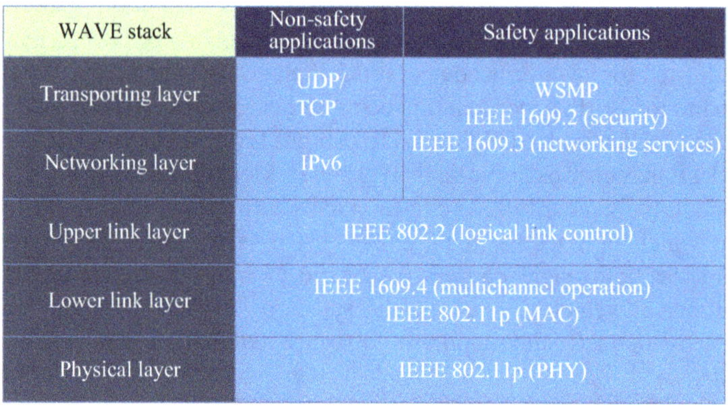

Figure 7.4 The DSRC/WAVE Protocol Stack and Associated Standards

The explicit multichannel nature of DSRC necessitates a concurrent multichannel operational scheme for safety and non-safety applications [15]. This need is facilitated with a MAC protocol extension by the IEEE 1609 working group, which deals with the standardization of the DSRC communication stack between the link layer and applications.

The IEEE 802.11p and IEEE 1609.x protocols combined form the wireless access in vehicular environments (WAVE) stack, and they aim to enable wireless communication between vehicles for safety (via the CCH) and other purposes (via the SCH). The complete WAVE/DSRC stack is presented at Figure 7.4.

7.3.2 WAVE Short Message Protocol

So there are two stacks supported by WAVE, one being the classic Internet Protocol version 6 (IPv6) and a proprietary one known as WAVE Short Message Protocol (WSMP). The reason for having two variations in the upper layers is to distinguish the messages as high-priority/time sensitive and less-demanding transmissions such as User Datagram Protocol (UDP) transactions.

Vehicular safety applications do not require big datagram lengths or complex packets to be transmitted rather than very strict probability of reception and little latency. The overhead is 11 bytes, when a typical UDP-IPv6 packet has a minimum overhead of 52 bytes [16]. WSMP enables sending short messages while directly manipulating the physical layer characteristics such as the transmission power and data rate so that nearby vehicles have a high probability of reception within a set time frame. A provider service ID field is similar to a port number in Transmission

WSM version 1 byte	Security type 1 byte	Channel number 1 byte	Data rate 1 byte	TX Power 1 byte	PSID 4 bytes	Length 2 bytes	DATA variable

Figure 7.5 The format of a WSMP packet

Control Protocol (TCP)/UDP, which acts as an identity and answers which application is a specific WSMP heading towards. To reduce latency, WSMP exchanges do not necessitate the formation of a BSS, which is a requirement for SCH exchanges. The WSMP format can be seen at Figure 7.5.

However, WSMP is not able to support the classic Internet applications or exchange of multimedia, and it does not need to, since such applications are more tolerant to delay or fluctuations in network performance. By supporting the IPv6 stack, which is open and already widely deployed, third-party internet services are easily deployable in a vehicular environment, and the cost of deployment would be significantly lower for private investors.

7.3.3 Control channel behaviour

The CCH is the one to facilitate safety communications through the exchange of safety messages, while having the following link layer characteristics:

- Most of the safety applications are based on **single hop communication** since they are much localized. This means that the basic DSRC communication design does not feature any networking (packet routing) capabilities, but there are scenarios where multi-hop communication is of use (warning for an accident/hazard along a highway). Some rebroadcast schemes for enhancing the broadcasting performance or reliability can be found in the literature [17] but are not considered as proper multi-hop.
- As mentioned already, safety communications are of **broadcast** nature, which means that it is targeted at vehicles for where they are rather than who they are. Additionally, channel access is not centrally managed in DSRC. Vehicular safety communication is fully distributed.
- A major concern for DSRC is that since all DSRC-enabled vehicles and infrastructure continuously broadcast beacon messages and event-triggered safety messages, such a system would require special design so that it can work reliably and efficiently in a large scale. Although safety communications are mostly single-hop, the system is unbounded which means that V2V communication can stretch to great distances, unlike a bounded system (cells in mobile telephony) [15].
- The CCH is to facilitate the exchange of safety messages, complying with WSMP. Occasionally, it is used for advertising non-safety applications (by RSUs) which take place in one of the SCHs. These are called WAVE Service Advertisement (WSA) messages. The receiving node would get informed of the existence of such applications and tune in the appropriate channel if it needs to make use of these. These advertisements are generally lightweight and their effect to the CCH's load is insignificant [15]. Consequently, the focus when designing the CCH characteristics is towards safety applications.

7.3.4 Message types

Two types of (safety) WSMP messages are to be sent through the CCH by every DSRC-enabled vehicle:

- **Periodic safety messages:** Cooperative awareness messages (CAMs) are broadcast status messages (beacons) containing the location, direction, velocity and other kinematic information of the transmitting vehicle. These messages are meaningful for little time, so that the receivers can predict the movement of the sender, and after a few seconds become irrelevant. RSUs also utilize these beacons for traffic light status and other applications. The beaconing interval is 100 ms ($F_{beacon} = 10$ Hz).
- **Event-triggered messages:** Changes in the vehicle dynamics (hard breaking) or RSU status activate the broadcasting of emergency messages containing safety information (i.e. road accident warning, unexpected breaking ahead, slippery road), called decentralized environmental notification messages.
- There are also **non-safety communications**, that can happen for file transfers (local map updates, infotainment) or transactions (toll collection) and others. These take place in the SCHs but are often advertised through WSA messages in the CCH, in which every DSRC-enabled vehicle is tuned in by default.

7.4 The IEEE 802.11p medium access control

Reliable transmission and reception of messages can be affected by packet collisions. Two or more transmitters within the range of the same receiver sending a packet simultaneously would lead to a packet collision at the receiver, and the receiver would not receive any message. To tackle this problem, a MAC protocol, which serves the purpose of allowing only one node to transmit at a time, would have to be designed and implemented. Two nearby nodes transmitting frames at the same time means that these frames will collide leading to wasted bandwidth. A MAC protocol is a set of rules defined in L2 by which the radio [12] (physical layer – L1) decides when to send data and when to defer from transmission.

Given that wireless vehicular networks are *ad hoc* in nature, TDMA, FDMA or CDMA are difficult to realize since some short of centralized control (AP) would be needed to dynamically allocate time slots, frequency channels or codes, respectively. In addition to the infrastructure-less nature of VANETs, the high degree of mobility makes these MAC protocols unsuitable for such networks [12]. Random access mechanisms are better suited to *ad hoc* networks, such as ALOHA or in the case of VANETs, CSMA.

7.4.1 Distributed coordination function

The *de facto* technique for sharing access to the medium among multiple nodes without central coordination in IEEE 802.11-based networks is the **distributed coordination**

function (DCF). It employs a CSMA/CA algorithm. DCF defines two access mechanisms to enable fair packet transmission, a two-way handshaking (basic mode) or a four-way handshaking (request-to-send/clear-to-send (RTS/CTS)).

Under the **basic access mechanism**, a node wishing to transmit would have to sense the channel for a DCF Interframe Space (IFS) (DIFS) interval. If the channel is found busy during the interval, the node would not transmit but instead wait for an additional DIFS interval plus a specific period of time known as the backoff interval, and then try again. If the channel was not found busy for a DIFS interval, the node would transmit.

Another optional mechanism for transmitting data packets is RTS/CTS reservation scheme. Small RTS/CTS packets are used to reserve the medium before large packets are transmitted.

7.4.2 Basic access mechanism

In a network like VANETs where many stations contend for the wireless medium, if multiple stations sense the channel and find it busy, they would also find it being idle virtually the same time and try to use it at that time instant. To avoid the collisions that would occur that very moment, as seen in Figure 7.6, every station would have to wait for a backoff interval, whose length is specified by the random backoff mechanism in DCF. This interval is picked randomly from the uniform distribution over the interval [0,CW] where CW is the current **contention window**. According to the IEEE 802.11p, the CW can be a number between $CW_{\min} = 3$ and $CW_{\max} = 255$ [18].

For every DIFS interval that the node senses the medium to be idle, the backoff timer is decreased. If the medium is used, the counter will freeze and resume when the channel is again idle for a DIFS interval. The station whose backoff timer expires (reaches 0) first will begin the transmission, and the other stations freeze their timers and defer transmission. Once the transmitting station completes transmission, the backoff process starts again and the remaining stations resume their backoff timers.

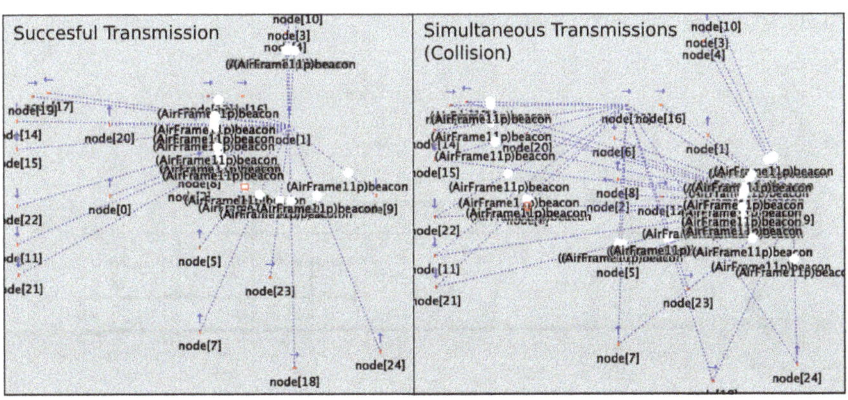

Figure 7.6 Collisions in a simulated environment

For unicast packet transmissions, in the case of a successful reception, the destination will send an ACK to the source node after a short IFS (SIFS), so it (the ACK) can be given priority (since SIFS<DIFS). If the source does not receive an ACK within a set time frame, it reactivates the sending process after the channel remains idle for an extended IFS. If two or more nodes decrease their backoff counter to 0 simultaneously, a collision occurs. For each retransmission attempt (because of collision and no ACK), the used CW is doubled, until it reaches $CW_{\max}$. Upon successful transmission, CW resets to $CW_{\min}$. The operation of CSMA/CA for both unicast and broadcast transmissions can be seen in Figure 7.7.

7.4.3 Binary exponential backoff

This mechanism of CW adaptation for unicast packets is called the binary exponential backoff (BEB) algorithm. It is the CA part of CSMA/CA and specifies that for every packet transmission, the station uniformly selects a random value for its backoff counter within $[0, W_i - 1]$, where W_i is the current CW and i is the number of failed attempts to transmit this single packet:

$$W_i = 2^i \times CW \quad \text{for} \quad i \in [0, m], \tag{7.1}$$

where the number of the backoff stages m is given by

$$m = \log_2\left(\frac{CW_{\max}}{CW_{\min}}\right) \tag{7.2}$$

At the first transmission attempt for a packet:

$$W_0 = CW_{\min} = CW. \tag{7.3}$$

If a unicast packet encounters a collision (meaning no ACK was received for a set time frame), $W_1 = 2 \times CW$. W_i is doubled every time a collision happens, until it reaches a $W_m = CW_{\max} = 2^m CW$. When $W_i = W_{\max}$, it maintains this value until a successful transmission (ACK received). W_i will be reset to $CW_{\min}$, and the process will start again for the next unicast packet.

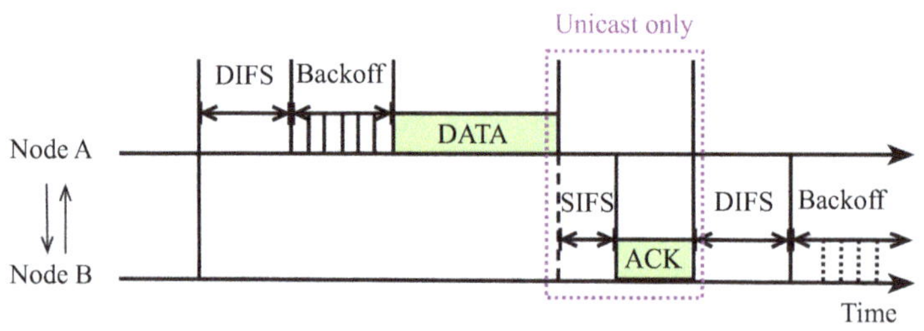

Figure 7.7 A CSMA/CA cycle for both unicast and broadcast cases. It manages channel access among transmitting nodes A and B

Two problems seem to appear with the BEB mechanism when trying to establish unicast communication among many highly mobile nodes. First, in dense wireless networks such as VANETs, there is higher probability that more than one node choose the same CW, resulting to collisions. Second, every time a collision occurs, the CW is doubled to avoid more collisions. But given that the network density for a VANET can vary a lot over small time periods because of high mobility, a node with a large CW (because of previous failed transmissions) will wait more than it needs to before transmitting under lighter network conditions. This will result in unnecessary delay.

7.4.4 RTS/CTS handshake

The hidden terminal problem, shown in Figure 7.8, appears when node C wants to transmit to node B but cannot hear that it is already occupied. This is due to node C not being within the transmission range of node A. The RTS/CTS scheme (also known as virtual carrier sensing) can help to reduce contention caused from this phenomenon. An RTS packet is transmitted first from the sender, containing the size of the upcoming, larger, data frame and the channel time which is required for it to be transmitted. In the case that the receiving node is free to receive the data frame, it sends a CTS packet back to the sender. The neighbouring nodes defer from the medium until the channel becomes free again. This mechanism is helpful when transmitting large data packets and tackles the hidden and exposed terminal problem, but in the case of *ad hoc* networks it was found [19] not to be as effective. The overhead associated with the exchange of RTS/CTS packets does not worth it when the network has many, highly mobile stations, especially when targeting low latency communication. Additionally, for VANETs, vehicle-stations positioned at the edge of a station's transmission range are not as depended on geo-significant safety information transmitted from that station, as much as the vehicles of close proximity to the transmitter.

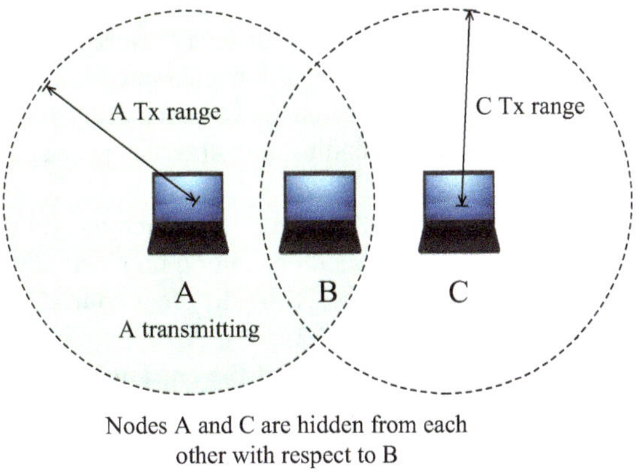

Figure 7.8 Hidden terminal problem

7.4.5 DCF for broadcasting

Frames exchanged in an 802.11 DCF-based network can be distinguished as unicast, if sent to a single destination node, or broadcast if sent to all available nodes within the transmission range. In both transmission cases, the MAC layer of the future-transmitting node receives a request to transmit from the upper layers of the networking stack, and then physical carrier sensing takes place to observe whether the channel is unoccupied by another transmitter. The channel being found idle for more than a DIFS interval means that the MAC layer of that station will begin the process of transmission, for either unicast or broadcast packets.

For the unicast case, the DCF backoff mechanism uses multiple backoff stages, as defined by (7.1) and (7.2). For every transmission that an ACK packet from the destination is not received in time or at all, the transmitter's CW is doubled, except from the last stage in which CW stays at maximum value.

For **broadcast packet transmission** though, the algorithm can have only a single backoff stage. The reason for this is that the 802.11 protocol does not require ACKs from the destination nodes for broadcast transmission, for practical reasons. All the receiving nodes sending an ACK on the reception of a message would impose an overhead, causing even more collisions (among all the ACK packets) and generally degrading the performance of the network for the given time.

So as usual, when the backoff counter expires for the first backoff stage, the message is going to be sent but will not be acknowledged, which means there is no definite way to know whether the packet actually reached the destination nodes. Collisions would be unrecoverable in this case [20], since no intelligent retransmission strategy is implemented by default for broadcasting. The different behaviours of the backoff algorithm for unicast and broadcast traffic also lead to different contention times [20].

7.4.6 Enhanced distributed channel access

When just the basic DCF scheme is employed, all nodes contend for access to the medium using the CSMA/CA algorithm with the same parameters. Data packets that are different regarding content, priority or delay-tolerance should be handled differently, and quality of service (QoS) should be guaranteed [21]. Real-time traffic information and collision-warning messages have strict delay requirements, while applications such as map data downloading and Internet browsing are more time tolerant.

In order to meet the different QoS requirements such as end-to-end delay and throughput, traffic should be differentiated depending on such requirements. A way of doing this service differentiation is by setting different contention parameters for different classes of data, as seen in Table 7.2.

The IEEE 802.11p/WAVE stack can adopt the **enhanced distributed channel access** from 802.11e in order to improve the QoS. It offers traffic classification through four access categories (ACs). When packets have different ACs, they contend internally and the winner will participate in external contention [22]. As shown on the table below, highly important messages (safety broadcasts) fall in AC3 which has the

Table 7.2 Contention parameters for different access categories in 802.11p

AC	Data class	CW_{min}	CW_{max}	AIFS
3	Safety related	3	7	2
2	Voice	3	7	3
1	Best effect	7	15	6
0	Background traffic	15	1,023	9

Table 7.3 Typical DSRC QoS requirements as seen in [12]

Applications	Packet size (bytes)/ bandwidth	Allowable latency (ms)	Network traffic type	Message range (m)	Priority
Intersection collision warning/avoidance	~100	~100	Event	300	Safety of life
Cooperative collision warning	~100/~10 kbps	~100	Periodic	50–300	Safety of life
Work zone warning	~100/~1 kbps	~1,000	Periodic	300	Safety
Transit vehicle signal priority	~100	~1,000	Event	300–1,000	Safety
Toll collection	~100	~50	Event	15	Non-safety
Service announcements	~100/2 kbps	~500	Periodic	0–90	Non-safety
Movie download (2 hours of MPEG 1): 10 min download time	>20 Mbps	N/A	N/A	0–90	Non-safety

lowest arbitrary IFS (AIFS) and CW size, so they are more likely to win the internal contention and affect transmission delay as little as possible (up to seven time slots for unicast and up to three time slots for broadcast transmissions). The QoS requirements for various vehicular networking applications can be found at Table 7.3, taken from [12].

7.5 Network traffic congestion in wireless vehicular networks

It is by now clear that the safety applications made possible through VANETs require a low end-to-end delay and high packet delivery ratio (PDR). Additionally, since the safety messages will be of broadcast nature, VANETs will be the first large-scale networks where communication is based on broadcast rather than on unicast messages. The choice of an IEEE 802.11-based technology for this kind of network raises some issues [23]. The MAC protocol in this family of standards is well known for its inability to cope with large-scale broadcast communications, since it was designed for a different use case, and it clearly favours unicast [20] communication.

Network traffic congestion in VANETs has a devastating impact on the performance of ITS applications. Given the large number of contending nodes, especially in an urban environment, it has been found [24] that the CSMA/CA algorithm, which is the basic medium access scheme in the IEEE 802.11, is not reliable enough due to high collision rates. This means channel congestion control and broadcast performance improvements of the 802.11p MAC are of particular concern and need to be addressed [15] in order to meet the QoS requirements of DSRC applications. The basic reason for this is the non-adaptation of the CW size depending on sensed network traffic. Work presented in [25] proves that throughput (derived from a simple Markov chains model) is diminished with an increased number of competing nodes exchanging broadcast packets.

The node density in a typical scenario can vary from very sparse connectivity to more than 150 cars/lane/km [23], so VANETs have to be able to accommodate the needs (channel-wise) of multiple simultaneous transmitters. The modifications brought by the IEEE 802.11p amendment focused on the physical layer, while the classic 802.11 MAC layer was enhanced for transmission of data outside BSS context which will contribute towards the scalability goal by removing the association and authentication overheads. But IEEE 802.11 was designed for unicast applications in mind, so it comes as no surprise that the CCH operating under 802.11p can be saturated solely by periodic broadcasting (beaconing), even for medium vehicle density [23].

7.5.1 Transmission power control

One idea on how to treat degrading performance on increasing vehicle density that has been extensively studied is limiting the number of contending nodes, which can be done by using TX power control mechanisms. When access to the wireless medium becomes difficult, lowering the transmission power of a station reduces the interference area [26]. WSMPs exploit this by providing the capability to set the transmission power on a per-packet basis. There are, however, some limitations on the minimum area that safety messages should reach.

7.5.2 Transmission rate control

Another solution, often combined with power control, is controlling the transmission time of a beacon. Since the packets' size is set by the application, only the data rate can be adjusted. Higher data rate translates into higher transmission probability [27], but also higher Signal-to-noise ratio (SNR) at the destination of the message, so the coverage area is reduced. This solution suffers from the same limitation as power control.

7.5.3 Adaptive backoff algorithms

A way to operate on maximum coverage area and still avoid collisions and degrading performance would be an adaptive backoff mechanism. With a high number of transmitting nodes, a large CW size is needed to avoid unnecessary collisions. On the other hand, when the traffic load of the network is low, a small CW size is needed so that potential senders can access the wireless medium with a short delay [18], thus making more efficient use of channel bandwidth. Additionally, the time the channel is idle because of nodes being in the backoff stage could be minimized. In an ideal situation, there would be zero idle time (which is essentially lost and is a synonym of bandwidth wastage) between messages with the exception of the DIFS [23].

7.6 Reinforcement learning-based channel access control

Machine-learning-based techniques have the potential to enter and improve every layer of the network stack for the IoT and other applications. The focus of this chapter is on RL [28] in the context of MAC for wireless V2V/V2X communication in the *ad-hoc* domain of vehicular networks. An adaptive backoff algorithm based on RL can help tackle the channel congestion when many stations are deployed, without reducing the transmission range or data rate, or knowing any details about the network beforehand.

RL is a general class of machine-learning algorithms fit for problems of sequential decision-making and control. It can be used as a parameter-perturbation/adaptive-control method for MDPs [29], a discrete time stochastic control formulation. It is based on the idea that if an action is followed by a satisfactory state of affairs, or by an improvement in the state of affairs (or a reward function), then the agent's tendency to produce that action is strengthened, i.e. reinforced. Specifically, we develop and evaluate a solution based on Q-learning, a much-used model-free RL algorithm that can solve MDPs with very little information from the dynamic VANET environment, but still reveals effective solutions regarding contention control for various network conditions. In addition, we employ a strategy for building self-improving Q-learning controllers that yield instant performance benefits since the vehicle-station's deployment and always strive for optimum operation while online.

7.6.1 Review of learning channel access control protocols

When it comes to relevant work which is focused specifically on the MAC layer issues, [30] uses the MDP formulation to design a MAC with deterministic backoff for virtualized IEEE 802.11 WLANs. For V2V exchanges, the work presented in [31] examines the IEEE 802.11p MAC regarding channel contention using the Markov model from [32] and proposes a passive contention estimation technique by observing the count of idle inter-frame slots.

RL is inspired by behaviourist psychology and deals with how software agents should take actions in an environment while aiming to maximize their cumulative reward. The problem, because of its generality, is studied in many disciplines such as game theory, control systems, IT, simulation-based optimization, statistics and genetic algorithms. There have been attempts to apply RL for optimizing the access control layer of wireless networks. The protocol in [33] is targeted on wireless sensor networks, optimizing battery-power node energy consumption. The protocol in [34] is targeted on wireless vehicular networks that operate on a unicast basis. It employs CW adaptation [35] which is a proven technique to improve the network contention because of interference in wireless networks. The premise is interesting, but the proposed IEEE 802.11p is a broadcast-based protocol. The current literature does not deal with the broadcasting issues within the context of contention resolution on the MAC level.

7.6.2 Markov decision processes

In RL, the learning agents can be studied mathematically by adopting the MDP formalism. An MDP is defined as a (S, A, P, R) tuple, where S stands for the set of possible states, A_s is the set of possible actions from state $s \in S$, $P_a(s, s')$ is the probability to transit from a state $s \in S$ to $s' \in S$ by performing an action $a \in A$. $R_a(s, s')$ is the reinforcement (or immediate reward), result of the transition from state s to state s' because of an action a, as seen in Figure 7.9. The decision policy π maps the state set to the action set, $\pi : S \rightarrow A$. Therefore, the MDP can be solved by discovering the optimal policy that decides the action $\pi(s) \in A$ that the agent will make when in state $s \in S$.

7.6.3 Q-learning

There are, though, many practical scenarios such as the channel access control problem studied in this work, for which the transition probability $P_{\pi(s)}(s, s')$ or the reward function $R_{\pi(s)}(s, s')$ are unknown, which makes it difficult to evaluate the policy π. Q-learning [36,37] is an effective and popular algorithm for learning from delayed reinforcement to determine an optimal policy π in the absence of transition probability. It is a form of model-free RL which provides agents the ability to learn how to act optimally in Markovian domains by experiencing the consequences of their actions, without requiring maps of these domains.

In Q-learning, the agent maintains a table of $Q[S, A]$, where S is the set of states and A is the set of actions. At each discrete time step $t = 1, 2, \ldots, \infty$, the agent observes the state $s_t \in S$ of the MDP, selects an action $a_t \in A$, receives the resultant reward r_t and observes the resulting next state $s_{t+1} \in S$. This experience (s_t, a_t, r_t, s_{t+1}) updates the Q-function at the observed state-action pair, thus providing the updated $Q(s_t, a_t)$. The algorithm, therefore, is defined by the function (1) that calculates the

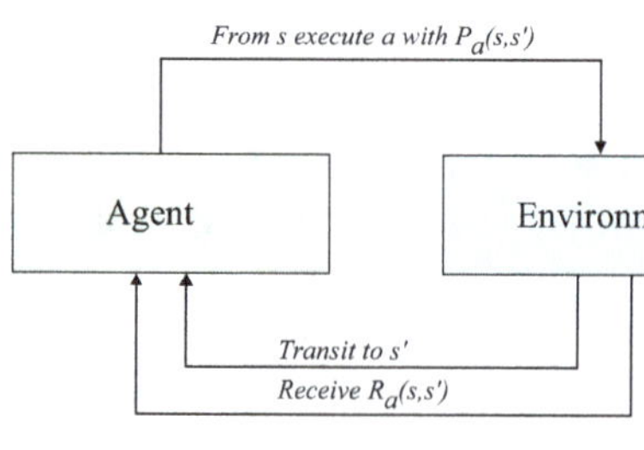

Figure 7.9 Abstract MDP model

quantity of a state-action (s, a) combination. The goal of the agent is to maximize its cumulative reward. The core of the algorithm is a value iteration update. It assumes the current value and makes a correction based on the newly acquired information, as in the following equation:

$$Q(s_t, a_t) \leftarrow Q(s_t, a_t) + \alpha \times [r_t + \gamma \times \max_{a_{t+1}} Q(s_{t+1}, a_{t+1}) - Q(s_t, a_t)], \tag{7.4}$$

where the discount factor γ models the importance of future rewards. A factor of $\gamma = 0$ will make the agent "myopic" or short-sighted by only considering current rewards, while a factor close to $\gamma = 1$ will make it strive for a high long-term reward. The learning rate α quantifies to what extent the newly acquired information will override the old information. An agent with $\alpha = 0$ will not learn anything, while with $\alpha = 1$, it would consider only the most recent information. The $\max_{a_{t+1} \in A} Q(s_{t+1}, a_{t+1})$ quantity is the maximum Q value among possible actions in the next state. In the following sections, we present employing (7.4) as a learning, self-improving, control method for managing channel access among IEEE 802.11p stations.

7.7 *Q*-learning MAC protocol

The adaptive backoff problem fits into the MDP formulation. RL is used to design a MAC protocol that selects the appropriate CW parameter based on gained experience from its interactions with the environment within an immediate communication zone. The proposed MAC protocol features a *Q*-learning-based algorithm that adjusts the CW size based on binary feedback from probabilistic rebroadcasts in order to avoid packet collisions.

7.7.1 The action selection dilemma

The state space S contains the discrete IEEE 802.11p-compatible CW values ranging from $CW_{min} = 3$ to $CW_{max} = 255$. The CW is adapted prior to every packet transmission by performing one of the following actions:

$$CW_{t+1} \xleftarrow{a \in \{(CW_t - 1)/2, CW_t, CW_t \times 2 - 1\}} CW_t\,. \tag{7.5}$$

RL differs from supervised learning in which correct input/output pairs are never presented, nor suboptimal actions are explicitly corrected. In addition, in RL there is a focus on online performance, which involves finding a balance between exploration of uncharted territory and exploitation of current knowledge. This in practice translates as a trade-off in how the learning agent in this protocol selects its next action for every algorithm iteration. It can either randomly pick an action from (7.5) (exploration) so that the algorithm can transit to a different (s, a) pair and get experience (reward) for it or follow a greedy strategy (exploitation) and choose the action with the highest Q-value for its current state given by

$$\pi(s) = \operatorname*{argmax}_a Q(s, a). \tag{7.6}$$

7.7.2 Convergence requirements

The RL algorithm's purpose is to converge to a (near) optimum output, in terms of CW. Watkins and Dayan [36] proved that Q-learning converges to the optimum action-values with probability 1 as long as all actions are repeatedly sampled in all states and the action-value pairs are represented discretely.

The greedy policy, with respect to the Q-values, tries to exploit continuously. However, since it does not explore all (s, a) pairs properly, it fails satisfying the first criterion. At the other extreme, a fully random policy continuously explores all (s, a) pairs, but it will behave suboptimally as a controller. An interesting compromise between the two extremes is the ε-greedy policy [28], which executes the greedy policy with probability $1 - \varepsilon$. This balancing between exploitation and exploration can guarantee convergence and often good performance.

The proposed protocol uses the ε-greedy strategy to focus the algorithm's exploration on the most promising CW trajectories. Specifically, it guarantees the first convergence criterion by forcing the agent to sample all (s, a) pairs over time with probability ε. Consequently, the proposed algorithmic implementation satisfies both convergence criteria, but further optimization is needed regarding convergence speed and applicability of the system.

In practice, the Q-learning algorithm converges under different factors, depending on the application and complexity. When deployed in a new environment, the agent should mostly explore and value immediate rewards and then progressively show its preference for the discovered (near) optimal actions $\pi(s)$ as it is becoming more sure of its Q estimates. This can be achieved via the decay function shown in the following equation:

$$\varepsilon = \alpha = 1 - \frac{N_{tx}}{N_{decay}} \quad \text{for} \quad 0 \leq N_{tx} \leq N_{decay}, \tag{7.7}$$

where N_{tx} is the number of transmitted broadcast packets and N_{decay} is a preset number of packets that sets the decay period. This decay function is necessary to guarantee convergence towards the last-known optimum policy in probabilistic systems such as the proposed contention-based MAC since there is no known optimum final state. By reducing the values of ε and α over time via (7.7), the agent is forced to progressively focus on exploitation of gained experience and strive for a high long-term reward. This way, when approaching the end of the decay period, the found (near) optimal states-CW/s are revealed.

7.7.3 A priori approximate controller

The above strategy can be used to get instant performance benefits, starting from the first transmission. This is done by preloading approximate controllers, pretrained for different transmitted bit rates and number of neighbours via (7.7), to the station's memory. These controllers define an initial policy that positively biases the search and accelerates the learning process.

The agent's objective in this phase is to quickly populate its Q-table with values (explore all the state-action pairs multiple times) and form an initial impression of the environment. The lookup Q-table is produced by encoding this knowledge (Q-values) for a set period of N_{decay} *a priori* and can be used as an initial approximate controller, which yields an instant performance benefit since the system is deployed.

Q-learning is an iterative algorithm so it implicitly assumes an initial condition before the first update occurs. Zero initial conditions are used the very first time the algorithm is trained on a set environment, except for some forbidden state-action pairs with large negative values, so it does not waste iterations in which it would try to increase/decrease the CW when it is already set on the upper/lower limit. The algorithm is also explicitly programmed to avoid performing these actions on exploration. The un-trained, initial Q-table is set as in (7.8), where the rows represent the possible states – CW sizes and columns stand for the action space:

$$\begin{array}{cc} & \begin{array}{cccc} \mathbf{CW} & (CW-1)/2 & CW & CW\times 2+1 \end{array} \\ \mathbf{Q_0}[7][3] = & \begin{array}{c} \mathbf{3} \\ \mathbf{7} \\ \mathbf{15} \\ \mathbf{31} \\ \mathbf{63} \\ \mathbf{127} \\ \mathbf{255} \end{array} \begin{pmatrix} -100 & 0 & 0 \\ 0 & 0 & 0 \\ 0 & 0 & 0 \\ 0 & 0 & 0 \\ 0 & 0 & 0 \\ 0 & 0 & 0 \\ 0 & 0 & -100 \end{pmatrix} \end{array} \tag{7.8}$$

Each station can employ various different learning, self-improving, controllers and use the appropriate one depending on a combination of sensed density and received bit rate. This is feasible because the station has the ability to sense the number of one-hop neighbours since they all transmit heart-beat, status packets periodically. It also does not have the memory constraints that typical sensor networks have. An example of a controller's table at the end of the ε decay period as in (7.7) can be seen in (7.9). The controller has been trained *a priori* with $\gamma = 0.7$ and a decay period lasting for 180 s in a 60-car network, where every car transmits 256 bytes every 100 ms. A trajectory leading to optimum/near-optimum CW/s is being formed (depending on past experience) by choosing the maximum Q-value for every CW-state, seen in bold font. The controller in (7.9) oscillates between the values 31 and 63 when exploiting the Q-table to find the optimum CW:

$$\begin{array}{cc} & \begin{array}{cccc} \mathbf{CW} & (CW-1)/2 & CW & CW * 2+1 \end{array} \\ \mathbf{Q}^{\pi}[7][3] \approx & \begin{array}{c} \mathbf{3} \\ \mathbf{7} \\ \mathbf{15} \\ \mathbf{31} \\ \mathbf{63} \\ \mathbf{127} \\ \mathbf{255} \end{array} \begin{pmatrix} -100 & -0.07218 & \mathbf{0.2388} \\ -0.076 & -0.0325 & \mathbf{0.6748} \\ 0.198 & 0.28012 & \mathbf{0.817} \\ 0.2896 & 0.2985 & \mathbf{0.4917} \\ \mathbf{0.4945} & 0.10115 & 0.2838 \\ \mathbf{0.2043} & -0.055 & -0.0218 \\ \mathbf{0.1745} & -0.86756 & -100 \end{pmatrix} \end{array} \tag{7.9}$$

Figure 7.10 shows the steps (in terms of CW size) the algorithm takes (exploration/pretraining) until it converges to an optimal value, or more specifically oscillates among two values (63–127) so that it can exploit (90% of the time) its knowledge and yield performance benefits. Figure 7.11 shows how the proposed

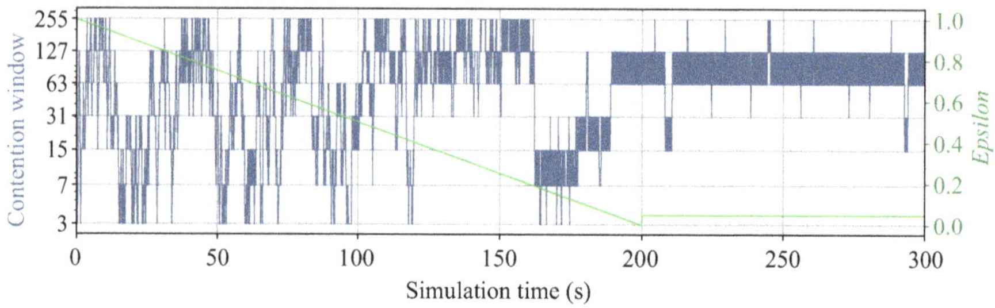

Figure 7.10 *Trace of CW over time for a station in a 100-car network. The first stage is the a priori controller training phase via (4) for 200 sec (or $N_{decay} = 2000$ original packets), then online stage for the remaining time, with an exploration to exploitation ratio of 1:9*

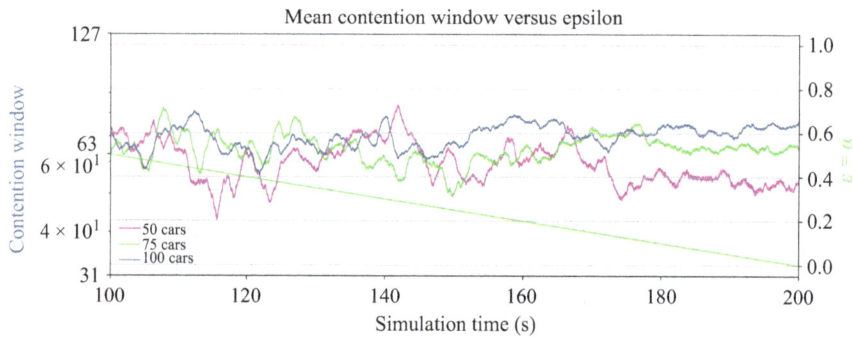

Figure 7.11 *Mean network-wide CW versus training time (second half) for networks of different densities using the Q-learning-based MAC*

MAC discovers the optimum CW size of the stations in three networks of different densities.

7.7.4 Online controller augmentation

While the pretrained, preloaded, approximate controller is useful for speeding up the learning process as well as getting an instant performance benefit, its drawback is that by default it is not adaptive to changes in the environment while online. The online efficiency of the *Q*-learning controller depends on finding the right balance between exploitation of the station's current knowledge and exploration for gathering new information. This means that the algorithm must sometimes perform actions other than the ones dictated by the current policy to update and augment that controller with new information.

While the station is online, exploratory action selection is performed less frequently ($\varepsilon = 0.1$) than in *a priori* learning (7.7) (ε starts from 1), primarily to compensate for modelling errors in the approximate controller. This means that the controller in its online operation uses the optimum *Q*-value 90% of the time and makes exploratory CW perturbations 10% of the time in order to gain new experience. In this way, the agent still has the opportunity to correct its behaviour based on new interactions with the VANET and corresponding rewards.

7.7.5 Implementation details

In RL, the only positive or negative reinforcement an agent receives upon acting so that it can learn to behave correctly in its environment comes in a form of a scalar reward signal. Taking advantage of the link capacity for maximum packet delivery (throughput) was of primary concern for this design, aiming to satisfy the requirements of V2V traffic (frequent broadcasting of kinematic and multimedia information). For this purpose, the reward function is based on the success of these transmissions. Reward r can be either 1 or -1 for successful (ACK) and failed transmissions (no ACK) correspondingly. A successful transmission from the same consecutive state – CW – is not given any reward. The pseudo-code in Algorithm 7.1 summarizes the operation of our proposed protocol.

Algorithm 7.1: *Q*-learning V2V MAC

Initialize $Q_0(CW, A)$ at $t_0 = 0$ ▷ as in (7.8)
procedure ACTION-SELECTION(CW_t) ▷ ε-greedy
    if $p_\varepsilon \leq \varepsilon$ **then**
        $a_{t+1} \leftarrow random[\frac{(CW_t-1)}{2}, CW_t, CW_t * 2 - 1]$
    else if $p_\varepsilon \geq 1 - \varepsilon$ **then**
        $a_{t+1} \leftarrow a_\pi$ ▷ Optimum a from (7.5)
    end if
    if A-priori Controller Learning **then**
        $\varepsilon = \alpha \rightarrow decay$ ▷ according to rule (7.7)
    else if On-line Learning **then**
        $\varepsilon = \alpha \rightarrow constant$
    end if
    $CW_{t+1} \leftarrow CW^{a_{t+1}}$
end procedure
TX Broadcast Packet: *MessageId* ▷ Transmit
procedure FEEDBACK(CW_{t+1}, a_{t+1}) ▷ Collect Reward
    Initialize: $RTT \leftarrow 0$ s
    if RX *MessageId* AND $RTT < 0.1$ s **then**
        if $a_t \neq (CW_{t+1} \longleftarrow CW_t)$ **then**
            $r_t \leftarrow 1$
        end if
    else if $RTT \geq 0.1$ s **then**
        $r_t \leftarrow -1$
    end if
end procedure
Update $Q(CW_{t+1}, a_{t+1})$ ▷ according to rule (7.4)
GOTO 2

The first step of the MAC protocol would be to set the default CW of the station to the minimum possible value, which is suggested by the IEEE 802.11p standard. After that, the node makes an exploratory move with probability ε (exploration) or picks the best known action to date (highest Q value) with probability $1 - \varepsilon$.

Received packet rebroadcasts can be used as ACKs since some will definitely be overheard from the source vehicle, even assuming that they move at the maximum speed limit. These rebroadcasts can happen for forwarding purposes, and they enhance the reliability of the protocol, since the original packet senders can detect collisions as well as provide a means to reward them if they succeed in broadcasting a packet. We use probabilistic rebroadcasting for simplicity, but various routing protocols can be used instead.

Every time a packet containing original information is transmitted, a timer is initiated which waits for a predefined time for an overheard retransmission of that packet, which will have the same *MessageId*. These broadcast packets are useful for a short lifetime, which is the period between refreshes. So a rebroadcast packet, received after that period, is not considered to be a valid ACK because the information will not be relevant any more, since the nodes in VANETs attempt to broadcast fresh information frequently (i.e. 1–10 Hz).

7.8 VANET simulation modelling

A VANET simulation has two main components; a **network component** as described above, which must have the capability to simulate the behaviour of communication networks as well as a **vehicular traffic component** which provides accurate enough mobility patterns for the nodes of such a network (vehicles/cars).

7.8.1 Network simulator

There are a few software environments for simulating a wireless network [38], of which OMNeT++ 4.6 is chosen for its many available models, maturity and advanced GUI capabilities. OMNeT++ [39] is a simulation platform written in C++ with a component-based, modular and extensible architecture.

The basic entities in OMNeT++ are simple modules implemented in C++. Compound modules can be built of simple modules as well as compound modules. These modules can be hosts, routers, switches or any other networking devices. Modules communicate with each other via message passing through gates. The connections from one gate to another can have various channel characteristics such as error/data rate or propagation delay.

Another important reason for choosing OMNeT++ to conduct simulation experiments is the availability of third party libraries containing many protocol implementations for wireless networks. The INET [40] framework version 3.2.3 is used for higher layer protocol implementations to achieve Internet connectivity for the OBUs. The VEINS 4.4 (Vehicles in Network Simulation) framework is used for its DSRC/IEEE 802.11p implementation and its ability to bind a network simulation

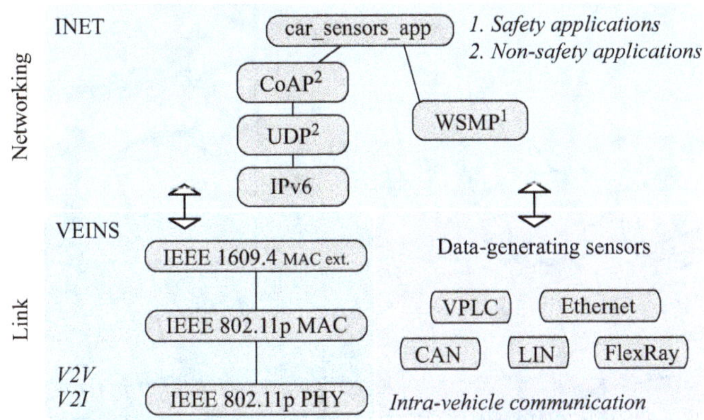

Figure 7.12 Network protocols used on all communication domains

with a live mobility simulation conducted by Simulation of Urban Mobility (SUMO) v0.25. Figure 7.12 shows the network protocols made available by OMNeT++ to enable in-vehicle and inter-vehicle communications.

7.8.2 Mobility simulator

Since vehicular traffic flow is very complex to model, researchers try to predict road traffic using simulations. A traffic simulation introduces models of transportation systems such as freeway junctions, arterial routes, roundabouts to the system under study. SUMO [41] is an open-source microscopic and continuous road traffic simulation package which enables us to simulate the car flow in a large road network such as the one in the city of Brighton. Microscopic traffic flow models, in contrast to macroscopic, simulate single vehicle units, taking under consideration properties such as position and velocity of individual vehicles.

7.8.3 Implementation

The simulation environment on which novel medium access algorithms are to be evaluated uses SUMO and open data to reproduce accurate car mobility [42]. The map is extracted off OpenStreetMap and converted to an XML file which defines the road network. Then random trips are generated from this road network file, and finally these trips are converted to routes and traffic flow. The resulting files are used in SUMO for live traffic simulation as depicted in Figure 7.13. The vehicles are dynamically generated with unique IDs shown in green labels.

Each node within OMNeT++, either mobile (car) or static (RSU) consists of a network interface that uses the 802.11p PHY and MAC, and the application layer that describes a basic safety message exchange and a mobility module. A car, chosen in random fashion, broadcasts a periodic safety message, much like the ones specified in the WSMP.

Figure 7.13 Large scale simulation in the city of Brighton

Listing 7.1 SUMO scripts and parameters to produce the needed XML files

```
//map file to road network XML
$netconvert --osm city.osm
//random trips from XML; source and
//destination edge weighted by length "-l"
$randomTrips.py -n city.net.xml -l -e 800 -o
city.trips.xml
//routes using Shortest Path computation
$duarouter -n city.net.xml -t city.trips.xml
-o city.rou.xml
```

As well as safety message exchange, connected cars can provide extra functionality and enable driving assisting and infotainment systems, such as downloading city map content from RSUs, exchanging video for extended driver vision or even uploading traffic information to the cloud towards an efficient traffic light system. The protocols used for such applications would be different from WSMP, such as the Internet protocols (IPv6, UDP) for the pervasiveness of IP-based applications. Figure 7.14 shows an example of V2V connectivity, where a car broadcasts a safety message to neighbouring cars within range.

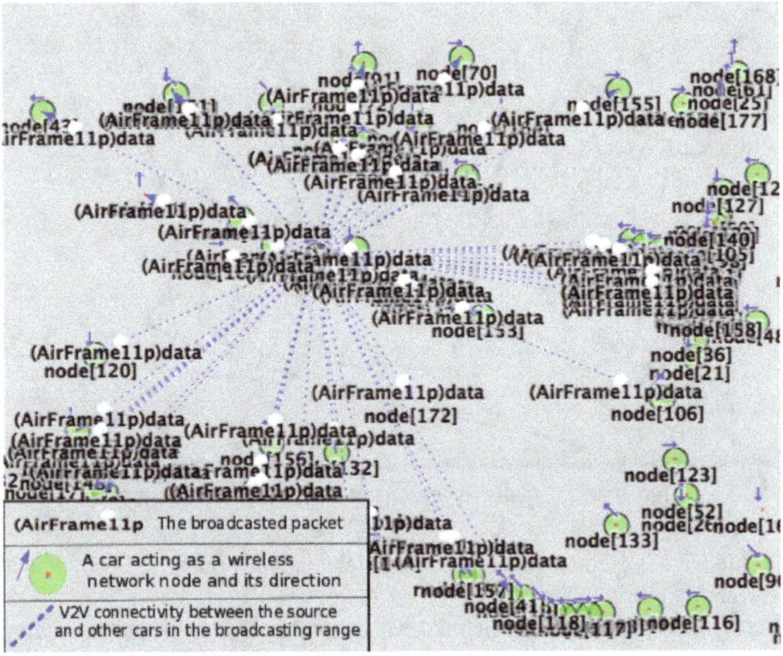

Figure 7.14 A car is broadcasting to neighbouring cars using IEEE 802.11p in OMNeT++

7.9 Protocol performance

The MAC method of the vehicular communication standard IEEE 802.11p has been simulated in a realistic vehicular traffic scenario with vehicle stations periodically broadcasting packets. In order to evaluate the performance of the novel proposed RL-based channel sharing protocol in comparison to the baseline IEEE 802.11p protocol, V2V simulations were carried out using OMNeT++ 5 simulator and the Veins framework. Realistic mobility simulation is achieved by using SUMO coupled with OMNeT++.

7.9.1 Simulation setup

All the cars within the area content for access to medium when trying to transmit a packet or rebroadcast a copy of one. Retransmission probability is set so that a proportion of nodes in the area of interest will rebroadcast the same information upon receipt (i.e. for 100 cars it is set at 2%). We collect most of our results within a specific ROI of ~600 m × 500 m within the University of Sussex campus and set the power to a high enough level within the DSRC limit, in order to not be influenced by border

Figure 7.15 Campus map used in network simulations

effects (hidden/exposed terminals). The artificial campus map used for simulations can be seen in Figure 7.15.

The achieved improvement on link-level contention was of primary concern, so a multitude of tests were run for a single hop scenario, with every node being within the range of the others. By eliminating the hidden terminal problem from the experiment and setting an infinite queue size, packet losses from collisions can be accurately measured. A multi-hop scenario is also presented, which makes the hidden terminal effect apparent in the performance of the network.

The simulation run time for the proposed MAC protocol consists of two stages, as seen in Figure 7.10. First is the approximate controller training stage, which lasts for $N_{decay} = 1{,}800$ transmitted packets (or 180 s with $f_b = 10$ Hz). Then follows the evaluation or online period which lasts for 120 s, in which the agent acts with an $\varepsilon = \alpha = 0.1$. During this time, we benchmark the effect of the trained controllers regarding network performance as well as keep performing some learning for the controller augmentation. For IEEE 802.11p simulations, only the evaluation stage is needed, which lasts for the same time.

All cars in the network are continuously transmitting broadcast packets, such as CAMs with a period $T_b = (1/f_b) = 100$ ms. The packets are transmitted using the highest priority, voice traffic (AC_VO) AC. In VANETs, the network density changes depending on location and time of the day. We test the performance of the novel MAC against the standard IEEE 802.11p protocol for different number of cars. The data rate is set at 6 Mbps so it can conveniently accommodate hundreds of vehicles within the DSRC communication range. Simulation parameters can be found at Table 7.4.

7.9.2 Effect of increased network density

The scalability of the MAC protocols is evaluated against a varying number of vehicles travelling in the simulated campus map shown in Figure 7.15. The packet size L_p

Table 7.4 Simulation parameters

Parameter	Value
Evaluation time	120 s
A priori training time	180 s
Channel frequency	5.9 GHz
Transmission rate	6 Mbps
Transmission power	1-hop: 100 mW, 2-hop: 40 mW
Packet size L_p	256 bytes
Backoff slot time	13 μs
Broadcasting frequency f_b	10 Hz
No of relays	≥2 cars (probabilistic)
Discount rate γ	0.7
Learning rate α	Training: (7.4), on-line: 0.1
Epsilon ε	training: (7.4), on-line: 0.1

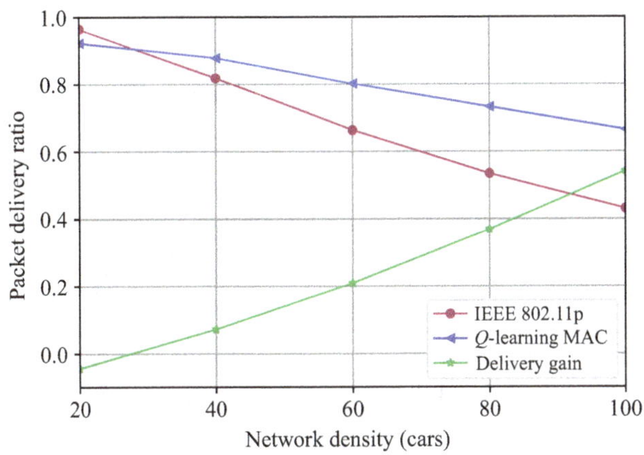

Figure 7.16 PDR versus network density for broadcasting of 256-byte packets with $f_b = 10$ Hz

used in this scenario is 256 bytes, and the broadcasting frequency f_b is set at 10 Hz. Figure 7.16 shows the increase in goodput when using this novel MAC protocol, expressed as a PDR. When using the standard IEEE 802.11p, PDR decreases in denser networks due to the increased collisions between data packets.

The PDR for the proposed Q-learning MAC is measured after the initial, exploratory phase (since the agent by then has gained significant experience). We observed a 37.5% increase in performance (original packets delivered) in a network formed of 80 cars when using the modified, "learning" MAC. There is a slight loss in performance (4%) for 20-car networks. In such sparse networks, the minimum CW is optimal, since with a big CW (waiting for more b time slots), transmission opportunities can be lost and the channel access delay will increase. When using our

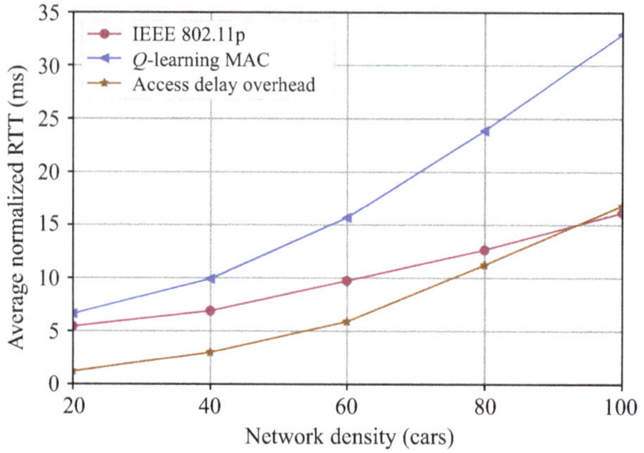

Figure 7.17 Packet Return Time (delay) versus network density for broadcasting of 256-byte packets with $f_b = 10$ Hz

learning protocol, the agent still explores larger CW levels 10% of the time ($\varepsilon = 0.1$), for better adaptability and augmentation of its initial controller. When the network density exceeds 40 cars, the proposed learning MAC performs much better regarding successful deliveries.

The round-trip time (RTT) shown in Figure 7.17 is defined as the length of time it takes for an original broadcast packet to be sent plus the length of time it takes for a rebroadcast of that packet to be received by the original sender. We can see that the increased CW of the learning MAC adds to the channel-access delay time. The worst case scenario simulated is for 100 simultaneous transceivers within the immediate range of each other, in which the average RTT doubles to 32.8 ms when using the Q-learning MAC. Given that both the transmission and heard retransmission are of the same packet size, we can assume that the mean packet delivery latency is 16.4 ms when using the learning MAC instead of 8 ms for baseline IEEE 802.11p, while PDR is improved by 54%.

7.9.3 Effect of data rate

We also examine the performance of both the standard and enhanced protocol for different data rates. PDR is measured for a network of 60 nodes without hidden terminals. The broadcasting frequency is set at $f_b = 10$ Hz, and the packet size L_p varies from 64 to 512 bytes, as seen in Figure 7.18. For 512 byte packets, the mean achieved goodput T_{avg} per IEEE 802.11p node from (7.10) is 16.925 kbps. For the same settings, each learning MAC station achieves 29.218 kbps on average, yielding to a 72.63% increase in goodput. It is clear that for larger packet transmissions the Q-learning-based protocol will be much faster and more reliable:

$$T_{avg} = L_p \times f_b \times 8\,\text{bit} \times PDR. \tag{7.10}$$

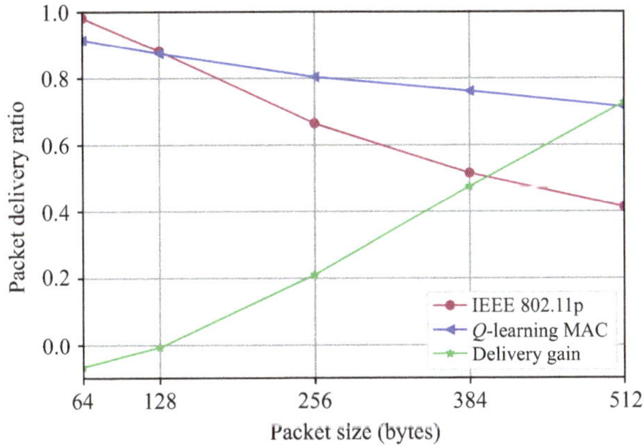

Figure 7.18 PDR versus packet size for 60 vehicles broadcasting with $f_b = 10$ Hz

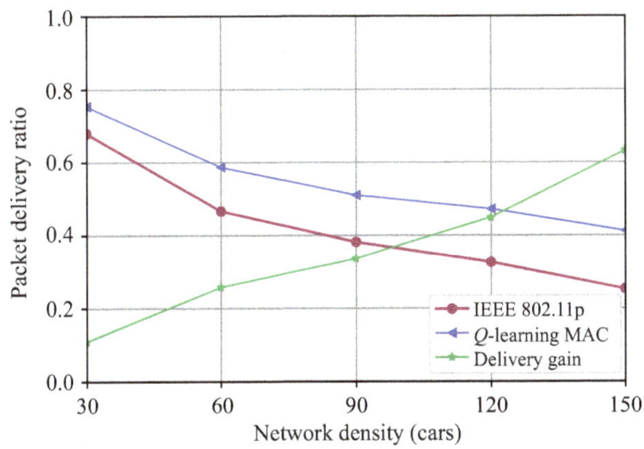

Figure 7.19 PDR versus network density for broadcasting of 256-byte packets with $f_b = 10$ Hz in a two-hop scenario

7.9.4 Effect of multi-hop

In a network without fixed topology, the most common way to disseminate information is to broadcast packets across the network. In VANETs, vehicles often cooperate to deliver data messages through multi-hop paths, without the need of centralized administration. In this scenario, we test the performance of the proposed protocol when attempting to transmit two hops away. We evaluate performance for two-hop transmissions by reducing the transmission power to 40 mW. As the network density increases, the proposed MAC offers a valid delivery benefit for vehicle-stations contenting for access on the same channel. The performance of both IEEE 802.11p and the proposed learning MAC regarding two-hop packet reception ratio is compared in Figure 7.19.

We see that because the hidden terminal phenomenon appears, the performance deteriorates compared to the single hop scenario, but the performance gain regarding packet delivery is still apparent when using Q-learning to adapt the backoff. Packets lost are not recovered since we are concerned with the performance of the link layer.

7.10 Conclusion

A contention-based MAC protocol for V2V/V2I transmissions was introduced in this chapter. It relies on Q-learning to discover the optimum CW by continuously interacting with the network. Simulations were developed to demonstrate the effectiveness of this learning-based MAC protocol. Results prove that the proposed method allows the network to scale better to increase network density and accommodate higher packet delivery rates compared to the IEEE 802.11p standard. This translates to more reliable packet delivery and higher system throughput, while maintaining acceptable delay levels. Future work will be focused on how the learning MAC responds to drastic changes in the networking environment via invoking the ε decay function while online as well as improving fairness and transmission latency.

References

[1] IEEE. IEEE Standard for Information Technology–Telecommunications and Information Exchange between Systems–Local and Metropolitan Area Networks–Specific Requirements Part 11:Wireless LAN Medium Access Control (MAC) and Physical Layer (PHY) Specifications Amendment 6: Wireless Access in Vehicular Environments, IEEE, pp. 1–51, 2010.

[2] Pressas A, Sheng Z, Ali F, *et al.* Contention-based learning MAC protocol for broadcast vehicle-to-vehicle communication. In: 2017 IEEE Vehicular Networking Conference (VNC); 2017. p. 263–270.

[3] Oliveira R, Bernardo L, and Pinto P. The influence of broadcast traffic on IEEE 802.11 DCF networks. Computer Communications. 2009;32(2): 439–452. Available from: http://www.sciencedirect.com/science/article/pii/S0140366408005847.

[4] Delgrossi L, and Zhang T. Vehicle Safety Communications: Protocols, Security, and Privacy; 2012. Available from: http://dx.doi.org/10.1002/9781118452189.ch3.

[5] Navet N, and Simonot-Lion F. In-vehicle communication networks – a historical perspective and review; 2013.

[6] Ku I, Lu Y, Gerla M, *et al.* Towards Software-Defined VANET: Architecture and Services; 2014.

[7] Achour I, Bejaoui T, and Tabbane S. Network coding approach for vehicle-to-vehicle communication: principles, protocols and benefits. In: 2014 22nd International Conference on Software, Telecommunications and Computer Networks, SoftCOM 2014; 2011. p. 154–159.

[8] Schoch E, Kargl F, and Weber M, Leinmuller T. Communication patterns in VANETs – topics in automotive networking. IEEE Communications Magazine. 2008;46(11):119–125.

[9] Liang W, Li Z, Zhang H, Wang S, and Bie R. Vehicular Ad Hoc Networks: Architectures, Research Issues, Methodologies, Challenges, and Trends. International Journal of Distributed Sensor Nctworks. 2015;11(8):745303.

[10] Al-Sultan S, Al-Doori MMM, Al-Bayatti AH, *et al.* A comprehensive survey on vehicular ad hoc network. Journal of Network and Computer Applications. 2014;37:380–392. Available from: http://dx.doi.org/10.1016/j.jnca.2013.02.036\nhttp://www.sciencedirect.com/science/article/pii/S108480451300074X.

[11] Faezipour M, Nourani M, Saeed A, *et al.* Progress and challenges in intelligent vehicle area networks. Communications of the ACM. 2012;55(2):90–100. Available from: http://doi.acm.org/10.1145/2076450.2076470.

[12] Xu Q, Mak T, Ko J, *et al.* Vehicle-to-vehicle safety messaging in DSRC. In: Proceedings of the first ACM Workshop on Vehicular Ad Hoc Networks – VANET '04; 2004. p. 19–28. Available from: http://portal.acm.org/citation.cfm?doid=1023875.1023879.

[13] Hameed Mir Z, and Filali F. LTE and IEEE 802.11p for vehicular networking: a performance evaluation. EURASIP Journal on Wireless Communications and Networking. 2014;2014(1):89. Available from: http://dx.doi.org/10.1186/1687-1499-2014-89.

[14] Zhang X, and Qiao D. Quality, reliability, security and robustness in heterogeneous networks. In: 7th International Conference on Heterogeneous Networking for Quality. Springer Publishing Company, Incorporated; 2012.

[15] Jiang D, Taliwal V, Meier A, *et al.* Design of 5.9GHz DSRC-based vehicular safety communication. IEEE Wireless Communications. 2006;13(5):36–43. Available from: http://dx.doi.org/10.1109/WC-M.2006.250356.

[16] Li YJ. An Overview of the DSRC / WAVE technology. In: Quality, Reliability, Security and Robustness in Heterogeneous Networks, Xi Z, Daji Q (eds.). Berlin: Springer 2012. p. 544–558.

[17] Xu Q, Mak T, Ko J, *et al.* Medium access control protocol design for vehicle – vehicle safety messages. IEEE Transactions on Vehicular Technology. 2007;56(2):499–518.

[18] Wu C, Ohzahata S, Ji Y, *et al.* A MAC protocol for delay-sensitive VANET applications with self-learning contention scheme. In: 2014 IEEE 11th Consumer Communications and Networking Conference, CCNC 2014; 2014. p. 438–443.

[19] Xu K, Gerla M, and Bae S. How effective is the IEEE 802.11 RTS/CTS handshake in ad hoc networks. In: Global Telecommunications Conference, 2002 GLOBECOM'02 IEEE; 2002; 1. p. 72–76.

[20] Oliveira R, Bernardo L, and Pinto P. Performance analysis of the IEEE 802.11 distributed coordination function with unicast and broadcast traffic. In: The 17th Annual IEEE International Symposium on Personal, Indoor and Mobile Radio Communications; 2006. p. 1–5.

[21] Xia X, Member N, Niu Z, *et al.* Enhanced DCF MAC scheme for providing differentiated QoS in ITS. In: Proceedings The 7th International IEEE Conference on Intelligent Transportation Systems (IEEE Cat No04TH8749); 2004. p. 280–285. Available from: http://ieeexplore.ieee.org/lpdocs/epic03/wrapper.htm?arnumber=1398911.

[22] Qiu HJF, Ho IWH, Tse CK, *et al.* A methodology for studying 802.11p VANET broadcasting performance with practical vehicle distribution. IEEE Transactions on Vehicular Technology. 2015;64(10):4756–4769.

[23] Stanica R, Chaput E, and Beylot AL. Enhancements of IEEE 802.11p protocol for access control on a VANET control channel. In: IEEE International Conference on Communications; 2011.

[24] Miao L, Djouani K, Wyk BJV, *et al.* Performance evaluation of IEEE 802.11p MAC protocol in VANETs safety applications. In: Wireless Communications and Networking Conference (WCNC), 2013 IEEE; 2013; i. p. 1663–1668.

[25] Choi J, So J, and Ko Y. Numerical analysis of IEEE 802.11 broadcast scheme in multihop wireless ad hoc networks. International Conference on Information Networking; 2005. Available from: http://www.springerlink.com/index/10.1007/b105584\nhttp://link.springer.com/chapter/10.1007/978-3-540-30582-8_1.

[26] Torrent-Moreno M, Mittag J, Santi P, *et al.* Vehicle-to-vehicle communication: Fair transmit power control for safety-critical information. IEEE Transactions on Vehicular Technology. 2009;58(7):3684–3703.

[27] Mertens Y, Wellens M, and Mahonen P. Simulation-based performance evaluation of enhanced broadcast schemes for IEEE 802.11-based vehicular networks. In: IEEE Vehicular Technology Conference; 2008. p. 3042–3046.

[28] Sutton RS, and Barto AG. Introduction to Reinforcement Learning. 1st ed. Cambridge, MA, USA: MIT Press; 1998.

[29] Bellman R. A Markovian Decision Process. Indiana University Mathematics Journal. 1957;6:679–684.

[30] Shoaei AD, Derakhshani M, Parsaeifard S, *et al.* MDP-based MAC design with deterministic backoffs in virtualized 802.11 WLANs. IEEE Transactions on Vehicular Technology. 2016;65(9):7754–7759.

[31] Tse Q, Si W, and Taheri J. Estimating contention of IEEE 802.11 broadcasts based on inter-frame idle slots. In: Proc. IEEE Conf. on Local Computer Networks – Workshops; 2013. p. 120–127.

[32] Bianchi G. Performance analysis of the IEEE 802.11 distributed coordination function. IEEE Journal on Selected Areas in Communications. 2000;18(3):535–547.

[33] Liu Z, and Elhanany I. RL-MAC: a QoS-aware reinforcement learning based MAC protocol for wireless sensor networks. In: Proc. IEEE Int. Conf. on Netw., Sens. and Control; 2006. p. 768–773.

[34] Wu C, Ohzahata S, Ji Y, *et al.* A MAC protocol for delay-sensitive VANET applications with self-learning contention scheme. In: Proc. IEEE Consumer Comm. and Netw. Conference; 2014. p. 438–443.

[35] Yang Q, Xing S, Xia W, *et al.* Modelling and performance analysis of dynamic contention window scheme for periodic broadcast in vehicular ad hoc networks. IET Communications. 2015;9(11):1347–1354.

[36] Watkins CJCH, and Dayan P. *Q*-learning. Machine Learning. 1992;8(3): 279–292. Available from: http://dx.doi.org/10.1007/BF00992698.

[37] Watkins CJCH. Learning from Delayed Rewards. Cambridge, UK: King's College; 1989. Available from: http://www.cs.rhul.ac.uk/~chrisw/new_thesis.pdf.

[38] Lessmann J, Janacik P, Lachev L, *et al.* Comparative study of wireless network simulators. In: Proc. of Seventh International Conference on Networking (ICN). IEEE; 2008. p. 517–523.

[39] Varga A, and Hornig R. An overview of the OMNeT++ simulation environment. In: Proc. of the 1st International Conference on Simulation Tools and Techniques for Communications, Networks and Systems & Workshops. SimuTools; 2008. p. 60:1–60:10. Available from: http://dl.acm.org/citation.cfm?id=1416222.1416290.

[40] INET Framework; 2010. [Online; accessed 21-June-2016]. http://inet.omnetpp.org.

[41] Behrisch M, Bieker L, Erdmann J, *et al.* SUMO – Simulation of Urban MObility – an overview. In: Proceedings of the 3rd International Conference on Advances in System Simulation (SIMUL'11); 2011;(c). p. 63–68. Available from: http://www.thinkmind.org/index.php?view=article&articleid=simul_2011_3_40_50150.

[42] Pressas A, Sheng Z, Fussey P, *et al.* Connected vehicles in smart cities: interworking from inside vehicles to outside. In: 2016 13th Annual IEEE International Conference on Sensing, Communication, and Networking (SECON); 2016. p. 1–3.

Chapter 8

Machine-learning-based perceptual video coding in wireless multimedia communications

Shengxi Li[1], *Mai Xu*[2], *Yufan Liu*[3], *and Zhiguo Ding*[4]

We present in this chapter the advantage of applying machine-learning-based perceptual coding strategies in relieving bandwidth limitation for wireless multimedia communications. Typical video-coding standards, especially the state-of-the-art high efficiency video coding (HEVC) standard as well as recent research progress on perceptual video coding, are included in this chapter. We further demonstrate an example that minimizes the overall perceptual distortion by modeling subjective quality with machine-learning-based saliency detection. We also present several promising directions in learning-based perceptual video coding to further enhance wireless multimedia communication experience.

8.1 Background

At present, multimedia applications, such as Facebook and Twitter, are becoming integral components in the daily lives of millions, leading to the explosion of big data. Among them, videos are one of the largest types of big data [1], thus posing a great challenge to the limited communication and storage resources. Meanwhile, due to more powerful camera hardware, their resolutions are significantly increasing, further intensifying the hunger on communication and storage resources. Aiming at overcoming this resource-hungry issue, a set of video-coding standards have been proposed to condense video data, e.g., MPEG-2 [2], MPEG-4 [3] VP9 [4], H.263 [5] and H.264/AVC [6].

Most recently, as the successor of H.264/AVC, HEVC [7] was formally approved in April, 2013. In HEVC, several new features, e.g., the quadtree-based coding

[1]Department of Electrical and Electronic Engineering, Imperial College London, UK
[2]Department of Electronic and Information Engineering, Beihang University, China
[3]Institute of Automation, Chinese Academy of Sciences, China
[4]School of Electrical and Electronic Engineering, The University of Manchester, UK

structure and intra-prediction modes with 33 directions,[1] were adopted. Consequently, the HEVC Main Still Picture (HEVC-MSP) profile [8], which is designed for still picture compression, achieves the best performance among all the state-of-the-art standards on image compression, with an approximately 10% (over VP9) to 40% (over JPEG) improvement in bit-rate savings [9]. However, all existing standards, including HEVC-MSP, primarily focus on removing statistical redundancy by adopting various techniques [10], e.g., intra-prediction and entropy coding. Further reducing statistical redundancy may help to improve coding efficiency, but at the cost of extremely high computational complexity.

Koch *et al.* [12] investigated that the bandwidth between the human eyes and brain is approximately 8 Mbps, which is far insufficient to process the visual input captured by millions of optical cells. Thus, the human eye is mostly at a quite low resolution, except for a small area at the fovea (visual angle of approximately 2°), which is called the region-of-interest (ROI) in the video-coding community. Meanwhile, as pointed out by [13], human ROIs are similar across different individuals. It is also well known [14] that the coding mechanism can be modified to cater to the human visual system (HVS) by moving bits from non-ROIs to ROIs to achieve better subjective quality. This is also illustrated in Figure 8.1(b) and (c). Perceptual video coding has received a great deal of research effort from 2000 onwards, due to its great potential in improving coding efficiency [15–18]. In H.263, a perceptual rate control scheme [15] was proposed. In this scheme, a perceptual sensitive weight map of conversational scene (i.e., scene with frontal human faces) is obtained by combining stimulus-driven (i.e., luminance adaptation and texture masking) and cognition-driven (i.e., skin colors) factors together. According to such a map, more bits are allocated to ROIs by reducing QP values in these regions. Afterwards, for H.264/AVC, a novel resource allocation method [16] was proposed to optimize the subjective rate–distortion (R–D)-complexity performance of conversational video coding, by improving the visual quality of face region extracted by the skin-tone algorithm. Moreover, Xu *et al.* [19] utilized a novel window model to characterize the relationship between the size of window and variations of picture quality and buffer occupancy, ensuring a better perceptual quality with less quality fluctuation. This model was advanced in [20] with an improved video quality metric for better correlation to the HVS. Most recently, in HEVC, the perceptual model of structural similarity (SSIM) has been incorporated for perceptual video coding [21]. Instead of minimizing mean squared error (MSE) and sum of absolute difference, SSIM is minimized [21] to improve the subjective quality of perceptual video coding in HEVC. However, through our investigation, the substantial low quality in non-ROIs may also significantly degrade image quality, as shown in Figure 8.1(d). Thus, how many bits should "move" from non-ROIs to ROIs, together with accurate ROI detection, is crucial for compression. In other words, we need to ensure that the detected ROIs are the regions that attract human attention, and then bit allocation needs to be optimized according to ROIs, targeting minimal overall perceptual distortion.

[1] Planar and DC are two other intra-prediction modes.

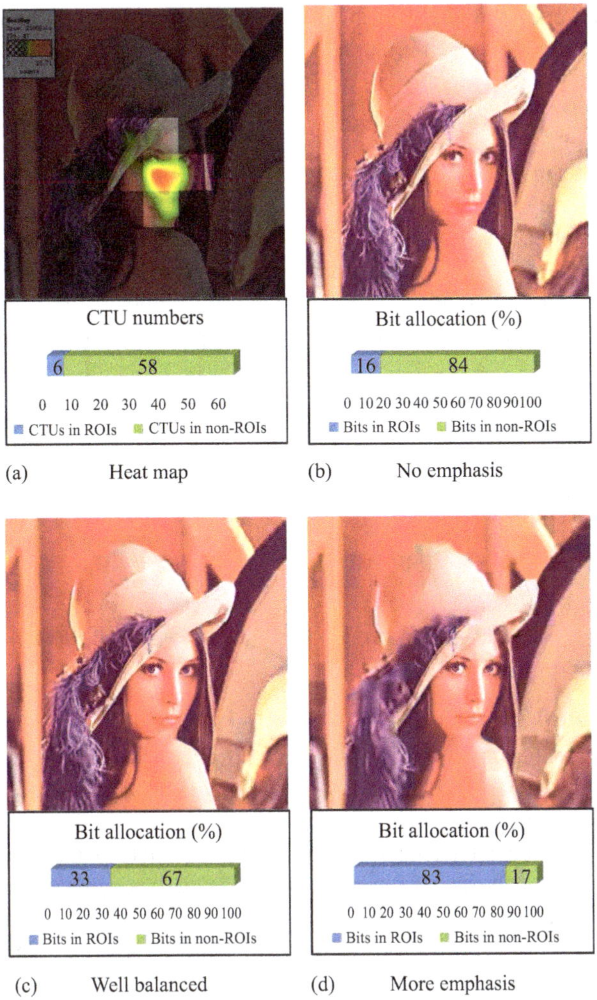

Figure 8.1 An example of HEVC-based compression for Lena image, with different bit allocation emphasis on ROIs. Note that (a) is the heat map of eye fixations; (b), (c) and (d) are compressed by HEVC-MSP at 0.1 bpp with no, well balanced and more emphasis on face regions. The difference mean opinion scores (DMOS) for (b), (c) and (d) are 63.9, 57.5 and 70.3, respectively [11]

The organization of this chapter is as follows. The literature review is first introduced in Section 8.2, from perspectives of perceptual models and incorporations in video coding. We then present in Section 8.3 the recursive Taylor expansion (RTE) method for optimal bit allocation toward the perceptual distortion and also provide rigorous proofs. The computational analysis of the proposed RTE method is introduced in

Section 8.4. For experimental validations, we first verify the proposed RTE method on compressing one single image/frame in Section 8.5, followed by the results on compressing video sequences in Section 8.6.

8.2 Literature review on perceptual video coding

Generally speaking, main parts of perceptual video coding are perceptual models, perceptual model incorporation in video coding and performance evaluations. Specifically, perceptual models, which imitate the output of the HVS to specify the ROIs and non-ROIs, need to be designed first for perceptual video coding. Second, on the basis of the perceptual models and existing video-coding standards, perceptual model incorporation in video coding from perceptual aspects needs to be developed to encode/decode the videos, mainly through removing their perceptual redundancy. Rather than incorporating perceptual model in video coding, some machine-learning-based image/video compression approaches have also proposed during the past decade.

8.2.1 Perceptual models

Perceptual models can be classified into two categories: either manual or automatic identifications.

8.2.1.1 Manual identification

This kind of perceptual models requires manual effort to distinguish important regions which need to be encoded with high quality. In the early years, Geisler and Perry [22] employed a foveated multi-resolution pyramid video encoder/decoder to compress each image of varying resolutions into five or six regions in real-time, using a pointing device. This model requires the users to specify which regions attract them most during the video transmission. Thus, this kind of models may lead to transmission and processing delay between the receiver and transmitter sides, when specifying the ROIs. Another way [23] is to specify ROIs before watching, hence avoiding the transmission and processing delay. However, considering the workload of humans, these models cannot be widely applied to various videos.

In summary, the advantage of manual identification models is the accurate detection of ROIs. However, as the cost, it is expensive and intractable to extensively apply these models due to the involvement of manual effort or hardware support. In addition, for the models of user input-based selection, there exists transmission and processing delay, thus making the real-time applications impractical.

8.2.1.2 Automatic identification

Just as its name implies, this category of perceptual models is to automatically recognize ROIs in videos, according to visual attention mechanisms. Therefore, visual attention models are widely used among various perceptual models. There are two classes of visual attention models: either bottom-up or top-down models. Itti's model [24] is one of the most popular bottom-up visual attention models in perceptual video coding. Mimicking processing in primate occipital and posterior parietal cortex, Itti's model integrates low-level visual cues, in terms of color, intensity, orientation, flicker and motion, to generate a saliency map for selecting ROIs [17].

The other class of visual attention models is top-down processing [14,18,25–29]. The top-down visual attention models are more frequently applied to video applications, since they are more correlated with human attractiveness. For instance, human face [16,18,26] is one of the most important factor that draws top-down attention, especially for conversational video applications. Beyond, a hierarchical perceptual model of face [18] has been established, endowing unequal importance within face region. However, abovementioned approaches are unable to figure out the importance of face region.

In this article, we quantify the saliency of face and facial features via learning the saliency distribution from the eye fixation data of training videos, via conducting the eye-tracking experiment. Then, after detecting face and facial features for automatically identifying ROI [18], the saliency map of each frame of encoded conversational video is assigned using the learnt saliency distribution. Although the same ROI is utilized in [18], the weight map of our scheme is more reasonable for the perceptual model for video coding, as it is in light of learnt distribution of saliency over face regions. Note that the difference between ROI and saliency is that the former refers to the place that may attract visual attention while the later refers to the possibility of each pixel/region to attract visual attention.

8.2.2 Incorporation in video coding

The existing incorporation schemes can be mainly divided into two aspects: model-based and learning-based approaches. The model-based approaches apply prior models or the above perceptual models to the existing video-coding approaches, while the learning-based approaches aim at discovering similarities among pixels or blocks to reduce redundancy in video coding.

8.2.2.1 Model-based approaches

One category of approaches called preprocessing is to control the nonuniform distribution of distortion before encoding [30–32]. A common way for preprocessing is spatial blurring [30,31]. For instance, the spatial blurring approach [30] separates the scene into foreground and background. The background is blurred to remove high-frequency information in spatial domain so that less bits are allocated to this region. However, this may cause obvious boundaries between the background and foreground.

Another category is to control the nonuniform distribution of distortion during encoding, therefore called embedded encoding [16,18,33–35]. As it is embedded into the whole coding process, this category of approaches is efficient in more flexibly compressing videos with different demands. In [16], Liu *et al.* established importance map at macro block (MB) level based on face detection results. Moreover, combining texture and nontexture information, a linear rate–quantization (R–Q) model is applied to H.264/AVC. Based on the importance map and R–Q model, the optimized

QP values are assigned to all MBs, which enhances the perceived visual quality of compressed videos. In addition, after obtaining the importance map, the other encoding parameters, such as mode decision and motion estimation search, are adjusted to provide ROIs with more encoding resources. Xu *et al.* [18] proposed a new weight-based unified *R–Q* (URQ) rate control scheme for compressing conversational videos, which assigns bits according to bpw, instead of bpp in conventional URQ scheme. Then, the quality of face regions is improved such that its perceived visual quality is enhanced. The scheme in [18] is based on the URQ model [36], which aims at establishing the relationship between bite-rate *R* and quantization parameters Q, i.e., *R–Q* relationship. However, since various flexible coding parameters and structures are applied in HEVC, *R–Q* relationship is hard to be precisely estimated [37]. Therefore, Lagrange multiplier λ [38], which stands for the slope of *R–D* curve, has been investigated. According to [37], the relationship between λ and *R* can be better characterized in comparison with *R–Q* relationships. This way, on the basis of *R*–λ model, the state-of-the-art *R*–λ rate control scheme [39] has better performance than the URQ scheme. Therefore, on the basis of the latest *R*–λ scheme, this article proposes a novel weight-based *R*–λ scheme to further improve the perceived video quality of HEVC.

8.2.2.2 Learning-based approaches

From the viewpoint of machine learning, the pixels or blocks from one image or several images may have high similarity. Such similarity can be discovered by machine-learning techniques and then utilized to decrease redundancy of video coding. For exploiting the similarity within an image/video, image inpainting has been applied in [40,41] to use the image blocks from spatial or temporal neighbors for synthesizing the unimportant content, which is deliberately deleted at the encoder side. As such, the bits can be saved as not encoding the missing areas of the image/video. Beyond, rather than predicting the missing intensity information in [40,41] , several approaches [42–45] have been proposed to learn to predict the color in an images using the color information of some representative pixels. Then, only representative pixels and grayscale image need to be stored, such that the image [43–45] or video [42] coding can be achieved. Most recently, the deep-learning technique has also been applied to reduce coding complexity via an early-terminated coding unit (CU) partition scheme [46].

For working on similarity across various images or frames of videos, dictionary learning has been developed to discover the inherent patterns of image blocks. Together with dictionary learning, sparse representation can be then used to effectively represent an image for image [47] or video coding [48], instead of conventional image transforms such as discrete cosine transform.

The above approaches primarily improve the fidelity of ROIs, but they may fail in ensuring the overall subjective quality, as extremely low quality on non-ROIs can also degrade the subjective quality. In the next section, we propose an approach to optimize the overall subjective quality, different from the above approaches that only increase bits in ROIs.

8.3 Minimizing perceptual distortion with the RTE method

In this section, we primarily focus on minimizing the perceptual distortion of one image/frame compression for clarify and propose a closed-form bit allocation approach to minimize the perceptual distortion [49]. It needs to point out that the approach can also be applied in perceptual video compression, which is to be presented in Section 8.6.

Specifically, the most recent work [17] has pointed out that eye-tracking weighted peak signal-to-noise ratio (EWPSNR), which is the combination of eye-tracking fixations and MSE, is highly correlated with subjective quality. Due to the unavailability of eye-tracking data, we utilize the saliency weighted PSNR (SWPSNR) instead as the perceptual distortion to approximate subjective quality. Automatic saliency detection is thus the first step of our approach for saliency-guided image compression. In our approach, we leverage on our most recent face saliency-detection method [50] for compressing face images and a latest saliency-detection method [51] for compressing other generic images. Note that face and non-face images are automatically classified using the face detector in [50]. Then, we propose a formulation to minimize perceptual distortion with reasonable bit allocation on compressed images. Unfortunately, it is intractable to obtain a closed-form solution to the proposed optimization formulation because the formulation is a high-order algebraic equation, and its non-integer exponents vary across different coding tree units (CTUs). We thus develop a new method, namely, RTE, to acquire the solution for optimal bit allocation in a closed-form manner. In the proposed RTE method, we iterate a third-order Taylor expansion to reach the optimal solution for bit allocation. We also develop an optimal bit reallocation process to alleviate the mismatch between the target and actual bits, while maintaining perceptual distortion optimization. We further verify via both theoretical and numerical analyses that little time cost is incurred by our approach.

We first transplant the R–λ RC approach [37] into HEVC-MSP in Section 8.3.1. Upon this, an optimization formulation is proposed in Section 8.3.2, which aims at maximizing the SWPSNR at a given bit rate for each image. The RTE method is then proposed in Section 8.3.3 to solve this formulation with a closed-form solution. In this way, the perceptual distortion can be minimized via bit allocation. In addition, we develop an optimal bit reallocation method in Section 8.3.4 to alleviate the mismatch between the target and actual bit rates.

8.3.1 Rate control implementation on HEVC-MSP

The latest R–λ approach is proposed in [37] for RC in HEVC. Since we concentrate on applying RC to image compression, the CTU level RC in one video frame is discussed here. Specifically, for HEVC, it has been verified that the hyperbolic model can better fit the R–D relationship [37]. Based on the hyperbolic model, an R–λ model is developed for bit allocation in the latest HEVC RC approach, where λ is the slope of the R–D relationship [38]. Assuming that d_i, r_i and λ_i represent the distortion, bits and *R–D* slope for the *i*th CTU, respectively, the *R–D* relationship and *R–λ* model are formulated as follows:

$$d_i = c_i r_i^{-k_i}, \tag{8.1}$$

and

$$\lambda_i = -\frac{\partial d_i}{\partial r_i} = c_i k_i \cdot r_i^{-k_i - 1}, \tag{8.2}$$

where c_i and k_i are the parameters that reflect the content of the *i*th CTU. In the *R–λ* approach [37], r_i is first allocated according to the predicted mean absolute difference, and then its corresponding λ_i is obtained using (8.2). By adopting a fitting relationship between λ_i and QP, the QPs of all CTUs within the frame can be estimated such that RC is achieved in HEVC. For more details, refer to [37].

However, for HEVC-MSP, c_i and k_i cannot be obtained when encoding CTUs. Thus, it is difficult to directly apply the *R–λ* RC approach to HEVC-MSP. In the work of [52], the sum of the absolute transformed differences (SATD), calculated by the sum of Hadamard transform coefficients, is utilized for HEVC-MSP. Specifically, the modified *R–λ* model is

$$\lambda_i = \alpha_i \left(\frac{s_i}{r_i} \right)^{\beta_i}, \tag{8.3}$$

where α_i and β_i are the constants for all CTUs and remain the same when encoding an image. Moreover, s_i denotes the SATD for the *i*th CTU, which measures the CTU texture complexity. Nevertheless, SATD is too simple to reflect image content, leading to an inaccurate *R–D* relationship during RC.

To avoid the above issues, we adopt a preprocessing process in calculating c_i and k_i. After pre-compressing, the pre-encoded distortion, bits and λ can be obtained for the *i*th CTU, which are denoted as $\bar{d}_i$, $\bar{r}_i$ and $\bar{\lambda}_i$, respectively. Then, the RC-related parameters, c_i and k_i, can be estimated upon (8.1) and (8.2) before encoding the *i*th CTU:

$$c_i = \frac{\bar{d}_i}{\left(\bar{r}_i^{-\bar{\lambda}_i \cdot \bar{r}_i / \bar{d}_i} \right)}, \tag{8.4}$$

and

$$k_i = \frac{\bar{\lambda}_i \cdot \bar{r}_i}{\bar{d}_i}. \tag{8.5}$$

With the estimated c_i and k_i, the RC of the *R–λ* approach [37] can be implemented in HEVC-MSP.

Here, a fast pre-compressing process is developed in our approach, which sets the maximum CU depth to 0 for all CTUs. We have verified that the fast pre-compressing process slightly increases the computational complexity by a 5% burden, which is slightly larger than the 3% of the SATD-based method [52]. However, this process is able to well reflect the *R–D* relationship, as to be verified Section 8.5.4.

8.3.2 Optimization formulation on perceptual distortion

The primary objective of this chapter is to minimize perceptual distortion for HEVC-based image compression. In our approach, the SWPSNR is applied to measure the perceptual distortion, as [53] has shown that SWPSNR is highly correlated with subjective quality. For SWPSNR, the pixel-wise saliency values need to be detected as the first step in our approach, and these values are used for weighting the MSE. In this chapter, we utilize two state-of-the-art saliency-detection methods for calculating SWPSNR. Specifically, the latest Boolean-map-based saliency (BMS) method [51] is applied in modeling SWPSNR for generic images. Furthermore, for face images, our most recent work [50] has better accuracy in saliency detection than the BMS method. Thus, when computing the SWPSNR of face images, we use the work of [50] to obtain the saliency values.

Here, we denote w_i as the average saliency value within the ith CTU. Meanwhile, we calculate distortion d_i by the sum of pixel-wise square error for the ith CTU. Then, based on d_i and w_i, the optimization on SWPSNR at a given target bit rate R can be formulated as

$$\min\left(\frac{\Sigma_{i=1}^{M} w_i d_i}{\Sigma_{i=1}^{M} w_i}\right) \quad \text{s.t.} \quad \Sigma_{i=1}^{M} r_i = R. \tag{8.6}$$

In (8.6), M denotes the number of CTUs in the image. By using the Lagrange multiplier λ, (8.6) can be turned to find the minimum value of R–D cost J [38], which is defined as

$$J = \left(\frac{\Sigma_{i=1}^{M} w_i d_i}{\Sigma_{i=1}^{M} w_i}\right) + \lambda \cdot (\Sigma_{i=1}^{M} r_i). \tag{8.7}$$

By setting the partial derivatives of (8.7) to zero, the minimum J can be found as follows[2]:

$$\begin{aligned}\frac{\partial J}{\partial r_i} &= \frac{\partial\left(\Sigma_{i=1}^{M} w_i d_i / \Sigma_{i=1}^{M} w_i + \lambda(\Sigma_{i=1}^{M} r_i)\right)}{\partial r_i} \\ &= \frac{w_i}{\Sigma_{i=1}^{M} w_i} \cdot \frac{\partial d_i}{\partial r_i} + \lambda \\ &= 0.\end{aligned} \tag{8.8}$$

Given (8.1) and (8.2), (8.8) is turned to

$$r_i = \left(\frac{\lambda \cdot \Sigma_{i=1}^{M} w_i}{c_i k_i w_i}\right)^{-(1/(k_i+1))} = \left(\frac{\widetilde{w}_i a_i}{\lambda}\right)^{b_i}, \tag{8.9}$$

where $a_i = c_i k_i$ and $b_i = (1/(k_i + 1))$ also reflect the image content for each CTU. Moreover, $\widetilde{w}_i = w_i/(\Sigma_{i=1}^{M} w_i)$ represents the visual importance for each CTU. Note that with our pre-compressing process, c_i and k_i can be obtained in advance. Thus, a_i

[2] It needs to point out that J in (8.7) is convex with regard to r_i and λ, which ensures the global minimum of the problem (8.7).

and b_i are available before encoding the image. Once λ is known, r_i can be estimated using (8.9) for achieving the minimum J.

Meanwhile, there also exists a constraint on bit rate, which is formulated as

$$\sum_{i=1}^{M} r_i = R. \tag{8.10}$$

According to (8.9) and (8.10), we need to find the "proper" λ and bit allocation r_i to satisfy the following equation:

$$\sum_{i=1}^{M} r_i = \sum_{i=1}^{M} \left(\frac{\widetilde{w}_i a_i}{\lambda}\right)^{b_i} = R. \tag{8.11}$$

After solving (8.11) to find the "proper" λ, the target bits can be assigned to each CTU with the maximum SWPSNR.

Unfortunately, since a_i and b_i vary across different CTUs, (8.11) cannot be solved by a closed-form solution. Next, the RTE method is proposed to provide a closed-form solution.

8.3.3 RTE method for solving the optimization formulation

For solving (8.11), we assume that $\widetilde{r}_i(\widetilde{\lambda})^{b_i} = (\widetilde{w}_i a_i)^{b_i}$, where $\widetilde{r}_i$ and $\widetilde{\lambda}$ are the estimated r_i and λ, respectively. Then, (8.11) can be rewritten as

$$\sum_{i=1}^{M} r_i = \sum_{i=1}^{M} \left(\frac{\widetilde{w}_i a_i}{\lambda}\right)^{b_i} = \sum_{i=1}^{M} \widetilde{r}_i \left(\frac{\widetilde{\lambda}}{\lambda}\right)^{b_i} = R. \tag{8.12}$$

From (8.12), we can see that once $\widetilde{\lambda} \rightarrow \lambda$, there exists $\widetilde{r}_i \rightarrow r_i$. As such, the optimization formulation of (8.11) can be solved in our approach. However, we do not know $\widetilde{\lambda}$ at the beginning. Meanwhile, λ of (8.12) is also unknown because it is intractable to find the closed-form solution to (8.11). Therefore, a chicken-and-egg dilemma exists between $\widetilde{\lambda}$ and λ. To solve this dilemma, a possible $\widetilde{\lambda}$ is initially set. In our RTE method, the picture λ (denoted as λ_{pic}) is chosen as the initial value of $\widetilde{\lambda}$ for quick convergence. It is calculated by the R–λ model at the picture level [37,52]:

$$\lambda_{pic} = \alpha_{pic} \left(\frac{s_{pic}}{R}\right)^{\beta_{pic}}, \tag{8.13}$$

where α_{pic} and β_{pic} are the fitted constants ($\alpha_{pic} = 6.7542$ and $\beta_{pic} = 1.7860$ in HM 16.0) and s_{pic} represents the SATD for the current picture. Recall that R denotes the target bits allocated to the currently encoded picture.

In the following, the RTE method is proposed to iteratively update $\widetilde{\lambda}$ for making $\widetilde{\lambda} \rightarrow \lambda$.

Specifically, we preliminarily apply Taylor expansion on $(\widetilde{\lambda}/\lambda)^{b_i}$ of (8.12), and then we discard the biquadratic and higher order terms. The process can be formulated as follows:

$$\begin{aligned} R &= \sum_{i=1}^{M} \widetilde{r}_i \left(\frac{\widetilde{\lambda}}{\lambda}\right)^{b_i} \\ &= \sum_{i=1}^{M} \widetilde{r}_i \left(1 + \frac{\ln(\widetilde{\lambda}/\lambda)}{1!} b_i + \cdots + \frac{(\ln(\widetilde{\lambda}/\lambda))^n}{n!} b_i^n + \cdots\right) \\ &\approx \sum_{i=1}^{M} \widetilde{r}_i \left(1 + \frac{\ln(\widetilde{\lambda}/\lambda)}{1!} b_i + \frac{(\ln(\widetilde{\lambda}/\lambda))^2}{2!} b_i^2 + \frac{(\ln(\widetilde{\lambda}/\lambda))^3}{3!} b_i^3\right) \end{aligned} \tag{8.14}$$

In the following equation, we use $\widehat{\lambda}$ to denote the approximation solution to (8.14) after discarding the biquadratic and higher order terms. Consequently, (8.12) can be approximated to be a cubic equation with variable $\ln \widehat{\lambda}$:

$$\begin{aligned} R &= \sum_{i=1}^{M} \widetilde{r}_i \left(1 + \frac{\ln(\widetilde{\lambda}/\widehat{\lambda})}{1!} b_i + \frac{(\ln(\widetilde{\lambda}/\widehat{\lambda}))^2}{2!} b_i^2 + \frac{(\ln(\widetilde{\lambda}/\widehat{\lambda}))^3}{3!} b_i^3\right) \\ &= \underbrace{-\sum_{i=1}^{M} \widetilde{r}_i \left(\frac{b_i^3}{6}\right)}_{A} \ln^3 \widehat{\lambda} + \underbrace{\sum_{i=1}^{M} \widetilde{r}_i \left(\frac{b_i^2}{2} + \frac{b_i^3}{2} \ln \widetilde{\lambda}\right)}_{B} \ln^2 \widehat{\lambda} \\ &\quad \underbrace{-\sum_{i=1}^{M} \widetilde{r}_i \left(b_i^2 \ln \widetilde{\lambda} + b_i + \frac{b_i^3}{2} \ln^2 \widetilde{\lambda}\right)}_{C} \ln \widehat{\lambda} \\ &\quad \underbrace{+\sum_{i=1}^{M} \widetilde{r}_i \left(1 + b_i \ln \widetilde{\lambda} + \frac{b_i^2}{2} \ln^2 \widetilde{\lambda} + \frac{b_i^3}{6} \ln^3 \widetilde{\lambda}\right)}_{D}. \end{aligned} \tag{8.15}$$

By applying the Shengjin formula [54], this cubic equation is evaluated to obtain the solution of $\widehat{\lambda}$ as

$$\widehat{\lambda} = e^{((-B-(\sqrt[3]{Y_1}+\sqrt[3]{Y_2}))/3A)},\ Y_{1,2} = BE + 3A\left(\frac{-F \pm \sqrt{F^2 - 4EG}}{2}\right), \tag{8.16}$$

where $E = B^2 - 3AC$, $F = BC - 9A(D - R)$ and $G = C^2 - 3B(D - R)$. Since $\Delta = F^2 - 4EG > 0$ in practical encoding, (8.16) has only one real solution [54]. Thus, the value of $\widehat{\lambda}$ is unique for optimizing bit allocation. After further removing the cubic-order term, (8.14) is turned to be a quadratic equation. We found that such a quadratic equation may have no real solution or two solutions. Meanwhile, using only one term may lead to large approximation error and slow convergence speed, while keeping more than four terms probably makes the polynomial equations on $\ln \widehat{\lambda}$ unsolvable.

Therefore, discarding the biquadratic and higher order terms of the Taylor expansion is the best choice for our approach.

However, due to the truncation of high-order terms in the Taylor expansion, $\widehat{\lambda}$ estimated by (8.16) may not be an accurate solution to (8.12). Fortunately, as proven in Lemma 8.1, $\widehat{\lambda}$ is more accurate[3] than $\widetilde{\lambda}$ when $\widetilde{\lambda} < \lambda$.

Lemma 8.1. *Consider $\lambda > \widetilde{\lambda} > 0$, $b_i > 0$, and $R > 0$ for (8.12). When the solution of λ to (8.12) is $\widehat{\lambda}$, the following inequality holds for $\widehat{\lambda}$:*

$$|\widehat{\lambda} - \lambda| < |\widetilde{\lambda} - \lambda|. \tag{8.17}$$

Proof. As can be seen in (8.15), $\widehat{\lambda}$ is the solution of λ to the third-order Taylor expansion on $\sum_{i=1}^{M} \widetilde{r}_i(\widetilde{\lambda}/\lambda)^{b_i}$. Hence, the following equation exists:

$$\begin{aligned} R &= \sum_{i=1}^{M} \widetilde{r}_i \left(\frac{\widetilde{\lambda}}{\lambda}\right)^{b_i} \\ &= \sum_{i=1}^{M} \widetilde{r}_i + \sum_{i=1}^{M} \widetilde{r}_i \frac{\ln(\widetilde{\lambda}/\widehat{\lambda})}{1!} b_i + \sum_{i=1}^{M} \widetilde{r}_i \frac{(\ln(\widetilde{\lambda}/\widehat{\lambda}))^2}{2!} b_i^2 + \sum_{i=1}^{M} \widetilde{r}_i \frac{(\ln(\widetilde{\lambda}/\widehat{\lambda}))^3}{3!} b_i^3. \end{aligned} \tag{8.18}$$

In fact, $0 < (\widetilde{\lambda}/\lambda)^{b_i} < 1$ holds for $0 < \widetilde{\lambda} < \lambda$ and $b_i > 0$. Besides, there exists $R = \sum_{i=1}^{M} \widetilde{r}_i(\widetilde{\lambda}/\lambda)^{b_i}$ in (8.18). Therefore, $\sum_{i=1}^{M} \widetilde{r}_i > R$ can be evaluated.

Next, assuming that $\widehat{\lambda} \le \widetilde{\lambda}$, we have $\ln(\widetilde{\lambda}/\widehat{\lambda}) \ge 0$. Due to $\sum_{i=1}^{M} \widetilde{r}_i > R$, $\ln(\widetilde{\lambda}/\widehat{\lambda}) \ge 0$, and $b_i > 0$, the inequality below holds:

$$\sum_{i=1}^{M} \widetilde{r}_i + \sum_{i=1}^{M} \widetilde{r}_i \frac{\ln(\widetilde{\lambda}/\widehat{\lambda})}{1!} b_i + \sum_{i=1}^{M} \widetilde{r}_i \frac{(\ln(\widetilde{\lambda}/\widehat{\lambda}))^2}{2!} b_i^2 + \sum_{i=1}^{M} \widetilde{r}_i \frac{(\ln(\widetilde{\lambda}/\widehat{\lambda}))^3}{3!} b_i^3 > R, \tag{8.19}$$

which is contradictory with (8.18). Therefore, it can be proven that $\widetilde{\lambda} < \widehat{\lambda}$. Then, given Lemma 8.2, $\widetilde{\lambda} < \widehat{\lambda} < \lambda$ can be obtained. As a result, $|\widehat{\lambda} - \lambda| < |\widetilde{\lambda} - \lambda|$ exists.

This completes the proof of Lemma 8.1. □

Lemma 8.2. *Consider $\widetilde{\lambda} > 0$, $\lambda > 0$, $b_i > 0$, $\lambda \neq \widetilde{\lambda}$ and $R > 0$ for (8.12). If $\widehat{\lambda}$ is the solution of λ to (8.12), then the following holds:*

$$\widehat{\lambda} < \lambda. \tag{8.20}$$

[3]It is obvious that $0 < b_i = 1/(k_i + 1) < 1$ and $R > 0$ in HEVC encoding.

Proof. Toward the Taylor expansion of $\sum_{i=1}^{M} \widetilde{r}_i(\widetilde{\lambda}/\lambda)^{b_i}$ in (8.12), we can obtain the following equations:

$$
\begin{aligned}
R &= \sum_{i=1}^{M} \widetilde{r}_i \left(\frac{\widetilde{\lambda}}{\lambda}\right)^{b_i} \\
&= \sum_{i=1}^{M} \widetilde{r}_i + \sum_{i=1}^{M} \widetilde{r}_i \frac{\ln(\widetilde{\lambda}/\widehat{\lambda})}{1!} b_i + \sum_{i=1}^{M} \widetilde{r}_i \frac{(\ln(\widetilde{\lambda}/\widehat{\lambda}))^2}{2!} b_i^2 + \sum_{i=1}^{M} \widetilde{r}_i \frac{(\ln(\widetilde{\lambda}/\widehat{\lambda}))^3}{3!} b_i^3 \\
&= \sum_{i=1}^{M} \widetilde{r}_i + \sum_{i=1}^{M} \widetilde{r}_i \frac{\ln(\widetilde{\lambda}/\lambda)}{1!} b_i + \sum_{i=1}^{M} \widetilde{r}_i \frac{(\ln(\widetilde{\lambda}/\lambda))^2}{2!} b_i^2 + \sum_{i=1}^{M} \widetilde{r}_i \frac{(\ln(\widetilde{\lambda}/\lambda))^3}{3!} b_i^3 \\
&\quad + \sum_{i=1}^{M} \widetilde{r}_i \frac{(\ln(\widetilde{\lambda}/\lambda))^4}{4!} b_i^4 + \sum_{i=1}^{M} \widetilde{r}_i \frac{(\ln(\widetilde{\lambda}/\lambda))^5}{5!} b_i^5 + \sum_{i=1}^{M} \widetilde{r}_i \frac{(\ln(\widetilde{\lambda}/\lambda))^6}{6!} b_i^6 + \cdots .
\end{aligned}
\tag{8.21}
$$

There exist two cases of $\widetilde{\lambda}$ and λ:

- For $\widetilde{\lambda} > \lambda > 0$ and $b_i > 0$, we can obtain $(\ln(\widetilde{\lambda}/\lambda)) \cdot b_i > 0$. It is known that (8.21) holds with $R > \sum_{i=1}^{M} \widetilde{r}_i > 0$ and $\ln(\widetilde{\lambda}/\lambda) > 0$ because of $(\widetilde{\lambda}/\lambda) > 1$. Thus, $\ln(\widetilde{\lambda}/\widehat{\lambda}) \geq \ln(\widetilde{\lambda}/\lambda) > 0$ exists such that $\widehat{\lambda} < \lambda$ can be achieved.
- For $\lambda > \widetilde{\lambda} > 0$ and $b_i > 0$, we have

$$
\sum_{i=1}^{M} \widetilde{r}_i \frac{(\ln(\widetilde{\lambda}/\lambda))^4}{4!} b_i^4 + \sum_{i=1}^{M} \widetilde{r}_i \frac{(\ln(\widetilde{\lambda}/\lambda))^5}{5!} b_i^5 + \sum_{i=1}^{M} \widetilde{r}_i \frac{(\ln(\widetilde{\lambda}/\lambda))^6}{6!} b_i^6 + \cdots > 0. \tag{8.22}
$$

Then, with (8.21), the following inequality exists:

$$
\begin{aligned}
&\sum_{i=1}^{M} \widetilde{r}_i \frac{\ln(\widetilde{\lambda}/\widehat{\lambda})}{1!} b_i + \sum_{i=1}^{M} \widetilde{r}_i \frac{(\ln(\widetilde{\lambda}/\widehat{\lambda}))^2}{2!} b_i^2 + \sum_{i=1}^{M} \widetilde{r}_i \frac{(\ln(\widetilde{\lambda}/\widehat{\lambda}))^3}{3!} b_i^3 \\
&\quad > \sum_{i=1}^{M} \widetilde{r}_i \frac{\ln(\widetilde{\lambda}/\lambda)}{1!} b_i + \sum_{i=1}^{M} \widetilde{r}_i \frac{(\ln(\widetilde{\lambda}/\lambda))^2}{2!} b_i^2 + \sum_{i=1}^{M} \widetilde{r}_i \frac{(\ln(\widetilde{\lambda}/\lambda))^3}{3!} b_i^3 .
\end{aligned}
\tag{8.23}
$$

Moreover, viewing $\widehat{\lambda}$ and λ as variable x, the inequality (8.23) can be analyzed by (8.24). The function of (8.24) monotonously decreases to 0 along with the increasing of variable x (until $x \leq \widetilde{\lambda}$):

$$
\sum_{i=1}^{M} \widetilde{r}_i + \sum_{i=1}^{M} \widetilde{r}_i \frac{\ln(\widetilde{\lambda}/x)}{1!} b_i + \sum_{i=1}^{M} \widetilde{r}_i \frac{(\ln(\widetilde{\lambda}/x))^2}{2!} b_i^2 + \sum_{i=1}^{M} \widetilde{r}_i \frac{(\ln(\widetilde{\lambda}/x))^3}{3!} b_i^3 . \tag{8.24}
$$

By combining (8.23) and (8.24), we can obtain $\widehat{\lambda} < \lambda$.

Therefore, $\widehat{\lambda} < \lambda$ holds for both cases. This completes the proof of Lemma 8.2. □

Remark 8.1. *Given Lemma 8.2, for the subsequent iterations of the RTE method,* $0 < \widetilde{\lambda} < \lambda$ *of Lemma 8.1 can be satisfied since the value of* $\widetilde{\lambda}$ *has been replaced by that of* $\widehat{\lambda}$*. Furthermore, as shown in Lemma 8.1, although both* $\widetilde{\lambda}$ *and* $\widehat{\lambda}$ *may be inaccurate for estimating* λ *in (8.11),* $\widehat{\lambda}$*, obtained through (8.12)–(8.16), is closer to* λ *than* $\widetilde{\lambda}$*. Therefore, we can iterate the Taylor expansion by using* $\widehat{\lambda}$ *as* $\widetilde{\lambda}$ *to the next iteration, which is the core of our RTE method. In this way, the closed-form solution* λ *can be obtained by iteratively estimating* $\widetilde{\lambda}$*.*

Our RTE method is summarized in Table 8.1. For each iteration, the convergence criterion is set according to the approximation error, $E_a < 10^{-10}$, where $E_a = |\Sigma_{i=1}^{M}\widetilde{r}_i - R|/R$. As analyzed in Section 8.4, the approximation error of our RTE method is able to converge to 10^{-10}, generally with no more than three iterations. In other words, after three or fewer iterations, the RTE method is able to reduce the difference between $\widetilde{\lambda}$ and λ to an extremely small range, meeting the convergence criterion. Thus, $\widetilde{\lambda}$ can be output as the closed-form solution to (8.12) (as well as (8.11)). Finally, we replace λ by $\widetilde{\lambda}$ in (8.9) to allocate the target bits to each CTU such that SWPSNR can be maximized.

The physical explanation for the fast convergence speed of our RTE method is as follows. Obviously, the approximation error for each iteration of the RTE method is largely related to $\ln(\widetilde{\lambda}/\lambda)$ in $((\ln(\widetilde{\lambda}/\lambda))^n/n!)b_i^n$ of (8.14). To reduce the value of $\ln(\widetilde{\lambda}/\lambda)$ for small approximation error, our RTE method utilizes a more accurate solution $\widehat{\lambda}$ after each iteration to replace $\widetilde{\lambda}$ for the next iteration, making $((\ln(\widetilde{\lambda}/\lambda))^n/n!)b_i^n$ decrease sharply. Therefore, such a replacement not only provides a more accurate input for the next iteration but also greatly reduces the values of the discarded terms and the approximation error. In this way, the convergence speed can be accelerated along with iterations. Moreover, keeping three terms for the Taylor expansion rather than other terms is solvable and also contributes to the fast convergence speed of our RTE method.

8.3.4 Bit reallocation for maintaining optimization

As we discussed in Section 8.3.3, bits are reasonably allocated in our approach to minimize perceptual distortion. However, in practical encoding, a slight difference

Table 8.1 The RTE method for solving (8.12) [11]

- **Input:** a_i, b_i, w_i for each encoding CTUs and target bits R.
- **Output:** reasonable bit allocation $\widetilde{r}_i$ for each CTU on maximizing SWPSNR.
- Initialize $\widetilde{\lambda}$ to be λ_{pic}.
- **While** $\widetilde{\lambda}$ does not meet the convergence criterion
  1 Calculate A, B, C and D of (8.15) with $\widetilde{\lambda}$.
  2 Obtain $\widehat{\lambda}$ estimated by (8.16).
  3 Update $\widetilde{\lambda}$ with the obtained $\widehat{\lambda}$.
  End
- Save the final $\widetilde{\lambda}$.
- Apply it to bit allocation $\widetilde{r}_i$ with (8.9).
- **Return** $\widetilde{r}_i$ for each CTU.

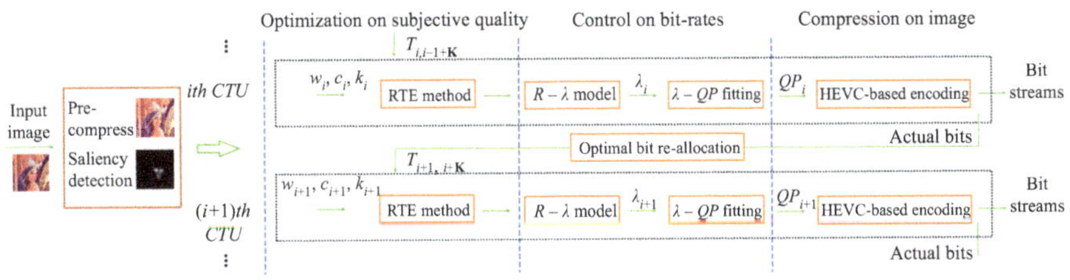

Figure 8.2 The procedure of our approach on minimizing perceptual distortion [11]

between the target and actual bits may exist for each CTU. This difference may degrade RC accuracy. To overcome this, we develop a bit reallocation process to accurately control bit rates, meanwhile maintaining the optimization for perceptual distortion.

Specifically, for compensating the bit-rate error after encoding the ith CTU, the target bits for the incoming K CTUs (denoted as $T_{i+1,i+K}$) are updated by

$$T_{i+1,i+K} = \sum_{j=i+1}^{j=i+K} \widetilde{r}_j + \underbrace{\left(\widehat{T} - \sum_{j=i+1}^{j=M} \widetilde{r}_j \right)}_{\text{bit-rate error}}. \tag{8.25}$$

In (8.25), $\widehat{T}$ is the remaining bits for encoding remaining CTUs, and $\widetilde{r}_j$ represents the target bits for the jth CTU by our RTE method. Recall that M denotes the total number of CTUs. Obviously, as seen from (8.25), the bit error is compensated during encoding the next K CTUs. Here, the RTE method of Section 8.3.3 is applied to reallocate $T_{i+1,i+K}$ to the next K CTUs. Note that we follow [52] and [37] to set $K = 4$, which means that bits are reassigned in the next four CTUs. Moreover, note that due to the fast convergence speed of our RTE method, the complexity increases little for the bit reallocation process.

Finally, we summarize our HEVC-based image compression approach in Figure 8.2. Specifically, we first transplant RC to HEVC-MSP with a simplified pre-compression process, and the saliency values are detected for the input image. Then, our RTE method obtains the target bits of each CTU, which can minimize perceptual distortion at a given bit rate. Next, the QP value of each CTU is estimated using the R–λ model and QP fitting. Note that the bits need to be reallocated in the following CTUs to bridge the gap between the target and actual bits. In addition, as to be verified in Section 8.4, little computational complexity cost is introduced in our RTE method, further highlighting the efficiency of our approach.

8.4 Computational complexity analysis

In this section, we primarily focus on the computational complexity of our approach. Since our approach adopts the RTE method to optimize perceptual distortion, the convergence speed of the RTE method is first discussed from both theoretical and numerical perspectives. In the numerical analysis, we also provide the practical computational time of our approach.

8.4.1 Theoretical analysis

For the theoretical analysis, we investigate the difference between $\widetilde{\lambda}$ and λ alongside the iterations of our RTE method. Here, we define $\Delta\lambda$ as the difference between $\widetilde{\lambda}$ and λ as

$$\Delta\lambda = \frac{\widetilde{\lambda} - \lambda}{\lambda}. \tag{8.26}$$

If $|\Delta\lambda| \to 0$, then it indicates that our RTE method is stably convergent. Therefore, we take into consideration $\Delta\lambda$ along with each iteration in our RTE method to analyze its convergence speed.

In practice, k_i (> 0) of (8.9) varies in a small range when encoding images using HEVC-MSP. Therefore, we assume that b_i ($0 < b_i = (1/(k_i + 1)) < 1$ in (8.9)) remains constant for simplicity. Based on this assumption, the convergence speed of our RTE method can be determined with Lemma 8.3.

Lemma 8.3. *Consider that $\widetilde{\lambda} > 0$, $\widehat{\lambda} > 0$, $\lambda > 0$, $R > 0$, and $\forall i$, $b_i = l \in (0,1)$. Recall that $\widetilde{\lambda}$ is the estimated λ of (8.12) before each iteration of our RTE method and that $\widehat{\lambda}$ is the solution of λ to (8.12) after each iteration of our RTE method. After each iteration in our RTE method, $\widetilde{\lambda}$ is replaced by $\widehat{\lambda}$. Then, there exists $|\Delta\lambda| \to 0$ along with iterations. Specifically, when $-0.9 < \Delta\lambda < 0$:*

$$|\Delta\lambda| < 0.04 \tag{8.27}$$

exists after two iterations.

Proof. Since $\widetilde{r}_i = (a_i/\widetilde{\lambda})^{b_i}$ and $r_i = (a_i/\lambda)^{b_i}$, we can obtain $\widetilde{r}_i = r_i \cdot (\lambda/\widetilde{\lambda})^{b_i}$. Then, by combining (8.14) and (8.15), we obtain the following equation:

$$\begin{aligned} R &= \sum_{i=1}^{M} \widetilde{r}_i \left(\frac{\widetilde{\lambda}}{\lambda}\right)^{b_i} \\ &= \sum_{i=1}^{M} \widetilde{r}_i + \sum_{i=1}^{M} \widetilde{r}_i \frac{\ln(\widetilde{\lambda}/\widehat{\lambda})}{1!} b_i + \sum_{i=1}^{M} \widetilde{r}_i \frac{(\ln(\widetilde{\lambda}/\widehat{\lambda}))^2}{2!} b_i^2 + \sum_{i=1}^{M} \widetilde{r}_i \frac{(\ln(\widetilde{\lambda}/\widehat{\lambda}))^3}{3!} b_i^3. \end{aligned} \tag{8.28}$$

Since $\forall i$, $b_i = l \in (0, 1)$ and $\sum_{i=1}^{M} r_i = R$, there exists $\sum_{i=1}^{M} \widetilde{r}_i = \sum_{i=1}^{M} r_i \cdot (\lambda/\widetilde{\lambda})^{b_i} = R \cdot (\lambda/\widetilde{\lambda})^{l}$. Next, we can rewrite (8.28) as

$$\begin{aligned} &\left(\frac{\lambda}{\widetilde{\lambda}}\right)^{l} \frac{R \cdot l^3}{3!} \cdot \left(\ln \frac{\widetilde{\lambda}}{\widehat{\lambda}}\right)^3 + \left(\frac{\lambda}{\widetilde{\lambda}}\right)^{l} \frac{R \cdot l^2}{2!} \cdot \left(\ln \frac{\widetilde{\lambda}}{\widehat{\lambda}}\right)^2 \\ &\quad + \left(\frac{\lambda}{\widetilde{\lambda}}\right)^{l} \frac{R \cdot l}{1!} \cdot \ln \frac{\widetilde{\lambda}}{\widehat{\lambda}} + R \cdot \left(\frac{\lambda}{\widetilde{\lambda}}\right)^{l} = R. \end{aligned} \tag{8.29}$$

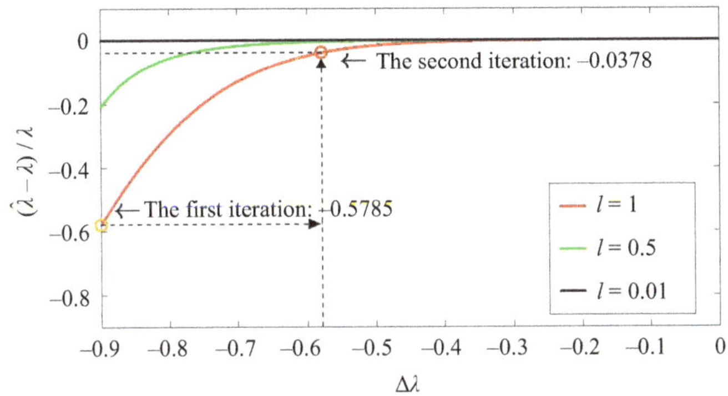

Figure 8.3 The relationship between $(\widehat{\lambda}-\lambda)/\lambda$ and $\Delta\lambda$ for each iteration. Note that the first (second) iteration shown in the figure represents the lowest convergence speed, i.e., when $l = 1$ and initial $|\Delta\lambda| = 0.9$ [11]

By solving this cubic equation, we can obtain:

$$\widehat{\lambda} = \widetilde{\lambda} \cdot e^{(1+2(\sqrt[3]{Z_1}+\sqrt[3]{Z_2}))/l}, \tag{8.30}$$

where

$$Z_1, Z_2 = -\frac{1}{8} + \frac{1}{8} \cdot \left(-3 \left(\frac{\widetilde{\lambda}}{\lambda} \right)^{l} + 2 \pm \sqrt{9 \left(\frac{\widetilde{\lambda}}{\lambda} \right)^{2l} - 6 \left(\frac{\widetilde{\lambda}}{\lambda} \right)^{l} + 2} \right). \tag{8.31}$$

Given (8.30), the relationship between $(\widehat{\lambda}-\lambda)/\lambda$ (which is $\Delta\lambda$ for the next iteration) and $\Delta\lambda$ (which is $\Delta\lambda$ for the current iteration) is illustrated in Figure 8.3. From this figure, we can determine that $|\Delta\lambda| \to 0$ in a quite fast speed. On the other hand, the convergence speed of $|\Delta\lambda| \to 0$ depends on l. When $l = 1$, the lowest convergence speed of $|\Delta\lambda|$ holds. In this case, $\Delta\lambda$ decreases at least to 0.58 for one iteration (when the largest initial $|\Delta\lambda| = 0.9$) and to 0.038 for two iterations. For other cases (e.g., $l = 0.5$ and $l = 0.01$), $|\Delta\lambda|$ decreases at a considerably faster speed. Therefore, $|\Delta\lambda| < 0.04$ exists after two iterations.

This completes the proof of Lemma 8.3. □

As proven in Lemma 8.2, $\widetilde{\lambda} < \lambda$ of our RTE method holds after the first iteration, which means that $\Delta\lambda \in (-1, 0)$. Moreover, we empirically found that $\Delta\lambda$ for all CTUs is restricted to $(-0.9, 0)$ after the first iteration in HEVC-MSP. Then, Lemma 8.3 indicates that $|\Delta\lambda|$ can be reduced to below 0.04 in at most three iterations, quickly approaching 0. This verifies the fast convergence speed of the RTE method in terms of $\Delta\lambda$. Next, we numerically evaluate the convergence speed of our RTE method in terms of E_a.

8.4.2 Numerical analysis

In this section, the numerical analysis of the convergence speed of our approach is presented. Specifically, we utilize the approximation error E_a to verify the convergence speed of the RTE method. Recall that $E_a = |\Sigma_{i=1}^{M}\widetilde{r}_i - R|/R$ (defined in Section 8.3.3). Figure 8.4 shows E_a versus RTE iterations when applying our approach to image compression in the HM 16.0 platform. As shown in this figure, with no more than three iterations, E_a reaches below 10^{-10}, thereby reflecting the fast convergence speed of our RTE method. This result is in accordance with the theoretical analysis of Section 8.4.1.

We further investigate the computational time for each iteration of the RTE method. As shown in Table 8.1, the computational time for each iteration is independent of the image content in our RTE method. Therefore, one image was randomly chosen from our test set, and the average time of one iteration of our RTE method was then recorded. The computer used for the test has an Intel Core i7-4770 CPU at

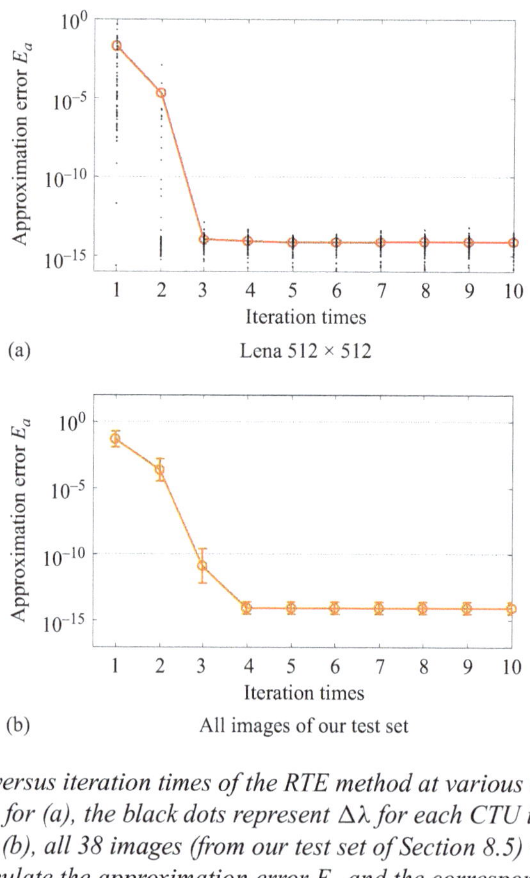

Figure 8.4 *E_a versus iteration times of the RTE method at various bit rates. Note that for (a), the black dots represent $\Delta\lambda$ for each CTU in* Lena *image. For (b), all 38 images (from our test set of Section 8.5) were used to calculate the approximation error E_a and the corresponding standard deviation along with the increasing iterations [11]*

3.4 GHz and 16 GB of RAM. From this test, we found out that one iteration of our RTE method only consumes approximately 0.0015 ms for each CTU. Since it takes at most three iterations to acquire the closed-form solution, the computational time for our RTE method is less than 0.005 ms.

Our approach consists of two parts: bit allocation and reallocation with the RTE method. For bit allocation, three iterations are sufficient for encoding one image, thus consuming at most 0.005 ms. For bit reallocation, the computational time depends on the number of CTUs of the image since each CTU requires at most three iterations to obtain the reallocated bits. For a $1,600 \times 1,280$ image, the computational time of our approach is approximately 2.5 ms because it includes 500 CTUs. This implies the negligible computational complexity burden of our approach.

8.5 Experimental results on single image coding

In this section, experimental results are presented to validate the performance of our approach. Specifically, the test and parameter settings for image compression are first presented in Section 8.5.1. Then, the *R–D* performance is evaluated in Section 8.5.2. In Section 8.5.3, the Bjontegaard delta bit rate (BD-rate) savings are provided to show how many bits can be saved in our approach for image compression. Then, the accuracy of bit-rate control is discussed in Section 8.5.4. Finally, the generalization of our approach is verified in Section 8.5.5.

8.5.1 Test and parameter settings

To evaluate the performance of our approach, we established a test set consisting of 38 images at different resolutions. Table 8.2 summarizes all 38 of these images in our test set. Among these images, ten images have faces, and the other images have no faces. Saliency for these images is first detected in our approach. Note that the face and non-face images are automatically recognized by using the face detector in [50]. Specifically, the face detector is first utilized to determine whether there is any face in the image. For the images with detected faces, we use [50] to predict saliency, and then we calculate SWPSNR as the optimization objective in our approach. Otherwise, [51] is utilized to predict saliency for SWPSNR for optimization.

Since the detected salient regions may deviate from the regions attracting human attention, in our experiments, we measure the EWPSNR of compressed images, which adopts the ground-truth eye fixations to weight MSE. The previous work of [17] has also verified that the EWPSNR is highly correlated with subjective quality. To obtain the ground-truth eye fixations[4] for measuring EWPSNR, 21 subjects (12 males and 9 females) with either corrected or uncorrected normal eyesight participated in our eye-tracking experiments by viewing all images of our test set. Note that only one among the 21 subjects was an expert who worked in the research field of saliency

[4]The ground-truth eye fixations, together with their corresponding images, can be obtained from our website at https://github.com/RenYun2016/TMM2016.

Table 8.2 Details of our test set [11]

Images	*Tourist*	*Golf*	*Travel*	*Doctor*	*Woman*	*Cafe*	*Bike*	*Picture01*	*Picture06*	*Picture10*	*Picture14*	*Picture30*
From	[50]				JPEG XR test set							
Face	√	√	√	√	√	×	×	×	×	×	×	×
Resolution	1,920×1,080				1,280×1,600							

Images	*Kodim01*	*Kodim02*	*Kodim03*	*Kodim05*	*Kodim06*	*Kodim07*	*Kodim08*	*Kodim11*	*Kodim12*	*Kodim13*	*Kodim14*	*Kodim15*	*Kodim16*	*Kodim20*	*Kodim21*	*Kodim22*	*Kodim23*	*Kodim24*
From	Kodak test set																	
Face	×	×	×	×	×	×	×	×	×	×	×	√	×	×	×	×	×	×
Resolution	768×512																	

Images	*Kodim04*	*Kodim09*	*Kodim10*	*Kodim17*	*Kodim18*	*Kodim19*	*Tiffany*	*Lena*
From	Kodak test set						Standard images	
Face	√	×	×	×	√	×	√	√
Resolution	512×768						512×512	

detection. The other 20 subjects did not have any background in saliency detection, and they were naive to the purpose of the eye-tracking experiment. Then, a Tobii TX60 eye tracker integrated with a monitor of a 23-in. LCD display was used to record the eye movement at a sample rate of 60 Hz. All subjects were seated on an adjustable chair at a distance of 60 cm from the monitor of the eye tracker. Before the experiment, the subjects were instructed to perform the 9-point calibration for the eye tracker. During the experiment, each image was presented in a random order and lasts for 4 s, followed by a 2-s black image for a drift correction. All subjects were asked to freely view each image. Overall, 9,756 fixations were collected for our 38 test images.

In our experiments, our approach was implemented in HM 16.0 with the MSP configuration profile. Then, the non-RC HEVC-MSP [9], also on the HM 16.0 platform, was utilized for comparison. The RC HEVC-MSP was also compared, the RC of which is mainly based on [52]. Note that both our approach and the RC HEVC-MSP have integrated RC to specify the bit rates, and the other parameters in the configuration profile were set by default, the same as those of the non-RC HEVC-MSP. To obtain the target bit rates, we encoded each image with the non-RC HEVC-MSP at six fixed *QP*s, the values of which are 22, 27, 32, 37, 42 and 47. Then, the target bit rates of our approach and the RC HEVC-MSP were set to be the actual bits obtained by the non-RC HEVC-MSP. As such, high ranges of visual quality for compressed images can be ensured.

8.5.2 Assessment on rate–distortion performance

Now, we assess the *R–D* performance of our approach and of the conventional non-RC and RC HEVC-MSP approaches. The *R–D* curves for face and non-face images are first plotted and analyzed. Subsequently, we present the results of image quality improvement of our approach at different *QP*s, which are measured by the EWPSNR and SWPSNR increase of our approach over the conventional approaches. Next, we evaluate how ROI detection accuracy affects the quality improvement in our approach. Finally, the subjective quality is evaluated by calculating the DMOS, as well as showing several compressed images.

R–D **curve:** The first ten figures of Figures 8.5 and 8.6 show the EWPSNR and PSNR versus bit rates for all ten face images of our test set. As shown in these figures, our approach is able to significantly improve the EWPSNR of compressed images, despite the slight decrease in PSNR. Consequently, subjective quality can be dramatically improved by our approach. Moreover, The last eight figures of Figures 8.5 and 8.6 show the curves of EWPSNR and PSNR versus bit rates for eight non-face images randomly selected from our test set. These figures show that our approach is also capable of achieving superior subjective quality for non-face images.

EWPSNR assessment: To quantify the *R–D* improvement of our approach, we tabulate in Table 8.3 the EWPSNR enhancement of our approach over conventional approaches. We have the following observations with regard to the EWPSNR enhancement. For face images, our approach achieves significant EWPSNR improvement, as the increase over the non-RC HEVC-MSP and RC HEVC-MSP is 2.31±1.23 dB and

Figure 8.5 EWPSNR and PSNR versus bit rates for our approach and the non-RC HEVC-MSP [11]

2.47±1.20 dB, respectively. In addition, the maximum increase of EWPSNR is 5.75 and 6.30 dB in our approach over the non-RC and RC HEVC-MSP approaches, respectively, whereas the minimum increase is 0.39 and 0.71 dB for these two approaches, respectively. For non-face images, the EWPSNR improvement of our approach reaches 1.49 dB on average compared with the RC HEVC-MSP approach, with a standard deviation of 0.70 dB. Compared to the non-RC HEVC-MSP approach, our approach enhances the EWPSNR by 1.21 dB on average, and the standard deviation of this enhancement is 0.61 dB. In a word, our approach dramatically improves the EWPSNR over the conventional approaches for both face and non-face images.

SWPSNR assessment: Since the optimization objective of our approach is to maximize SWPSNR, we further report in Table 8.3 the SWPSNR improvement of our approach over the conventional approaches. As shown in Table 8.3, our approach

Figure 8.6 EWPSNR and PSNR versus bit rates for our approach and the RC HEVC-MSP [11]

also achieves significant improvements in SWPSNR at different *QP*s. Specifically, compared with RC HEVC-MSP, our approach achieves an SWPSNR improvement over all images, with up to a 4.14 dB SWPSNR enhancement for face images and up to a 2.14 dB enhancement for non-face images. On average, for non-face images, our approach increases the SWPSNR by 0.72 and 1.00 dB over non-RC and RC HEVC-MSP, respectively. For face images, a more average SWPSNR gain is obtained by our approach, which has 1.56 and 1.67 dB increase over non-RC and RC HEVC-MSP.

Influence of ROI detection accuracy: Now, we investigate how the ROI detection accuracy influences the results of quality improvement in our approach. To this end, we further implement our approach using EWPSNR (instead of SWPSNR) as the optimization objective, which means that ROI detection is of 100% accuracy when

Table 8.3 EWPSNR and SWPSNR improvement of our approach over non-RC and RC HEVC-MSP approaches, for the 38 images [11]

		Face				**Non-face**			
		SWPSNR improvement		**EWPSNR improvement**		**SWPSNR improvement**		**EWPSNR improvement**	
		Avg. ± Std.	**Max./Min.**	**Avg. ± Std.**	**Max./Min.**	**Avg. ± Std.**	**Max./Min.**	**Avg. ± Std.**	**Max./Min.**
$QP=47$	Over non-RC	1.10 ± 0.47	2.05/0.44	1.55 ± 0.79	2.93/0.39	0.44 ± 0.19	0.95/0.14	0.71 ± 0.43	1.91/0.04
	Over RC	1.19 ± 0.52	2.21/0.65	1.67 ± 0.86	2.87/0.71	0.90 ± 0.40	1.84/0.25	1.15 ± 0.55	2.51/0.24
$QP=42$	Over non-RC	1.21 ± 0.43	1.83/0.39	1.71 ± 0.79	2.84/0.47	0.58 ± 0.23	1.17/0.18	0.92 ± 0.42	2.13/0.15
	Over RC	1.43 ± 0.55	2.43/0.55	1.99 ± 0.80	2.98/1.07	0.97 ± 0.45	1.74/0.31	1.34 ± 0.62	2.74/0.23
$QP=37$	Over non-RC	1.29 ± 0.38	1.95/0.72	1.92 ± 0.93	3.64/0.67	0.71 ± 0.29	1.23/0.25	1.16 ± 0.46	2.21/0.31
	Over RC	1.42 ± 0.50	2.40/0.90	2.16 ± 0.92	3.56/0.80	1.00 ± 0.50	1.96/0.25	1.47 ± 0.65	2.83/0.51
$QP=32$	Over non-RC	1.51 ± 0.51	2.48/0.67	2.23 ± 1.08	4.20/1.04	0.81 ± 0.34	1.32/0.24	1.35 ± 0.54	2.40/0.27
	Over RC	1.57 ± 0.52	2.49/0.95	2.38 ± 1.10	4.18/0.91	0.99 ± 0.50	1.99/0.21	1.56 ± 0.68	2.90/0.36
$QP=27$	Over non-RC	1.90 ± 0.73	3.26/0.79	2.85 ± 1.37	5.41/1.66	0.86 ± 0.36	1.48/0.33	1.49 ± 0.61	2.66/0.10
	Over RC	2.01 ± 0.65	3.14/1.01	2.98 ± 1.25	5.16/1.73	0.97 ± 0.47	2.13/0.36	1.58 ± 0.73	2.77/0.23
$QP=22$	Over non-RC	2.38 ± 0.92	4.14/1.26	3.60 ± 1.21	5.75/2.17	0.92 ± 0.38	1.54/0.40	1.62 ± 0.69	3.07/0.12
	Over RC	2.42 ± 1.05	4.14/1.17	3.65 ± 1.21	6.30/2.07	1.15 ± 0.51	2.14/0.39	1.85 ± 0.82	3.60/0.08
Overall	Over non-RC	1.56 ± 0.73	4.14/0.39	2.31 ± 1.23	5.75/0.39	0.72 ± 0.34	1.54/0.14	1.21 ± 0.61	3.07/0.04
	Over RC	1.67 ± 0.76	4.14/0.55	2.47 ± 1.20	6.30/0.71	1.00 ± 0.47	2.14/0.21	1.49 ± 0.70	3.60/0.08

Table 8.4 EWPSNR difference (dB) of our approach after replacing SWPSNR with EWPSNR as the optimization objective [11]

QP	47	42	37	32	27	22	Overall
Face	0.72	0.77	0.67	0.66	0.57	0.45	0.64
Non-face	0.70	0.77	0.84	0.92	0.98	1.01	0.87

compressing images using our approach. Specifically, Table 8.4 shows the EWPSNR difference averaged over all 38 test images when replacing SWPSNR with EWPSNR as the optimization objective in our approach. This reflects the influence of ROI detection accuracy on the quality improvement of our approach. We can see from Table 8.4 that the EWPSNR of our approach can be enhanced by 0.64 and 0.87 dB on average for face and non-face images after replacing SWPSNR by EWPSNR as the optimization objective. Thus, visual quality can be further improved in our approach when ROI detection is more accurate.

Subjective quality evaluation: Next, we compare our approach with the non-RC HEVC-MSP using DMOS. Note that the DMOS of the RC HEVC-MSP is not evaluated in our test because it produces even worse visual quality than the non-RC HEVC-MSP. The DMOS test was conducted by the means of single stimulus continuous quality score, which is processed by Rec. ITU-R BT.500 to rate the subjective quality. The total number of subjects involved in the test is 12, consisting of 6 males and 6 females. Here, a Sony BRAVIA XDV-W600, with a 55-in. LCD, was utilized for displaying the images. The viewing distance was set to be four times the image height for rational evaluation. During the experiment, each image was displayed for 4 s, and the order in which the images were displayed was random. Then, the subjects were asked to rate after each image was displayed, i.e., excellent (100–81), good (80–61), fair (60–41), poor (40–21) and bad (21–0). Finally, DMOS was computed to qualify the difference in subjective quality between the compressed and uncompressed images.

The DMOS results for the face images are tabulated in Table 8.5. Smaller values of DMOS indicate better subjective quality. As shown in Table 8.5, our approach has considerably better subjective quality than the non-RC HEVC-MSP at all bit rates. Note that for all images, the DMOS values of our approach at QP $=$ 47 are almost equal to those of the non-RC HEVC-MSP at QP $=$ 42, which approximately doubles the bit rates of QP $=$ 47. This indicates that a bit rate reduction of nearly half can be achieved in our approach. This result is also in accordance with the ~40% BD-rate saving of our approach (to be discussed in Section 8.5.3). We further show in Figure 8.7 *Lena* and *Kodim*18 compressed by our and the other two approaches. Obviously, our approach, which incorporates the saliency-detection method of [50], is able to significantly meliorate the visual quality over face regions (that humans mainly focus on). Consequently, our approach yields significantly better subjective quality than the non-RC and RC HEVC-MSP for face images.

In addition, the DMOS results of those eight non-face images are listed in Table 8.6. Again, our approach is considerably superior to the non-RC HEVC-MSP

Table 8.5 DMOS results for face images between our approach and the non-RC HEVC-MSP [11]

		Tourist	**Golf**	**Travel**	**Doctor**	**Woman**	**Kodim15**	**Kodim04**	**Kodim18**	**Tiffany**	**Lena**
$QP=47$	Bits (bpp)	0.04	0.02	0.04	0.02	0.04	0.03	0.03	0.05	0.03	0.05
	Our	**57.2**	**58.0**	**56.9**	**56.5**	**61.4**	**64.5**	**68.9**	**55.0**	**59.2**	**57.5**
	Non-RC	74.3	69.6	69.1	63.9	78.4	70.1	73.9	66.3	67.6	63.9
$QP=42$	Bits (bpp)	0.08	0.03	0.10	0.03	0.13	0.06	0.06	0.16	0.06	0.09
	Our	**45.0**	**50.0**	**42.7**	**47.8**	**43.9**	**50.7**	**53.6**	**43.1**	**43.1**	**47.9**
	Non-RC	58.5	56.3	53.7	52.1	61.3	61.2	61.9	56.9	54.1	55.5
$QP=32$	Bits (bpp)	0.27	0.08	0.36	0.10	0.56	0.29	0.31	0.76	0.26	0.28
	Our	**28.1**	**35.2**	**26.1**	**34.1**	**28.9**	**30.0**	**30.0**	**20.8**	**27.1**	**36.9**
	Non-RC	36.4	42.0	34.0	42.3	36.0	38.7	38.8	28.5	30.2	44.0

Note: The bold values mean the best subjective quality per test QP and test image.

Figure 8.7 Subjective quality of Lena and Kodim18 images at both 0.05 bpp (QP = 47) for three approaches [11]: (a) human fixations, (b) non-RC HEVC-MSP, (c) RC HEVC-MSP and (d) our

approach at all bit rates. Moreover, Figure 8.8 shows two images *Kodim*06 and *Kodim*07 compressed by our approach and by the other two approaches. From this figure, we can see that our approach improves the subjective quality of compressed images, as the fixated regions are with higher quality.

8.5.3 Assessment of BD-rate savings

It is interesting to investigate how many bits can be saved when applying our approach to image compression. In our experiments, BD-rates were calculated for this investigation. To calculate the BD-rates, the six different bit rates, each of which corresponds to one fixed QP (among $QP = 22, 27, 32, 37, 42$, and 47), were all utilized. Since the above section has shown that the EWPSNR is more effective than the PSNR for evaluating subjective quality, the EWPSNRs of each image at six bit rates were measured as the distortion metric. Given the bit rates and their corresponding EWPSNRs, the BD rate of each image was achieved. Then, the BD-rate savings of our approach can be obtained, with the non-RC or RC HEVC-MSP as an anchor.

Table 8.7 reports the BD-rate savings of our approach averaged over all 38 images of our test set. As shown in this table, a 24.3% BD-rate saving is achieved in our approach for all images over the non-RC HEVC-MSP. The BD-rate saving of our approach increases to 27.7%, when compared with the RC HEVC-MSP. In Table 8.7, the results of BD-rate savings for face and non-face images are also listed. Accordingly, we can see that our approach is able to save 39.1% and 42.5% BD-rates over non-RC and RC HEVC-MSP, respectively. Note that compared with non-face images, face images witness more gains in our approach. It is probably due to the fact

Table 8.6 DMOS results for non-face images between our approach and the non-RC HEVC-MSP [11]

		Bike	**Picture14**	**Kodim02**	**Kodim06**	**Kodim07**	**Kodim10**	**Kodim16**	**Kodim24**
$QP = 47$	Bits (bpp)	0.07	0.04	0.02	0.04	0.05	0.03	0.02	0.06
	Our	**53.3**	**59.6**	**65.5**	**62.0**	**56.8**	**63.0**	**71.1**	**67.1**
	Non-RC	57.2	63.1	69.9	72.1	67.0	68.1	79.2	70.2
$QP = 42$	Bits (bpp)	0.14	0.10	0.04	0.12	0.10	0.08	0.06	0.17
	Our	**36.8**	**50.3**	**50.0**	**52.7**	**50.1**	**54.5**	**56.2**	**55.4**
	Non-RC	38.9	54.2	53.4	57.6	56.3	58.7	62.1	59.3
$QP = 32$	Bits (bpp)	0.49	0.40	0.26	0.60	0.33	0.28	0.36	0.71
	Our	**30.3**	**31.7**	**33.5**	**34.8**	**36.3**	**34.7**	**35.6**	**32.6**
	Non-RC	30.8	32.6	35.2	35.6	38.0	37.9	40.8	33.8

Note: The bold values mean the best subjective quality per test QP and test image.

(a) (b) (c) (d)

Figure 8.8 Subjective quality of Kodim06 and Kodim07 image at 0.04 and 0.05 bpp (QP = 47) for three approaches [11]: (a) human fixations, (b) non-RC HEVC-MSP, (c) RC HEVC-MSP and (d) our

Table 8.7 BD-rate savings and encoding time ratio of our approach over non-RC and RC HEVC-MSP [11]

	Over non-RC HEVC-MSP	**Over RC HEVC-MSP**
Face images (%)	39.18	42.50
Non-face images (%)	18.98	22.43
All generic images (%)	24.30	27.72
Encoding time (%)	108.3	105.2

that human faces are more consistent than other objects in attracting human attention. Meanwhile, in our approach, the saliency of face images can be better predicted than that of non-face images. Consequently, the ROI-based compression of face images by our approach is more effective in satisfying human perception, resulting in larger improvements in EWPSNR, BD-rate savings and DMOS scores.

As a result of BD-rate saving, the computational time of our approach increases, which is also reported in Table 8.7. Specifically, our approach increases the encoding time by approximately 8% and 5% over non-RC and RC HEVC-MSP, respectively. The computational time of our approach mainly comes from three parts, i.e., saliency detection, pre-compression and RTE optimization. As discussed above (Sections 8.3.1 and 8.4.2), our pre-compression process slightly increases the computational cost by ~3%, while our RTE method consumes negligible computational time. Besides, saliency detection, which is the first step in our approach, consumes ~2% extra time.

8.5.4 Assessment of control accuracy

The control accuracy is another factor in evaluating the performance of RC-related image compression. Here, we compare the control accuracy of our approach and of the RC HEVC-MSP over all images in our test set. Since the bit reallocation process is developed in our approach to bridge the gap between the target and actual bits, the control accuracy of our approach with and without the bit reallocation process is also compared. In the following, the control accuracy is evaluated from two aspects: the CTU level and the image level.

For the evaluation of control accuracy at the CTU level, we compute the bit-rate error of each CTU, i.e., the absolute difference between target and actual bits assigned to one CTU. Then, Figure 8.9 demonstrates the heat maps of bit-rate errors at the CTU level averaged over all images with the same resolutions from the Kodak and JPEG XR sets. The heat maps of our approach and of the RC HEVC-MSP are both shown in Figure 8.9. It can easily be observed that our approach ensures a considerably smaller bit-rate error for almost all CTUs when compared with the RC HEVC-MSP. Note that the accurate rate control at the CTU level is meaningful because it ensures that the bit consumption follows the amount that it is allocated, satisfying the subjective *R–D* optimization formulation of (8.6). As a result, the bits in our approach can be accurately assigned to ROIs with optimal subjective quality. In contrast, the conventional RC HEVC-MSP normally accumulates redundant bits at the end of image bitstreams, resulting in poor performance in *R–D* optimization.

For the evaluation of control accuracy at the image level, the bit-rate error, defined as the absolute difference between the target and actual bits of the compressed image, is worked out. Figure 8.10 shows the bit-rate errors of all 38 images from our test set in terms of maximum, minimum, average and standard deviation values. As shown in this figure, our approach achieves smaller bit-rate error than the RC HEVC-MSP from the aspects of mean, standard deviation, maximum and minimum values. This verifies the effectiveness of our approach in RC and also makes our approach more practical because the accurate bit allocation of our approach well meets the bandwidth or storage requirements. Furthermore, Figure 8.10 shows that the bit-rate error significantly increases from 1.43% to 6.91% and also dramatically fluctuates once bit reallocation is disabled in our approach. This indicates the effectiveness of the bit-reallocation process in our approach. Note that because a simple reallocation process is also adopted in the RC HEVC-MSP, the bit-rate errors of RC HEVC-MSP are also much smaller than those of our approach without bit reallocation.

In summary, our approach has more accurate RC at both the CTU and image levels compared to the RC HEVC-MSP.

8.5.5 Generalization test

To verify the generalization of our approach, we further compare our approach and conventional approaches on 112 raw images from 3 test sets grouped into 4 categories, i.e., 22 face images, 41 non-face images, 4 graphics images and 45 aerial images. The resolutions of these images range from 256×256 to $7{,}216 \times 5{,}408$. The experimental results on these 112 images are reported in Table 8.8, including the mean, standard deviation, maximum and minimum values of SWPSNR[3] as well as bit-rate errors. Due to space limitation, this table only shows the results of compression at $QP = 32$ and the overall results of compression at $QP = 22$, 27, 32, 37, 42 and 47.

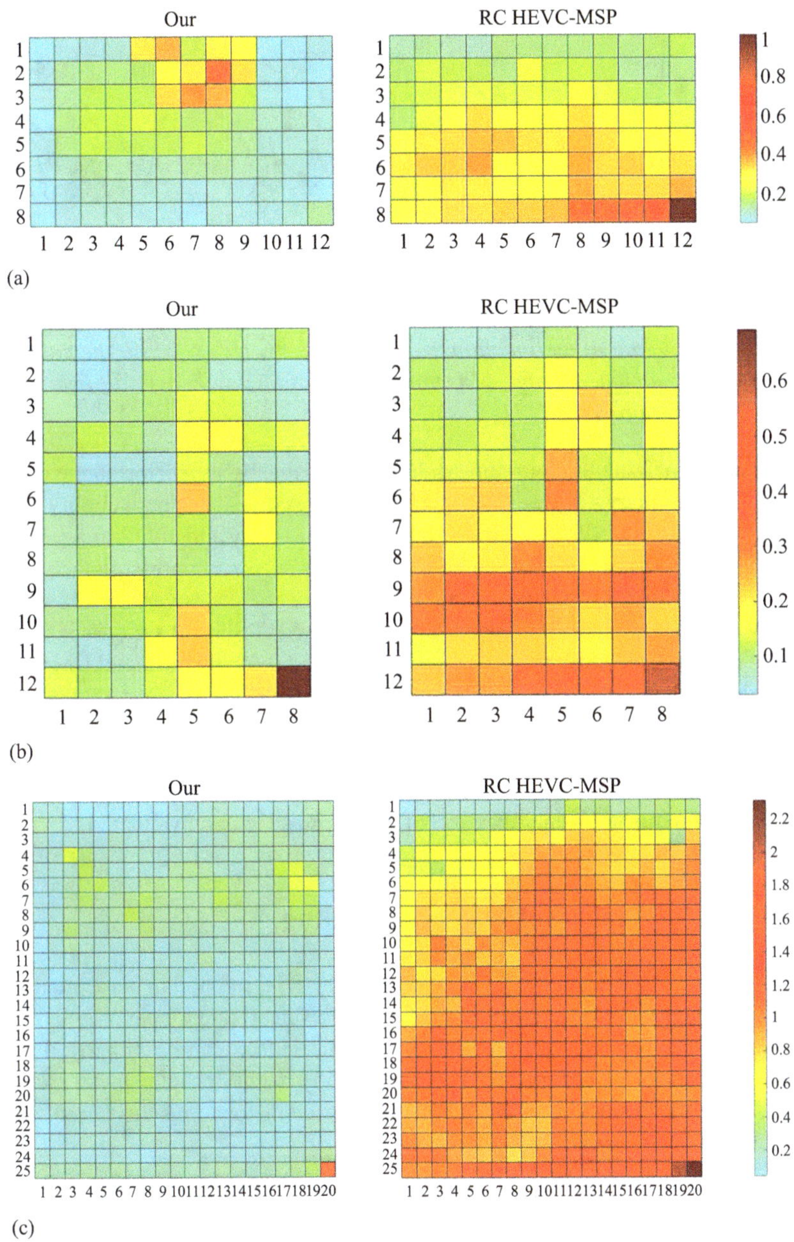

Figure 8.9 *Heat maps of bit-rate errors at CTU level for our approach and RC HEVC-MSP. Each block in this figure indicates the bit-rate error of one CTU. Note that the bit-rate errors are obtained via averaging all images compressed by our and the RC HEVC MSP at six different bit rates (corresponding to QP = 22, 27, 32, 37, 42, 47) [11]. (a) Kodak 768 × 512, (b) Kodak 512 × 768, (c) JPEG XR 1,280 × 1,600*

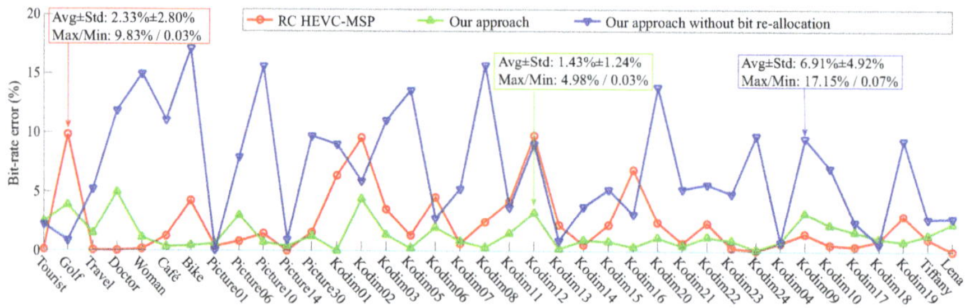

Figure 8.10 The bit-rate errors of each single image for our approach with and without bit reallocation, as well as the RC HEVC-MSP. The maximum, minimum, average and standard deviation values over all images are also provided [11]

As shown in Table 8.8, our approach still dramatically outperforms the conventional approaches across different categories of images in terms of both quality and RC error. Specifically, the SWPSNR improvement on the newly added 112 images is similar to that on the above 38 test images. In particular, when compressing face images at 6 QPs, our approach has 1.50 ± 0.84 dB SWPSNR increase over the conventional RC HEVC-MSP. Moreover, the average increase in SWPSNR at six QPs is 0.75 dB for non-face images, 0.83 dB for graphic images and 0.60 dB for aerial images. For control accuracy, the average bit-rate errors of our approach stabilize at 1.84%–3.74% across different categories, while the conventional RC approach in HEVC fluctuates from 4.08% to 12.40% on average with an even larger standard deviation. This result validates that our approach can achieve a stable and accurate RC, compared to RC HEVC-MSP. Finally, the generalization of our approach can be validated.

8.6 Experimental results on video coding

In this section, we present an implementation of the RTE method in perceptual video coding, i.e., to optimize the subjective quality at a given bit rate for panoramic videos. Several saliency-detection methods for panoramic videos can be employed in our framework [55,56]. We adopt the S-PSNR of [55] in this chapter as an example. Based on the definition of S-PSNR, the sphere-based distortion is the sum of square error between pixels sampled from sphere:

$$d_i = \sum_{n \in \mathbf{C}_i} (S(x_n, y_n) - S'(x_n, y_n))^2, \tag{8.32}$$

where $\mathbf{C}_i$ is the set of pixels belonging to the ith CTU. Therefore, at target bit rate R, the optimization on S-PSNR can be formulated by

$$\min \left(\sum_{i=1}^{M} d_i \right) \quad \text{s.t.} \sum_{i=1}^{M} r_i = R. \tag{8.33}$$

Table 8.8 Performance improvement of our approach over non-RC and RC HEVC-MSP approaches, for 112 test images belonging to different categories [11]

QP	**SWPSNR improvement (dB)**		**Face**	**Non-face**	**Graphics**	**Aerial**	**All**
32	Over non-RC	Avg. ± Std.	1.14 ± 0.03	0.54 ± 0.40	0.51 ± 0.28	0.35 ± 0.30	0.58 ± 0.50
		Max./Min.	2.82/0.11	1.60/0.02	0.81/0.14	1.04/0.00	2.82/0.00
	Over RC	Avg. ± Std.	1.40 ± 0.76	0.73 ± 0.49	0.65 ± 0.13	0.59 ± 0.61	0.80 ± 0.66
		Max./Min.	2.85/0.17	1.80/0.01	0.83/0.53	2.97/0.00	2.97/0.00
All	Over non-RC	Avg. ± Std.	1.25 ± 0.71	0.53 ± 0.45	0.50 ± 0.21	0.30 ± 0.33	0.58 ± 0.58
		Max./Min.	3.30/0.01	3.35/0.00	0.90/0.14	2.28/0.01	3.35/0.00
	Over RC	Avg. ± Std.	1.50 ± 0.84	0.75 ± 0.59	0.83 ± 0.68	0.60 ± 0.52	0.84 ± 0.71
		Max./Min.	4.59/0.01	3.13/0.01	2.88/0.06	2.97/0.01	4.59/0.01
	Bit-rate error (%)		**Face**	**Non-face**	**Graphics**	**Aerial**	**Overall**
32	RC HEVC-MSP	Avg. ± Std.	2.40 ± 2.76	3.53 ± 9.11	6.43 ± 9.80	6.93 ± 9.09	4.78 ± 8.39
		Max./Min.	10.9/0.06	53.65/0.01	20.99/0.47	35.07/0.02	53.65/0.01
	Our	Avg. ± Std.	2.72 ± 2.62	2.80 ± 4.15	1.89 ± 1.69	1.63 ± 3.34	2.28 ± 3.51
		Max./Min.	12.11/0.36	25.45/0.03	4.42/0.85	20.38/0.06	25.45/0.03
All	RC HEVC-MSP	Avg. ± Std.	4.08 ± 5.51	7.96 ± 15.64	11.96 ± 21.20	12.40 ± 16.29	9.12 ± 15.07
		Max./Min.	33.61/0.04	98.81/0.00	86.00/0.12	69.12/0.00	98.81/0.00
	Our	Avg. ± Std.	3.37 ± 3.63	3.74 ± 5.71	2.17 ± 1.85	1.84 ± 3.03	2.85 ± 4.37
		Max./Min.	25.79/0.10	39.32/0.01	7.00/0.31	21.39/0.00	39.32/0.00

In (8.33), r_m is the assigned bits at the mth CTU, and M is the total number of CTUs in the current frame. To solve the above formulation, a Lagrange multiplier λ is introduced, and (8.33) can be converted to an unconstrained optimization problem:

$$\min_{\{r_i\}_{i=1}^{M}} \quad J = \sum_{i=1}^{M} (d_i + \lambda r_i). \tag{8.34}$$

Here, we define J as the value of R–D cost. By setting derivative of (8.34) to zero, minimization on J can be achieved by

$$\begin{aligned} \frac{\partial J}{\partial r_i} &= \frac{\partial \left(\sum_{i=1}^{M} (d_i + \lambda r_i) \right)}{\partial r_i} \\ &= \frac{\partial d_i}{\partial r_i} + \lambda \\ &= 0. \end{aligned} \tag{8.35}$$

Next, we need to model the relationship between distortion d_i and bit rate r_i, for solving (8.35). Note that d_i and r_i are equivalent to S-MSE and bpp divided by the number of pixels in a CTU, respectively. Similar to [37], we use the hyperbolic model to investigate the relationship between sphere-based distortion S-MSE and bit-rate bpp, on the basis of four encoded panoramic video sequences. Figure 8.11 plots the fitting R–D curves using the Hyperbolic model, for these four sequences. In this figure, bpp is calculated by

$$\text{bpp} = \frac{R}{f \times W \times H}, \tag{8.36}$$

where f means frame rate, and W and H stand for width and height of video, respectively. Figure 8.11 shows that the Hyperbolic model is capable of fitting on the relationship between S-MSE [55] and bpp, and R-square for the fitting curves of four sequences are all more than 0.99. Therefore, the hyperbolic model is used in our RC scheme as follows:

$$d_i = c_i \cdot (r_i)^{-k_i}, \tag{8.37}$$

where c_m and k_m are the parameters of the hyperbolic model that can be updated for each CTU using the same way as [11].

The above equation can be rewritten by

$$-\frac{\partial d_i}{\partial r_i} = c_i \cdot k_i \cdot r_i^{(-k_i-1)}. \tag{8.38}$$

Given (8.35) and (8.38), the following equation holds:

$$r_i = \left(\frac{c_i k_i}{\lambda} \right)^{(1/(k_i+1))}. \tag{8.39}$$

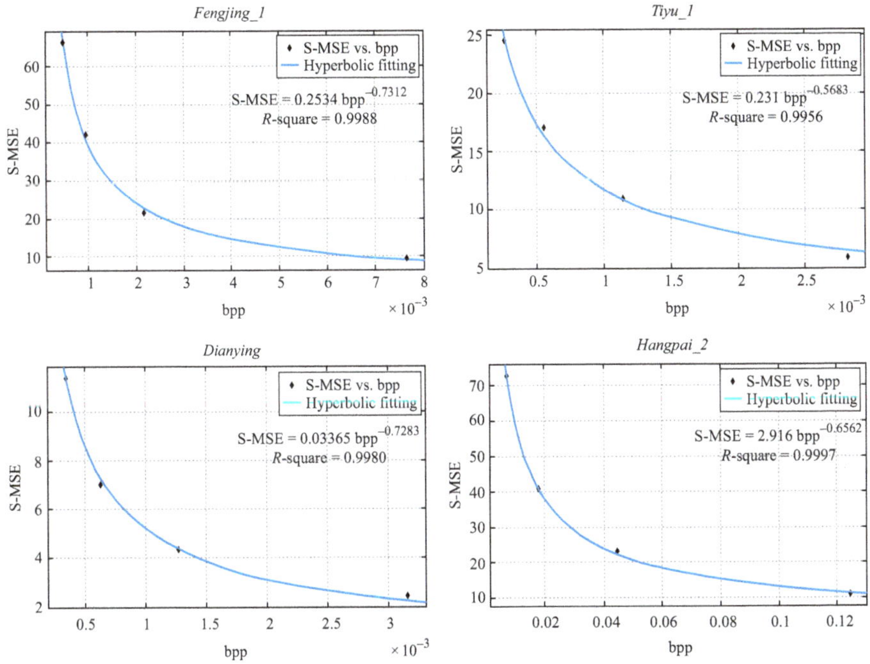

Figure 8.11 *R–D fitting curves using the hyperbolic model. Note that these four sequences are encoded by HM 15.0 with the default low delay P profile. The bit rates are set as the actual bit rates when compressing at four fixed QP (27, 32, 37, 42), to be described in Section 8.6.1.1 [55]*

Moreover, according to (8.33), we have the following constraint:

$$\sum_{i=1}^{M} r_i = R. \tag{8.40}$$

Upon (8.39) and (8.40), the bit allocation for each CTU can be formulated as follows:

$$\sum_{i=1}^{M} r_i = \sum_{i=1}^{M} \left(\frac{c_i \cdot k_i}{\lambda} \right)^{(1/(k_i+1))} = R. \tag{8.41}$$

Therefore, once (8.41) is solved, target bit r_i can be obtained for each CTU, with maximization on S-PSNR. In this chapter, we apply the RTE method [49] in solving (8.41) with the closed-form solution.

After obtaining the optimal bit-rate allocation, quantization parameter (QP) of each CTU can be estimated using the method of [37]. Figure 8.12 summarizes the overall procedure of our RC scheme for panoramic video coding. Note that our RC scheme is mainly applicable for the latest HEVC-based panoramic video coding, and

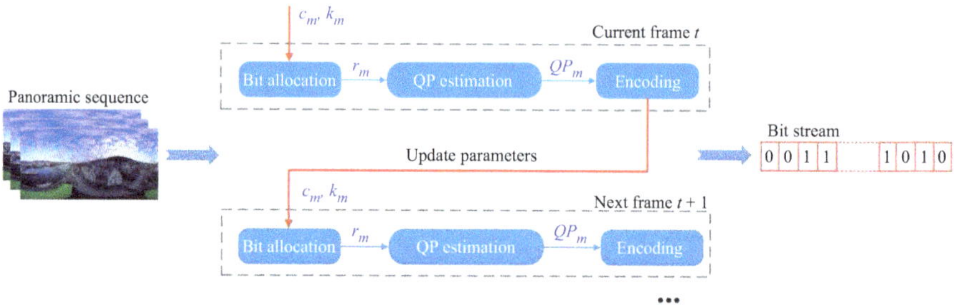

Figure 8.12 The framework of the proposed RC scheme for panoramic video coding [55]

it can be extended to other video-coding standards by reinvestigating the hyperbolic model of bit rate and distortion.

8.6.1 Experiment

In this section, experiments are conducted to validate the effectiveness of our RC scheme. Section 8.6.1.1 presents the settings for our experiments. Section 8.6.1.2 evaluates our approach from aspects of *R–D* performance, BD-rate and Bjontegaard delta S-PSNR (BD-PSNR). Section 8.6.1.3 discusses the RC accuracy of our scheme.

8.6.1.1 Settings

Due to space limitation, eight panoramic video sequences at 4K are chosen from the test set of IEEE 1857 working group in our experiments. They are shown in Figure 8.13. These sequences are all at 30 fps with duration of 10 s. Figure 8.13 shows that the contents of these sequences, which vary from indoor to outdoor scenes and contain people and landscapes. Then, these panoramic video sequences are compressed by the HEVC reference software HM-15.0. Here, we implement our RC scheme in HM-15.0, and then compare our scheme with the latest *R*–λ RC scheme [37] that is default RC setting of HM-15.0. For HM-15.0, the Low Delay P setting is applied with the configuration file *encoder_lowdelay_P_main.cfg*. The same as [37], we first compress panoramic video sequences using the conventional HM-15.0 at four fixed *QP*s, which are 27, 32, 37 and 42. Then, the obtained bit rates are used to set the target bit rates of each sequence for both our and conventional [37] schemes. It is worth pointing out that we only compare with the state-of-the-art RC scheme [37] of HEVC for 2D video coding, since there exists no RC scheme for panoramic video coding.

8.6.1.2 Evaluation on *R–D* performance

***R–D* curves.** We compare the *R–D* performance of our and the conventional RC [37] schemes using S-PSNR in Y channel. We plot in Figure 8.14 the *R–D* curves of all test panoramic video sequences, for both our and the conventional RC schemes.

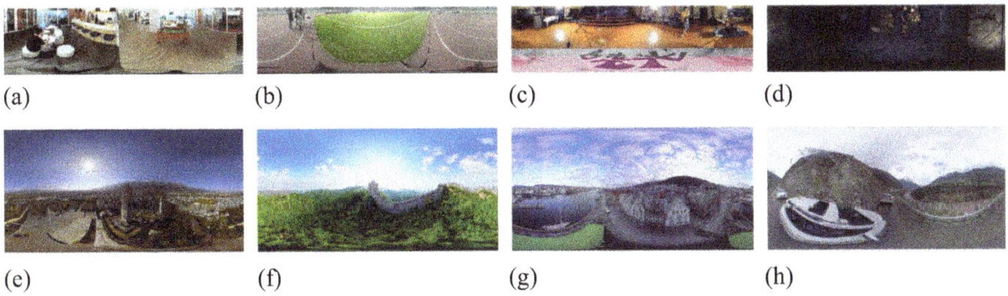

Figure 8.13 Selected frames from all test panoramic video sequences [55]: (a) Fengjing_1 (4,096 × 2,048), (b) Tiyu_1 (4,096 × 2,048), (c) Yanchanghui_2 (4,096 × 2,048), (d) Dianying (4,096 × 2,048), (e) Hangpai_1 (4,096 × 2,048), (f) Hangpai_2 (4,096 × 2,048), (g) AerialCity (3,840 × 1,920), (h) DrivingInCountry (3,840 × 1,920)

We can see from these *R–D* curves that our scheme achieves higher S-PSNR than [37] at the same bit rates, for all test sequences. Thus, our RC scheme is superior to [37] in *R–D* performance.

BD-PSNR and BD-rate. Next, we quantify *R–D* performance in terms of BD-PSNR and BD-rate. Similar to the above *R–D* curves, we use S-PSNR in Y channel for measuring BD-PSNR and BD-rate. Table 8.9 reports the BD-PSNR improvement of our scheme over [37]. As can be seen from this table, our scheme averagely improves 0.1613 dB in BD-PSNR over [37]. Such improvement is mainly because our scheme aims at optimizing S-PSNR, while [37] deals with optimization on PSNR. Table 8.9 also tabulates the BD-rate saving of our RC scheme with [37] being an anchor. We can see that our RC scheme is able to save 5.34% BD-rate in average, when compared with [37]. Therefore, our scheme has potential in relieving the bandwidth-hungry issue posed by panoramic videos.

Subjective quality. Furthermore, Figure 8.15 shows visual quality of one selected frame of sequence *Dianying*, encoded by HM-15.0 with our and conventional RC schemes at the same bit rate. We can observe that our scheme yields better visual quality than [37], with smaller blurring effect and less artifacts. For example, both regions of fingers and light generated by our scheme is much more clearer than those by [37]. Besides, the region of the leg encoded with our RC scheme has less blurring effect, compared to [37]. In summary, our scheme outperforms [37] in *R–D* performance, evaluated by *R–D* curves, BD-PSNR, BD-rate and subjective quality.

8.6.1.3 Evaluation on RC accuracy

Now, we evaluate the RC accuracy of our scheme. For such evaluation, Table 8.10 illustrates the error rate of actual bit rate with respect to target bit rate, for both our and the conventional RC [37] schemes. We can see from this table that the average RC error rate is less than 1‰, comparable to the error rate of [37]. Besides, the maximum error rate for our RC scheme is 3.02‰ for sequence *Tiyu_1*, and the error rate of [37]

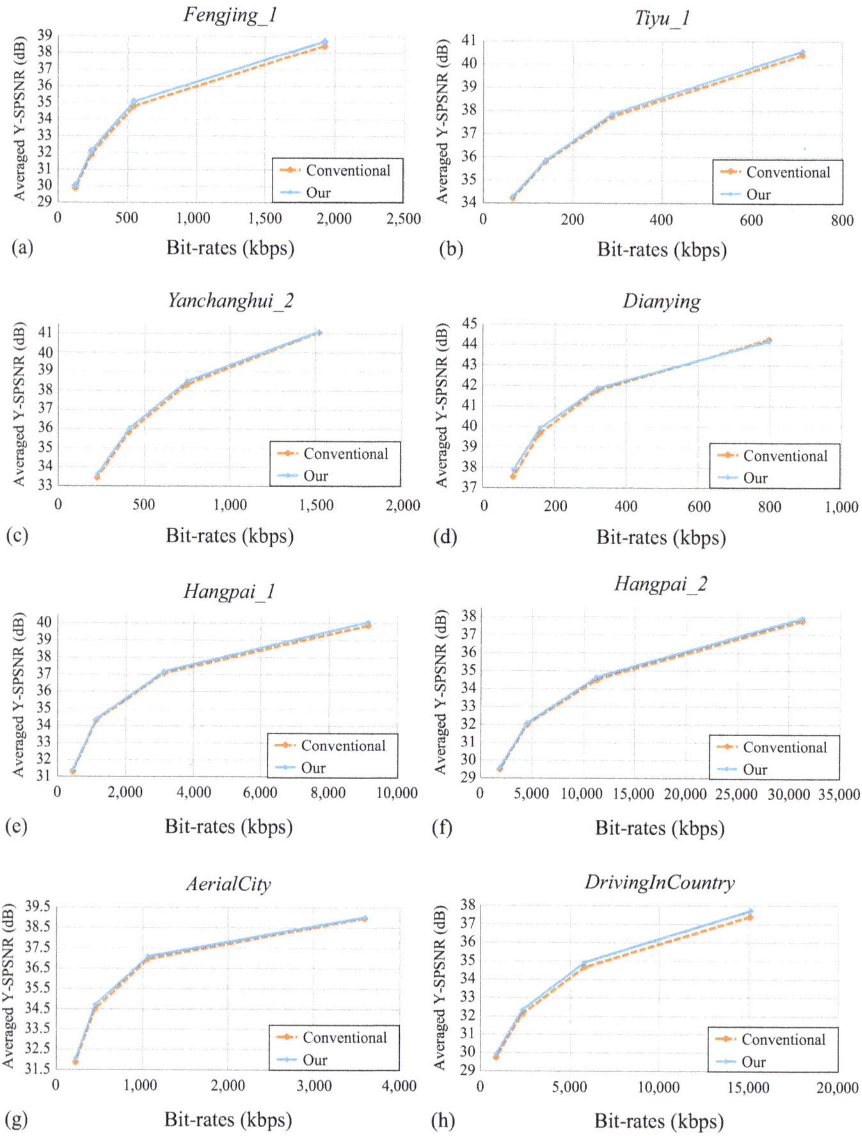

Figure 8.14 R–D curves of all test sequences compressed by HM-15.0 with our and conventional RC [37] schemes [55]: (a) Fengjing_1, *(b)* Tiyu_1, *(c)* Yanchanghui_2, *(d)* Dianying, *(e)* Hangpai_1, *(f)* Hangpai_2, *(g)* AerialCity, *(h)* DrivingInCountry

Table 8.9 BD-rate saving and BD-PSNR enhancement for each test panoramic video sequence [55]

Name	***Fengjing_1***	***Tiyu_1***	***Yanchanghui_2***	***Dianying***	***Hangpai_1***	***Hangpai_2***	***AerialCity***	***DrivingInCountry***	**Average**
BD-rate saving (%)	−7.63	−4.39	−3.96	−4.81	−3.87	−4.04	−5.41	−8.63	−5.34
BD-PSNR (dB)	0.2527	0.1155	0.1619	0.1441	0.1143	0.1197	0.1356	0.2464	0.1613

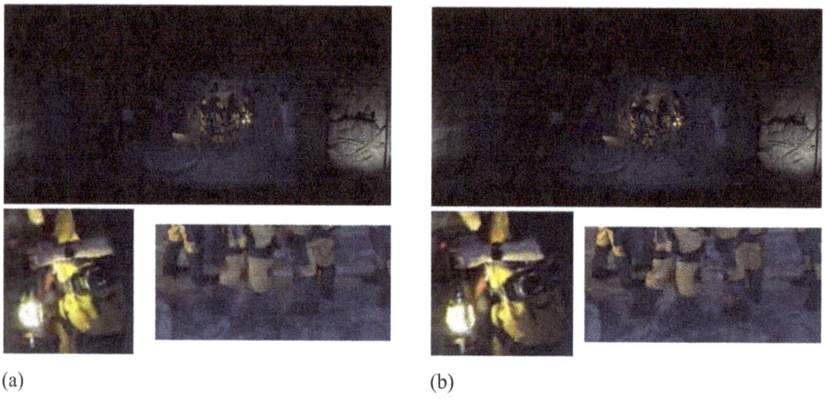

Figure 8.15 Visual quality of Dianying *compressed at 158 kbps by HM-15.0 with our and conventional RC [37] schemes. Note that this figure shows the 68th frame of compressed Dianying [55]. (a) Conventional RC scheme and (b) our scheme*

is up to 1.37‰. Although the RC accuracy of our scheme is smaller than [37], it is rather high and very close to 100% accuracy. Therefore, our scheme is effective and practical for controlling bit rate of HEVC-based panoramic video coding. More importantly, our RC is capable of improving RC performance for panoramic video coding.

8.7 Conclusion

In this chapter, we have proposed a novel HEVC-based compression approach that minimizes the perceptual distortion. Benefiting from the state-of-the-art saliency detection, we developed a formulation to minimize perceptual distortion, which maintains properly high quality at regions that attract attention. Then, the RTE method was proposed as a closed-form solution to our formulation with little extra time for minimizing perceptual distortion, followed by the bit allocation and reallocation process. Consequently, we validated our approach in experiments of compressing both images and videos.

There are two possible directions for future work: (1) our approach only takes into account the visual attention in improving the subjective quality of compressed images/videos. In fact, other factors of the HVS, e.g., JND, may also be integrated into our approach for perceptual compression. (2) Our approach in its present form only concentrates on minimizing perceptual distortion according to the predicted visual attention of uncompressed frames. However, the distribution of visual attention may be influenced by the distortion of compressed frames in reverse. A long-term

Table 8.10 S-PSNR improvement and RC accuracy of our RC scheme, compared with the conventional scheme [37,55]

Name	Fixed *QP*	RC error (‰) (conventional)	RC error (‰) (our)	S-PSNR (dB) improvement	Name error (‰)	Fixed *QP*	RC error (‰) (conventional)	RC error (‰) (our)	S-PSNR (dB) improvement	Name	Fixed *QP*	RC error (‰) (conventional)	RC error (‰) (our)	S-PSNR (dB) improvement
Fengjing_1	27	0.07	0.12	0.27	*Tiyu_1*	27	0.04	0.04	0.18	*Yanchanghui_2*	27	0.74	0.34	0.03
	32	0.06	0.05	0.28		32	0.46	0.08	0.13		32	0.34	0.52	0.17
	37	0.32	0.20	0.23		37	0.01	1.98	0.07		37	0.25	0.96	0.19
	42	0.31	0.23	0.15		42	1.37	3.02	0.10		42	0.54	1.89	0.20
Dianying	27	0.02	1.68	−0.07	*Hangpai_1*	27	0.05	0.19	0.19	*Hangpai_2*	27	0.02	0.18	0.15
	32	0.00	2.00	0.11		32	0.10	0.29	0.11		32	0.45	0.27	0.14
	37	0.26	0.49	0.25		37	0.06	1.46	0.09		37	0.42	0.50	0.09
	42	0.70	0.04	0.32		42	0.12	0.10	0.12		42	0.01	0.78	0.09
AerialCity	27	0.17	0.95	0.06	*Driving InCountry*	27	0.02	2.75	0.34	Average	27	0.14	0.78	0.14
	32	0.28	0.74	0.12		32	0.04	0.24	0.27		32	0.21	0.52	0.17
	37	0.06	1.18	0.20		37	0.02	0.52	0.21		37	0.17	0.91	0.17
	42	0.09	4.43	0.18		42	0.05	1.24	0.19		42	0.40	1.47	0.17
Overall average		0.23	0.92	0.16										

goal of perceptual compression should thus include the loop between visual attention and perceptual distortion over compressed images/videos.

References

[1] Chen S-C, Li T, Shibasaki R, Song X, and Akerkar R. Call for papers: Multimedia: the biggest big data. Special Issue of IEEE Transactions on Multimedia. 2015;17(9):1401–1403.
[2] Haskell BG, Puri A, and Netravali AN. Digital Video: An Introduction to MPEG-2. New York:Kluwer Academic Publishers; 1997.
[3] Vetro A, Sun H, and Wang Y. MPEG-4 rate control for multiple video objects. IEEE Transactions on Circuits and Systems for Video Technology. 1999;9(1): 186–199.
[4] Bankoski J, Bultje RS, Grange A, *et al.* Towards a next generation open-source video codec. In: IS&T/SPIE Electronic Imaging. International Society for Optics and Photonics; 2013. p. 866606-1-14.
[5] Cote G, Erol B, Gallant M, *et al.* H. 263+: Video coding at low bit rates. IEEE Transactions on Circuits and Systems for Video Technology. 1998;8(7): 849–866.
[6] Wiegand T, Sullivan GJ, Bjontegaard G, *et al.* Overview of the H.264/AVC video coding standard. IEEE Transactions on Circuits and Systems for Video Technology. 2003;13(7):560–576.
[7] Sullivan GJ, Ohm JR, Han WJ, *et al.* Overview of the high efficiency video coding (HEVC) standard. IEEE Transactions on Circuits and Systems for Video Technology. 2012;22(12):1649–1668.
[8] Lainema J, Bossen F, Han WJ, *et al.* Intra coding of the HEVC standard. IEEE Transactions on Circuits and Systems for Video Technology. 2012;22(12):1792–1801.
[9] Nguyen T, and Marpe D. Objective performance evaluation of the HEVC main still picture profile. IEEE Transactions on Circuits and Systems for Video Technology. 2014;25(5):790–797.
[10] Lee J, and Ebrahimi T. Perceptual video compression: a survey. IEEE Journal of Selected Topics in Signal Processing. 2012;6(6):684–697.
[11] Li S, Xu M, Ren Y, *et al.* Closed-form optimization on saliency-guided image compression for HEVC-MSP. IEEE Transactions on Multimedia. 2018;20(1):155–170.
[12] Koch K, McLean J, Segev R, *et al.* How much the eye tells the brain. Current Biology. 2006;16(14):1428–1434.
[13] Wandell BA. Foundations of vision. Sinauer Associates. 1995.
[14] Doulamis N, Doulamis A, Kalogeras D, *et al.* Low bit-rate coding of image sequences using adaptive regions of interest. IEEE Transactions on Circuits and Systems for Video Technology. 1998;8(8):928–934.
[15] Yang X, Lin W, Lu Z, *et al.* Rate control for videophone using local perceptual cues. IEEE Transactions on Circuits and Systems for Video Technology. 2005;15(4):496–507.

[16] Liu Y, Li ZG, and Soh YC. Region-of-interest based resource allocation for conversational video communication of H.264/AVC. IEEE Transactions on Circuits and Systems for Video Technology. 2008;18(1):134–139.

[17] Li Z, Qin S, and Itti L. Visual attention guided bit allocation in video compression. Image and Vision Computing. 2011;29(1):1–14.

[18] Xu M, Deng X, Li S, *et al.* Region-of-interest based conversational HEVC coding with hierarchical perception model of face. IEEE Journal of Selected Topics on Signal Processing. 2014;8(3):475–489.

[19] Xu L, Zhao D, Ji X, *et al.* Window-level rate control for smooth picture quality and smooth buffer occupancy. IEEE Transactions on Image Processing. 2011;20(3):723–734.

[20] Xu L, Li S, Ngan KN, *et al.* Consistent visual quality control in video coding. IEEE Transactions on Circuits and Systems for Video Technology. 2013;23(6):975–989.

[21] Rehman A, and Wang Z. SSIM-inspired perceptual video coding for HEVC. In: Multimedia and Expo (ICME), 2012 IEEE International Conference on. IEEE; 2012. p. 497–502.

[22] Geisler WS, and Perry JS. A real-time foveated multi-resolution system for low-bandwidth video communication. In: Proceedings of the SPIE: The International Society for Optical Engineering. vol. 3299; 1998. p. 294–305.

[23] Martini MG, and Hewage CT. Flexible macroblock ordering for context-aware ultrasound video transmission over mobile WiMAX. International Journal of Telemedicine and Applications. 2010;2010:6.

[24] Itti L. Automatic foveation for video compression using a neurobiological model of visual attention. IEEE Transactions on Image Processing. 2004;13(10):1304–1318.

[25] Chi MC, Chen MJ, Yeh CH, *et al.* Region-of-interest video coding based on rate and distortion variations for H. 263+. Signal Processing: Image Communication. 2008;23(2):127–142.

[26] Cerf M, Harel J, Einhäuser W, *et al.* Predicting human gaze using low-level saliency combined with face detection. Advances in Neural Information Processing Systems. 2008;20:241–248.

[27] Saxe DM, and Foulds RA. Robust region of interest coding for improved sign language telecommunication. IEEE Transactions on Information Technology in Biomedicine: A Publication of the IEEE Engineering in Medicine and Biology Society. 2002;6(4):310–316.

[28] Sun Y, Ahmad I, Li D, *et al.* Region-based rate control and bit allocation for wireless video transmission. IEEE Transactions on Multimedia. 2006;8(1): 1–10.

[29] Chi MC, Yeh CH, and Chen MJ. Robust region-of-interest determination based on user attention model through visual rhythm analysis. IEEE Transactions on Circuits and Systems for Video Technology. 2009;19(7): 1025–1038.

[30] Cavallaro A, Steiger O, and Ebrahimi T. Semantic video analysis for adaptive content delivery and automatic description. IEEE Transactions on Circuits and Systems for Video Technology. 2005;15(10):1200–1209.

[31] Boccignone G, Marcelli A, Napoletano P, *et al.* Bayesian integration of face and low-level cues for foveated video coding. IEEE Transactions on Circuits and Systems for Video Technology. 2008;18(12):1727–1740.

[32] Karlsson LS, and Sjostrom M. Improved ROI video coding using variable Gaussian pre-filters and variance in intensity. In: Image Processing, 2005. ICIP 2005. IEEE International Conference on. vol. 2. IEEE; 2005. p. II–313.

[33] Chai D, and Ngan KN. Face segmentation using skin-color map in videophone applications. IEEE Transactions on Circuits and Systems for Video Technology. 1999;9(4):551–564.

[34] Wang M, Zhang T, Liu C, *et al.* Region-of-interest based dynamical parameter allocation for H.264/AVC encoder. In: Picture Coding Symposium, 2009. PCS 2009. IEEE; 2009. p. 1–4.

[35] Chen Q, Zhai G, Yang X, *et al.* Application of scalable visual sensitivity profile in image and video coding. In: Circuits and Systems, 2008. ISCAS 2008. IEEE International Symposium on. IEEE; 2008. p. 268–271.

[36] Choi H, Yoo J, Nam J, *et al.* Pixel-wise unified rate-quantization model for multi-level rate control. Journal of Selected Topics in Signal Processing. 2013;7(6):1112–1123.

[37] Li B, Li H, Li L, *et al.* λ Domain based rate control for high efficiency video coding. IEEE Transactions on Image Processing. 2014;23(9):3841–3854.

[38] Sullivan GJ, and Wiegand T. Rate-distortion optimization for video compression. IEEE Signal Processing Magazine. 1998;15(6):74–90.

[39] Li B, Li H, Li L, *et al.* Rate control by R-lambda model for HEVC. Document: JCTVC-K0103, Joint Collaborative Team on Video Coding; 2012 Oct.

[40] Liu D, Sun X, Wu F, *et al.* Image compression with edge-based inpainting. IEEE Transactions on Circuits and Systems for Video Technology. 2007;17(10):1273–1287.

[41] Xiong H, Xu Y, Zheng YF, *et al.* Priority belief propagation-based inpainting prediction with tensor voting projected structure in video compression. IEEE Transactions on Circuits and Systems for Video Technology. 2011;21(8):1115–1129.

[42] Cheng L, and Vishwanathan S. Learning to compress images and videos. In: Proceedings of the 24th International Conference on Machine Learning. ACM; 2007. p. 161–168.

[43] He X, Ji M, and Bao H. A unified active and semi-supervised learning framework for image compression. In: Computer Vision and Pattern Recognition, IEEE Conference on. IEEE; 2009. p. 65–72.

[44] Levin A, Lischinski D, and Weiss Y. Colorization using optimization. In: ACM Transactions on Graphics (TOG). vol. 23. ACM; 2004. p. 689–694.

[45] Kavitha E, and Ahmed MA. A machine learning approach to image compression. International Journal of Technology in Computer Science and Engineering. 2014;1(2):70–81.

[46] Xu M, Li T, Wang Z, *et al.* Reducing complexity of HEVC: a deep learning approach. IEEE Transactions on Image Processing. 2018;27(10):5044–5059.

[47] Xu M, Li S, Lu J, *et al.* Compressibility constrained sparse representation with learnt dictionary for low bit-rate image compression. IEEE Transactions on Circuits and Systems for Video Technology. 2014;24(10):1743–1757.
[48] Sun Y, Xu M, Tao X, *et al.* Online dictionary learning based intra-frame video coding. Wireless Personal Communications. 2014;74(4):1281–1295.
[49] Li S, Xu M, Wang Z, *et al.* Optimal bit allocation for CTU level rate control in HEVC. IEEE Transactions on Circuits and Systems for Video Technology. 2017;27(11):2409–2424.
[50] Xu M, Ren Y, and Wang Z. Learning to predict saliency on face images. In: Proc. ICCV; 2015. p. 3907–3915.
[51] Zhang J, and Sclaroff S. Saliency detection: a Boolean map approach. In: Proc. ICCV; 2013. p. 153–160.
[52] Marta Karczewicz XW. Intra frame rate control based on SATD. Document: JCTVC-M0257, Joint Collaborative Team on Video Coding; 2013 Apr.
[53] Wang Z, and Li Q. Information content weighting for perceptual image quality assessment. IEEE Transactions on Image Processing. 2011;20(5):1185–1198.
[54] Fan S. A new extracting formula and a new distinguishing means on the one variable cubic equation. Nature Science Journal of Hainan Teachers College. 1989;2:91–98.
[55] Liu Y, Xu M, Li C, *et al.* A novel rate control scheme for panoramic video coding. In: Multimedia and Expo (ICME), 2017 IEEE International Conference on. IEEE; 2017. p. 691–696.
[56] Xu M, Song Y, Wang J, *et al.* Modeling Attention in Panoramic Video: A Deep Reinforcement Learning Approach. arXiv preprint arXiv:171010755. 2017.

Chapter 9
Machine-learning-based saliency detection and its video decoding application in wireless multimedia communications

Mai Xu[1], Lai Jiang[1], and Zhiguo Ding[2]

Saliency detection has been widely studied to predict human fixations, with various applications in wireless multimedia communications. For saliency detection, we argue that the state-of-the-art high-efficiency video-coding (HEVC) standard can be used to generate the useful features in compressed domain. Therefore, this chapter proposes to learn the video-saliency model, with regard to HEVC features. First, we establish an eye-tracking database for video-saliency detection. Through the statistical analysis on our eye-tracking database, we find out that human fixations tend to fall into the regions with large-valued HEVC features on splitting depth, bit allocation, and motion vector (MV). In addition, three observations are obtained from the further analysis on our eye-tracking database. Accordingly, several features in HEVC domain are proposed on the basis of splitting depth, bit allocation, and MV. Next, a support vector machine (SVM) is learned to integrate those HEVC features together, for video-saliency detection. Since almost all video data are stored in the compressed form, our method is able to avoid both the computational cost on decoding and the storage cost on raw data. More importantly, experimental results show that the proposed method is superior to other state-of-the-art saliency-detection methods, either in compressed or uncompressed domain.

9.1 Introduction

According to the study on the human visual system (HVS) [1], when a person looks at a scene, he/she may pay much visual attention on a small region (the fovea) around a point of eye fixation at high resolution. The other regions, namely, the peripheral regions, are captured with little attention at low resolutions. As such, humans are able to avoid the processing of tremendous visual data. Visual attention is therefore

[1] School of Electronic and Information Engineering, Beihang University, China
[2] School of Electrical and Electronic Engineering, The University of Manchester, UK

a key to perceive the world around humans, and it has been extensively studied in psychophysics, neurophysiology, and even computer vision societies [2]. Saliency detection is an effective way to predict the amount of human visual attention attracted by different regions in images/videos. Most recently, saliency detection has been widely applied in wireless multimedia communications and other computer vision tasks, such as, object detection [3,4], object recognition [5], image retargeting [6], image-quality assessment [7], and image/video compression [8,9].

In earlier time, some heuristic saliency-detection methods are developed according to the understanding of the HVS. Specifically, in light of the HVS, Itti and Koch [10] found out that the low-level features of intensity, color, and orientation are efficient in detecting saliency of still images. In their method, center-surround responses in those feature channels are established to yield the conspicuity maps. Then, the final saliency map can be obtained by linearly integrating conspicuity maps of all three features. For detecting saliency in videos, Itti *et al.* [11] proposed to add two dynamic features (i.e., motion and flicker contrast) into Itti's image saliency model [10]. Later, other advanced heuristic methods [12–18] have been proposed for modeling video saliency.

Recently, data-driven methods [19–24] have emerged to learn the visual attention models from the ground-truth eye-tracking data. Specifically, Judd *et al.* [19] proposed to learn a linear classifier of SVM from training data for image saliency detection, based on several low, middle, and high-level features. For video-saliency detection, most recently, Rudoy *et al.* [23] have proposed a novel method to predict saliency by learning the conditional saliency map from human fixations over a few consecutive video frames. This way, the inter-frame correlation of visual attention is taken into account, such that the accuracy of video-saliency detection can be significantly improved. Rather than free-view saliency detection, a probabilistic multitask learning method was developed in [21] for the task-driven video-saliency detection, in which the "stimulus-saliency" functions were learned from the eye-tracking data as the top-down attention models.

HEVC [25] was formally approved as the state-of-the-art video-coding standard in April 2013. It achieves double coding efficiency improvement over the preceding H.264/AVC standard. Interestingly, we found out that the state-of-the-art HEVC encoder can be explored as a feature extractor to efficiently predict video saliency. As shown in Figure 9.1, the HEVC domain features on splitting depth, bit allocation, and MV for each coding tree unit (CTU), are highly correlated with the human fixations. The statistical analysis of Section 9.3.2 verifies such high correlation. Therefore, we develop several features in our method for video-saliency detection, which are based on splitting depths, bit allocation, and MVs in HEVC domain. It is worth pointing out that most videos exist in the form of encoded bitstreams and the features related to entropy and motion have been well exploited by video coding at the encoder side. Since [2] has argued that entropy and motion are very effective in video-saliency detection, our method utilizes these well-exploited HEVC features (splitting depth, bit allocation, and MV) at the decoder side to achieve high accurate detection on video saliency.

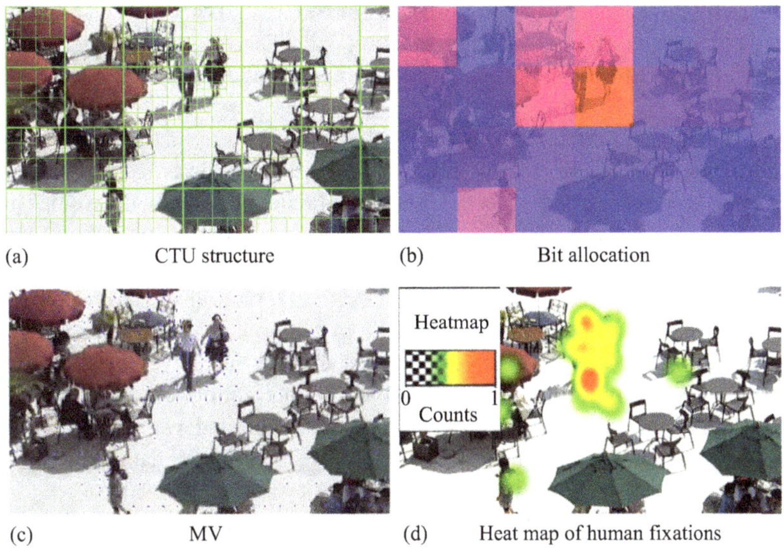

Figure 9.1 An example of HEVC domain features and heat map of human fixations for one video frame. Parts (a), (b), and (c) are extracted from the HEVC bitstream of video BQSquare (resolution: 416 × 240) at 130 kbps. Note that in (c) only the MVs that are larger than 1 pixel are shown. Part (d) is the heat map convolved with a 2D Gaussian filter over fixations of 32 subjects

Generally speaking, the main motivation of using HEVC features in our saliency detection method is 2-fold: (1) our method takes advantage from sophisticated encoding of HEVC, to effectively extract features for video-saliency detection. Our experimental results in this chapter also show that the HEVC features are quite effective in video-saliency detection. (2) Our method can efficiently detect video saliency from HEVC bitstreams without completely decoding the videos, thus saving both the computational time and storage. Consequently, our method is generally more efficient than the aforementioned video-saliency detection methods at pixel domain (or called uncompressed domain), which have to decode the bitstreams into raw data. Such efficiency is also validated by our experiments.

There are only a few methods [26–28] proposed for detecting video saliency in compressed domain of previous video-coding standards. Among these methods, the block-wise discrete cosine transform (DCT) coefficients and MVs are extracted in MPEG-2 [26] and MPEG-4 [27]. Bit allocation of H.264/AVC is exploited for saliency prediction in [28]. However, all above methods do not take full advantage of the sophisticated features of the modern HEVC encoder, such as CTU splitting [29] and R–λ bit allocation [30]. More importantly, all methods of [26–28] fail to find out the precise impact of each compressed domain feature on attracting visual attention.

In fact, the relationship between compressed domain features and visual attention can be learned from the ground-truth eye-tracking data. Thereby, this chapter proposes to learn the visual attention model of videos with regard to the well-explored HEVC features.

Similar in spirit, the latest work of [31] also makes use of HEVC features for saliency detection. Despite conceptually similar, our method is greatly different from [31] in two aspects. From the aspect of feature extraction, our method develops pixel-wise HEVC features, while [31] directly uses block-based HEVC features with deeper decoding (e.g., inverse DCT). Instead of going deeper, our method develops shallow decoded HEVC features with sophisticated design of temporal and spatial difference on these features, more unrestrictive than [31]. In addition, camera motion is detected and then removed in our HEVC features, such that our features are more effective in predicting attention. From the aspect of feature integration, compared with [31], our method is data driven, in which a learning algorithm is developed to bridge the gap between HEVC features and video saliency. Meanwhile, our data-driven method benefits from thorough analysis of our established eye-tracking database.

Specifically, the main contributions of this chapter are listed in the following:

- We establish an eye-tracking database on viewing 33 raw videos of the latest data sets, with the thorough analysis and observations on our database.
- We propose several saliency-detection features in HEVC domain, according to the analysis and observations on our established eye-tracking database.
- We develop a data-driven method for video-saliency detection, with respect to the proposed HEVC features.

The rest of this chapter is organized as follows: in Section 9.2, we briefly review the related work on video-saliency detection. In Section 9.3, we present our eye-tracking database as well as the analysis and observations on our database. In light of such analysis and observations, Section 9.4 proposes several HEVC features for video-saliency detection. Section 9.5 outlines our learning-based method, which is based on the proposed HEVC features. Section 9.6 shows the experimental results to validate our method. Finally, Section 9.7 concludes this chapter.

9.2 Related work on video-saliency detection

9.2.1 Heuristic video-saliency detection

For modeling saliency of a video, a great number of methods [11–18] have been proposed. Itti *et al.* [11] started the initial work of video-saliency detection, by adding two dynamic features of motion and flicker contrast into Itti's image saliency model [10]. Later, a novel term called *surprise* was defined in [14] to measure how the visual change attracts human observers. With the new term *surprise*, [14] developed a Bayesian framework to calculate the Kullback–Leibler (KL) divergence between spatiotemporal posterior and prior beliefs, for predicting video saliency. Some other Bayesian-framework-related methods, e.g., [15], were also proposed for video-saliency detection. Most recently, some advanced video-saliency detection methods [16–18] have been proposed. To be more specific, Guo *et al.* [16] applied phase spectrum of quaternion Fourier transform (PQFT) on four feature channels (two color channels, one intensity channel, and one motion channel) to detect video saliency. Lin *et al.* [18] utilized earth mover's distance to measure the center-surround difference in spatiotemporal receptive filed, for producing the dynamic saliency maps of videos. Inspired by sparse representation, Ren *et al.* [17] proposed to explore the movement of a target patch for temporal saliency detection of videos. In their method, the movement of the target patch can be estimated by finding the minimal reconstruction error of sparse representation regarding the patches of neighboring frames. In addition to temporal saliency detection, the center-surround contrast needs be modeled for spatial saliency detection. This is achieved through sparse representation with respect to neighboring patches.

In fact, top-down visual cues play an important role in determining the saliency of a scene. Thereby, the top-down visual attention models have been studied in [32,33] for predicting the saliency of dynamic scenes in a video. In [32], Pang *et al.* proposed to integrate the top-down information of eye-movement patterns (i.e., passive and active states [13]) for video saliency detection. In [33], Wu and Xu found out that the high-level features, such as face, person, car, speaker, and flash, may attract extensive human attention. Thus, these high-level features are integrated with the bottom-up model [16] for saliency detection of news videos.

However, the understanding of the HVS is still in its infancy, and saliency detection thus has a long way to go yet. In fact, we may rethink saliency detection by taking advantage of the existing video-coding techniques. Specifically, the video-coding standards have evolved for almost three decades, with HEVC being the latest one. The evolution of video coding adopts several elegant and effective techniques to produce several sophisticated features, for continuously improving coding efficiency. For example, the state-of-the-art HEVC standard introduced fractional sample interpolation to represent MVs with quarter-sample precision, thus being able to precisely model object motions. Moreover, HEVC proposes to partition CTUs into smaller blocks using the tree structure and quadtree-like signaling [29], which can well reflect the texture complexity of video frames. On the other hand, the HEVC features, which are generated by the sophisticated process of the latest HEVC techniques, may be explored for efficient video-saliency detection.

9.2.2 Data-driven video-saliency detection

During the past decade, data-driven methods have emerged as a possible way to learn video-saliency model from ground-truth eye-tracking data, instead of the study on the HVS. The existing data-driven video-saliency detection can be further divided into task-driven [13,21,22,34,35] and free-view [20,23,24,36] methods.

For task-driven video-saliency detection, Peter and Itti [13] proposed to incorporate the computation on signatures of each video frame. Then, a regression classifier is learned from the subjects' fixations on playing video games, which associates the different classes of signatures (seen as gist) with the gaze patterns of task-driven attention. Combined with 12 multi-scale bottom-up features, [13] has high accuracy in task-driven saliency detection. Most recently, a dynamic Bayesian network method [35] has been proposed for learning top-down visual attention model of playing video games. Besides the task of playing video games, a data-driven method [34] on video-saliency detection was proposed with the dynamic consistency and alignment models, for the task of action recognition. In [34], the proposed models are learned from the task-driven human fixations on large-scale dynamic computer vision databases like Hollywood-2 [37] and UCF Sports [38]. In [21], Li *et al.* developed a probabilistic multitask learning method to include the task-related attention models for video-saliency detection. The "stimulus-saliency" functions are learned from the eye-tracking database, as the top-down attention models to some typical tasks of visual search. As a result, [21] is "good at" video-saliency detection in multiple tasks, more generic than other methods that focus on single visual task. However, all task-driven saliency-detection methods can only deal with the specific tasks.

For free-view video-saliency detection, Kienzle *et al.* [20] proposed a nonparametric bottom-up method to model video saliency, via learning the center-surround texture patches and temporal filters from the eye-tracking data. Recently, Lee *et al.* [24] have proposed to extract the spatiotemporal features, i.e., rarity, compactness, center prior, and motion, for the bottom-up video-saliency detection. In their bottom-up method, all extracted features are combined together by an SVM, which is learned from the training eye-tracking data. In addition to the bottom-up model, Hua *et al.* [36] proposed to learn the middle-level features, i.e., gists of a scene, as the top-down cue for both video and image-saliency detection. Most recently, Rudoy *et al.* [23] have proposed to detect the saliency of a video, by simulating the way that humans watch the video. Specifically, a visual attention model is learned to predict the saliency map of a video frame, given the fixation maps from the previous frames. As such, the inter-frame dynamics of gaze transitions can be taken into account during video-saliency detection.

As aforementioned, this chapter mainly concentrates on utilizing the HEVC features for video-saliency detection. However, there is a gap between HEVC features and human visual attention. From data-driven perspective, machine learning can be utilized in our method to investigate the relationship between HEVC features and visual attention, according to eye-tracking data. Thus, this chapter aims at learning an SVM classifier to predict saliency of videos using the features from HEVC domain.

9.3 Database and analysis

9.3.1 Database of eye tracking on raw videos

In this chapter, we conducted the eye-tracking experiment to obtain fixations on viewing videos of the latest test sets. Here, all 33 raw videos from the test sets [9,39], which have been commonly utilized for evaluating HEVC performance, were included in our eye-tracking experiment. We further conducted the extra experiment to obtain the eye-tracking data on watching all videos of our database compressed by HEVC at different quality. Through the data analysis, we found that visual attention is almost unchanged when videos are compressed at high or medium quality (more than 30 dB). This is consistent with the result of [40]. Compared with the conventional databases (e.g., SFU [41] and DIEM [42]), the utilization of these videos benefits from the state-of-the-art test sets in providing videos with diverse resolutions and content. For the resolution, the videos vary from 1080p (1,920 × 1,080) to 240p (416 × 240). For the content, the videos include sport events, surveillance, video conferencing, video games, videos with the subscript, etc.

In our eye-tracking experiment, all videos are with YUV 4:2:0 sampling. Here, the resolutions of the videos in Class A of [39] were down-sampled to be 1,280 × 800, as the screen resolution of the eye tracker can only reach to 1,920 × 1,080. Other videos were displayed in their original resolutions. In our experiment, the videos were displayed in a random manner at their default frame rates to reduce the influence of video-playing order on the eye-tracking results. Besides, a blank period of 5 seconds was inserted between two consecutive videos, so that the subjects can have a proper rest time to avoid eye fatigue.

There were a total of 32 subjects (18 male and 14 female, aging from 19 to 60) involved in our eye-tracking experiment. These subjects were selected from the campuses of Beihang University and Microsoft Research Asia. All subjects have either corrected or uncorrected normal eyesight. Note that only two subjects were experts, who are working in the research field of saliency detection. The other 30 subjects did not have any research background in video-saliency detection, and they were also native to the purpose of our eye-tracking experiment.

The eye fixations of all 32 subjects over each video frame were recorded by a Tobii TX300 eye tracker at a sample rate of 300 Hz. The eye tracker is integrated with a monitor of 23-inch LCD screen, and the resolution of the monitor was set to be 1,920 × 1,080. All subjects were seated on an adjustable chair at a distance of around 60 cm from the screen of the eye tracker, ensuring that their horizontal sight is in the center of the screen. Before the experiment, subjects were instructed to perform the 9-point calibration for the eye tracker. Then, all subjects were asked to free-view each video. After the experiment, 392,163 fixations over 13,020 frames of 33 videos were collected. Here, the eye fixations of all subjects and the corresponding MATLAB® code for our eye-tracking database are available online: https://github.com/remega/video_database.

9.3.2 Analysis on our eye-tracking database

Figure 9.1 has shown that the HEVC features, i.e., splitting depth, bit allocation, and MV, are effective in predicting human visual attention. It is therefore interesting to statistically analyze the correlation between these HEVC features and visual attention. From now on, we concentrate on the statistical analysis on our eye-tracking database to show the effectiveness of the HEVC features on the prediction of visual attention. This is a new finding, which reveals the correlation between HEVC features and visual attention.

For all videos of our database, the features on splitting depth, bit allocation, and MV were extracted from the corresponding HEVC bitstreams. Then, the maps of these features were generated for each video frame. Note that the configuration to generate the HEVC bitstreams can be found in Section 9.6. Afterwards, a 2D Gaussian filter was applied to all three feature maps of each video frame. For each feature map, after sorting pixels in the descending order of their feature values, the pixels were equally divided into ten groups according to the values of corresponding features. For example, the group of 0%–10% stands for the set of pixels, the features of which rank top 10%. Finally, the number of fixations belonging to each group was counted upon all 33 videos in our database.

We show in Figure 9.2 the percentages of eye fixations belonging to each group, in which the values of the corresponding HEVC features decrease alongside the groups. From this figure, we can find out that extensive attention is drawn by the regions with large-valued HEVC features, especially for the feature of bit allocation. For example, about 33% fixations fall into the regions of top 10% high-valued feature of bit allocation, whereas the percentage of those hitting the bottom 10% is much less than 2%. Hence, the HEVC features on splitting depth, bit allocation, and MV, are explored for video-saliency detection in our method (Section 9.4).

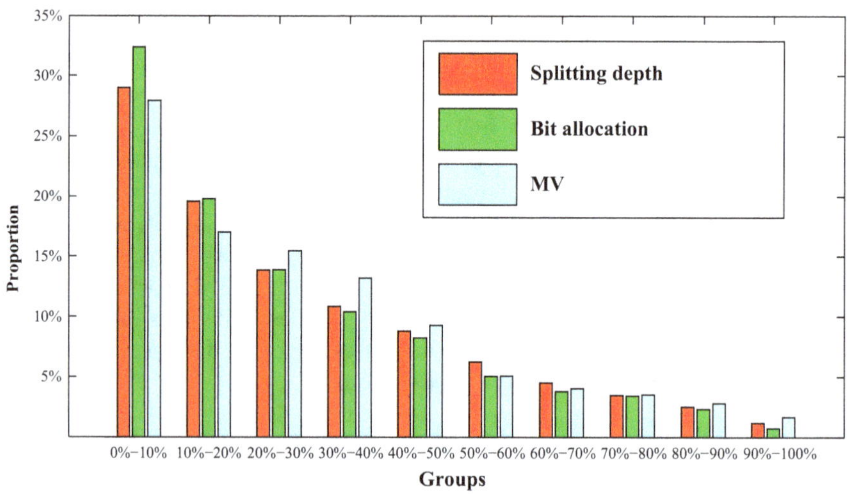

Figure 9.2 The statistical results for fixations belong to different groups of pixels, in which values of the corresponding HEVC features are sorted in the descending order. Here, all 392,163 fixations of 33 videos are used for the analysis. In this figure, the horizontal axis indicates the groups of pixels. For example, 0%–10% means that the first group of pixels, the features of which rank top 10%. The vertical axis shows the percentage of fixations that fall into each group

9.3.3 Observations from our eye-tracking database

Beyond the analysis of our eye-tracking database, we verify some other factors on attracting human attention, with the following three observations. These observations provide insightful guide for developing our saliency-detection method.

Observation 9.1: *Human fixations lag behind the moving or new objects in a video by some microseconds.*

In Figure 9.3, we show the frames of videos *BasketballDrive* and *Kimono* with the corresponding heat maps of human fixations. The first row of this figure reveals that the visual attention falls behind the moving object, as the fixations trail the moving basketball. In particular, the distance between the basketball and fixations becomes large, when the basketball moves at high speed. Besides, the second row of Figure 9.3 illustrates that the human fixations lag behind the new appearing objects by a few frames. It is because the human fixations still stay in the location of the salient region in previous frames, even when the scene has been changed. This completes the analysis of Observation 9.1.

Observation 9.2: *Human fixations tend to be attracted by the new objects appearing in a video.*

It is intuitive that visual attention is probably to be attracted by the objects newly emerging in a video. It is thus worth analyzing the influence of the object emergence on human visual attention. Figure 9.4 shows the heat maps of fixations on several frames selected from videos *vidyo1* and *ParkScene*. Note that a person appears in the door from the 553th frame of the video *vidyo1*, and that a person riding bicycle arises from the 64th frame of the video *ParkScene*. From Figure 9.4, one may observe that once a new object appears in the video, it probably attracts a huge amount of visual

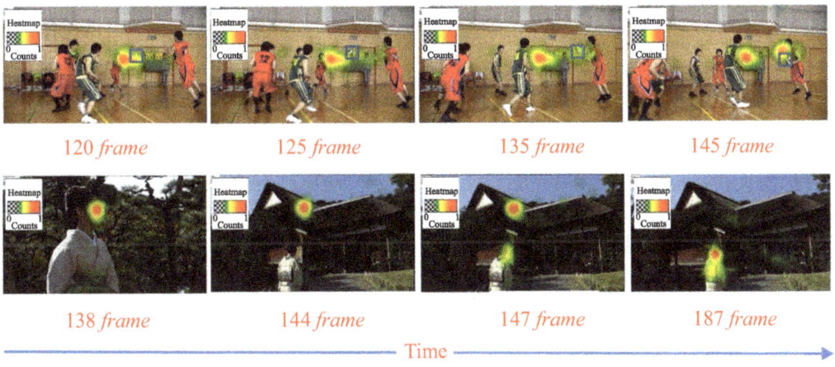

Figure 9.3 Illustration of Observation 9.1. This figure shows the heat maps of human fixations of all 32 subjects on several selected frames of videos BasketballDrive and Kimono. In BasketballDrive, the green box is drawn to locate the moving basketball

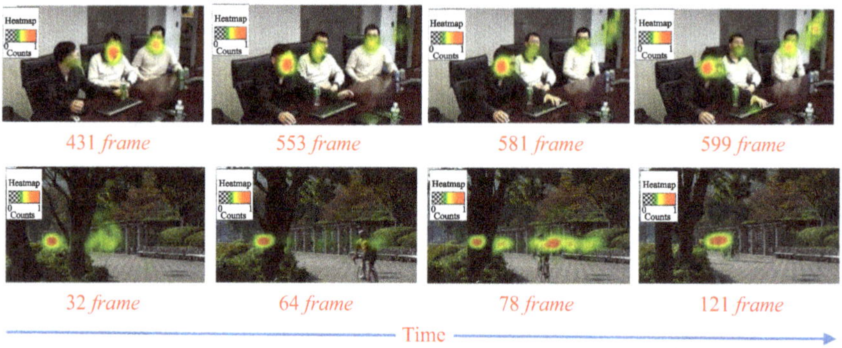

Figure 9.4 *Illustration of Observation 9.2. This figure shows the heat maps of visual attention of all 32 subjects, over several selected frames of videos vidyo1 and ParkScene*

Figure 9.5 *Illumination of Observation 9.3. This figure shows the map of human fixations of all 32 subjects, over a selected frame of video PeopleOnStreet. Note that in the video a lot of visual attention is attended to the old man, who pushes a trolley and walks in the opposite direction of the crowd*

attention. This completes the analysis of Observation 9.2. Note that there exists the lag of human fixations, as the door is still fixated on when the person has left. This also satisfies Observation 9.1.

Observation 9.3: *The object, which moves in the opposite direction of the surrounding objects, is possible to receive extensive fixations.*

The previous work [10] has verified that the human fixations on still images are influenced by the center-surround features of color and intensity. Actually, the center-surround feature of motions also has an important effect on attracting visual attention. As seen from Figure 9.5, the old man with a trolley moves in the opposite direction of the surrounding crowd, and he attracts the majority of visual attention. Therefore, this suggests that the object moving in the opposite direction to its surround (i.e., it is with large center-surround motion) may receive extensive fixations. This completes the analysis of Observation 9.3.

9.4 HEVC features for saliency detection

In this section, we mainly focus on exploring the features in HEVC domain, which can be used to efficiently detect video saliency. As analyzed above, three HEVC features, i.e., splitting depth, bit allocation, and MV, are effective in predicting video saliency. Therefore, they are worked out as the basic features for video saliency detection, to be presented in Section 9.4.1. Note that the camera motion has to be removed for the MV feature, with an efficient algorithm developed in Section 9.4.1. Based on the three basic HEVC features, the features on temporal and spatial difference are discussed in Sections 9.4.2 and 9.4.3, respectively.

9.4.1 Basic HEVC features

Splitting depth. The CTU partition structure [29], a new technique introduced by HEVC, can offer more flexible block sizes in video coding. In HEVC, the block sizes range from 64 × 64 to 8 × 8. In other words, the splitting depth varies from 0 (=64 × 64 block size) to 3 (=8 × 8 block size). In HEVC, rather than raw pixels, the residual of each coding block is encoded, which reflects spatial texture in intra-frame prediction and temporal variation in inter-frame prediction. Consequently, in intra-frame prediction, splitting depth of each CTU can be considered to model spatial saliency. In inter-frame prediction, splitting depth of each coding block can be used to model temporal saliency. Since Section 9.3.2 has demonstrated that most fixations fall into groups with high-valued splitting depths, the splitting depth of each CU is applied as a basic HEVC feature in video-saliency detection.

Let d_{ij}^k be the normalized splitting depth of pixel (i,j) at the kth frame. First, the splitting depths of all CUs need to be extracted from HEVC bitstreams. Then, we assume that the splitting depth of each pixel is equivalent to that of its corresponding CU. Afterwards, all splitting depths should be normalized by the maximal splitting depth in each video frame. At last, all normalized d_{ij}^k can be yielded as one basic feature of our method.

Bit allocation. Since the work of [30] is a state-of-the-art rate control scheme for HEVC, it has been embedded into the latest HEVC reference software (HM 16.0) for assigning bits to different CTUs. In the work of [30], the rate-distortion was optimized in each video frame, such that the CTUs with high-information were generally encoded by more bits. It has been argued in [2] that high-information regions attract extensive visual attention. Thus, the bits, allocated by [30] in HEVC, are considered a basic feature, modeling spatial saliency in intra-frame prediction and temporal saliency in inter frame prediction. Specifically, Section 9.3.2 has shown that visual attention is highly correlated with the bit allocation of each CTU. Thereby, bit per pixel (bpp) is extracted from HEVC bitstreams, toward saliency detection. Let b_{ij}^k denote the normalized bpp of pixel (i,j) at the kth frame. Here, the bpp is achieved via averaging all consumed bits in the corresponding CTU. Next, the bpp is normalized to be b_{ij}^k in each video frame, and it is then included as one of basic HEVC features to detect saliency.

MV. In video coding, MV identifies the location of matching prediction unit (PU) in the reference frame. In HEVC, MV is sophisticatedly developed to indicate motion between neighboring frames. Intuitively, MV can be used to detect video saliency, as motion is an obvious cue [16] of salient regions. This intuition has also been verified by the statistical analysis of Section 9.3.2. Therefore, MV is extracted as a basic HEVC feature in our method.

During video coding, MV is accumulated by two factors: the camera motion and object motion. It has been pointed out in [43] that in a video, moving objects may receive extensive visual attention, while static background normally draws little attention. It is thus necessary to distinguish moving objects and static background. Unfortunately, MVs of static background may be as large as moving objects, due to the camera motion. On the other hand, although temporal difference of MVs is able to make camera motion negligible for static background, it may also miss the moving objects. Therefore, the camera motion has to be removed from calculated MVs to estimate object motion for saliency detection.

Figure 9.6 shows that the camera motion can be estimated to be the dominant MVs in a video frame. In this chapter, we therefore develop a voting algorithm to estimate the motion of camera. Assuming that $\mathbf{m}_{ij}^k$ is the two-dimensional MV of pixel (i, j) at the kth frame, the dominant camera motion $\mathbf{m}_c^k$ in this frame can be determined in the following way.

First, the static background $\mathbf{S}_b^k$ is roughly extracted to be

$$\mathbf{S}_b^k = \left\{ (i,j) | d_{ij}^k \cdot b_{ij}^k < \frac{1}{|\mathbf{I}^k|} \sum_{(i',j') \in \mathbf{I}^k} d_{i'j'}^k \cdot b_{i'j'}^k \right\}, \tag{9.1}$$

(a) Without camera motion (b) With camera motion

Figure 9.6 An example of MV values of all PUs in (a) a frame with no camera motion and (b) a frame with right-to-left camera motion. Note that the MVs are extracted from HEVC bitstreams. In (a) and (b), the dots stand for the origin of each MV, and the blue lines indicate the intensity and angle of each MV. It can be seen that in (a) there is no camera motion, as most MV values are close to zero, whereas the camera motion in (b) is from right to left according the most MV values for the kth frame $\mathbf{I}^k$ (with $|\mathbf{I}^k|$ pixels). It is because the static background is generally with less splitting depth and bit allocation than the moving foreground objects. Then, the azimuth $a(\mathbf{m}_c^k)$ for the dominant camera motion can be calculated via voting all MV angles in the background $\mathbf{S}_b^k$ as

$$\max \text{hist}\left(\bigcup_{i,j\in\mathbf{S}_b^k} a(\mathbf{m}_{ij}^k)\right), \tag{9.2}$$

where $a(\mathbf{m}_{ij}^k)$ is the azimuth for MV $\mathbf{m}_{ij}^k$, and hist($\cdot$) is the azimuth histogram of all MVs. In this chapter, 16 bins with equal angle width ($= 360°/16 = 22.5°$) are applied for the histogram. After obtaining $a(\mathbf{m}_c^k)$, radius $r(\mathbf{m}_c^k)$ for the camera motion needs to be calculated via averaging over all MVs from the selected bin of $a(\mathbf{m}_c^k)$. Finally, the camera motion of each frame can be achieved upon $a(\mathbf{m}_c^k)$ and $r(\mathbf{m}_c^k)$. For justification, we show in Figure 9.7 some subjective results of the camera motion estimated by our voting algorithm (in yellow arrows), as well as the annotated ground truth of camera motion (in blue arrows). As can be seen from this figure, our algorithm is capable of accurately estimating the camera motion. See Appendix for more justification on the estimation of camera motion.

Next, in order to track the motion of objects, all MVs obtained in HEVC domain need to be processed to remove the estimated camera motion. All processed MVs should be then normalized in each video frame, denoted as $\hat{\mathbf{m}}_{ij}^k$. Since it has been argued in [16] that visual attention is probably attracted by moving objects, $\|\hat{\mathbf{m}}_{ij}^k\|_2$ is utilized as one of the basic HEVC features to predict video saliency.

Figure 9.7 The results of camera motion estimation, yielded by our voting algorithm. The first six videos are with some extended camera motion, whereas the last one is without any camera motion. In the frames of the second row, the yellow and blue arrows represent the estimated and manually annotated vectors of the camera moving from frames of the first row to frames of the second row, respectively. Similarly, the yellow and blue arrows in the frames of the third row show the camera motion from frames of the second row to the third row. Refer to Appendix for the way of annotating ground-truth camera motion

9.4.2 Temporal difference features in HEVC domain

As revealed in Observation 9.2, humans tend to fixate on the new objects appearing in a video. In fact, the new appearing or moving objects in the video also leads to large temporal difference of HEVC features in colocated regions of neighboring frames. Hence, the temporal difference features, which quantify the dissimilarity of splitting depth, bit allocation and MV across neighboring frames, are developed as novel HEVC features in our method. However, the temporal difference in colocated region across video frames refers to the sum of object motion and camera motion. It has been figured out in [43] that moving objects attract extensive visual attention, whereas camera motion receives little attention. Therefore, when developing temporal difference features, camera motion needs to be removed to compensate object motion (to be discussed in the following).

Specifically, let us first look at the way on estimating temporal difference of splitting depths. For pixel (i, j) at the kth frame, $\Delta_t d_{ij}^k$ is defined as the difference value of splitting depth across neighboring frames. It can be calculated by averaging the weighted difference values of the splitting depths over all previous frames:

$$\Delta_t d_{ij}^k = \frac{\sum_{l=1}^{k} \exp(-(l^2/\sigma_d^2))\|d_{ij}^k - d_{ij}^{k-l}\|_1}{\sum_{l=1}^{k} \exp(-(l^2/\sigma_d^2))}, \tag{9.3}$$

where parameter σ_d controls the weights on splitting depth difference between two frames. In (9.3), d_{ij}^{k-l} is the splitting depth of pixel (i, j) at the $(k - l)$th frame. After considering the camera motion with our voting algorithm, we assume that $(i^{k,l}, j^{k,l})$ is the pixel at the $(k - l)$th frame matching to pixel (i, j) at the kth frame. To remove the influence of the camera motion, we replace d_{ij}^{k-l} in (9.3) by $d_{i^{k,l}j^{k,l}}^{k-l}$. Then, (9.3) is rewritten to be

$$\Delta_t d_{ij}^k = \frac{\sum_{l=1}^{k} \exp(-(l^2/\sigma_d^2))\|d_{ij}^k - d_{i^{k,l}j^{k,l}}^{k-l}\|_1}{\sum_{l=1}^{k} \exp(-(l^2/\sigma_d^2))}. \tag{9.4}$$

After calculating (9.4), $\Delta_t d_{ij}^k$ needs to be normalized in each video frame, as one of temporal difference features in HEVC domain.

Furthermore, the bpp difference across neighboring frames is also regarded as a feature for saliency detection. Let $\Delta_t b_{ij}^k$ denote the temporal difference of the bpp at pixel (i, j) between the currently processed kth frame and its previous frames. Similar to (9.4), $\Delta_t b_{ij}^k$ can be obtained by

$$\Delta_t b_{ij}^k = \frac{\sum_{l=1}^{k} \exp(-(l^2/\sigma_b^2))\|b_{ij}^k - b_{i^{k,l}j^{k,l}}^{k-l}\|_1}{\sum_{l=1}^{k} \exp(-(l^2/\sigma_b^2))}, \tag{9.5}$$

where σ_b decides the weights of the bpp difference between frames. In (9.5), with the compensated camera motion, $b_{i^{k,l}j^{k,l}}^{k-l}$ is the bpp for pixel $(i^{k,l}, j^{k,l})$ at the $(k - l)$th frame, which matches to pixel (i, j) at the kth frame.

Finally, the temporal difference of MV is also taken into account, by adopting the similar way presented above. Recall that $\hat{\mathbf{m}}_{ij}^k$ is the extracted MV of each pixel, with the camera motion being removed. Since $\hat{\mathbf{m}}_{ij}^k$ is a 2D vector, ℓ_2-norm operation is applied to compute the temporal difference of MVs (denoted by $\Delta_t \hat{m}_{ij}^k$) as follows:

$$\Delta_t \hat{m}_{ij}^k = \frac{\sum_{l=1}^{k} \exp(-(l^2/\sigma_m^2)) \| \hat{\mathbf{m}}_{ij}^k - \hat{\mathbf{m}}_{i^{k,l} j^{k,l}}^{k-l} \|_2}{\sum_{l=1}^{k} \exp(-(l^2/\sigma_m^2))}. \tag{9.6}$$

In (9.6), we can use parameter σ_m to determine the weights of MV difference between two frames. Moreover, $\hat{\mathbf{m}}_{i^{k,l} j^{k,l}}^{k-l}$ is the MV value for pixel $(i^{k,l}, j^{k,l})$ at the $(k-l)$th frame, which is the colocated pixel of (i, j) at the kth frame, after the camera motion is removed by our voting algorithm.

9.4.3 Spatial difference features in HEVC domain

However, the above features are not sufficient to model saliency in a video, since some smooth regions may stand out from complicated background for drawing attention (like a salient smooth ball appearing in grass land). Generally speaking, the basic features of splitting depth and bit allocation in a smooth region are significantly different from those in its surrounding background. Thus, we here develop spatial difference features for saliency detection. In addition, according to Observation 9.3, the object moving in the opposite direction to the nearby objects may result in extensive visual attention. Actually, the dissimilarity of object motion can be measured by the spatial difference of MVs between neighboring PUs. Hence, the spatial difference of all three basic features is incorporated into our method, as given below.

Recall that $\mathbf{I}^k$ is the kth video frame, and that d_{ij}^k, b_{ij}^k, and $\mathbf{m}_{ij}^k$ denote the splitting depth, bit allocation, and MV , respectively, for pixel (i, j) of this video frame. For the spatial difference of MV, the camera motion has to be removed in each $\mathbf{m}_{ij}^k$, defined by $\hat{\mathbf{m}}_{ij}^k$. Then, we have

$$\begin{cases} \Delta_s d_{ij}^k = \dfrac{\sum_{(i',j')\in\mathbf{I}^k} \exp(-(((i'-i)^2 + (j'-j)^2)/\xi_d^2)) \| d_{i'j'}^k - d_{ij}^k \|_1}{\sum_{(i',j')\in\mathbf{I}^k} \exp(-(((i'-i)^2 + (j'-j)^2)/\xi_d^2))} \\ \Delta_s b_{ij}^k = \dfrac{\sum_{(i',j')\in\mathbf{I}^k} \exp(-(((i'-i)^2 + (j'-j)^2)/\xi_b^2)) \| b_{i'j'}^k - b_{ij}^k \|_1}{\sum_{(i',j')\in\mathbf{I}^k} \exp(-(((i'-i)^2 + (j'-j)^2)/\xi_b^2))} \\ \Delta_s \hat{m}_{ij}^k = \dfrac{\sum_{(i',j')\in\mathbf{I}^k} \exp(-(((i'-i)^2 + (j'-j)^2)/\xi_m^2)) \| \hat{\mathbf{m}}_{i'j'}^k - \hat{\mathbf{m}}_{ij}^k \|_2}{\sum_{(i',j')\in\mathbf{I}^k} \exp(-(((i'-i)^2 + (j'-j)^2)/\xi_m^2))}, \end{cases} \tag{9.7}$$

to compute the spatial difference of splitting depth, bit allocation, and MV. As in the above equations, ξ_d, ξ_b, and ξ_m are the parameters to control the spatial weighting of each feature.

Finally, all nine features in HEVC domain can be achieved in our saliency detection method. Since all the proposed HEVC features are block wise, the block-to-pixel refinement is required to obtain smooth feature maps. For the block-to-pixel refinement, a 2D Gaussian filter is applied to three basic features. In this chapter, the dimension and standard deviation of the Gaussian filter are tuned to be

$(2h/15) \times (2h/15)$ and $(h/30)$, where h is the height of the video. It is worth mentioning that the above features on spatial and temporal difference are explored in compressed domain with the block-to-pixel refinement, while the existing methods compute contrast features in pixel domain (e.g., in [10,11]). Additionally, unlike the existing methods, the camera motion is estimated and removed when calculating the feature contrast in our method. Despite being simple and straightforward, these features are effective and efficient, as evaluated in experiment section.

Figure 9.8 summarizes the procedure of HEVC feature extraction in our saliency detection method. As seen from Figure 9.8, the maps of nine features have been obtained, based on splitting depth, bit allocation, and MV of HEVC bitstreams. We argue that one single feature is not capable enough [2] but has different impact on saliency detection. We thus integrate the maps of all nine features with the learned weights. For more details, refer to the next section.

9.5 Machine-learning-based video-saliency detection

This section mainly concentrates on learning an SVM classifier to detect video saliency, using the abovementioned nine HEVC features. The framework of our learning-based method is summarized in Figure 9.9. As shown in this figure, given the HEVC bitstreams, all HEVC features need to be extracted and calculated. Then, the saliency map of each single video frame is yielded by combining the HEVC features with C-support vector classification (C-SVC), which is a kind of nonlinear SVM classifier. Here, the C-SVC classifier is learned from the ground-truth human fixations of training videos. At last, a simple forward smoothing filter is applied to the yielded saliency maps across video frames, outputting the final video-saliency maps. More details about our learning-based method are to be discussed in the following.

9.5.1 Training algorithm

In our method, the nonlinear C-SVC [44], a kind of SVM, is trained as the binary classifier to decide if each pixel can attract attention, according to the proposed HEVC features. First, for the binary classifier, both positive and negative samples need to be obtained from the training set, in which the positive samples mean the pixels attracting fixations, and negative samples indicate the pixels without any visual attention. Next, three basic HEVC features of each training sample are extracted from the HEVC bitstreams, and then other spatial and temporal features are computed upon the corresponding basic features. Let $\{(\mathbf{f}_n, l_n)\}_{n=1}^N$ be those training samples, where $\mathbf{f}_n$ is the vector of the nine HEVC features for the nth training sample, and $l_n \in \{-1, 1\}$ is the class label indicating whether the sample is positive ($l_n = 1$) or negative ($l_n = -1$). Finally, the C-SVC for saliency detection can be worked out, via solving the following optimization problem:

$$\begin{aligned} &\min_{\mathbf{w},b,\{\beta_n\}_{n=1}^N} \frac{1}{2}\|\mathbf{w}\|_2^2 + C\sum_{n=1}^N \beta_n \\ &\text{s.t.} \quad \forall n, \quad l_n(\mathbf{w}^T \cdot \phi(\mathbf{f}_n) + b) \geq 1 - \beta_n, \beta_n \geq 0. \end{aligned} \tag{9.8}$$

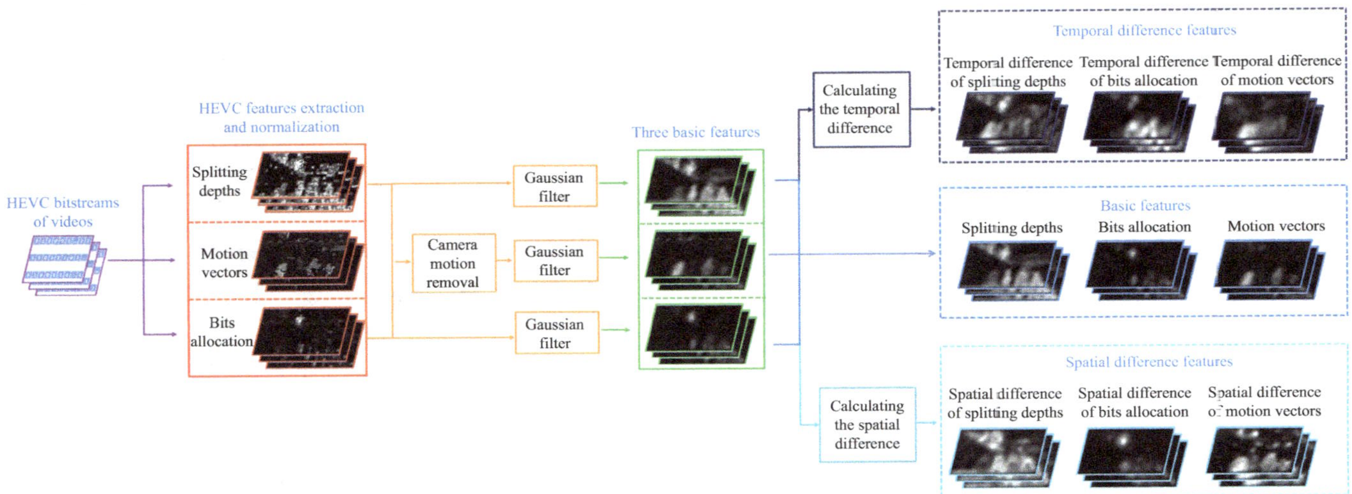

Figure 9.8 Framework of our HEVC feature extractor for video-saliency detection

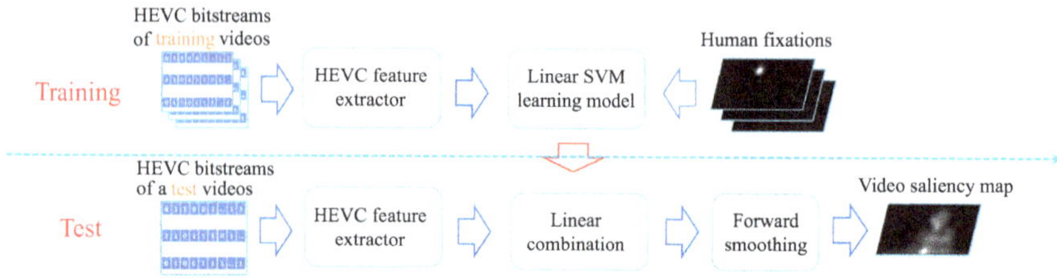

Figure 9.9 Framework of our learning-based method for video-saliency detection with HEVC features. For the HEVC feature extractor, refer to Figure 9.8

In (9.8), $\mathbf{w}$ and b are the parameters to be learned for maximizing the margin between positive and negative samples, and β_n is a nonnegative slack variable evaluating the degree of classification error of $\mathbf{f}_n$. In addition, C balances the trade-off between the error and margin. Function $\phi(\cdot)$ transforms the training vector of HEVC features $\mathbf{f}_n$ to higher dimensional space. Then, $\mathbf{w}$ can be seen as the linear combination of transformed vectors:

$$\mathbf{w} = \sum_{m=1}^{N} \lambda_m l_m \cdot \phi(\mathbf{f}_m), \tag{9.9}$$

where λ_m is the Lagrange multiplier to be learned. Then, the following holds:

$$\begin{aligned} \mathbf{w}^T \cdot \phi(\mathbf{f}_n) &= \left(\sum_{m=1}^{N} \lambda_m l_m \cdot \phi(\mathbf{f}_m) \right)^T \cdot \phi(\mathbf{f}_n) \\ &= \sum_{m=1}^{N} \lambda_m l_m \cdot \langle \phi(\mathbf{f}_m), \phi(\mathbf{f}_n) \rangle. \end{aligned} \tag{9.10}$$

Note that $\langle \phi(\mathbf{f}_m), \phi(\mathbf{f}_n) \rangle$ indicates the inner product of $\phi(\mathbf{f}_m)$ and $\phi(\mathbf{f}_n)$. To calculate (9.10), a kernel of radial based function (RBF) is introduced:

$$K(\mathbf{f}_m, \mathbf{f}_n) = \langle \phi(\mathbf{f}_m), \phi(\mathbf{f}_n) \rangle = \exp(-\gamma \|\mathbf{f}_m - \mathbf{f}_n\|_2^2), \tag{9.11}$$

where γ (> 0) stands for the kernel parameter. Here, we utilize the above RBF kernel due to its simplicity and effectiveness. When training the C-SVC for saliency detection, the penalty parameter C in (9.8) is set to 2^{-3}, and γ of the RBF kernel is tuned to be 2^{-15}, such that the trained C-SVC is rather efficient in detecting saliency. Finally, $\mathbf{w}$ and b can be worked out in the trained C-SVC as the model of video-saliency detection, to be discussed below.

9.5.2 Saliency detection

To detect the saliency of test videos, all nine HEVC features are integrated together using the learned $\mathbf{w}$ and b of our C-SVC classifier. Then, the saliency map $\mathbf{S}_k$ for each single video frame can be yielded by

$$\mathbf{S}_k = \mathbf{w}^T \cdot \phi(\mathbf{F}_k) + b, \tag{9.12}$$

where $\mathbf{F}_k$ defines the pixel-wise matrix of nine HEVC features at the kth video frame. Note that $\mathbf{w}$ in (9.12) is one set of weights for the binary classifier of C-SVC, which have been obtained using the above training algorithm.

Since Observation 9.1 offers a key insight that visual attention may lag behind the moving or new appearing objects, a forward smoothing filter is developed in our method to take into account the saliency maps of previous frames. Mathematically, the final saliency map $\hat{\mathbf{S}}_k$ of the kth video frame is calculated by the forward smoothing filter as follows:

$$\hat{\mathbf{S}}_k = \frac{1}{\lceil t \cdot \mathrm{fr} \rceil} \sum_{k'=k-\lceil t \cdot \mathrm{fr} \rceil+1}^{k} \mathbf{S}_{k'}, \tag{9.13}$$

where t (> 0) is the time duration[1] of the forward smoothing, and fr is the frame rate of the video. Note that a simple forward smoothing filter of (9.13) is utilized here, since we mainly concentrate on extracting and integrating features for saliency detection. Some advanced tracking filters may be applied, instead of the forward smoothing filter in our method, for further improving the performance on saliency detection. To model visual attention on video frames, the final saliency maps need to be smoothed with a 2D Gaussian filter, which is in addition to the one for each single feature map (as shown in Figure 9.8). Note that the 2D Gaussian filter here shares the same parameters as those for feature maps.

9.6 Experimental results

In this section, we present the experimental results on video-saliency detection to validate the performance of our method. Section 9.6.1 shows the settings of our method, and Section 9.6.2 discusses the parameter selection in our method. Sections 9.6.3 and 9.6.4 compare the saliency detection results by our and other seven methods, over our and other two public databases, respectively. For comparing the accuracy of saliency detection, receiver operating characteristic (ROC) curves, the equal error rate (EER), the area under ROC curve (AUC), normalized scanpath saliency (NSS), linear correlation coefficient (CC), and KL were measured on the saliency maps generated by our and other seven methods. Section 9.6.5 evaluates the performance of our method at different working conditions. In Section 9.6.6, we demonstrate the effectiveness of each single HEVC feature in saliency detection.

9.6.1 Setting on encoding and training

HEVC configuration. Before saliency detection, the bitstreams of both training and test videos were generated by the HEVC encoder, for extracting features. In our experiments, the HEVC reference software HM 16.0 (JCT-VC; http://hevc.hhi.fraunhofer.de/) was used as the HEVC encoder. Then, the HEVC

[1]We found out through experiments that $t = 0.3$ s makes the saliency detection accuracy highest. So, time duration t of our forward smoothing was set to be 0.3 in Section 9.6.

bitstreams of all 33 videos in our database were produced for both training and test. In HM 16.0, the low delay (LD) P main configuration was chosen. In addition, the latest R–λ rate control scheme [30] was enabled in HM 16.0. Since the test videos are with diverse content and resolutions, we followed the way of [30] to set the bit rates the same as those at fixed QPs. The CTU size was set to 64×64 and maximum CTU depth was 3 to allow all possible CTU partition structures for saliency detection. Each group of pictures (GOP) was composed of 4 P frames. Other encoding parameters were set by default, using the common *encoder_lowdelay_P_main.cfg* configuration file of HM.

Other working conditions. The implementation of our method in random access (RA) configuration is to be presented in Section 9.6.5. The rate control of RA in HM 16.0 is also enabled. In our experiments, we set all other parameters of RA via the *encoder_randomaccess_main.cfg* file. Note that the GOP of RA is 8 B frames for HM 16.0. Section 9.6.5 further presents the saliency detection results of our method for the bitstreams of ×265, which is more practical than the HM encoder from the aspects of encoding and decoding time.[2] Here, ×265 v1.8 encoder, embedded in the latest FFmpeg, was applied. For ×265, both LD and RA were tested. In ×265, the bit rates were chosen using the same way as we applied for HM 16.0. The GOP structure is 4 P frames for LD and four frames (BBBP) for RA. Other parameters were all set by default in the FFmpeg with the ×265 codec. It is worth pointing out that the ×265 codec was used to extract features from the bitstreams encoded by ×265, while the features of HM 16.0 bitstreams were extracted by the software of HM 16.0.

Training setting. In order to train the C-SVC, our database of Section 9.3.1 was divided into nonoverlapping sets. For the fair evaluation, 3-fold cross validation was conducted in our experiments, and the averaged results are reported in Sections 9.6.2 and 9.6.3. Specifically, our database was equally partitioned into three nonoverlapping sets. Then, two sets were used as training data, and the remaining set was retained for validating saliency detection. The cross-validation process is repeated by 3-fold, with each of the three sets being used exactly once as the validation data. In the training set, 3 pixel of each video frame were randomly selected from top 5% salient regions of ground-truth fixation maps as the positive samples. Similarly, 3 pixel of each video frame were further chosen from bottom 70% salient regions as negative samples. Then, both positive and negative samples were available in each cross validation, to train the C-SVC with (9.8).

9.6.2 Analysis on parameter selection

In HEVC, the bit allocation, splitting depth, and MV of each CTU may change along with increased or decreased bit rates. Therefore, we analyze the performance of our method with regard to the videos compressed at different bit rates. Since the resolutions of test videos vary from 416×240 to $1,920 \times 1,080$, there is an issue on

[2]It takes around 100 s for HM to encode a 1080p video frame, in a PC with Intel Core i7-4770 CPU and 16 GB RAM. By contrast, ×265 adopts parallel computing and fast methods to encode videos, such that real-time 4K HEVC encoding can be achieved by ×265.

finding bit rates suitable for all videos to ensure proper visual quality. To solve such an issue, we follow [30] in setting the bit rates of each video for rate control the same as those of fixed QPs. Then, we report in Figure 9.10 the AUC, CC, and NSS results of our method at different bit rates. Note that the bit rates averaged over all 33 videos are shown, varying from 2,068 to 100 kbps. Figure 9.10 shows that our method achieves the best performance in terms of CC and NSS, when the averaged bit rate of rate control is 430 kbps (equal to those of fixed QP = 37). Therefore, such bit-rate setting is used for the following evaluation. Figure 9.10 also shows that the bit rates have slight impact on the overall performance of our method in terms of AUC, NSS, and CC. The minimum values of AUC, NSS, and CC are above 0.82, 1.52, and 0.41, respectively, at different bit rates, which are superior to all other methods reported in Section 9.6.3. Besides, one may observe from Figure 9.10 that the saliency detection accuracy of some HEVC features is fluctuating when the bit rate is changed. Hence, this figure suggests that our saliency detection should not rely

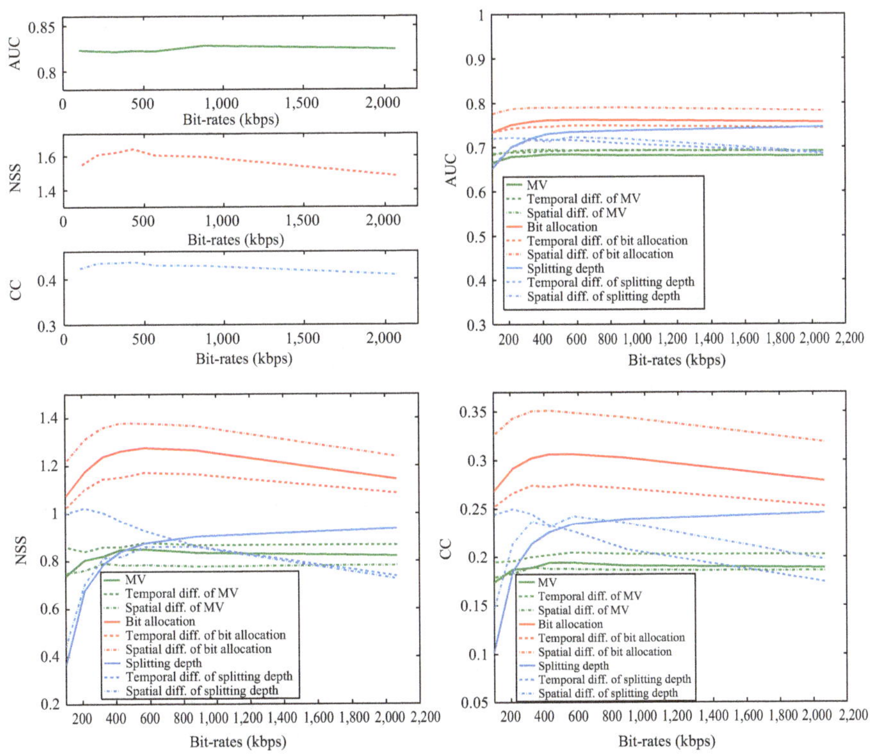

Figure 9.10 Performance comparison of our method (first column) and our single features (second to fourth columns) at different bit rates. The bit rates of each video in our rate control are the same as those of fixed QPs, i.e., QP = 27, 32, 35, 37, 39, 42, and 47. Here, the bit rates averaged over all 33 videos are shown in the horizontal axis

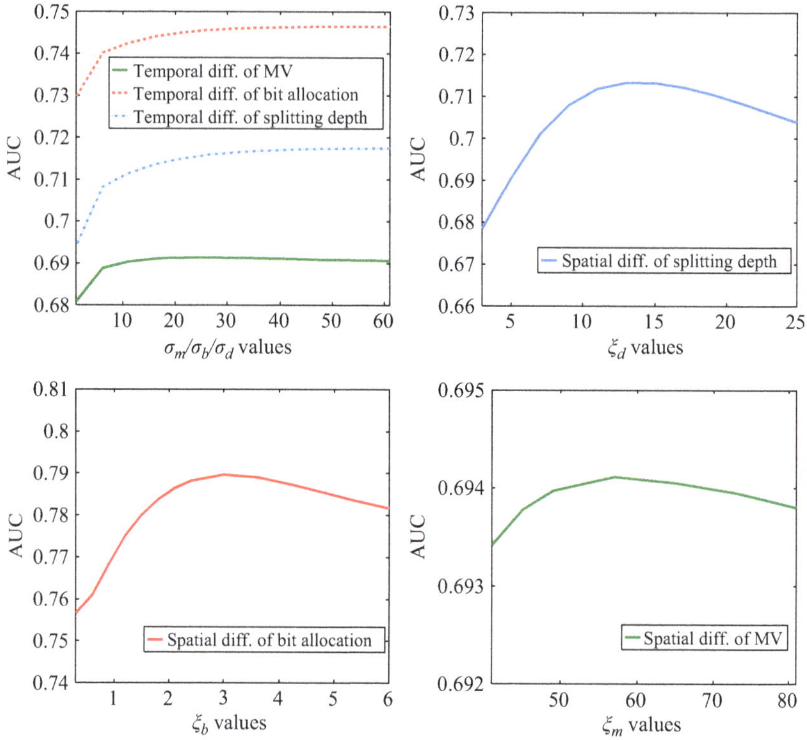

Figure 9.11 Saliency detection performance of each single feature at different parameter settings. Note that only the AUC is utilized here to evaluate the saliency detection performance. For other metrics (e.g., NSS and CC), similar results can be found for choosing the optimal values of parameters

on a single feature. On the contrary, the combination of all features is robust across various bit rates, implying the benefit of applying the C-SVC in learning to integrate all HEVC features for saliency detection.

Next, we analyze the parameters of our saliency detection method. When computing the spatial difference features through (9.7), parameters ξ_d, ξ_b, and ξ_m have been all traversed to find the optimal values. The results are shown in Figure 9.11. As can be seen in this figure, parameters ξ_d, ξ_b, and ξ_m should be set to 13, 3, and 57 for optimizing saliency detection results. In addition, the saliency detection accuracy of temporal difference features almost reaches the maximum, when σ_d, σ_b, and σ_m of (9.4), (9.5), and (9.6) are equivalent to 46, 46, and 26, respectively. Finally, we achieve the optimal parameter selection for the following evaluation (i.e., $\xi_d = 13$, $\xi_b = 3$, $\xi_m = 57$, $\sigma_d = 46$, $\sigma_b = 46$, and $\sigma_m = 26$).

The effectiveness of the center bias in saliency detection has been verified in [45], as humans tend to pay more attention on the center of the image/video than the

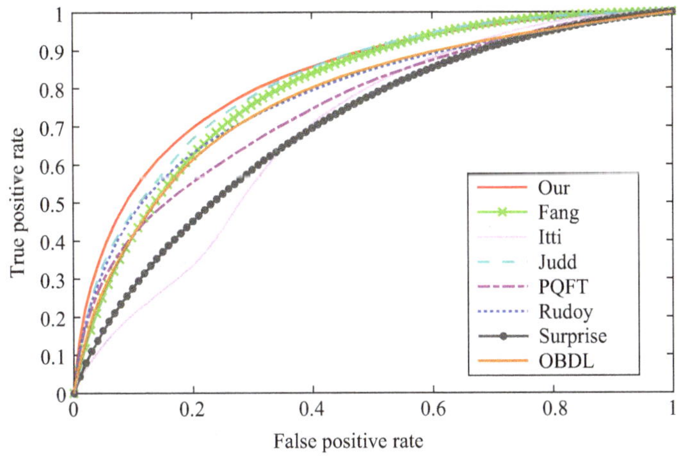

Figure 9.12 ROC curves of saliency detection by our and other state-of-the-art methods. Note that the results are averaged over frames of all test videos of 3-fold cross validation

surround. In this chapter, we follow [45] to impose the same center bias map **B** to both our and other compared methods, for fair comparison. Specifically, the center bias is based on the Euclidean distance of each pixel to video frame center (i_c, j_c) as follows:

$$\mathbf{B}(i,j) = 1 - \frac{\sqrt{(i-i_c)^2 + (j-j_c)^2}}{\sqrt{i_c^2 + j_c^2}}, \tag{9.14}$$

where $\mathbf{B}(i, j)$ is the center bias value at pixel (i, j). Then, the detected saliency maps of all methods are weighted by the above center bias maps.

9.6.3 Evaluation on our database

In this section, we evaluate the saliency detection accuracy of our method, in comparison with other seven state-of-the-art methods,[3] i.e., Itti's model [10], Bayesian surprise [14], Judd *et al.* [19], PQFT [16], Rudoy *et al.* [23], Fang *et al.* [27], and OBDL [28]. Note that 3-fold cross validation was applied in our database for evaluation, and the saliency detection accuracy was averaged over the frames of all test videos of 3-fold cross validation. Furthermore, the saliency maps of some selected video frames are provided for each cross validation, to show the subjective saliency detection results of our and other methods.

ROC curves. The ROC curves of our and other seven methods are shown in Figure 9.12, to evaluate the accuracy of saliency detection in predicting human fixations. As can be seen in this figure, our method generally has higher true positive

[3]In our experiments, we directly used the codes by the authors to implement all methods except Fang *et al.* [27], which was realized by ourselves as the code is not available online.

rates than others at the same false positive rates. In a word, the ROC curves illustrate the superior performance of our method in saliency detection.

AUC and EER. In order to quantify the ROC curves, we report in Table 9.1 the AUC and EER results of our and other seven state-of-the-art methods. Here, both mean and standard deviation are provided for the AUC and EER results of all test video frames of 3-fold cross validation. This table shows that our method performs better than all other seven methods. Specifically, there are 0.026 and 0.038 enhancement of AUC, over Fang *et al.* [27] and OBDL [28], respectively, which also work in compressed domain. The EER of our method has 0.028 and 0.036 decrease, compared with compressed domain methods of [27,28]. Smaller EER means that there is a lower misclassifying probability in our method when the false positive rate equals to the false negative rate. The possible reasons for the improvement of our method are (1) the new compressed domain features (i.e., CTU structure and bit allocation) are developed in light of the latest HEVC standard; (2) the camera motion has been removed in our method; (3) the learning mechanism is incorporated into our method to bridge the gap between HEVC features and human visual attention. Besides, our method outperforms uncompressed domain learning-based methods [19,23], with 0.007 and 0.038 improvement in AUC as well as 0.009 and 0.029 reduction in EER. This verifies the effectiveness of the newly proposed features in compressed domain, which benefit from the well-developed HEVC standard. However, since extensive high and middle level features are applied in [19], there is little AUC improvement (around 0.007) of our method over [19]. Generally speaking, our method outperforms all other seven methods, which are in compressed or uncompressed domain.

NSS, CC, and KL. Now, we concentrate on the comparison of NSS, CC, and KL metrics to evaluate the accuracy of saliency detection on all test videos. The averaged results (with their standard deviation) of NSS, CC, and KL, by our and other seven state-of-the-art methods, are also reported in Table 9.1. Note that the method with a higher value of NSS, CC or KL can better predict the human fixations. Again, it can be seen from Table 9.1 that our method improves the saliency detection accuracy over all other methods, in the terms of NSS, CC, and KL. Moreover, the improvement of NSS, CC, and KL, especially CC, is much larger than that of AUC.

Saliency maps. Figure 9.13 shows the saliency maps of four randomly selected test videos, detected by our and other seven methods, as well as the ground-truth human fixation maps. Note that the results of only one frame for each video are shown in these figures. From these figures, one may observe that in comparison with all other seven methods, our method is capable of well locating the saliency regions in a video frame, much closer to the maps of human fixations. In summary, the subjective results here, together with the objective results above, demonstrate that our method is superior to other state-of-the-art methods in our database.

Computational time. For time-efficiency evaluation, the computational time of our and other methods has been recorded[4] and listed in Table 9.2. We can see from this table that our method ranks third in terms of computational speed, only

[4]All methods were run in the same environment: MATLAB 2012b at a computer with Intel Core i7-4770 CPU@3.4 GHz and 16 GB RAM.

Table 9.1 The averaged accuracy of saliency detection by our and other seven methods, in mean (standard deviation) of all test videos of 3-fold cross validation over our database

	Our	**Itti [10]**	**Surprise [14]**	**Judd [19]**	**PQFT [16]**	**Rudoy [23]**	**Fang [27]**	**OBDL [28]**
AUC	**0.823**(0.071)	0.688(0.066)	0.752(0.083)	0.816(0.065)	0.750(0.084)	0.785(0.100)	0.797(0.073)	0.785(0.086)
NSS	**1.658**(0.591)	0.445(0.464)	1.078(0.739)	1.427(0.440)	1.300(0.529)	1.401(0.708)	1.306(0.560)	1.511(0.825)
CC	**0.438**(0.133)	0.119(0.098)	0.272(0.156)	0.387(0.111)	0.311(0.121)	0.386(0.186)	0.370(0.133)	0.352(0.166)
KL	**0.300**(0.086)	0.104(0.043)	0.183(0.086)	0.285(0.076)	0.239(0.076)	0.269(0.111)	0.266(0.081)	0.236(0.111)
EER	**0.241**(0.075)	0.365(0.051)	0.305(0.075)	0.250(0.064)	0.307(0.074)	0.270(0.094)	0.269(0.071)	0.277(0.098)

Note: The bold values indicate the best saliency prediction results in the table.

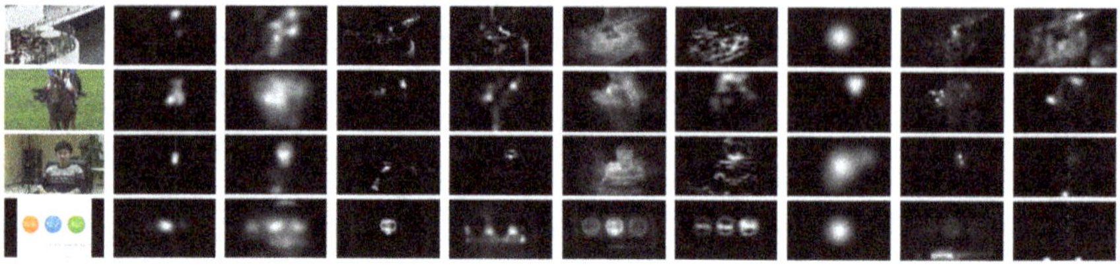

(a) (b) (c) (d) (e) (f) (g) (h) (i) (j)

Figure 9.13 Saliency maps of four videos selected from the first time of our cross-validation experiments. The maps were yielded by our and other seven methods as well the ground-truth human fixations. Note that the results of only one frame are shown for each selected video: (a) input, (b) human, (c) our, (d) Itti, (e) surprise, (f) Judd, (g) PQFT, (h) Rudoy, (i) Fang, (j) OBDL

Table 9.2 Computational time per video frame averaged over our database for our and other seven methods

	Our	**Itti**	**Surprise**	**Judd**	**PQFT**	**Rudoy**	**Fang**	**OBDL**
Time (s)	3.1	1.6	40.6	23.9	0.5	98.5	15.4	5.8

slower than Itti [10] and PQFT [16]. However, as discussed above, the performance of Itti and PQFT is rather inferior compared with other methods, and their saliency detection accuracy is much lower than that of our method. In summary, our method has high time efficiency with effective saliency prediction performance. The main reason is that our method benefits from the modern HEVC encoder and the learning mechanism, thus not wasting much time on exploiting saliency-detection features. We further transplanted our method into C++ program on the VS.net platform to figure out its potential in real-time implementation. After the transplantation, our method consumes averaged 140 ms per frame over all videos of our database and achieves real-time detection for 480p videos at 30 frame per second (fps). It is worth pointing out that some speeding-up techniques, like parallel computing, may further reduce the computational time of our method for real-time saliency detection of high-resolution videos.

9.6.4 Evaluation on other databases

For evaluating the generalization of our method, we compared our and other seven methods on all videos of SFU [41] and DIEM [42], which are two widely used databases. In DIEM, the first 300 frames of each video were tested for matching the length of videos in SFU and our databases. Here, all 33 videos of our database were selected for training the C-SVC classifier. Table 9.3 presents the saliency detection

Table 9.3 Mean (standard deviation) values for saliency detection accuracy of our and other methods over SFU and DIEM databases

SFU								
	Our	**Itti [10]**	**Surprise [14]**	**Judd [19]**	**PQFT [16]**	**Rudoy [23]**	**Fang [27]**	**OBDL [28]**
AUC	**0.83**(0.06)	0.70(0.07)	0.65(0.12)	0.77(0.07)	0.72(0.08)	0.79(0.08)	0.80(0.07)	0.80(0.07)
NSS	**1.42**(0.34)	0.27(0.36)	0.47(0.58)	1.05(0.33)	0.86(0.45)	1.38(0.57)	1.23(0.40)	1.36(0.57)
CC	**0.49**(0.11)	0.09(0.09)	0.16(0.17)	0.37(0.10)	0.29(0.14)	0.46(0.16)	0.42(0.12)	0.44(0.16)
KL	**0.28**(0.07)	0.09(0.03)	0.13(0.08)	0.18(0.06)	0.19(0.07)	0.25(0.10)	0.24(0.07)	0.27(0.09)
EER	**0.24**(0.06)	0.34(0.06)	0.32(0.09)	0.29(0.06)	0.28(0.06)	0.26(0.07)	0.26(0.07)	0.26(0.06)
DIEM								
	Our	**Itti [10]**	**Surprise [14]**	**Judd [19]**	**PQFT [16]**	**Rudoy [23]**	**Fang [27]**	**OBDL [28]**
AUC	**0.86**(0.07)	0.77(0.07)	0.75(0.12)	0.75(0.09)	0.79(0.08)	0.80(0.11)	0.80(0.09)	0.79(0.12)
NSS	**1.82**(0.65)	0.54(0.67)	0.93(0.91)	0.99(0.40)	1.28(0.75)	1.48(0.91)	1.23(0.57)	1.62(1.01)
CC	**0.49**(0.14)	0.13(0.12)	0.23(0.19)	0.29(0.11)	0.30(0.15)	0.41(0.22)	0.35(0.14)	0.39(0.22)
KL	**0.37**(0.10)	0.10(0.06)	0.18(0.13)	0.20(0.07)	0.25(0.11)	0.29(0.14)	0.28(0.10)	0.30(0.13)
EER	**0.21**(0.07)	0.28(0.07)	0.29(0.10)	0.31(0.08)	0.26(0.07)	0.25(0.10)	0.25(0.08)	0.26(0.11)

Note: The bold values indicate the best saliency prediction results in the table.

Table 9.4 Comparison to the results reported in [23]

	Our	PQFT [16]	Rudoy [23]
Median shuffled-AUC	**0.74**	0.68	0.72

Note: The bold values indicate the best saliency prediction results in the table.

accuracy of our and other methods over the SFU and DIEM databases. Again, our method performs much better than others in terms of all five metrics. Although the C-SVC was trained on our database, our method still significantly outperforms all seven conventional methods over other databases.

Although above results were mainly upon the codes by their authors, it is fairer to compare with the results reported in their literatures. However, it is hard to find the literatures reporting the results of all seven methods on a same database. Due to this, we only compare to the reported results of the method with top performance. We can see from Tables 9.1 and 9.3 that among all methods we compared, Rudoy [23] generally ranks highest in our, SFU, and DIEM databases. Thus, we implemented our method on the same database as Rudoy [23] (also the DIEM database), and then we compared the results of our method to those of seven PQFT [16] and Rudoy [23], which were reported in [23]. The comparison is provided in Table 9.4. Note that the comparison is in terms of median shuffled-AUC, as shuffled version of AUC was measured with median values available in [23]. Note that shuffled-AUC is much smaller than AUC, due to the removed center bias prior. We can see from Table 9.4 that our method again performs better than [16,23].

9.6.5 Evaluation on other work conditions

For further assessing the generalization of our method, we extended the implementation of our method at different HEVC working conditions. The working conditions include HM 16.0 and ×265 v1.8 encoders, at both LD and RA configurations. We have discussed the parameter settings of these working conditions in Section 9.6.1. The rate control at these working conditions was also enabled, with the bit rates the same as above.

Figure 9.14 compares the saliency detection performance of our method applied to HM and ×265 encoders with LD and RA configurations. The performance is evaluated in the terms of AUC, CC, NSS, and KL, averaged over all videos of the three databases, i.e., our, SFU, and DIEM databases. The results of Rudoy [23] and Fang [27] are also provided in this figure as the reference. As seen from Figure 9.14, although our method in RA performs a bit worse than that in LD, it is much superior to other state-of-the-art methods. We can further see from Figure 9.14 that the performance of our method slightly decreases, when using ×265 bitstreams instead of HM bitstreams. Such a slight decrease is probably due to the simplified process of ×265 over HM. More importantly, when applied to ×265 bitstreams, our method still significantly outperforms other methods. In summary, our method is robust to different working conditions.

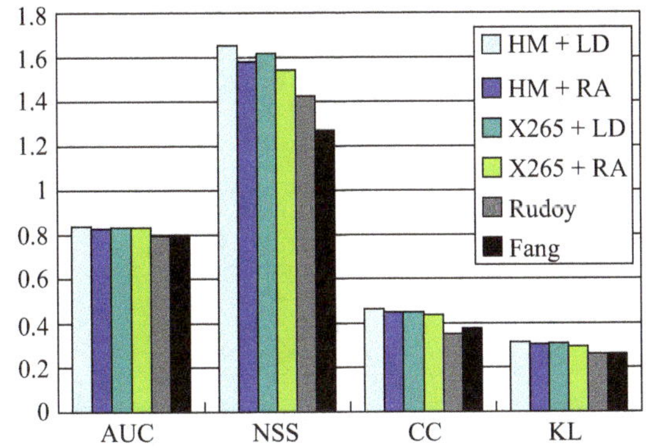

Figure 9.14 Performance of our method at different working conditions, compared with Rudoy [23] and Fang [27]. The performance is assessed in terms of AUC, NSS, CC, and KL, averaged over all videos of our, SFU, and DIEM databases

9.6.6 Effectiveness of single features and learning algorithm

It is interesting to investigate the effectiveness of each HEVC feature in our method. We utilized each single feature of our method to detect saliency of all 33 videos from our database. Since the learning process is not required when evaluating each feature of our method, all 33 videos of our database were tested here without any cross validation. In Table 9.5, we tabulate the saliency detection accuracy of each single feature, measured by AUC, NSS, CC, KL, and EER. This table shows that the AUC results of all nine HEVC features in our method are significantly better than that of random hit, the AUC of which is 0.5. This confirms that the HEVC encoder can be utilized as an effective feature extractor for saliency detection. Besides, it can be clearly observed from this table that the accuracy of bit-allocation-related features ranks the highest among all features. Therefore, we can conclude that the bit allocation of HEVC is rather effective in saliency detection, compared to other HEVC features.

Furthermore, Figure 9.15 evaluates the robustness of each single feature across various working conditions (HM+LD, HM+RA, ×265+LD and ×265+RA). Here, the evaluation is performed on AUC averaged all 33 videos of our database. We can see that the AUC of each single feature, especially the features of splitting depth, varies at different working conditions. This implies that each single feature relies on the working conditions. Benefitting from the machine-learning power of the C-SVC (presented in Section 9.5), the performance of combining all features is significantly more robust than a single feature as shown in Figure 9.15. Since the splitting depth is least robust across various working conditions, we plot in Figure 9.15 the AUC values of integrating six features (excluding spitting depth related features). It shows that the integration of six features underperforms the integration of all nine features for all working conditions. Thus, we can validate that the features of spitting depth are still able to improve the overall performance of our method at various working conditions.

Table 9.5 Mean (standard deviation) values for saliency detection accuracy by each single feature of our method, averaged over the frames of all 33 test videos

	Basic			Temporal difference			Spatial difference		
	Depth	**Bit**	**MV**	**Depth**	**Bit**	**MV**	**Depth**	**Bit**	**MV**
AUC	0.73(0.10)	0.76(0.09)	0.68(0.11)	0.72(0.09)	0.75(0.09)	0.69(0.10)	0.71(0.10)	0.79(0.08)	0.69(0.12)
NSS	0.84(0.49)	1.26(0.72)	0.85(0.67)	0.97(0.55)	1.15(0.63)	0.86(0.61)	0.82(0.50)	1.38(0.70)	0.78(0.62)
CC	0.23(0.12)	0.31(0.15)	0.19(0.15)	0.23(0.12)	0.27(0.14)	0.20(0.15)	0.23(0.13)	0.35(0.15)	0.19(0.15)
KL	0.19(0.09)	0.24(0.10)	0.19(0.09)	0.22(0.08)	0.24(0.09)	0.19(0.08)	0.19(0.08)	0.27(0.09)	0.12(0.09)
EER	0.27(0.08)	0.29(0.09)	0.35(0.09)	0.33(0.08)	0.30(0.08)	0.34(0.09)	0.33(0.09)	0.27(0.09)	0.35(0.02)

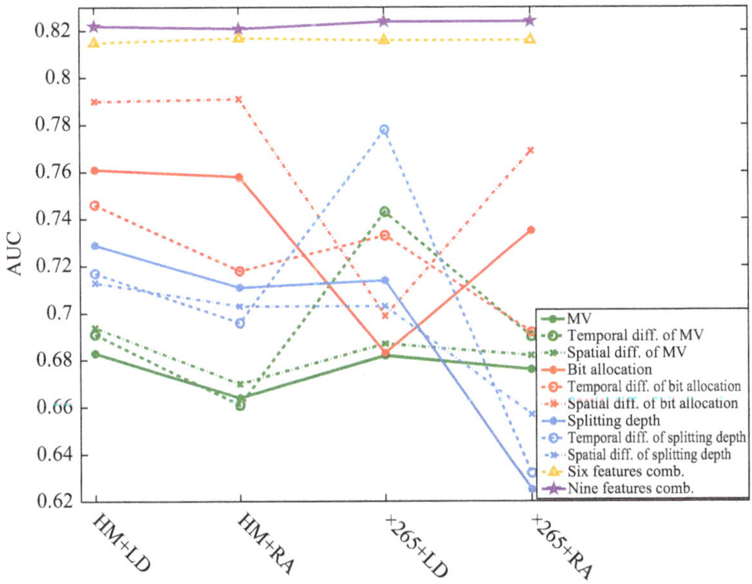

Figure 9.15 AUC curves of saliency detection by each single feature and feature combination. Six comb. and nine comb. mean the results of saliency detection by six features (excluding features of splitting depth) and by all six features, respectively. Similar results can be found for other metrics, e.g., CC

Table 9.6 The averaged accuracy of saliency detection by our method with C-SVC and equal weight

	AUC	NSS	CC	KL	EER
C-SVC	0.823(0.071)	1.658(0.591)	0.438(0.133)	0.300(0.086)	0.241(0.075)
Equal weight	0.775(0.087)	1.268(0.546)	0.330(0.129)	0.247(0.083)	0.279(0.084)

Finally, it is necessary to verify the effectiveness of the C-SVC learning algorithm in our method, since it bridges the gap between the proposed HEVC features and saliency. Provided that the learning algorithm is not incorporated, equal weighting is a common way for feature integration (e.g., in [10]). Table 9.6 compares saliency detection results of our method with the C-SVC learning algorithm and with equal weighting. As can be seen in this table, the C-SVC produces significantly better results in all metrics, compared with the equal weight integration. This indicates the effectiveness of the learning algorithm applied in our method for saliency detection.

9.7 Conclusion

In this chapter, we found out that the state-of-the-art HEVC encoder is not only efficient in video coding but also effective in providing the useful features in saliency detection. Therefore, this chapter has proposed a novel method for learning to detect video saliency with several HEVC features. Specifically, to facilitate the study on video-saliency detection, we first established an eye-tracking database on viewing 33 uncompressed videos from test sets commonly used for HEVC evaluation. The statistical analysis on our database revealed that human fixations tend to fall into the regions with the high-valued HEVC features of splitting depth, bit allocation, and MV. Besides, three observations were also found from our eye-tracking database. According to the analysis and observations, we proposed to extract and then compute several HEVC features, on the basis of splitting depth, bit allocation, and MV. Next, we developed the C-SVC, as a nonlinear SVM classifier, to learn the model of video saliency with regard to the proposed HEVC features. Finally, the experimental results verified that our method outperforms other state-of-the-art saliency detection methods, in terms of ROC, EER, AUC, CC, NSS, and KL metrics.

In the reality of wireless multimedia communications, almost all videos exist in the form of bitstreams, generated by video-coding techniques. Since HEVC is the latest video-coding standard, there is no doubt that the HEVC bitstreams will be prevalent in the near future. Accordingly, our method, performed in HEVC domain, is more practicable over other state-of-the-art uncompressed domain methods, as both time and storage complexity on decompressing videos can be saved.

References

[1] Matin E. Saccadic suppression: a review and an analysis. Psychological Bulletin. 1974;81(12):899–917.

[2] Borji A, and Itti L. State-of-the-art in visual attention modeling. IEEE Transactions on Pattern Analysis and Machine Intelligence. 2013;35(1): 185–207.

[3] Butko NJ, and Movellan JR. Optimal scanning for faster object detection. In: Proc. CVPR; 2009. p. 2751–2758.

[4] Wang W, Shen J, and Porikli F. Saliency-aware geodesic video object segmentation. In: Proc. CVPR; 2015. p. 3395–3492.

[5] Gao D, Han S, and Vasconcelos N. Discriminant saliency, the detection of suspicious coincidences, and applications to visual recognition. IEEE Transactions on Pattern Analysis and Machine Intelligence. 2009;31(6):989–1005.

[6] Rubinstein M, Gutierrez D, Sorkine O, *et al.* A comparative study of image retargeting. ACM Transactions on Graphics (TOG). 2010;29(6):160: 01–10.

[7] Engelke U, Kaprykowsky H, Zepernick H, *et al.* Visual attention in quality assessment. IEEE Signal Processing Magazine. 2011;28(6):50–59.

[8] Hadizadeh H, and Bajic IV. Saliency-aware video compression. IEEE Transactions on Image Processing. 2014;23(1):19–33.

[9] Xu M, Deng X, Li S, *et al.* Region-of-interest based conversational HEVC coding with hierarchical perception model of face. IEEE Journal of Selected Topics on Signal Processing. 2014;8(3):475–489.

[10] Itti L, Koch C, and Niebur E. A model of saliency-based visual attention for rapid scene analysis. IEEE Transactions on Pattern Analysis and Machine Intelligence. 1998;20(11):1254 1259.

[11] Itti L, Dhavale N, and Pighin F. Realistic avatar eye and head animation using a neurobiological model of visual attention. Optical Science and Technology. 2004;64:64–78.

[12] Harel J, Koch C, and Perona P. Graph-based visual saliency. In: Proc. NIPS; 2006. p. 545–552.

[13] Peters RJ, and Itti L. Beyond bottom-up: Incorporating task-dependent influences into a computational model of spatial attention. In: In Proc. CVPR; 2007. p. 1–8.

[14] Itti L, and Baldi P. Bayesian surprise attracts human attention. Vision Research. 2009;49(10):1295–1306.

[15] Zhang L, Tong MH, and Cottrell GW. SUNDAy: Saliency using natural statistics for dynamic analysis of scenes. In: Annual Cognitive Science Conference; 2009. p. 2944–2949.

[16] Guo C, and Zhang L. A novel multiresolution spatiotemporal saliency detection model and its applications in image and video compression. IEEE Transactions on Image Processing. 2010;19(1):185–198.

[17] Ren Z, Gao S, Chia LT, *et al.* Regularized feature reconstruction for spatio-temporal saliency detection. IEEE Transactions on Image Processing. 2013;22(8):3120–3132.

[18] Lin Y, Tang YY, Fang B, *et al.* A visual-attention model using earth mover's distance-based saliency measurement and nonlinear feature combination. IEEE Transactions on Pattern Analysis and Machine Intelligence. 2013;35(2):314–328.

[19] Judd T, Ehinger K, Durand F, *et al.* Learning to predict where humans look. In: Proc. ICCV; 2009. p. 2106–2113.

[20] Kienzle W, Schölkopf B, Wichmann FA, *et al.* How to find interesting locations in video: A spatiotemporal interest point detector learned from human eye movements. In: Pattern Recognition. vol. 4713; 2007. p. 405–414.

[21] Li J, Tian Y, Huang T, *et al.* Probabilistic multi-task learning for visual saliency estimation in video. International Journal of Computer Vision. 2010;90(2):150–165.

[22] Mathe S, and Sminchisescu C. Dynamic eye movement datasets and learnt saliency models for visual action recognition. In: Proc. ECCV; 2012. p. 842–856.

[23] Rudoy D, Goldman DB, Shechtman E, *et al.* Learning video saliency from human gaze using candidate selection. In: Proc. CVPR; 2013. p. 1147–1154.

[24] Lee SH, Kim JH, Choi KP, *et al.* Video saliency detection based on spatiotemporal feature learning. In: Proc. ICIP; 2014. p. 1120–1124.

[25] Sullivan GJ, Ohm J, Han W, *et al.* Overview of the high efficiency video coding (HEVC) standard. IEEE Transactions on Circuits and Systems for Video Technology. 2012;22(12):1649–1668.

[26] Muthuswamy K, and Rajan D. Salient motion detection in compressed domain. IEEE Signal Processing Letters. 2013;20(10):996–999.

[27] Fang Y, Lin W, Chen Z, *et al.* A video saliency detection model in compressed domain. IEEE Transactions on Circuits and Systems for Video Technology. 2014;24(1):27–38.

[28] Hossein Khatoonabadi S, Vasconcelos N, Bajic IV, *et al.* How many bits does it take for a stimulus to be salient? In: CVPR; 2015. p. 5501–5510.

[29] Sullivan GJ, and Baker RL. Efficient quadtree coding of images and video. IEEE Transactions on Image Processing. 1994;3(3):327–331.

[30] Li B, Li H, Li L, *et al.* Domain rate control algorithm for high efficiency video coding. IEEE Transactions on Image Processing. 2014;23(9):3841–3854.

[31] Shanableh T. Saliency detection in MPEG and HEVC video using intra-frame and inter-frame distances. Signal, Image and Video Processing. 2016;10(4):703–709.

[32] Pang D, Kimura A, Takeuchi T, *et al.* A stochastic model of selective visual attention with a dynamic Bayesian network. In: Proc. ICME; 2008. p. 1073–1076.

[33] Wu B, and Xu L. Integrating bottom-up and top-down visual stimulus for saliency detection in news video. Multimedia Tools and Applications. 2014;73(3):1053–1075.

[34] Borji A, Ahmadabadi MN, and Araabi BN. Cost-sensitive learning of top-down modulation for attentional control. Machine Vision and Applications. 2011;22(1):61–76.

[35] Borji A, Sihite DN, and Itti L. What/where to look next? Modeling top-down visual attention in complex interactive environments. IEEE Transactions on Systems, Man, and Cybernetics: Systems. 2014;44(5):523–538.

[36] Hua Y, Zhao Z, Tian H, *et al.* A probabilistic saliency model with memory-guided top-down cues for free-viewing. In: Proc. ICME; 2013. p. 1–6.

[37] Marszalek M, Laptev I, and Schmid C. Actions in context. In: Proc. CVPR; 2009. p. 2929–2936.

[38] Rodriguez MD, Ahmed J, and Shah M. Action MACH a spatio-temporal maximum average correlation height filter for action recognition. In: Proc. CVPR; 2008. p. 1–8.

[39] Ohm JR, Sullivan GJ, Schwarz H, *et al.* Comparison of the coding efficiency of video coding standards—including high efficiency video coding (HEVC). IEEE Transactions on Circuits and Systems for Video Technology. 2012;22(12):1669–1684.

[40] Le Meur O, Ninassi A, Le Callet P, *et al.* Do video coding impairments disturb the visual attention deployment?. Signal Processing: Image Communication. 2010;25(8):597–609.

[41] Hadizadeh H, Enriquez MJ, and Bajić IV. Eye-tracking database for a set of standard video sequences. IEEE Transactions on Image Processing. 2012;21(2):898–903.

[42] Mital PK, Smith TJ, Hill RL, *et al.* Clustering of gaze during dynamic scene viewing is predicted by motion. Cognitive Computation. 2011;3(1):5–24.

[43] Itti L. Automatic foveation for video compression using a neurobiological model of visual attention. IEEE Transactions on Image Processing. 2004;13(10):1304–1318.
[44] Chang CC, and Lin CJ. LIBSVM: A library for support vector machines. ACM Transactions on Intelligent Systems and Technology. 2011;2(3):1–27.
[45] Duan L, Wu C, Miao J, *et al.* Visual saliency detection by spatially weighted dissimilarity. In: Computer Vision and Pattern Recognition (CVPR), 2011 IEEE Conference on. IEEE; 2011. p. 473–480.

Chapter 10

Deep learning for indoor localization based on bimodal CSI data

Xuyu Wang[1] and Shiwen Mao[2]

In this chapter, we incorporate deep learning for indoor localization utilizing channel state information (CSI) with commodity 5 GHz Wi-Fi. We first introduce the state-of-the-art deep-learning techniques including deep autoencoder network, convolutional neural network (CNN), and recurrent neural network (RNN). We then present a deep-learning-based algorithm to leverage bimodal CSI data, i.e., average amplitudes and estimated angle of arrivals (AOA), for indoor fingerprinting. The proposed scheme is validated with extensive experiments. Finally, we discuss several open research problems for indoor localization based on deep-learning techniques.

10.1 Introduction

The proliferation of mobile devices has fostered great interest in indoor-location-based services, such as indoor navigation, robot tracking in factories, locating workers on construction sites, and activity recognition [1–8], all requiring accurately identifying the locations of mobile devices indoors. The indoor environment poses a complex radio-propagation channel, including multipath propagation, blockage, and shadow fading, and stimulates great research efforts on indoor localization theory and systems [9]. Among various indoor-localization schemes, *Wi-Fi-based fingerprinting* is probably one of the most widely used. In fingerprinting, a database is first built with data collected from a thorough measurement of the field in the off-line training stage. Then, the position of a mobile user can be estimated by comparing the newly received test data with that in the database. A unique advantage of this approach is that no extra infrastructure needs to be deployed.

Many existing fingerprinting-based indoor-localization systems use *received signal strength* (RSS) as fingerprints, due to its simplicity and low hardware requirement [10,11]. For example, radar is one of the first RSS-based fingerprinting systems

[1]Department of Computer Science, California State University, United States
[2]Department of Electrical and Computer Engineering, Auburn University, United States

that incorporate a deterministic method for location estimation [10]. For higher accuracy, Horus, another RSS-based fingerprinting scheme, adopts a probabilistic method based on K-nearest neighbor (KNN) [9] for location estimation [11]. The performance of RSS-based schemes is usually limited by two inherent shortcomings of RSS. First, due to the multipath effect and shadow fading, the RSS values are usually highly diverse, even for consecutively received packets at the same position. Second, RSS value only reflects the coarse channel information, since it is the sum of the powers of all received signals.

Unlike RSS, CSI represents fine-grained channel information, which can now be extracted from several commodity Wi-Fi network interface cards (NIC), e.g., Intel Wi-Fi Link 5300 NIC [12], the Atheros AR9390 chipset [13], and the Atheros AR9580 chipset [14]. CSI consists of subcarrier-level measurements of orthogonal frequency division multiplexing (OFDM) channels. It is a more stable representation of channel characteristics than RSS. Several CSI-based fingerprinting systems have been proposed and shown to achieve high localization accuracy [15,16]. For example, the fine-grained indoor fingerprinting system (FIFS) [15] uses a weighted average of CSI values over multiple antennas. To fully exploit the diversity among the multiple antennas and subcarriers, DeepFi [16] learns a large amount of CSI data from the three antennas and 30 subcarriers with an autoencoder. These CSI-based schemes only use the amplitude information of CSI, since the raw phase information is extremely random and not directly usable [17].

Recently, for the Intel 5300 NIC in 2.4 GHz, two effective methods have been proposed to remove the randomness in raw CSI phase data. In [18], the measured phases from 30 subcarriers are processed with a linear transformation to mitigate the random phase offsets, which is then employed for passive human-movement detection. In [17], in addition to the linear transformation, the difference of the sanitized phases from two antennas is obtained and used for line-of-sight (LOS) identification. Although both approaches can stabilize the phase information, the mean value of phase will be zero (i.e., lost) after such processing. This is actually caused by the firmware design of the Intel 5300 NIC when operating on the 2.4 GHz band [19]. To address this issue, Phaser [19] proposes to exploit CSI phase in 5 GHz Wi-Fi. Phaser constructs an AOA pseudospectrum for phase calibration with a single Intel 5300 NIC. These interesting works motivate us to explore effectively cleansed phase information for indoor fingerprinting with commodity 5 GHz Wi-Fi.

In this chapter, we investigate the problem of fingerprinting-based indoor localization with commodity 5 GHz Wi-Fi. We first present three hypotheses on CSI amplitude and phase information for 5 GHz OFDM channels. *First*, the average amplitude over two antennas is more stable over time for a fixed location than that from a single antenna as well as RSS. *Second*, the phase difference of CSI values from two antennas in 5 GHz is highly stable. Due to the firmware design of Intel 5300 NIC, the phase differences of consecutively received packets form four clusters when operating in 2.4 GHz. Such ambiguity makes measured phase difference unusable. However, we find this phenomenon does not exist in the 5 GHz band, where all the phase differences concentrate around one value. We further design a simple multi-radio hardware for phase calibration which is different from the technique in [19] that uses AOA pseudospectrum search with a high computation complexity, to calibrate phase in single Intel 5300 NIC. As a result, the randomness from the time and frequency difference between the transmitter and receiver, and the unknown phase offset can all be removed, and stable phase information can be obtained. *Third*, the calibrated phase difference in 5 GHz can be translated into AOA with considerable accuracy when there is a strong LOS component. We validate these hypotheses with both extensive experiments and simple analysis.

We then design BiLoc, bimodal deep learning for indoor localization with commodity 5 GHz Wi-Fi, to utilize the three hypotheses in an indoor fingerprinting system [20]. In BiLoc, we first extract raw amplitude and phase data from the three antennas, each with 30 subcarriers. We then obtain bimodal data, including average amplitudes over pairs of antennas and estimated AOAs, with the calibration procedure discussed above. In the off-line training stage, we adopt an autoencoder with three hidden layers to extract the unique channel features hidden in the bimodal data and propose to use the weights of the deep network to store the extracted features (i.e., fingerprints). To reduce the computational complexity, we propose a greedy learning algorithm to train the deep network in a layer-by-layer manner with a restricted Boltzmann machine (RBM) model. In the online test stage, bimodal test data is first collected for a mobile device. Then a Bayesian probability model based on the radial basis function (RBF) is leveraged for accurate online position estimation.

In the rest of this chapter, preliminaries on deep learning for indoor localization is introduced Section 10.2. Then, the three hypotheses are given in Section 10.3. We present the BiLoc system in Section 10.4 and validate its performance in Section 10.5. Section 10.6 discusses future research problems for indoor localization, and Section 10.7 concludes this chapter.

10.2 Deep learning for indoor localization

With the rapid growth of computation platforms like Tensorflow, Caffe, and Torch [21], deep learning has been widely applied in a variety of areas such as object recognition, natural-language processing, computer vision, robotics, automated vehicles, and artificial intelligence (AI) games [22]. Compared with shallow machine-learning algorithms, such as support vector machine (SVM) and KNN, deep learning is a branch of machine learning, which implements nonlinear transformations with multiple hidden layers and has high-level data abstractions. In addition, deep learning can train the weights and bias of the network with a huge quantity of data for improving classification performance and data-representation capability, which includes unsupervised and supervised learning with different deep-learning models [23]. In this chapter, three different deep-learning frameworks are discussed below for indoor localization problems.

10.2.1 Autoencoder neural network

A deep autoencoder neural network is an unsupervised learning, which can produce output data that is a de-noised input data. Moreover, it is also used to extract data

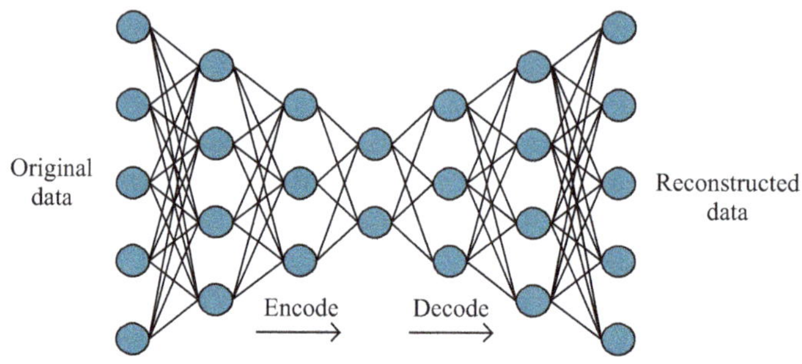

Figure 10.1 Autoencoder

features or reduce the size of data, which is more powerful than principal component analysis-based methods because of its nonlinear transformations with multiple hidden layers. Figure 10.1 shows the architecture of the deep autoencoder neural network. For training, a deep autoencoder neural network has three stages including pretraining, unrolling, and fine-tuning [24]. In the pretraining stage, each neighboring set of two layer is considered as an RBM, is denoted as a bipartite undirected graphical model. Then, a greedy algorithm is used to train the weights and biases for a stack of RBMs. In the unrolling stage, the deep autoencoder network is unrolled to obtain the reconstructed input data. Finally, the fine-tuning phase employs the *backpropagation* (BP) algorithm for training the weights in the deep autoencoder network by minimizing the loss function (i.e., the error).

The first work that applies a deep autoencoder to indoor localization is DeepFi [16,25], which is a deep autoencoder network-based indoor fingerprinting method with CSI amplitudes. For every training location, the deep autoencoder network is trained to obtain a set of weights and biases, which are used as fingerprints for the corresponding locations. For online test, the true location is estimated based on the Bayesian scheme. The experimental results show that the mean distance error in a living room environment and a laboratory environment is 1.2 and 2.3 m, respectively. In addition, PhaseFi [26,27] is proposed to use CSI calibrated phase, which still incorporates a deep autoencoder networks for indoor localization. Moreover, deep autoencoder networks are used for device-free indoor localization [28,29]. The denoising autoencoder-based indoor localization with Bluetooth Low Energy (BLE) is also used to provide 3-D localization [30]. In this chapter, we consider deep autoencoder networks for indoor localization using bimodal CSI data.

10.2.2 Convolutional neural network

CNN is also a useful deep-learning architecture, which has been successfully used in computer vision and activity recognition [23,31]. In 1998, LeCun proposed LeNet-5 [32], which is the first architecture of CNN. Figure 10.2 shows the CNN framework, which includes the convolutional layers, subsampling layers, and fully connected layers.

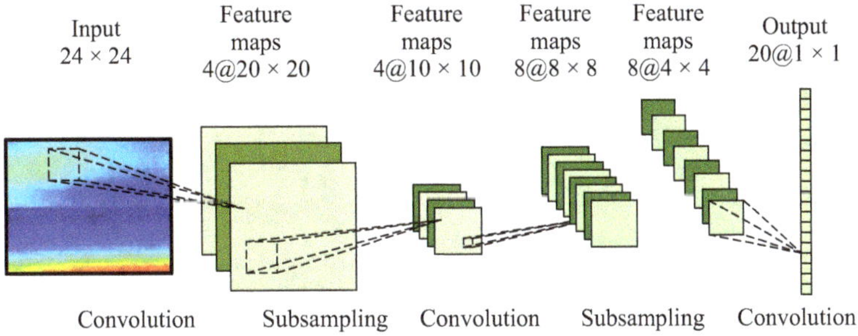

Figure 10.2 CNN

The convolutional layer can obtain feature maps within local regions in the previous layer's feature maps with linear convolutional filters, which is followed by nonlinear activation functions. The subsampling layer is to decrease the resolution of the feature maps by downsampling over a local neighborhood in the feature maps of the previous layer, which is invariant to distortions in the input data [33]. The feature maps in the previous layer are pooled over a local temporal neighborhood using the mean pooling function. Other operations such as the sum or max pooling function can also be incorporated in the subsampling layer.

After the convolutional and subsampling layers, there is a fully connected layer, which is a basic neural network with one hidden layer, to train the output data. Moreover, a loss function is used to measure the difference between the true location label and the output of CNN, where the squared error or cross entropy is used as loss function for training the weights. Currently, an increasing number of CNN models are proposed, such as AlexNet [31] and ResNet [34]. AlexNet is a larger and more complex model, where Max pooling and rectified linear unit (ReLU) nonlinear activation function are used in the model [35]. Moreover, dropout regularization is used to handle the overfitting problem. ResNet was proposed by Microsoft, where the residual block includes a direct path between the input and output, and the batch normalization technique is used to avoid diminishing or exploding of the gradient. ResNet is a 152 layers residual learning framework, which won the ILSVRC 2015 classification competition [31].

For indoor localization problems, the CiFi [33,36] system leverages the constructed images with estimated AOA values with commodity 5 GHz Wi-Fi for indoor localization. This system demonstrates that the performance of the localization has outperformed several existing schemes, like FIFS and Horus. Motivated by ResNet, the ResLoc [37] system uses bimodal CSI tensor data to train a *deep residual sharing learning*, which can achieve the best performance among deep-learning-based localization methods using CSI. CSI amplitude is also used to obtain CSI images for indoor localization [38]. In addition, input images by using received signal strength indicator (RSSI) of Wi-Fi signals are leveraged to train a CNN model [39,40]. CNN has also been used for TDoA-based localization systems, which can estimate nonlinearities in the signal propagation space but also predict the signal for multipath effects [41].

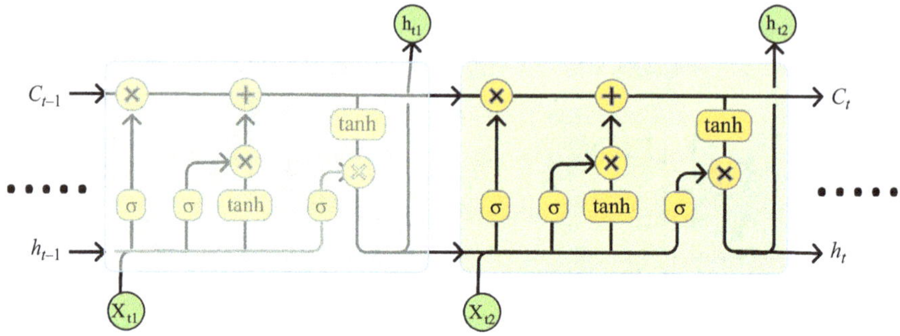

Figure 10.3 LSTM

10.2.3 Long short-term memory

To process variable-length sequence inputs, RNNs are proposed, where long range dependencies can be captured using the feedback loop in the recurrent layer. However, the dependencies also makes it hard to train an RNN, because of diminishing or exploding of the gradient of the loss function. Long short-term memory (LSTM) is proposed to handle the above problem, which has been widely applied for sequence data processing [42].

For the LSTM algorithm in Figure 10.3, the input gate i decides how much new information will be exploited in the current memory cell, the forget gate f controls how much information will be removed from the old memory cell, and the output gate o determines how much data will be output based on the current memory cell c. In addition, the sigmoid function σ can control how much information can be updated and the hyperbolic tangent function tanh can create new candidate values g. Thus, unlike RNN, LSTM can handle long-term dependency and has better data representation ability, and has been employed for speech recognition, machine translation, and time-series problems.

The recently proposed DeepML system uses a two-layer LSTM network for a higher learning and representation ability on exploiting magnetic and light sensor data for indoor localization, which can achieve submeter level localization accuracy [43]. LSTM can be used for sequence-based localization problems with other signals. We have also applied LSTM to wheat moisture level detection [44] and forecasting of renewable energy generation [45].

10.3 Preliminaries and hypotheses

10.3.1 Channel state information preliminaries

OFDM is widely used in wireless network standards, such as Wi-Fi (e.g., IEEE 802.11a/g/n), where the total spectrum is partitioned into multiple orthogonal subcarriers, and wireless data is transmitted over the subcarriers using the same modulation and coding scheme to mitigate frequency selective fading. Leveraging the device driver for off-the-shelf NICs, e.g., the Intel 5300 NIC, we can extract CSI for each received packet, that is a fine-grained physical layer (PHY) information. CSI reveals the channel characteristics experienced by the received signal such as the multipath effect, shadow fading, and distortion.

With OFDM, the Wi Fi channel at the 5 GHz band can bc considcrcd as a narrowband flat fading channel. In the frequency domain, the channel model can be expressed as

$$\vec{Y} = \mathrm{CSI} \cdot \vec{X} + \vec{N}, \tag{10.1}$$

where $\vec{Y}$ and $\vec{X}$ denote the received and transmitted signal vectors, respectively, $\vec{N}$ is the additive white Gaussian noise (AWGN), and CSI represents the channel's frequency response, which can be computed from $\vec{Y}$ and $\vec{X}$.

Although a Wi-Fi receiver uses an OFDM system with 56 subcarriers for a 20 MHz channel, the Intel 5300 NIC can report 30 out of 56 subcarriers. The channel frequency response of subcarrier i, CSI_i, is a complex value, that is,

$$\mathrm{CSI}_i = \mathscr{I}_i + j\mathscr{Q}_i = |\mathrm{CSI}_i| \exp(j\angle\mathrm{CSI}_i), \tag{10.2}$$

where $\mathscr{I}_i$ and $\mathscr{Q}_i$ are the in-phase component and quadrature component, respectively; $|\mathrm{CSI}_i|$ and $\angle\mathrm{CSI}_i$ are the amplitude response and phase response of subcarrier i, respectively.

10.3.2 Distribution of amplitude and phase

In general, both $\mathscr{I}_i$ and $\mathscr{Q}_i$ can be modeled as i.i.d. AWGN of variance σ^2. The amplitude response is $|\mathrm{CSI}_i| = \sqrt{\mathscr{I}_i^2 + \mathscr{Q}_i^2}$, which follows a Rician distribution when there is a strong LOS component [46]. The probability distribution function (PDF) of the amplitude response is given by

$$f(|\mathrm{CSI}_i|) = \frac{|\mathrm{CSI}_i|}{\sigma^2} \times \exp\left(-\frac{|\mathrm{CSI}_i|^2 + |\mathrm{CSI}_0|^2}{2\sigma^2}\right) \times I_0\left(\frac{|\mathrm{CSI}_i| \cdot |\mathrm{CSI}_0|}{\sigma^2}\right), \tag{10.3}$$

where $|\mathrm{CSI}_0|$ is the amplitude response without noise, $I_0(\cdot)$ is the zeroth-order modified Bessel function of the first kind. When the signal-to-noise ratio (SNR) is high, the PDF $f(|\mathrm{CSI}_i|)$ will converge to the Gaussian distribution as $\mathscr{N}(\sqrt{|\mathrm{CSI}_0|^2 + \sigma^2}, \sigma^2)$ [46].

The phase response of subcarrier i is computed by $\angle\mathrm{CSI}_i = \arctan(\mathscr{Q}_i/\mathscr{I}_i)$ [46]. The phase PDF is given by

$$\begin{aligned} f(\angle\mathrm{CSI}_i) = {} & \frac{1}{2\pi} \exp\left(-\frac{|\mathrm{CSI}_0|^2}{2\sigma^2}\right) \left(1 + \frac{|\mathrm{CSI}_0|}{\sigma}\sqrt{2\pi}\cos(\angle\mathrm{CSI}_i)\right. \\ & \left. \times \exp\left(\frac{|\mathrm{CSI}_0|^2\cos^2(\angle\mathrm{CSI}_i)}{2\sigma^2}\right)\left(1 - Q\left(\frac{|\mathrm{CSI}_0|\cos(\angle\mathrm{CSI}_i)}{\sigma}\right)\right)\right), \end{aligned}$$

where $Q(\cdot)$ is the Q-function. In the high SNR regime, the PDF $f(\angle\mathrm{CSI}_i)$ also converges to a Gaussian distribution as $\mathscr{N}\left(0, (\sigma/|\mathrm{CSI}_0|)^2\right)$ [46]. The distribution of amplitude and phase of the subcarriers would be useful to guide the design of localization algorithms.

10.3.3 Hypotheses

We next present three important hypotheses about the CSI data on 5 GHz OFDM channels, which are demonstrated and tested with our measurement study and theoretical analysis.

10.3.3.1 Hypothesis 1

The average CSI amplitude value of two adjacent antennas for the 5 GHz OFDM channel is highly stable for a fixed location.

We find CSI amplitude values exhibit great stability for continuously received packets at a given location. Figure 10.4 presents the cumulative distribution functions (CDF) of the standard deviations (STD) of (i) the normalized CSI amplitude averaged over two adjacent antennas, (ii) the normalized CSI amplitude from a single antenna, and (iii) the normalized RSS amplitude from a single antenna, for 90 positions. At each position, 50 consecutive packets are received by the Intel 5300 NIC operating on the 5 GHz band. It can be seen that 90% of the testing positions are blow 10% of the STD in the case of averaged CSI amplitudes, while the percentage is 80% for the case of single antenna CSI and 70% for the case of single antenna RSS. Thus, averaging over two adjacent antennas can make CSI amplitude highly stable for a fixed location with 5 GHz OFDM channels. We conduct the measurements over a long period of time, including midnight and business hours. No obvious difference in the stability of CSI is observed over different times, while RSS values exhibit large variations even for the same position. This finding motivates us to use average CSI amplitudes of two adjacent antennas as one of the features of deep learning in the BiLoc design.

Recall that the PDF of the amplitude response of a single antenna is Gaussian in the high SNR regime. Assuming that the CSI values of the two antennas are i.i.d.

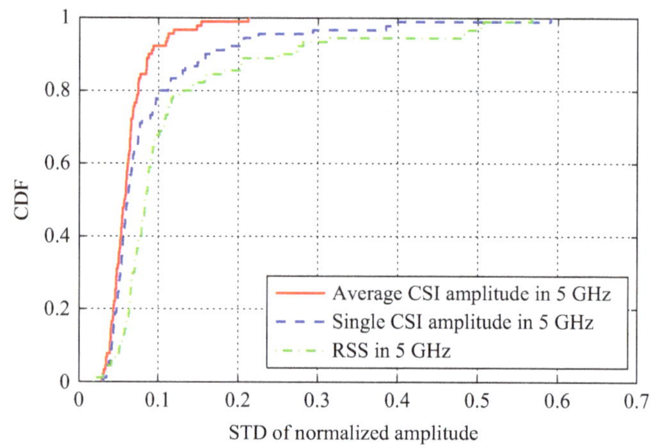

Figure 10.4 CDF of the standard deviations of normalized average CSI amplitude, a single CSI amplitude, and a single RSS in the 5 GHz OFDM channel for 90 positions

(true when the antennas are more than a half wavelength apart [17]), the average CSI amplitudes also follow the Gaussian distribution, as $\mathcal{N}(\sqrt{|\mathrm{CSI}_0|^2+\sigma^2}, \sigma^2/2)$, but with a smaller variance. This proves that stability can be improved by averaging CSI amplitudes over two antennas [47] (as observed in Figure 10.4). We consider the average CSI amplitudes over two antennas instead of three antennas or CSI amplitudes from only one antenna, and BiLoc employs bimodal data, including estimated AOAs and average amplitudes. This requires that we use the same number of nodes as the input for the deep network.

10.3.3.2 Hypothesis 2

The difference of CSI phase values between two antennas of the 5 GHz OFDM channel is highly stable, compared to that of the 2.4 GHz OFDM channel.

Although the CSI phase information is also available from the Intel 5300 NIC, it is highly random and cannot be directly used for localization, due to noise and the unsynchronized time and frequency of the transmitter and receiver. Recently, two useful algorithms are used to remove the randomness in CSI phase. The first approach is to make a linear transform of the phase values measured from the 30 subcarriers [18]. The other one is to exploit the phase difference between two antennas in 2.4 GHz and then remove the measured average [17]. Although both methods can stabilize the CSI phase in consecutive packets, the average phase value they produce is always near zero, which is different from the real phase value of the received signal.

Switching to the 5 GHz band, we find the phase difference becomes highly stable. In Figure 10.5, we plot the measured phase differences of the 30 subcarriers between two antennas for 200 consecutively received packets in the 5 GHz (in blue) and 2.4 GHz (in red) bands. The phase difference of the 5 GHz channel varies between [0.5, 1.8], which is considerably more stable than that of the 2.4 GHz channel (varies between $[-\pi, \pi]$). To further illustrate this finding, we plot the measured phase differences on the fifth subcarrier between two antennas using polar coordinates in Figure 10.6. We find that all the 5 GHz measurements concentrate around 30°, while the 2.4 GHz measurements form four clusters around 0°, 90°, 180°, and 270°. We conjecture that this may be caused by the firmware design of the Intel 5300 NIC when operating on the 2.4 GHz band, which reports the phase of channel modulo $\pi/2$ rather than 2π on the 5 GHz band [19]. Comparing to the ambiguity in the 2.4 GHz band, the highly stable phase difference in the 5 GHz band could be very useful for indoor localization.

As in Hypothesis 1, we also provide an analysis to validate the observation from the experiments. Let $\angle\widehat{CSI}_i$ denote the measured phase of subcarrier i, which is given by [14,48]:

$$\angle\widehat{CSI}_i = \angle\mathrm{CSI}_i + (\lambda_p + \lambda_s)m_i + \lambda_c + \beta + Z, \tag{10.4}$$

where $\angle\mathrm{CSI}_i$ is the true phase; Z is the measurement noise; β is the initial phase offset because of the phase-locked loop; m_i is the subcarrier index of subcarrier i; λ_p, λ_s, and λ_c are phase errors from the packet boundary detection (PBD); the

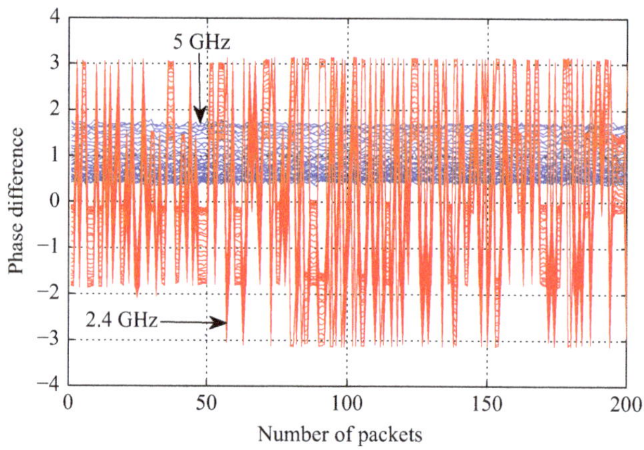

Figure 10.5 The measured phase differences of the 30 subcarriers between two antennas for 200 consecutively received packets in the 5 GHz (blue) and 2.4 GHz (red) bands

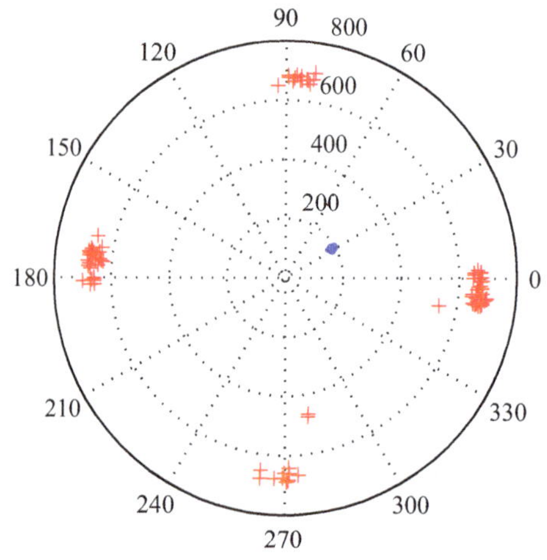

Figure 10.6 The measured phase differences of the fifth subcarrier between two antennas for 200 consecutively received packets in the 5 GHz (blue dots) and 2.4 GHz (red crosses) bands

sampling frequency offset and central frequency offset, respectively [48], which are expressed by

$$\begin{cases} \lambda_p = 2\pi \dfrac{\Delta t}{N} \\ \lambda_s = 2\pi \left(\dfrac{T' - T}{T} \right) \dfrac{T_s}{T_u} n \\ \lambda_c = 2\pi \, \Delta f \, T_s n, \end{cases} \tag{10.5}$$

where Δt is the PBD delay, N is the fast Fourier transform (FFT) size, T' and T are the sampling periods from the receiver and the transmitter, respectively, T_u is the length of the data symbol, T_s is the total length of the data symbol and the guard interval, n is the sampling time offset for current packet, Δf is the center frequency difference between the transmitter and receiver. It is noticed that we cannot obtain the exact values about Δt, $(T' - T)/T$, n, Δf, and β in (10.4) and (10.5). Moreover, λ_p, λ_s, and λ_c vary for different packets with different Δt and n. Thus, the true phase $\angle \text{CSI}_i$ cannot be derived from the measured phase value.

However, note that the three antennas of the Intel 5300 NIC use the same clock and the same down-converter frequency. Consequently, the measured phases of subcarrier i from two antennas have identical packet detection delay, sampling periods, and frequency differences (and the same m_i) [19]. Thus the measured phase difference on subcarrier i between two antennas can be approximated as

$$\Delta \angle \widehat{CSI}_i = \Delta \angle \text{CSI}_i + \Delta \beta + \Delta Z, \tag{10.6}$$

where $\Delta \angle \text{CSI}_i$ is the true phase difference of subcarrier i, $\Delta \beta$ is the unknown difference in phase offsets, which is in fact a constant [19], and ΔZ is the noise difference. We find that $\Delta \angle \widehat{CSI}_i$ is stable for different packets because of (10.6) where Δt and n are cancelled.

In the high SNR regime, the PDF of the phase response of subcarrier i for each of the antennas is $\mathcal{N}(0, (\sigma/|\text{CSI}_0|)^2)$. Due to the independent phase responses, the measured phase difference of subcarrier i is also Gaussian with $\mathcal{N}(\Delta\beta, 2\sigma^2(1 + 1/|\text{CSI}_0|^2))$. Note that although the variance is higher comparing to the true-phase response, the uncertainty from the time and frequency differences is removed, leading to much more stable measurements (as shown in Figure (10.6)).

10.3.3.3 Hypothesis 3

The calibrated phase difference in 5 GHz can be translated into the AOA with considerable accuracy when there is a strong LOS component.

The measured phase difference on subscriber i can be translated into an estimation of AOA, as

$$\theta = \arcsin\left(\frac{\Delta \angle \widehat{CSI}_i \lambda}{2\pi d}\right), \tag{10.7}$$

where λ is the wavelength and d is the distance between the two antennas (set to $d = 0.5\lambda$ in our experiments). Although the measured phase difference $\Delta \angle \widehat{CSI}_i$ is highly stable, we still wish to remove the unknown phase offset difference $\Delta\beta$ to further reduce the error of AOA estimation. For commodity Wi-Fi devices, the only existing approach for a single NIC, to the best of our knowledge, is to search for $\Delta\beta$ within an AOA pseudospectrum in the range of $[-\pi, \pi]$, which, however, has a high time complexity [19].

In this chapter, we design a simple method to remove the unknown phase offset difference $\Delta\beta$ using two Intel 5300 NICs. As in Figure 10.7, we use one Intel 5300 NIC as transmitter and the other as receiver, while a *signal splitter* is used to route signal from antenna 1 of the transmitter to antennas 1 and 2 of the receiver through cables

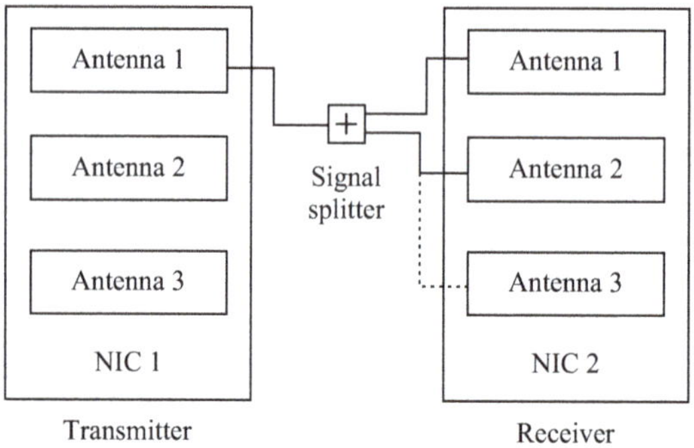

Figure 10.7 The multi-radio hardware design for calibrating the unknown phase offset difference $\Delta\beta$

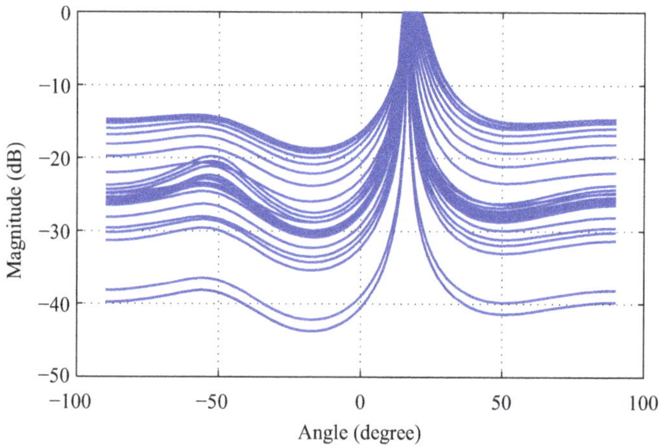

Figure 10.8 The estimated AOAs from the 30 subcarriers using the MUSIC algorithm, while the real AOA is 14°

of the same length. Since the two antennas receive the same signal, the true phase difference $\Delta\angle\text{CSI}_i$ of subcarrier i is zero. We can thus obtain $\Delta\beta$ as the measured phase offset difference between antennas 1 and 2 of the receiver. We also use the same method to calibrate antennas 2 and 3 of the receiver, to obtain the unknown phase offset difference between them as well. We find that the unknown phase offset difference is relatively stable over time.

Having calibrated the unknown phase offset differences for the three antennas, we then use the MUSIC algorithm for AOA estimation [49]. In Figure 10.8, the AOA estimation using MUSIC with the calibrated phase information for the 30 subcarriers is plotted for a high SNR signal with a known incoming direction of 14°. We can see that the peak occurs at around 20° in Figure 10.8, indicating an AOA estimation error of about 6°.

We can obtain the true incoming angle with MUSIC when the LOS component is strong. To deal with the case with strong NLOS paths (typical in indoor environments), we adopt a deep network with three hidden layers to learn the estimated AOAs and the average amplitudes of adjacent antenna pairs as fingerprints for indoor localization. As input to the deep network, the estimated AOA is obtained as follows:

$$\theta = \arcsin\left(\left(\Delta \angle \widehat{CSI}_i - \Delta\beta\right) \frac{\lambda}{2\pi d} \right) + \frac{\pi}{2}, \tag{10.8}$$

where $\Delta\beta$ is measured with the proposed multi-radio hardware experiment. The estimated AOA is in the range of $[0, \pi]$.

10.4 The BiLoc system

10.4.1 BiLoc system architecture

The overall architecture of BiLoc is illustrated in Figure 10.9. The BiLoc design uses only one access point and one mobile device, each equipped with an Intel 5300 NIC, servicing as receiver and transmitter, respectively. All the communications are on the

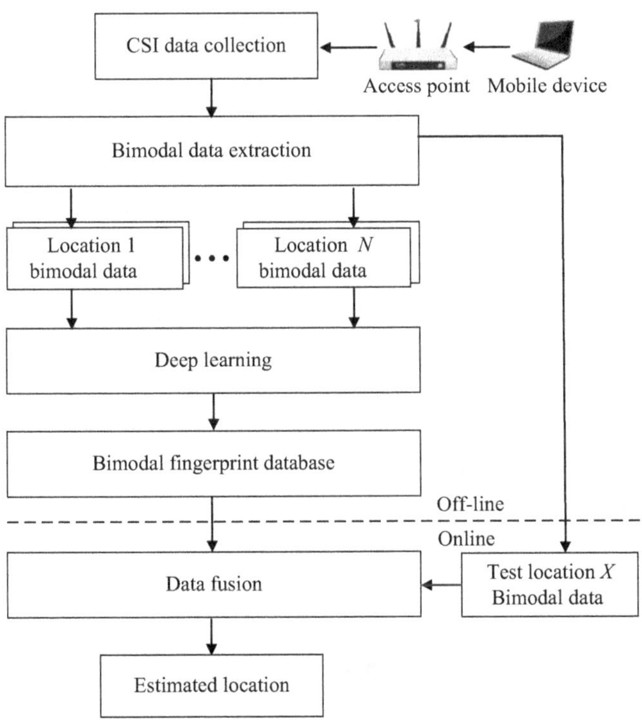

Figure 10.9 The BiLoc system architecture

5 GHz band. The Intel 5300 NIC has three antennas; at each antenna, we can read CSI data from 30 subcarriers. Thus we can collect 90 CSI data for every received packet. We then calibrate the phase information of the received CSI data using our multi-radio hardware design (see Figure 10.7). Both the estimated AOAs and average amplitudes of two adjacent antennas are used as location features for building the fingerprint database.

A unique feature of BiLoc is its bimodal design. With the three receiving antennas, we can obtain two groups of data: (i) 30 estimated AOAs and 30 average amplitudes from antennas 1 and 2 and (ii) that from antennas 2 and 3. BiLoc utilizes estimated AOAs and average amplitudes for indoor fingerprinting for two main reasons. First, these two types of CSI data are highly stable for any given position. Second, they are usually complementary to each other under some indoor circumstances. For example, when a signal is blocked, the average amplitude of the signal will be significantly weakened, but the estimated AOA becomes more effective. On the other hand, when the NLOS components are stronger than the LOS component, the average amplitude will help to improve the localization accuracy.

Another unique characteristic of BiLoc is the use of deep learning to produce feature-based fingerprints from the bimodal data in the off-line training stage, which is quite different from the traditional approach of storing the measured data as fingerprints. Specifically, we use the weights in the deep network to represent the features-based fingerprints for every position. By obtaining the optimal weights with the bimodal data on estimated AOAs and average amplitudes, we can establish a bimodal fingerprint database for the training positions. The third feature of BiLoc is the probabilistic data fusion approach for location estimation based on received bimodal data in the online test stage.

10.4.2 Off-line training for bimodal fingerprint database

In the off-line stage, BiLoc leverages deep learning to train and store the weights to build a bimodal fingerprint database, which is a deep autoencoder that involves three phases: pretraining, unrolling, and fine-tuning [24]. In the pretraining phase, a deep network with three hidden layers and one input layer is used to learn the bimodal data. We denote h^i as the hidden variable with K_i nodes at layer i, $i = 1, 2, 3$, and h^0 as the input data with K_0 nodes at the input layer. Let the average amplitude data be v^1 and the estimated AOA data be v^2. To build the bimodal fingerprint database, we set $h^0 = v^1$ and $h^0 = v^2$ for databases 1 and 2, respectively, each of which is a set of optimal weights. We denote W_1, W_2, and W_3 as the weights between input data and the first hidden layer, the first, and second hidden layers, and the second and third hidden layers, respectively.

We define $\Pr(h^0, h^1, h^2, h^3)$ as the probabilistic generative model for the deep network. To derive the optimal weights, we maximize the marginal distribution of the input data for the deep network, which is given by

$$\max_{\{W_1,W_2,W_3\}} \sum_{h^1} \sum_{h^2} \sum_{h^3} \Pr(h^0, h^1, h^2, h^3). \tag{10.9}$$

Because of the large number of nodes and the complex model structure, it is difficult to find the optimal weights for the input data with the maximum likelihood method. To reduce the computational complexity, BiLoc utilizes a greedy learning algorithm to train the weights *layer by layer* based on a stack of RBMs [50]. We consider an RBM as a bipartite undirected graphical model [50] with joint distribution $\Pr(h^{i-1}, h^i)$, as

$$\Pr(h^{i-1}, h^i) = \frac{\exp(-\mathbb{E}(h^{i-1}, h^i))}{\sum_{h^{i-1}} \sum_{h^i} \exp(-\mathbb{E}(h^{i-1}, h^i))}, \tag{10.10}$$

where $\mathbb{E}(h^{i-1}, h^i)$ denotes the free energy between layer $(i-1)$ and layer i, which is given by

$$\mathbb{E}(h^{i-1}, h^i) = -b^{i-1}h^{i-1} - b^i h^i - h^{i-1} W_i h^i, \tag{10.11}$$

where b^{i-1} and b^i are the biases for the units of layer $(i-1)$ and that of layer i, respectively. To obtain the joint distribution $\Pr(h^{i-1}, h^i)$, the *CD-1 algorithm* is used to approximate it as [50]:

$$\begin{cases} \Pr(h^{i-1}|h^i) = \prod_{j=1}^{K_{i-1}} \Pr(h_j^{i-1}|h^i) \\ \Pr(h^i|h^{i-1}) = \prod_{j=1}^{K_i} \Pr(h_j^i|h^{i-1}), \end{cases} \tag{10.12}$$

where $\Pr(h_j^{i-1}|h^i)$ and $\Pr(h_j^i|h^{i-1})$ are given by the sigmoid belief network as follows:

$$\begin{cases} \Pr(h_j^{i-1}|h^i) = \left(1 + \exp(-b_j^{i-1} - \sum_{t=1}^{K_i} W_i^{j,t} h_t^i)\right)^{-1} \\ \Pr(h_j^i|h^{i-1}) = \left(1 + \exp(-b_j^i - \sum_{t=1}^{K_{i-1}} W_i^{j,t} h_t^{i-1})\right)^{-1}. \end{cases} \tag{10.13}$$

We propose a greedy algorithm to train the weights and biases for a stack of RBMs. First, with the CD-1 method, we use the input data to train the parameters $\{b^0, b^1, W_1\}$ of the first layer RBM. Then, the parameters $\{b^0, W_1\}$ are frozen, and we sample from the conditional probability $\Pr(h^1|h^0)$ to train the parameters $\{b^1, b^2, W_2\}$ of the second layer RBM. Next, we freeze the parameters $\{b^0, b^1, W_1, W_2\}$ of the first and second layers and then sample from the conditional probability $\Pr(h^2|h^1)$ to train the parameters $\{b^2, b^3, W_3\}$ of the third layer RBM. In order to train the weights and biases of each RBM, we use the CD-1 method to approximate them. For the layer i RBM model, we estimate $\hat{h}^{i-1}$ by sampling from the conditional probability $\Pr(h^{i-1}|h^i)$; by sampling from the conditional probability $\Pr(h^i|\hat{h}^{i-1})$, we can estimate $\hat{h}^i$. Thus, the parameters are updated as follows:

$$\begin{cases} \Delta W_i = \varepsilon(h^{i-1}h^i - \hat{h}^{i-1}\hat{h}^i) \\ \Delta b^i = \varepsilon(h^i - \hat{h}^i) \\ \Delta b^{i-1} = \varepsilon(h^{i-1} - \hat{h}^{i-1}), \end{cases} \tag{10.14}$$

where ε is the step size.

After the pretraining phase, we obtain the near-optimal weights for the deep network. We then unroll the deep network with *forward propagation* to obtain the reconstructed input data in the unrolling phase. Finally, in the fine-tuning phase, the

BP algorithm is used to train the weights in the deep network according to the error between the input data and the reconstructed input data. The optimal weights are obtained by minimizing the error. In BiLoc, we use estimated AOAs and average amplitudes as input data and obtain two sets of optimal weights for the bimodal fingerprint database.

10.4.3 Online data fusion for position estimation

In the online phase, we adopt a probabilistic approach to location estimation based on the bimodal fingerprint database and the bimodal test data. We derive the posteriori probability $\Pr(l_i|v^1, v^2)$ using Bayes' law as

$$\Pr(l_i|v^1, v^2) = \frac{\Pr(l_i)\Pr(v^1, v^2|l_i)}{\sum_{j=1}^{N}\Pr(l_j)\Pr(v^1, v^2|l_j)}, \tag{10.15}$$

where N is the number of reference locations, l_i is the ith reference location in the bimodal fingerprint database, and $\Pr(l_i)$ is the prior probability that the mobile device is considered to be at the reference location l_i. Without loss of generality, we assume that $\Pr(l_i)$ is uniformly distributed. The posteriori probability $\Pr(l_i|v^1, v^2)$ becomes:

$$\Pr(l_i|v^1, v^2) = \frac{\Pr(v^1, v^2|l_i)}{\sum_{j=1}^{N}\Pr(v^1, v^2|l_j)}. \tag{10.16}$$

In BiLoc, we approximate $\Pr(v^1, v^2|l_i)$ with an RBF in the similar form of a Gaussian function, to measure the degree of similarity between the reconstructed bimodal data and the test bimodal data, given by

$$\Pr(v^1, v^2|l_i) = \exp\left(-(1-\rho)\frac{\|v^1 - \hat{v}^1\|}{\eta_1\sigma_1} - \rho\frac{\|v^2 - \hat{v}^2\|}{\eta_2\sigma_2}\right), \tag{10.17}$$

where $\hat{v}^1$ and $\hat{v}^2$ are the reconstructed average amplitude and reconstructed AOA, respectively; σ_1 and σ_2 are the variance of the average amplitude and estimated AOA, respectively; η_1 and η_2 are the parameters of the variance of the average amplitude and estimated AOA, respectively; and ρ is the ratio for the bimodal data.

For the (10.17), the average amplitudes $\hat{v}^1$ and the estimated AOAs $\hat{v}^2$ are as the input of deep network, where the different nodes of the input can express the different CSI channels. Then, by employing the test data $\hat{v}^1$ and $\hat{v}^2$, we compute the reconstructed average amplitude $\hat{v}^1$ and reconstructed AOA $\hat{v}^2$ based on databases 1 and 2, respectively, which is used to compute the likelihood function $\Pr(v^1, v^2|l_i)$.

The location of the mobile device can be finally estimated as a weighted average of all the reference locations, which is given by

$$\hat{l} = \sum_{i=1}^{N}\Pr(l_i|v^1, v^2)\cdot l_i. \tag{10.18}$$

10.5 Experimental study

10.5.1 Test configuration

We present our experimental study with BiLoc in the 5 GHz band in this section. In the experiments, we use a desktop computer as an access point and a Dell laptop as a mobile device, both equipped with an Intel 5300 NIC. In fact, we use the desktop computer instead of the commodity routers that are not equipped with the Intel 5300 NIC nowadays. Our implementation of BiLoc is executed on the Ubuntu desktop 14.04 LTS OS for both the access point and mobile device. We use Quadrature Phase Shift Keying (QPSK) modulation and a 1/2 coding rate for the OFDM system. For the access point, it is set in monitor model, and the distance between two adjacent antennas is $d = 2.68$ cm, which is half of a wavelength for the 5 GHz band. For the mobile device, it transmits packets at 100 packets per second using only one antenna in injection mode. By using packet-injection technique based on LORCON version 1, 5 GHz CSI data can be obtained. Then, we extract bimodal data for training and test stages as described in Section 10.4.2.

We also implement three representative schemes from the literature, i.e., Horus [11], FIFS [15], and DeepFi [16]. For a fair comparison, all the schemes use the same measured dataset captured in the 5 GHz band to estimate the location of the mobile device. We conduct extensive experiments with the schemes in the following two representative indoor environments:

Computer laboratory: This is a $6 \times 9\,\text{m}^2$ computer laboratory, a cluttered environment with metal tables, chairs, and desktop computers, blocking most of the LOS paths. The floor plan is shown in Figure 10.10, with 15 chosen training positions (marked as red squares) and 15 chosen test positions (marked as green dots). The distance

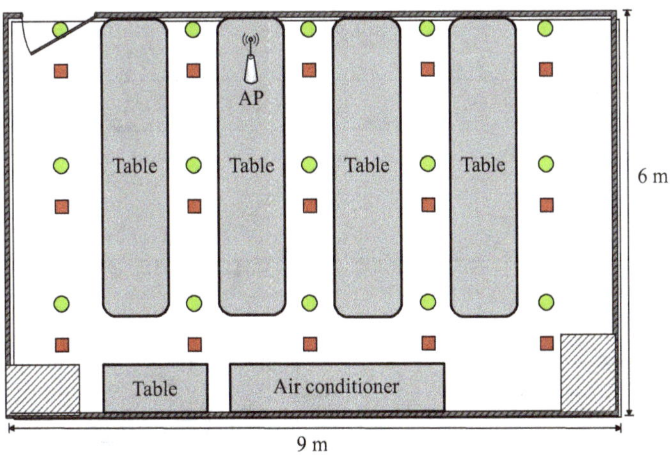

Figure 10.10 Layout of the computer laboratory: training positions are marked as red squares and testing positions are marked as green dots

between two adjacent training positions is 1.8 m. The single access point is put close to the center of the room. We collect bimodal data from 1,000 packet receptions for each training position, and from 25 packet receptions for each test position. The deep network used for this scenario is configured as $\{K_1 = 150, K_2 = 100, K_3 = 50\}$. Also, the ratio ρ for the bimodal data is set as 0.5.

Corridor: This is a 2.4×24 m^2 corridor, as shown in Figure 10.11. In this scenario, the AP is placed at one end of the corridor, and there are plenty of LOS paths. Ten training positions (red squares) and ten test positions (green dots) are arranged along a straight line. The distance between two adjacent training positions is also 1.8 m. We also collect bimodal data from 1,000 packets for each training position and from 25 packets for each test position. The deep network used for this scenario is configured as $\{K_1 = 150, K_2 = 100, K_3 = 50\}$. Also, the ratio ρ for the bimodal data is set as 0.1.

10.5.2 Accuracy of location estimation

Tables 10.1 and 10.2 present the mean and STD of localization errors, and the execution time of the four schemes for the two scenarios, respectively. In the laboratory

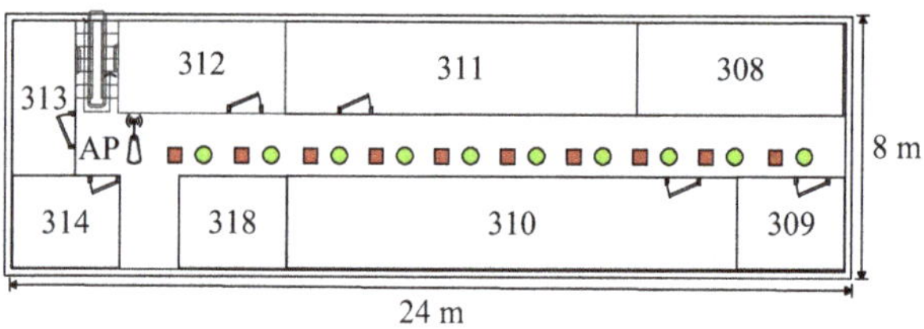

Figure 10.11 Layout of the corridor: training positions are marked as red squares and testing positions are marked as green dots

Table 10.1 Mean/STD error and execution time of the laboratory experiment

Algorithm	**Mean error (m)**	**Std. dev. (m)**	**Mean execution time (s)**
BiLoc	1.5743	0.8312	0.6653
DeepFi	2.0411	1.3804	0.3340
FIFS	2.7151	1.0805	0.2918
Horus	3.0537	1.0623	0.2849

Table 10.2 Mean/STD errors and execution time of the corridor experiment

Algorithm	**Mean error (m)**	**Std. dev. (m)**	**Mean execution time (s)**
BiLoc	2.1501	1.5420	0.5440
DeepFi	2.8953	2.5665	0.3707
FIFS	4.4296	3.4256	0.2535
Horus	4.8000	3.5242	0.2505

environment, BiLoc achieves a mean error of 1.5743 m and an STD error of 0.8312 m across the 15 test points. In the corridor experiment, because only one access point is used for this larger space, BiLoc achieves a mean error of 2.1501 m and an STD error of 1.5420 m across the ten test points. BiLoc outperforms the other three benchmark schemes with the smallest mean error, as well as with the smallest STD error, i.e., being the most stable scheme in both scenarios. We also compare the onlinc tcst time of all the schemes. Due to the use of bimodal data and the deep network, the mean executing time of BiLoc is the highest among the four schemes. However, the mean execution time is 0.6653 s for the laboratory case and 0.5440 s for the corridor case, which are sufficient for most indoor localization applications.

Figure 10.12 presents the CDF of distance errors of the four schemes in the laboratory environment. In this complex propagation environment, BiLoc has 100% of the test positions with an error under 2.8 m, while DeepFi, FIFS, and Horus have about 72%, 52%, and 45% of the test positions with an error under 2.8 m, respectively. For a much smaller error of 1.5 m, the percentage of test positions having a smaller error are 60%, 45%, 15%, and 5% for BiLoc, DeepFi, FIFS, and Horus, respectively. BiLoc achieves the highest precision among the four schemes, due to the use of bimodal CSI data (i.e., average amplitudes and estimated AOAs). In fact, when the amplitude of a signal is strongly influenced in the laboratory environment, the estimated AOA can be utilized to mitigate this effect by BiLoc. However, the other schemes-based solely on CSI or RSS amplitudes will be affected.

Figure 10.13 presents the CDF of distance errors of the four schemes for the corridor scenario. Only one access point is used at one end for this 24 m long corridor, making it hard to estimate the location of the mobile device. For BiLoc, more than 90% of the test positions have an error under 4 m, while DeepFi, FIFS, and Horus have about 70%, 60%, and 50% of the test positions with an error under 4 m, respectively. For a tighter 2 m error threshold, BiLoc has 60% of the test positions with an error below this threshold, while it is 40% for the other three schemes. For the corridor

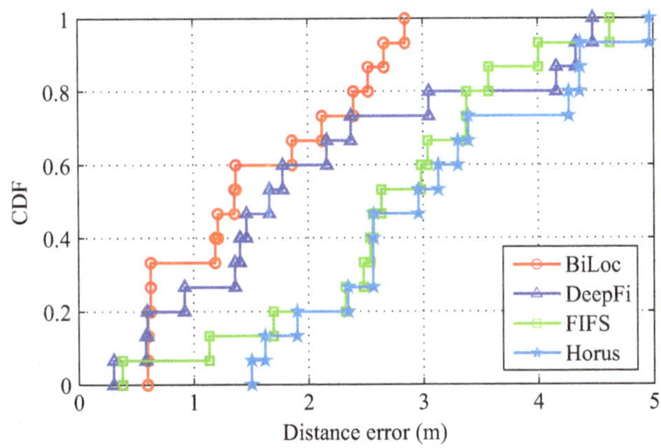

Figure 10.12 CDF of localization errors in 5 GHz for the laboratory experiment

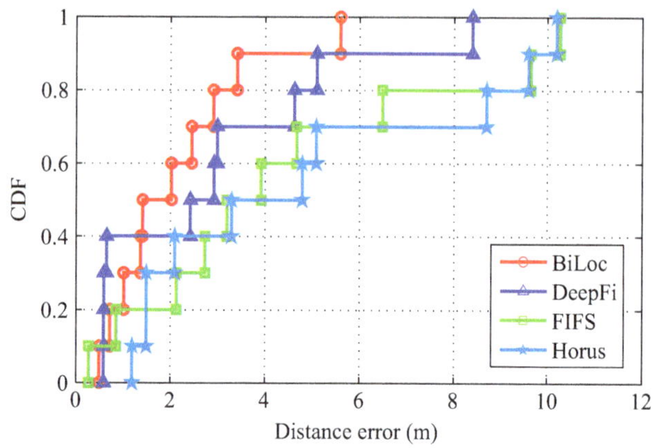

Figure 10.13 CDF of localization errors in 5 GHz for the corridor experiment

scenario, BiLoc mainly utilizes the average amplitudes of CSI data, because the estimated AOAs are similar for all the training/test positions (recall that they are aligned along a straight line with the access point at one end). This is a challenging scenario for differentiating different test points and the BiLoc mean error is 0.5758 m higher than that of the laboratory scenario.

10.5.3 2.4 versus 5 GHz

We also compare the 2.4 and 5 GHz channels with the BiLoc scheme. For a fair comparison, we conduct the experiments at night, because the 2.4 GHz band is much more crowded than the 5 GHz band during the day.

Figure 10.14 presents the CDF of localization errors in the 2.4 and 5 GHz band in the laboratory environment, where both average amplitudes and estimated AOAs are effectively used by BiLoc for indoor localization. We can see that for BiLoc, about 70% of the test positions have an error under 2 m in 5 GHz, while 50% of the test positions have an error under 2 m in 2.4 GHz. In addition, the maximum errors in 2.4 and 5 GHz are 6.4 and 2.8 m, respectively. Therefore, the proposed BiLoc scheme achieves much better performance in 5 than 2.4 GHz. In fact, the phase difference between two antennas in 2.4 GHz exhibits great variations, which lead to lower localization accuracy. This experiment also validates our Hypothesis 2.

10.5.4 Impact of parameter ρ

Recall that the parameter ρ is used to trade-off the impacts of average amplitudes and estimated AOAs in location estimation as in (10.17). We examine the impact of ρ on localization accuracy under the two environments. With BiLoc, we use bimodal data for online testing, and ρ directly influences the likelihood probability $\Pr(v^1, v^2|l_i)$ (10.17), which in turn influences the localization accuracy.

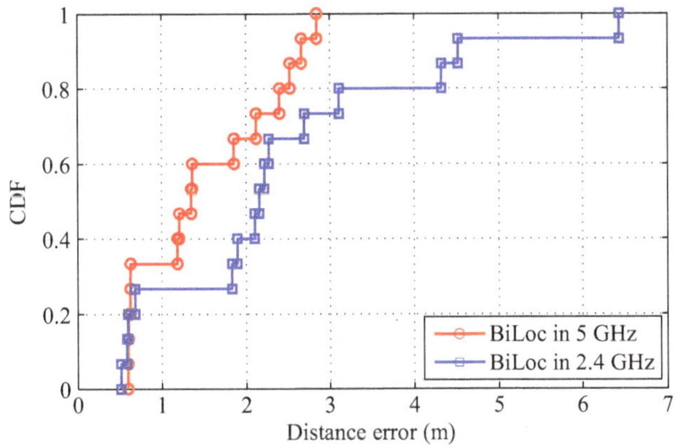

Figure 10.14 CDF of localization errors in 5 and 2.4 GHz for the laboratory experiment

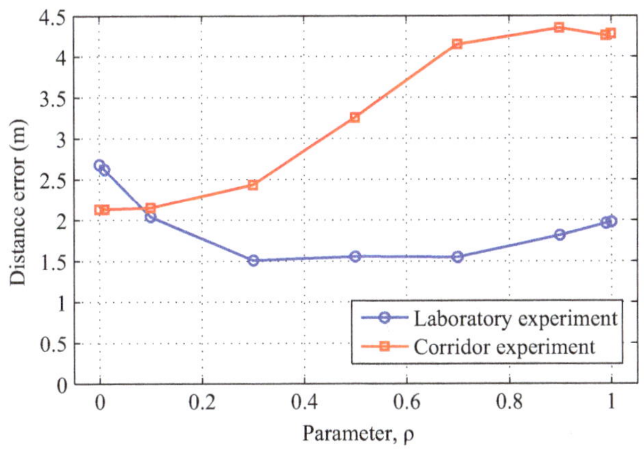

Figure 10.15 Mean localization errors versus parameter, ρ, for the laboratory and corridor experiments

Figure 10.15 presents the mean localization errors for increasing ρ for the laboratory and corridor experiments. In the laboratory experiment, when ρ is increased from 0 to 0.3, the mean error decreases from 2.6 to 1.5 m. Furthermore, the mean error remains around 1.5 m for $\rho \in [0.3, 0.7]$, and then increases from 1.5 to 2 m when ρ is increased from 0.6 to 1. Therefore, BiLoc achieves its minimum mean error for $\rho \in [0.3, 0.7]$, indicating that both average amplitudes and estimated AOAs are useful for accurate location estimation. Moreover, BiLoc has higher localization accuracy with the mean error of 1.5 m, compared with individual modality such as the estimated AOAs with that of 2.6 m or the average amplitudes with that of 2.0 m.

In the corridor experiment, we can see that the mean error remains around 2.1 m when ρ is increased from 0 to 0.1. When ρ is further increased from 0.1 to 1, the mean error keeps on increasing from 2.1 to about 4.3 m. Clearly, in the corridor experiment, the estimated AOAs provide similar characteristics for deep learning and are not useful for distinguishing the positions. Therefore, BiLoc should mainly use the average amplitudes of CSI data for better accuracy. These experiments provide some useful guidelines on setting the ρ value for different indoor environments.

10.6 Future directions and challenges

10.6.1 New deep-learning methods for indoor localization

This chapter has discussed three deep-learning technologies including autoencoder, CNN, LSTM for fingerprinting-based indoor localization. With the rapid growth in the AI field, new deep-learning approaches are proposed for mainly handling computer vision problems, such as robust object recognition and detection, data generation, as well as the Go game. For example, generative adversarial network (GAN) can be used for generating new data samples; deep reinforcement learning has been leveraged for AlphaGo; deep Gaussian process can be utilized for improving the robustness of object detection. In fact, these new deep-learning methods can be also used for solving basic indoor localization problems such as radio map constructions, environment change, and devices calibration. For example, deep reinforcement learning [51] can be used for improving localization performance and reduce cost. Moreover, Bayesian deep learning such as deep Gaussian process [52,53] has high robustness for environment noise, which can be exploited for radio map construction, and mitigating environment changes and devices calibration. Moreover, GAN can be incorporated for building radio map and increasing the number of training data samples. In addition, compressed deep learning [54] by using pruning and quantization can be considered for resource limited mobile devices. Thus, we can implement deep-learning models on smartphones in addition to servers for indoor localization.

10.6.2 Sensor fusion for indoor localization using deep learning

In this chapter, we have proposed bimodal CSI data for indoor localization. In fact, multiple sensor data sources can be fused for improving indoor localization performance. Traditionally, sequence models such as Kalman filter, particle filter, hidden Markov model, and conditional random field can fuse Wi-Fi and inertial sensor data on smartphones for indoor localization, which requires for obtaining the sequence data from continuing smartphone movement. Deep-learning techniques can improve the performance of indoor localization using multimodal sequence data. For example, LSTM method can be leveraged for indoor localization using sequence RSS or CSI data, which also fuse multimodal data for improving the localization accuracy. Considering Wi-Fi and magnetic sensor data from smartphone, we can integrate them into a large data matrix as input to LSTM for indoor localization. In fact, Wi-Fi and magnetic sensor data are complementary to each other. For example, because of lower resolution of Wi-Fi signals, only using Wi-Fi RSS values cannot obtain better performance at close locations, while magnetic sensor data at such positions is greatly different. LSTM can effectively fuse them for indoor localization [43]. In addition, an integrated CNN and LSTM model can be used for Wi-Fi RSS or CSI images data, which can be easily created from different access points or different subcarriers. In fact, the LSTM model can be combined with other deep-learning models such as autoencoder, GAN, deep reinforcement learning, Bayesian model for different localization problems such as radio map construction, device calibration, and environment change. For sensor data fusion for indoor localization, different sensor data sources should be normalized and aligned [23].

10.6.3 Secure indoor localization using deep learning

For wireless-fingerprinting-based indoor localization, security becomes increasingly important, where wireless signals are susceptible to eavesdropping, distributed denial-of-service (DDoS) attacks, and bad data injection [55]. Specially, for crowd-sourcing-based indoor localization, fingerprints are from different devices at different times, which greatly exposes the security problem. For attacker models, there are three general scenarios for RSS fingerprinting-based localization [56]. First, the attacker does not know the true RSS fingerprints and injects fake RSS data at random. Second, the attacker knows legitimate RSS fingerprints and add noise to them. Third, the attacker can change the mapping between RSS fingerprints and positions. For defense models, they can consider the temporal correlation and spatial correlation within RSS Traces against different attackers. In fact, deep learning can study the feature of Localization signals to address the above security problems. Deep learning can consider different data features from multiple paths of wireless signals to classify eavesdropping, DDoS attack, or bad data injection for fingerprinting-based indoor localization.

On the other hand, deep-learning security is also an important problem, which mainly focuses on how to recognize adversarial data and clean RSS data. Deep learning will have bad performance with adversarial data, which can be obtained by adding small noise into clear RSS data. Thus, adversarial data should be recognized before implementing indoor localization systems based on deep learning, thus guaranteeing good localization performance. In addition, privacy persevering deep learning can be used for indoor localization problems, which can protect user location privacy information.

10.7 Conclusions

In this chapter, we proposed a bimodal deep-learning system for fingerprinting-based indoor localization with 5 GHz commodity Wi-Fi NICs. First, the state-of-the-art deep-learning techniques including deep autoencoder network, CNN, and LSTM were introduced. We then extracted and calibrated CSI data to obtain bimodal CSI data, including average amplitudes and estimated AOAs, which were used in both the off-line and online stages. The proposed scheme was validated with extensive experiments. We concluded this chapter with a discussion of future directions and challenges for indoor localization problems using deep learning.

Acknowledgments

This work is supported in part by the US NSF under Grants ACI-1642133 and CNS-1702957, and by the Wireless Engineering Research and Education Center (WEREC) at Auburn University.

References

[1] Wang Y, Liu J, Chen Y, *et al.* E-eyes: Device-free location-oriented activity identification using fine-grained WiFi signatures. In: Proc. ACM Mobicom'14. Maui, HI; 2014. p. 617–628.

[2] Zhang D, Zhao S, Yang LT, *et al.* NextMe: Localization using cellular traces in internet of things. IEEE Transactions on Industrial Informatics. 2015;11(2):302–312.

[3] Derr K, and Manic M. Wireless sensor networks node localization for various industry problems. IEEE Transactions on Industrial Informatics. 2015;11(3): 752–762.

[4] Abu-Mahfouz A, and Hancke GP. Distance bounding: A practical security solution for real-time location systems. IEEE Transactions on Industrial Informatics. 2013;9(1):16–27.

[5] Pak J, Ahn C, Shmaliy Y, *et al.* Improving reliability of particle filter-based localization in wireless sensor networks via hybrid particle/FIR filtering. IEEE Transactions on Industrial Informatics. 2015;11(5):1089–1098.

[6] Ivanov S, Nett E. Localization-based radio model calibration for fault-tolerant wireless mesh networks. IEEE Transactions on Industrial Informatics. 2013;9(1):246–253.

[7] Lee S, Kim B, Kim H, *et al.* Inertial sensor-based indoor pedestrian localization with minimum 802.15.4a configuration. IEEE Transactions on Industrial Informatics. 2011;7(3):455–466.

[8] Wu B, and Jen C. Particle filter based radio localization for mobile robots in the environments with low-density WLAN APs. IEEE Transactions on Industrial Electronics. 2014;61(12):6860–6870.

[9] Liu H, Darabi H, Banerjee P, *et al.* Survey of wireless indoor positioning techniques and systems. IEEE Transactions on Systems, Man, and Cybernetics, Part C. 2007;37(6):1067–1080.

[10] Bahl P, Padmanabhan VN. Radar: An in-building RF-based user location and tracking system. In: Proc. IEEE INFOCOM'00. Tel Aviv, Israel; 2000. p. 775–784.

[11] Youssef M, and Agrawala A. The Horus WLAN location determination system. In: Proc. ACM MobiSys'05. Seattle, WA; 2005. p. 205–218.

[12] Halperin D, Hu WJ, Sheth A, *et al.* Predictable 802.11 packet delivery from wireless channel measurements. In: Proc. ACM SIGCOMM'10. New Delhi, India; 2010. p. 159–170.

[13] Sen S, Lee J, Kim KH, and Congdon P. Avoiding multipath to revive inbuilding WiFi localization. In: Proc. ACM MobiSys'13. Taipei, Taiwan; 2013. p. 249–262.

[14] Xie Y, Li Z, and Li M. Precise power delay profiling with commodity WiFi. In: Proc. ACM Mobicom'15. Paris, France; 2015. p. 53–64.

[15] Xiao J, Wu K, Yi Y, *et al.* FIFS: Fine-grained indoor fingerprinting system. In: Proc. IEEE ICCCN'12. Munich, Germany; 2012. p. 1–7.

[16] Wang X, Gao L, Mao S, *et al.* DeepFi: Deep learning for indoor fingerprinting using channel state information. In: Proc. WCNC'15. New Orleans, LA; 2015. p. 1666–1671.

[17] Wu C, Yang Z, Zhou Z, *et al.* PhaseU: Real-time LOS identification with WiFi. In: Proc. IEEE INFOCOM'15. Hong Kong, China; 2015. p. 2038–2046.

[18] Qian K, Wu C, Yang Z, *et al.* PADS: Passive detection of moving targets with dynamic speed using PHY layer information. In: Proc. IEEE ICPADS'14. Hsinchu, Taiwan; 2014. p. 1–8.

[19] Gjengset J, Xiong J, McPhillips G, *et al.* Phaser: Enabling phased array signal processing on commodity WiFi access points. In: Proc. ACM Mobicom'14. Maui, HI; 2014. p. 153–164.

[20] Wang X, Gao L, and Mao S. BiLoc: Bi-modal deep learning for indoor localization with commodity 5 GHz WiFi. IEEE Access. 2017;5:4209–4220.

[21] Abadi M, Barham P, Chen J, *et al.* Tensorflow: A system for large-scale machine learning. In: OSDI. vol. 16; 2016. p. 265–283.

[22] Mohammadi M, Al-Fuqaha A, Sorour S, *et al.* Deep learning for IoT big data and streaming analytics: A survey. IEEE Communications Surveys & Tutorials. 2018;20(4):2923–2960.

[23] Wang X, Wang X, and Mao S. RF sensing in the Internet of Things: A general deep learning framework. IEEE Communications Magazine. 2018;56(9): 62–67.

[24] Hinton GE, and Salakhutdinov RR. Reducing the dimensionality of data with neural networks. Science. 2006;313(5786):504–507.

[25] Wang X, Gao L, Mao S, *et al.* CSI-based fingerprinting for indoor localization: A deep learning approach. IEEE Transactions on Vehicular Technology. 2017;66(1):763–776.

[26] Wang X, Gao L, and Mao S. PhaseFi: Phase fingerprinting for indoor localization with a deep learning approach. In: Proc. GLOBECOM'15. San Diego, CA; 2015.

[27] Wang X, Gao L, and Mao S. CSI phase fingerprinting for indoor localization with a deep learning approach. IEEE Internet of Things Journal. 2016;3(6): 1113–1123.

[28] Chen X, Ma C, Allegue M, *et al.* Taming the inconsistency of Wi-Fi fingerprints for device-free passive indoor localization. In: INFOCOM 2017-IEEE Conference on Computer Communications, IEEE. IEEE; 2017. p. 1–9.

[29] Wang J, Zhang X, Gao Q, *et al.* Device-free wireless localization and activity recognition: A deep learning approach. IEEE Transactions on Vehicular Technology. 2017;66(7):6258–6267.

[30] Xiao C, Yang D, Chen Z, *et al.* 3-D BLE indoor localization based on denoising autoencoder. IEEE Access. 2017;5:12751–12760.

[31] Krizhevsky A, Sutskever I, and Hinton GE. ImageNet classification with deep convolutional neural networks. In: Advances in Neural Information Processing Systems; 2012. p. 1097–1105.

[32] LeCun Y, Bottou L, Bengio Y, *et al.* Gradient-based learning applied to document recognition. Proceedings of the IEEE. 1998;86(11):2278–2324.

[33] Wang X, Wang X, and Mao S. CiFi: Deep convolutional neural networks for indoor localization with 5 GHz Wi-Fi. In: Proc. IEEE ICC 2017. Paris, France; 2017. p. 1–6.

[34] He K, Zhang X, Ren S, *et al.* Deep residual learning for image recognition. In: Proceedings of the IEEE Conference on Computer Vision and Pattern Recognition; 2016. p. 770–778.

[35] Nair V, and Hinton GE. Rectified linear units improve restricted Boltzmann machines. In: Proceedings of the 27th International Conference on Machine Learning (ICML-10); 2010. p. 807–814.

[36] Wang W, Wang X, and Mao S. Deep convolutional neural networks for indoor localization with CSI images. IEEE Transactions on Network Science and Engineering; 2018; Early Access.

[37] Wang X, Wang X, and Mao S. ResLoc: Deep residual sharing learning for indoor localization with CSI tensors. In: Proc. IEEE PIMRC 2017. Montreal, Canada; 2017.

[38] Chen H, Zhang Y, Li W, *et al.* ConFi: Convolutional neural networks based indoor WiFi localization using channel state information. IEEE Access. 2017;5:18066–18074.

[39] Mittal A, Tiku S, and Pasricha S. Adapting convolutional neural networks for indoor localization with smart mobile devices. In: Proceedings of the 2018 on Great Lakes Symposium on VLSI. ACM; 2018. p. 117–122.

[40] Zhang T, and Yi M. The enhancement of WiFi fingerprint positioning using convolutional neural network. In: DEStech Transactions on Computer Science and Engineering (CCNT); 2018.

[41] Niitsoo A, Edelhäußer T, and Mutschler C. Convolutional neural networks for position estimation in TDoA-based locating systems. In: Proc. 9th Intl. Conf. Indoor Positioning and Indoor Navigation, (Nantes, France); 2018. p. 1–8.

[42] Gers FA, Schmidhuber J, and Cummins F. Learning to forget: Continual prediction with LSTM. Neural Computation. 2000;12(10):2451–2471.

[43] Wang X, Yu Z, and Mao S. DeepML: Deep LSTM for indoor localization with smartphone magnetic and light sensors. In: Proc. IEEE ICC 2017. Kansas City, MO; 2018.

[44] Yang W, Wang X, Cao S, *et al.* Multi-class wheat moisture detection with 5 GHz Wi-Fi: A deep LSTM approach. In: Proc. ICCCN 2018. Hangzhou, China; 2018.

[45] Wang Y, Shen Y, Mao S, *et al.* LASSO & LSTM integrated temporal model for short-term solar intensity forecasting. IEEE Internet of Things Journal. In press.

[46] Akbar MB, Taylor DG, and Durgin GD. Amplitude and phase difference estimation bounds for multisensor based tracking of RFID tags. In: Proc. IEEE RFID'15. San Diego, CA; 2015. p. 105–112.

[47] Kleisouris K, Chen Y, Yang J, *et al.* The impact of using multiple antennas on wireless localization. In: Proc. IEEE SECON'08. San Francisco, CA; 2008. p. 55–63.

[48] Speth M, Fechtel S, Fock G, *et al.* Optimum receiver design for wireless broadband systems using OFDM—Part I. IEEE Transactions on Communications. 1999;47(11):1668–1677.

[49] Schmidt R. Multiple emitter location and signal parameter estimation. IEEE Transactions on Antennas and Propagation. 1986;34(3):276–280.

[50] Bengio Y, Lamblin P, Popovici D, *et al.* Greedy layer-wise training of deep networks. In: Proc. Adv. Neural Inform. Proc. Syst. 19. Vancouver, Canada; 2007. p. 153–160.

[51] Mnih V, Kavukcuoglu K, Silver D, *et al.* Human-level control through deep reinforcement learning. Nature. 2015;518(7540):529.

[52] Shi J, Chen J, Zhu J, *et al.* ZhuSuan: A library for Bayesian deep learning. arXiv preprint arXiv:170905870. 2017.

[53] Wang X, Wang X, Mao S, *et al.* DeepMap: Deep Gaussian process for indoor radio map construction and location estimation. In: Proc. IEEE GLOBECOM 2018. Abu Dhabi, United Arab Emirates; 2018.

[54] Han S, Liu X, Mao H, *et al.* EIE: Efficient inference engine on compressed deep neural network. In: International Conference on Computer Architecture (ISCA); 2016.

[55] Shokri R, Theodorakopoulos G, Troncoso C, *et al.* Protecting location privacy: optimal strategy against localization attacks. In: Proceedings of the 2012 ACM conference on Computer and communications security. ACM; 2012. p. 617–627.

[56] Li T, Chen Y, Zhang R, *et al.* Secure crowd-sourced indoor positioning systems. In: IEEE INFOCOM'18; 2018.

Chapter 11
Reinforcement-learning-based wireless resource allocation

Rui Wang[1]

In wireless systems, radio resource management (RRM) is a necessary approach to improve the transmission efficiency. For example, the base stations (BSs) of cellular networks can optimize the selection of downlink receiving users in each frame and the transmission power for them according to channel state information (CSI), such that the total throughput is maximized. The RRM of various kinds are usually represented by mathematical optimization problems, with an objective to be optimized (e.g., throughput or delay) and some constraints on limited resources (e.g., transmission time, frequency or power). The mathematical modeling procedure from the needs of resource allocation to optimization problems is usually referred to as *"problem formulation."*

In this chapter, we shall focus on the formulation of RRM via Markov decision process (MDP). Convex optimization has been widely used in the RRM within a short-time duration, where the wireless channel is assumed to be quasi-static. These problems are usually referred to as *deterministic optimization* problems. On the other hand, MDP is an elegant and powerful tool to handle the resource optimization of wireless systems in a longer timescale, where the random transitions of system and channel status are considered. These problems are usually referred to as *stochastic optimization* problems. Particularly, MDP is suitable for the joint optimization between physical and media-access control (MAC) layers. Based on MDP, reinforcement learning is a practical method to address the optimization without *a priori* knowledge of system statistics. In this chapter, we shall first introduce some basics on stochastic approximation, which serves as one basis of reinforcement learning, and then demonstrate the MDP formulations of RRM via some case studies, which require the knowledge of system statistics. Finally, some approaches of reinforcement learning (e.g., *Q*-learning) are introduced to address the practical issue of unknown system statistics.

11.1 Basics of stochastic approximation

Stochastic approximation is a general iterative method to solve some stochastic fixed-point problems or optimization problems without the knowledge of statistics in the

[1] Department of Electrical and Electronic Engineering, The Southern University of Science and Technology, China

problem. Mathematically, these kinds of problems are not well defined. They may refer to some systems with unknown random behavior. For example, the wireless transmitter wants to make sure that the average receiving signal-to-interference-plus-ratio (SINR) is above certain quality level; however, the interference level at the receiver is hard to predict without its statistics. Clearly, this problem cannot be solved unless more information can be collected. In the transmission protocol design, the receiver can estimate the receiving interference level and report it to the transmitter periodically, so that the transmitter can adjust its power and guarantee an acceptable average SINR level. Hence, the procedure of problem-solving includes not only calculation but also system observation. Stochastic approximation is such an online learning and adapting procedure, which collects the information from each observation and finally converges to the solution.

11.1.1 Iterative algorithm

In this section, we use one example of deterministic fixed-point problem to demonstrate the structure of iterative solution, which is widely used in solving the problems without close-form solutions. The exemplary fixed-point problem is provided below.

Problem 11.1 (Deterministic fixed-point problem). *Find an x such that*

$$f(x) = 0, \tag{11.1}$$

where $f(x)$ is a monotonically increasing function.

Providing the expression of $f(x)$, this problem may be solved analytically. For example, $x = \log_{10} a$ when $f(x) = 10^x - a$ and a is a positive constant. Nevertheless, the following iterative algorithm is useful when the explicit expression of x cannot be derived.

Iterative algorithm for Problem 11.1

Let n be the index of iteration and x_n be the median value of x in the nth iteration, the solution of (11.1) can be achieved as follows:

1. Initialize the iteration index n as $n = 0$, and the value of x as x_0.
2. Update the median value of x as

$$x_{n+1} = x_n - \gamma_{n+1} f(x_n), \tag{11.2}$$

   where γ_n is the step size of iteration. Let $n = n + 1$.
3. Let ε be a threshold for terminating the iteration. The algorithm stops if $|x_{n+1} - x_n| < \varepsilon$ or goes to Step 2 otherwise.

Strictly speaking, the solution for Problem 11.1 may not be unique; the above algorithm is to find one feasible solution if it exists. There are a number of choices

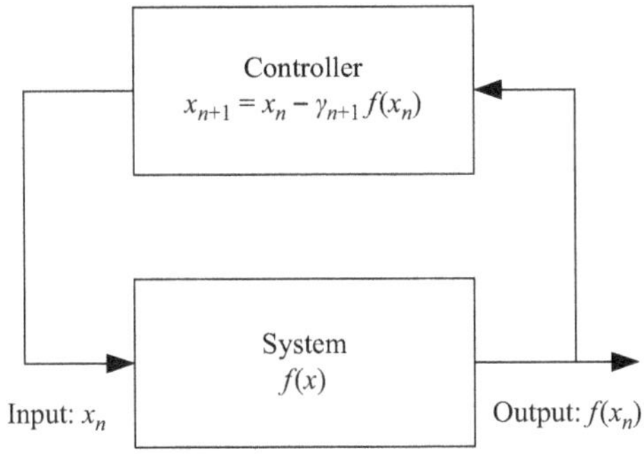

Figure 11.1 Block diagram for the iterative algorithm of Problem 11.1

on the step size $\{\gamma_n | \forall n = 1, 2, \ldots\}$. For the Newton's method (also known as the Newton–Raphson method), the step size is

$$\gamma_n = \frac{1}{f'(x_n)},$$

where $f'(x_n)$ is the first-order derivative of $f(x)$ at $x = x_n$. For the case that $f'(x_n)$ cannot be obtained, one more general choice of step size is the harmonic series

$$\gamma_n = \frac{1}{n}.$$

An intuitive explanation on using the harmonic series as iteration step size is provided below:

- Note that this series is monotonically decreasing. When x_n is close to the solution, smaller step size is better for fine adjustment.
- Note that $\sum_{n=1}^{+\infty}(1/n) = +\infty$, the incremental update of $-\gamma_{n+1} f(x)$ is not negligible as long as $f(x) \neq 0$. Hence the algorithm could drive x_n to the solution of $f(x) = 0$.

A block diagram of the iterative algorithm is illustrated in Figure 11.1, where $f(x)$ may be an observation of certain system with input x. A controller, with the objective of $f(x) = 0$, collects the observation of $f(x)$ and update the value of x.

11.1.2 Stochastic fixed-point problem

The problem introduced in the previous section assumes that the function $f(x)$ can be accurately estimated when an input value of x is provided. However, this may not be the case in some applications. For example, suppose that Y is a random variable whose distribution is unknown, and we want to find an appropriate x to satisfy the following equation:

$$\mathbb{E}[x - Y] = 0. \tag{11.3}$$

This is not a well-defined problem if there is no further information of Y, and its solution can only be obtained by learning its statistics. This section will provide a general iterative solution for such stochastic fixed-point problem, which is referred to as stochastic approximation.

Consider a general expression of stochastic fixed-point problem. Let $f(x, Y)$ be a function of variable x and random variable Y. $f(x, Y)$ can be treated as the output of a system, which depends on the input parameter x and random internal statue Y. In some applications, people may be interested in the expected output of the system. Hence, it is defined that

$$\bar{f}(x) = \mathbb{E}f(x, Y) = \int f(x, y)p(y)dy, \tag{11.4}$$

where $p(y)$ is the PDF of random variable Y. The fixed-point problem to be solved becomes.

Problem 11.2 (stochastic fixed-point problem). *Find x such that*

$$\bar{f}(x) = 0, \tag{11.5}$$

where $\bar{f}(x)$ and its realization $f(x, Y)$ satisfy the following conditions:

- *Suppose $\bar{f}(\theta) = 0$. Given x, there exists a finite positive constant δ such that*

$$\bar{f}(x) \leq -\delta \text{ for } x < \theta, \qquad \bar{f}(x) \geq \delta \text{ for } x > \theta. \tag{11.6}$$

- *There exists a finite constant c such that*

$$\Pr[|f(x, Y)| \leq c] = 1. \tag{11.7}$$

The condition (11.6) guarantees that the value of $\bar{f}(x)$ can be used to update the variable x: x should be decreased when $\bar{f}(x)$ is positive, and vice versa. The condition (11.7) assures that each realization $f(x, Y)$ can be adopted to evaluate its expectation $\bar{f}(x)$. According to the method of stochastic approximation, the solution of Problem 11.2 is described below.

Iterative algorithm for Problem 11.2

Let x_n be the nth input value of x, and Y_n be the nth realization of random variable Y. The solution of (11.5) can be achieved as follows:

1. Initialize the iteration index n as $n = 0$, and the value of x as x_0.
2. Update the input value of x as

$$x_{n+1} = x_n - \gamma_{n+1} f(x_n, Y_n), \tag{11.8}$$

   where γ_n is the step size of iteration. Let $n = n + 1$.
3. Let ε be a threshold for terminating the iteration. The algorithm stops if $|x_{n+1} - x_n| < \varepsilon$ or goes to Step 2 otherwise.

The convergence of the above algorithm is summarized below.

Theorem 11.1. *Let $e_n = \mathbb{E}(x_n - \theta)^2$. If $\gamma_n = 1/n$, the conditions of (11.6) and (11.7) are satisfied, then*

$$\lim_{n \to +\infty} e_n = 0.$$

Please refer to the Theorem 11.1 of [1] for the rigorous proof. Some insights on the convergence property are provided here. In fact, $f(x_n, Y_n)$ can be treated as one estimation of $\bar{f}(x_n)$ with estimation error Z_n. Thus:

$$f(x_n, Y_n) = \bar{f}(x_n) + Z_n, \quad \forall n. \tag{11.9}$$

Since $\mathbb{E} f(x_n, Y_n) = \bar{f}(x_n)$, we know that $\mathbb{E}[Z_n] = 0$. From the iterative update equation (11.8), it can be derived that

$$\begin{aligned}
x_1 &= x_0 - \gamma_1 f(x_0, Y_0) = x_0 - [\bar{f}(x_0) + Z_0] \\
x_2 &= x_1 - \gamma_2 f(x_1, Y_1) = x_0 - [\bar{f}(x_0) + Z_0] - \frac{1}{2}[\bar{f}(x_1) + Z_1] \\
&\ldots \\
x_n &= x_{n-1} - \gamma_n f(x_{n-1}, Y_{n-1}) = x_0 - \sum_{i=0}^{n-1} \frac{1}{i+1} \bar{f}(x_i) - \sum_{i=0}^{n-1} \frac{1}{i+1} Z_i \\
&= \underbrace{x_0 - \sum_{i=0}^{k-1} \frac{1}{i+1} [Z_i + \bar{f}(x_i)]}_{x_k} - \sum_{i=k}^{n-1} \frac{1}{i+1} \bar{f}(x_i) - \sum_{i=k}^{n-1} \frac{1}{i+1} Z_i.
\end{aligned} \tag{11.10}$$

Hence, the convergence of $\{x_n\}$ is discussed as follows:

1. The last term of (11.10), $\sum_{i=k}^{n-1} (1/(i+1)) Z_i$, can be treated as the noise of iteration. Because without it, the iteration becomes:

$$x_n = x_k - \sum_{i=k}^{n-1} \frac{1}{i+1} \bar{f}(x_i), \tag{11.11}$$

   which is the solution algorithm for the deterministic fixed point $\bar{f}(x) = 0$. The expectation of the noise is zero, i.e., $\mathbb{E}\left[\sum_{i=k}^{n-1} (1/(i+1)) Z_i\right] = 0$.
2. The variance of noise $\sum_{i=k}^{n-1} (1/(i+1)) Z_i$ are analyzed as follows:

$$Var\left[\sum_{i=k}^{n-1} \frac{1}{i+1} Z_i\right] = \sum_{i=k}^{n-1} \frac{1}{(i+1)^2} Var(Z_i) \leq \sum_{i=k}^{n-1} \frac{1}{(i+1)^2} \sigma_Z^2, \tag{11.12}$$

   where $\sigma_Z^2 = \max_i Var(Z_i)$.
3. $\sum_{i=k}^{\infty} (1/(i+1)^2) \sigma_Z^2$ is finite. Moreover, for arbitrary $\varepsilon > 0$, there exists an integer K such that $\sum_{i=K}^{\infty} (1/(i+1)^2) \sigma_Z^2 < \varepsilon$.

4. Let $n \to \infty$ on both sides of (11.10), we have

$$x_\infty = x_k - \sum_{i=k}^{\infty} \frac{1}{i+1}\bar{f}(x_i) - \sum_{i=k}^{\infty} \frac{1}{i+1} Z_i. \tag{11.13}$$

Note that $\sum_{i=k}^{\infty}(1/(i+1))Z_i$ is generally finite according to the above discussion. If $\bar{f}(x_i)$ is always greater than certain positive value, $\sum_{i=k}^{\infty}(1/(i+1))\bar{f}(x_i)$ will drive x_∞ to negative infinity, which will lead to negative $\bar{f}(x_i)$. Thus, the convergence can be intuitively understood according to the above contradiction.

11.2 Markov decision process: basic theory and applications

In wireless systems, the transmission time and spectrum are usually organized as frames. For example, every 1 ms in the long-term evolution (LTE) system is organized as one subframe, and every ten subframes constitutes a frame. The wireless channel is usually assumed to be quasi-static within one subframe or even one frame. In LTE systems, the CSI can be estimated in every subframe, which is used to decode the current subframe or determine the transmission parameters of the following subframes. The transmission resource allocation in one subframe can be formulated as various optimization problems. For example, one possible problem formulation is to jointly optimize the uplink or downlink transmission time and power among multiple mobile users, such that the overall throughput of the subframe is maximized. In fact, many resource optimization problems share the similar structure, which usually consists of the following:

- **System state:** A number of parameters specifying the system status, which can be estimated and notified to the scheduler, e.g., coefficients of large-scale and small-scale fading.
- **Scheduling action:** A number of transmission or receiving parameters can be adjusted, e.g., transmission time and power.
- **Objective:** A function measuring the utility or cost within a time slot, where the system state is assumed to be quasi-static. For example, the overall throughput in an LTE subframe. The objective is usually a function of system state and scheduling action.
- **Constraints:** A number of limitations on the transmission resources within the aforementioned time slot, e.g., the total resource elements (symbols) for data transmission in an LTE subframe. There might be a number of constraints on different types of transmission resource. Each constraint is usually a function of scheduling action; some of them may also depend on system state.

As a result, the resource allocation problems can be generally formulated by the following optimization problem:

$$\begin{aligned} \max_{\textit{Action}} \ \textit{or} \ \min_{\textit{Action}} \quad & \text{Objective}\,(\textit{state, action}) \\ \text{subject to} \quad & \text{Constraints}(\textit{state, action}) \in \textit{System affordable region}. \end{aligned}$$

These type of problems aim at finding the optimal values of some transmission or receiving parameters for each time slot (i.e., action). It implies that there is a scheduler, who observes the system state in each time slot, solves the above optimization problem, and uses the solution in transmission. In many of such optimization problems, the optimization action in one time slot does not affect that of the followings. Hence, we shall refer to this type of problems as the *"single stage"* optimization in the remaining of this chapter. The following is an example of single stage optimization formulation.

Example 11.1 (Multi-carrier power allocation). *Suppose that there is one point-to-point OFDM link with N_F subcarriers, and their channel gains in one certain time slot are denoted by $\{h_i|i = 1,2,\ldots,N_F\}$. Let p_i $(i = 1,2,\ldots,N_F)$ be the transmission power on the ith subcarrier. One typical power allocation problem is to determine the transmission power on each subcarrier $\{p_i|i = 1,2,\ldots,N_F\}$ such that the overall throughput is maximized, which can be formulated as follows:*

- *System state: $\{h_i|i = 1,2,\ldots,N_F\}$.*
- *Action: $\{p_i|i = 1,2,\ldots,N_F\}$.*
- *Objective: $\sum_{i=1}^{N_F} \log_2(1 + ((p_i\|h_i\|^2)/\sigma_z^2))$.*
- *Constraint: $\sum_{i=1}^{N_F} p_i \leq P$, where P is the peak transmission power.*

Hence, the overall optimization problem can be written as

$$\max_{\{p_i|i=1,2,\ldots,N_F\}} \sum_{i=1}^{N_F} \log_2\left(1 + \frac{p_i\|h_i\|^2}{\sigma_z^2}\right)$$

$$\text{subject to} \sum_{i=1}^{N_F} p_i \leq P,$$

where σ_z^2 is the power of noise. This problem can be solved by the well-known water-filling algorithm. Note that this is the single stage optimization, since there is no connection between the optimization in different time slots.

However, the above single stage formulation is not powerful enough to address all the wireless-resource-allocation problems, especially when the scope of optimization is extended to the MAC layer and larger timescale. From the MAC layer point of view, the BS maintains one queue for each active downlink mobile user. If the BS schedules more transmission resource to one user in certain time slot, the traffic load for this user in the following time slots can be relieved. Hence, the scheduling action in one time slot can affect that of the following ones, and a joint optimization along multiple time slots becomes necessary. The difficulty of such joint optimization is that the wireless channel is time varying. Hence, the scheduler cannot predict the channel of the following time slots and, of course, cannot determine their scheduling parameters in advance. We shall refer to this type of problems as the *"multistage"* optimization in the remaining of this chapter. Due to the uncertainty on future system state, its differences from the single-stage optimization problem are as follows:

- Instead of calculating the values for scheduling action, we should provide a mapping from arbitrarily possible system state to the corresponding scheduling action, so that the system can work properly in all possible situations. This mapping is called *policy*, thus:

$$\textit{Policy} \; : \; \textit{System State} \rightarrow \textit{Scheduling Action}. \tag{11.14}$$

- The expectation should be taken on the objective and constraints (if any) since these functions depend on random system state.

In this section, the MDP is introduced to formulate and solve this kind of multi-stage optimization problem. In order to bring up the basic principle without struggling with the mathematical details, this section is only about the discrete-time MDP with finite state and action spaces, and some mathematical proof is neglected. For the readers who are interested in a overwhelm and rigorous discussion on MDP optimization theory, please refer to [2,3].

11.2.1 Basic components of MDP

As illustrated in Figure 11.2, there are three basic components in an MDP, namely, system state, control or scheduling action, and cost function. Sometimes, people may want to maximize some utility function, this is equivalent to minimize the inverse of utility function, which can be treated as cost function. System state, denoted as $\mathbf{s}$, consists of the set of parameters, which uniquely specify the system at any time instance. The set of all the possible values of system state is called the *state space*, denoted as $\mathscr{S}$. In this chapter, we consider the case that the cardinality of state space

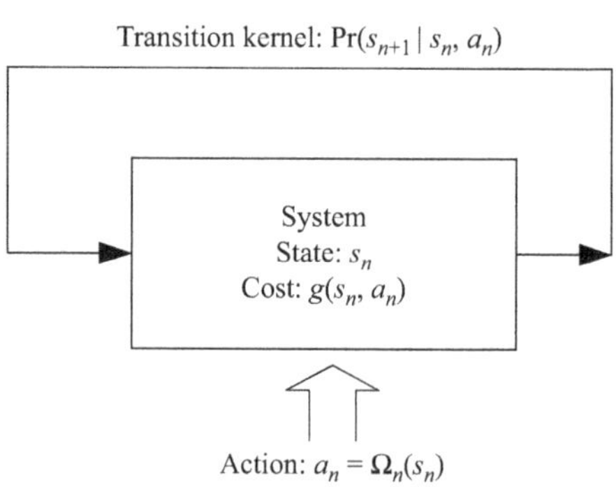

Figure 11.2 Block diagram for Markov decision process

$\mathscr{S}$, denoted as $|\mathscr{S}|$, is finite, or it can be compressed to finite. Given the state $\mathbf{s}$, there is a controller, which is able to manipulate the system by adjusting a set of control parameters $\mathbf{a}$. This set of control parameters is called the control action, and the set of all the possible choices of control actions are named as *action space*, denoted as $\mathscr{A}$. In Example 11.1 (it can be treated as a trivial case of MDP), the system state and control action are the CSI and the transmission powers of all the subcarriers, respectively. Thus,

$$\mathbf{s} = \{h_i | i = 1, 2, \ldots, N_F\}$$

and

$$\mathbf{a} = \{p_i | i = 1, 2, \ldots, N_F\}.$$

We focus on discrete-time MDP, where subscript of notations is used to indicate the stage index. For example, at the tth stage, the system state and control action are denoted as $\mathbf{s}_t$ and $\mathbf{a}_t$, respectively. The control action $\mathbf{a}_t$ is uniquely determined by the system state $\mathbf{s}_t$, the mapping from system state to control action is named as control policy. Let $\Omega_t : \mathscr{S} \rightarrow \mathscr{A}$ be the control policy at the tth stage, i.e.:

$$\Omega_t(\mathbf{s}_t) = \mathbf{a}_t, \quad \forall \mathbf{s}_t \in \mathscr{S}. \tag{11.15}$$

In MDP, the state of a system evolves with time in a Markovian way: providing the current system state and control action, the distribution of next system state is independent of other historical states or actions. In other words, the evolving of system state is a Markov chain, providing the control policy at each stage. Given the current (say the tth stage) system state $\mathbf{s}_t$ and control action $\mathbf{a}_t$, the distribution of next system state, $\Pr(\mathbf{s}_{t+1} | \mathbf{s}_t, \mathbf{a}_t)$, is called the state transition probability or *transition kernel*.

The expense of control action is measured by the cost function. The cost of the system at the tth stage is a function of $\mathbf{s}_t$ and $\mathbf{a}_t$, which is denoted as $g_t(\mathbf{s}_t, \mathbf{a}_t)$. In fact, g_t can be a random variable given $\mathbf{s}_t$ and $\mathbf{a}_t$, i.e., $g_t(\mathbf{s}_t, \mathbf{a}_t, \xi_t)$ where ξ_t for different t are independent variables. To simplify the elaboration, we focus on the form of $g_t(\mathbf{s}_t, \mathbf{a}_t)$ in the following discussion. Note that the cost function can be homogeneous or heterogeneous along the time line. Particularly, g_t can be different with respect to stage index t for an MDP with finite number of stages. However, when it is extended to the optimization over infinite number of stages, g_t should usually be homogeneous and subscript of stage index can be removed.

What are optimized in an MDP are not some parameters but a policy, which maps from the system state to control (scheduling) action. Thus, the solution of an MDP is a "function" rather than the values for some parameters. In Example 11.1, the water-filling algorithm can be used to figure out the values of the transmission powers on all the subcarriers. This is a physical-layer point of view. The following example shows that the if the scope of resource allocation is extended to MAC layer, the value optimization will transfer to policy optimization, which can be formulated as an MDP.

Example 11.2 (Multi-carrier power allocation: view from MAC-layer). *Similar to Example 11.1, a point-to-point OFDM link with N_F subcarriers is considered. In MAC layer, a transmission queue is maintained, which accepts packets from upper layer and delivers them via physical layer. Suppose that each packet consists of B information bits, and the number of arrival packets in each frame follows the Poisson distribution with expectation λ, i.e.:*

$$\Pr\left[\textit{The number of arrival packets in one frame} = n\right] = \frac{\lambda^n e^{-\lambda}}{n!}. \tag{11.16}$$

The packet departure in the MAC layer is determined by the physical layer transmission. Let $q(t)$ be the number of packets waiting to be transmitted in the tth frame, the queue dynamics can be represented by

$$q(t+1) = \max\{0, q(t) - d(t)\} + c(t), \tag{11.17}$$

where $d(t)$ and $c(t)$ are the numbers of departure and arrival packets in the tth frame.

In physical layer, let $\{h_i(t)|i = 1, 2, \ldots, N_F\}$ be the CSI of all the subcarriers in the tth frame, and $\{p_i(t)|i = 1, 2, \ldots, N_F\}$ be the corresponding power allocation. The number of packets can be delivered in the tth frame is

$$d(t) = \left\lfloor \frac{\sum_{i=1}^{N_F} \log_2\left(1 + ((p_i(t)\|h_i(t)\|^2)/\sigma_z^2)\right)}{B} \right\rfloor. \tag{11.18}$$

Clearly, larger transmission power will lead to larger departure rate of the transmission queue. However, some systems may have the following concerns on the average power consumption, which is particularly for battery-powered device:

$$\lim_{T\to+\infty} \mathbb{E}\left[\frac{1}{T}\sum_{t=1}^{T}\sum_{i=1}^{N_F} p_i(t)\right] \le \overline{P}, \tag{11.19}$$

where $\overline{P}$ is the average power constraint.

When the queue in the MAC layer is considered, the maximum throughput used in the Example 11.1 may not be suitable as the scheduling objective. One reasonable objective is the minimum average delay, which measures the average time duration for one packet from arrival to departure. Let:

$$\overline{Q} = \lim_{T\to+\infty} \mathbb{E}\left[\frac{1}{T}\sum_{t=1}^{T} q(t)\right] \tag{11.20}$$

be the average queue length at the transmitter, the average delay $\overline{W}$ is given below according to the Little's Law [4]:

$$\overline{W} = \frac{\overline{Q}}{\lambda} = \lim_{T\to+\infty} \mathbb{E}\left[\frac{1}{T}\sum_{t=1}^{T}\frac{q(t)}{\lambda}\right]. \tag{11.21}$$

Therefore, one possible problem formulation is to minimize the average packet delay while satisfying the average power constraint. Thus,

$$\min_{\{p_i(t)|\forall i,t\}} \quad \overline{W}$$

subject to (11.19).

In Example 11.1, the scheduling in different frames is independent. In other words, one does not need to worry about the impact of current frame scheduling on the future frames. However, when the average transmission delay $\overline{W}$ is the objective with the average power constraint, the scheduling in the current frame will affect that of the future frames. For example, the current frame may be scheduled with a power level greater than $\overline{P}$, which consumes the power budget of the following frames. Hence, it becomes meaningless to consider the resource optimization in one single frame (as Example 11.1), and the scope of optimization is the whole time line. As a result, what should be optimized is a mapping from CSI and queue length (also called queue state information or QSI) to the power allocation. Thus it is a "function," rather than some "variables."

Three forms of MDP formulation will be discussed in the following: first of all, we introduce the finite-horizon MDP, where the number of stages for joint optimization is finite. Then, we move to the infinite-horizon MDP, where two cost functions are considered: namely, average cost and discounted cost.

11.2.2 Finite-horizon MDP

In this section, we focus on the optimization along a fixed number of stages, say T stages. The overall cost function, denoted as $\overline{G}$, can be written as

$$\overline{G}\Big(\{\Omega_n|n = 1, 2, \ldots, T\}\Big) = \mathbb{E}\left[\sum_{t=1}^{T} g_t(\mathbf{s}_t, \mathbf{a}_t)\right], \tag{11.22}$$

where $\mathbf{a}_t = \Omega_t(\mathbf{s}_t)$. The expectation in the above equation is with respect to the randomness of the system state at the first stage and the state transition given the control action. Note that with the expectation on random system state, the overall cost function $\overline{G}$ depends on the control policies used in all the stages. With the objective of minimizing $\overline{G}$, the problem of finite-horizon MDP is described below.

Problem 11.3 (Finite-horizon MDP). *Find the optimal control polices for each stage, denoted as* $\{\Omega_t^*|t = 1, 2, \ldots, T\}$*, such that the overall cost* $\overline{G}$ *is minimized, i.e.:*

$$\{\Omega_t^*|t = 1, 2, \ldots, T\} = \arg \min_{\{\Omega_t|t=1,2,\ldots,T\}} \overline{G}\Big(\{\Omega_t|t = 1, 2, \ldots, T\}\Big). \qquad (11.23)$$

It is worth to highlight that in finite-horizon MDP, the optimal scheduling policies in different stages are usually different. For example, the policy design of the first stage should jointly consider the cost of the current stage and the potential cost of the following $N - 1$ stages, whereas the policy design of the last stage only needs to care the cost of the current stage.

In order to elaborate the solution structure of Problem 11.3, we first define the following cost-to-go function:

$$V_t(\mathbf{s}_t) = \min_{\{\Omega_k|k=t,t+1,\ldots,T\}} \mathbb{E}\left[\sum_{k=t}^{T} g_k(\mathbf{s}_k, \mathbf{a}_k)\right], \quad \forall t, \mathbf{s}_t, \qquad (11.24)$$

which is the average cost from the tth stage to the last one given the system state in the tth time slot $\mathbf{s}_t$. The cost-to-go function V_t is usually named as *value function*. It is straightforward to see that they satisfy the following iterative expressions:

$$V_t(\mathbf{s}_t) = \min_{\mathbf{a}_t}\left[g_t(\mathbf{s}_t, \mathbf{a}_t) + \sum_{\mathbf{s}_{t+1}} \Pr(\mathbf{s}_{t+1}|\mathbf{s}_t, \mathbf{a}_t)V_{t+1}(\mathbf{s}_{t+1})\right], \quad \forall t, \mathbf{s}_t. \qquad (11.25)$$

Equation (11.25) is usually referred to as the *Bellman equation*. It provides important insights on the solution of Problem 11.3. Given the system state at the tth stage $\mathbf{s}_t$, the optimal control action minimizing the right-hand-side of (11.25), denoted as $\Omega_t^*(\mathbf{s}_t)$, is obviously the optimal control action for system state $\mathbf{s}_t$ at the tth stage, i.e.:

$$\Omega_t^*(\mathbf{s}_t) = \arg\min_{\mathbf{a}_t}\left[g_t(\mathbf{s}_t, \mathbf{a}_t) + \sum_{\mathbf{s}_{t+1}} \Pr(\mathbf{s}_{t+1}|\mathbf{s}_t, \mathbf{a}_t)V_{t+1}(\mathbf{s}_{t+1})\right], \quad \forall t, \mathbf{s}_t. \quad (11.26)$$

Hence, in order to obtain the optimal control policy at the tth stage, it is necessary to first figure out the value function V_{t+1} for all possible next state. It implies that before calculating the optimal policy, a *backward recursion* for evaluating V_T, V_{T-1}, ..., V_1 sequentially is required, which is usually referred to as *value iteration* (VI). The VI algorithm for finite-horizon MDP is elaborated below.

VI algorithm for finite-horizon MDP

The value functions for finite-horizon MDP, as defined in (11.24), can be evaluated by the following steps:

1. Calculate the value function V_T for the last stage by

$$V_T(\mathbf{s}_T) = \min_{\mathbf{a}_T} g_T(\mathbf{s}_T, \mathbf{a}_T), \quad \forall \mathbf{s}_T. \qquad (11.27)$$

2. For $t = T - 1, T - 2, \ldots, 1$, calculate the value function V_t according to (11.25) sequentially.

Note that the value functions are calculated from the last stage to the first one. This is because of iterative structure as depicted in the Bellman equation (11.25). As a summary, the procedure to obtain the optimal control policy for the finite-horizon MDP can be described below.

- **Off-line VI:** Before running the system, the controller should evaluate the value functions for all the possible system states and all the stages. Their values can be stored in a table.
- **Online scheduling:** When the system is running, the controller should identify the system state, solve the corresponding Bellman equation, and apply the optimal action.

Hence, the solution raises both computation and memory requirements to the controller, whose complexities are proportional to the size of state space $|\mathscr{S}|$ and the number of stages T. In the following, we shall demonstrate the application of finite-horizon MDP via the multi-carrier power allocation problem.

11.2.2.1 Case study: multi-carrier power allocation via finite-horizon MDP

Suppose that there is one OFDM transmitter which wants to deliver a file of B bits to the receiver within T frames. The number of subcarriers is N_F. The transmitter is a battery-powered device, and it tries to save the transmission energy as much as possible by exploiting the channel temporal diversity in the T frames. One possible formulation for transmission scheduling at the transmitter is provided below:

- **System state:** In the tth frame ($t = 1, 2, \ldots, T$), the system state $\mathbf{s}_t$ is uniquely specified by the CSI of all the subcarriers $\{h_i(t)|i = 1, 2, \ldots, N_F\}$ and the number of remaining bits at the transmitter $q(t)$, which is usually called QSI. Thus:

$$\mathbf{s}_t = \Big\{\{h_i(t)|i = 1, 2, \ldots, N_F\}, q(t)\Big\}. \tag{11.28}$$

- **Control policy:** The control action in the tth frame ($t = 1, 2, \ldots, T$) is the power allocation on all the subcarriers, i.e.,

$$\mathbf{a}_t = \{p_i(t)|i = 1, 2, \ldots, N_F\}.$$

  Then the control policy in the tth frame, denoted as Ω_t, can be written as

$$\Omega_t(\mathbf{s}_t) = \mathbf{a}_t, \quad \forall t, \mathbf{s}_t. \tag{11.29}$$

- **Transition kernel:** The block fading channel model is considered, and the CSIs in different frames are i.i.d. distributed. Therefore, the transition kernel can be rewritten as

$$\Pr(\mathbf{s}_{t+1}|\mathbf{s}_t, \mathbf{a}_t) = \Pr\Big(\{h_i(t+1)|i = 1, 2, \ldots, N_F\}\Big) \Pr\Big(q(t+1)|\mathbf{s}_t, \mathbf{a}_t\Big), \tag{11.30}$$

where

$$q(t+1) = \max\{0, q(t) - d(t)\}, \tag{11.31}$$

and

$$d(t) = \left\lfloor \sum_{i=1}^{N_F} \log_2 \left(1 + \frac{p_i(t)\|h_i(t)\|^2}{\sigma_z^2}\right) \right\rfloor \tag{11.32}$$

is the number of bits transmitted in the tth frame. Hence, given $\mathbf{s}_t$ and $\mathbf{a}_t$, $q(t+1)$ is uniquely determined.

- **Cost:** In the tth frame ($t = 1, 2, \ldots, T$), the cost of the system is the total power consumption, i.e.:

$$g_t(\mathbf{s}_t, \mathbf{a}_t) = \sum_{i=1}^{N_F} p_i(t), \quad \forall t = 1, 2, \ldots, T. \tag{11.33}$$

Due to the randomness of the channel fading, a penalty is added in case there are some remaining bits after T frames transmission (penalty on the remaining bits in the $(T+1)$th frame). Hence, the following cost is introduced for the $(T+1)$th frame:

$$g_{T+1}(\mathbf{s}_{T+1}, \mathbf{a}_{T+1}) = w\, q(T+1), \tag{11.34}$$

where w is the weight for the penalty and $q(T+1)$ is the number of remaining bits after T frames. Note that there is no control action in the $(T+1)$th frame and $\mathbf{a}_{T+1}$ is introduced simply for notation consistency.

As a result, the optimization of transmission resource allocation can be written as the following finite-horizon MDP:

$$\min_{\{\Omega_t | t=1,2,\ldots,T\}} \mathbb{E}\left[\sum_{t=1}^{T+1} g_t(\mathbf{s}_t, \mathbf{a}_t)\right] = \min_{\{\Omega_t | t=1,2,\ldots,T\}} \mathbb{E}\left[\sum_{t=1}^{T}\sum_{i=1}^{N_F} p_i(t) + w\, q(T+1)\right]. \tag{11.35}$$

The expectation is because that $p_i(t)$ $(\forall i, t)$ and $q(T+1)$ are random due to channel fading. It can be observed that the choice of weight w may have strong impact on the scheduling policy: small weight leads to conservative strategy (try to save energy) and large weight makes the transmitter aggressive.

The Bellman equation for the above MDP is given in (11.25), where $V_{T+1}(\mathbf{s}_{T+1}) = w\, q(T+1)$ can be calculated directly. However, because the space of CSI is continuous and infinite, it is actually impossible to evaluate the other value functions. Note that the CSI is i.i.d. distributed among different frames, the expectation on CSI can be taken on both sides of the Bellman equation, which can be written as

$$\begin{aligned}\overline{V}_t(q(t)) &= \mathbb{E}_h\, V_t(\mathbf{s}_t)\\ &= \mathbb{E}_h \min_{\mathbf{a}_t}\left[g_t(\mathbf{s}_t,\mathbf{a}_t) + \sum_{\mathbf{s}_{t+1}} \Pr(\mathbf{s}_{t+1}|\mathbf{s}_t,\mathbf{a}_t)V_{t+1}(\mathbf{s}_{t+1})\right]\\ &= \mathbb{E}_h \min_{\mathbf{a}_t}\left[g_t(\mathbf{s}_t,\mathbf{a}_t) + \sum_{\mathbf{s}_{t+1}} \Pr\left(\{h_i(t+1)|\forall i\}\right)V_{t+1}(\mathbf{s}_{t+1})\Pr(q(t+1)|\mathbf{s}_t,\mathbf{a}_t)\right]\\ &= \mathbb{E}_h \min_{\mathbf{a}_t}\left[g_t(\mathbf{s}_t,\mathbf{a}_t) + \overline{V}_{t+1}(q(t+1))\right],\end{aligned} \tag{11.36}$$

where $\mathbb{E}_h$ denotes the expectation over CSI. Therefore, an equivalent Bellman equation with compressed system state is obtained, whose value function $\overline{V}_t$ $(t = 1, 2, \ldots, T)$ depends only on the QSI. The dependence of CSI is removed from the value function, which is mainly due to the nature of i.i.d. distribution. As a result, the state space is reduced from infinite to finite, and a practical solution becomes feasible.

The off-line VI can be applied to compute the new value function $\overline{V}_t$ for all states and stages (off-line VI), which is given below:

- Initialize the value function of the $(T + 1)$th stage as

$$\overline{V}_{T+1}(q(T+1)) = w\, q(T+1). \tag{11.37}$$

- For $t = T, T-1, \ldots, 1$, evaluate the value function according to (11.36). Note that there is an expectation with respect to channel fading, the Monte Carlo method can be used to calculate the value function numerically by generating sufficient number of CSI realizations according to its distribution (e.g., Rayleigh fading).

With the value functions, the optimal online scheduling when the system is running can be derived in each stage according to

$$\Omega_t^*(\mathbf{s}_t) = \mathbf{a}_t^* = \arg\min_{\mathbf{a}_t}\left[g_t(\mathbf{s}_t,\mathbf{a}_t) + \overline{V}_{t+1}(q(t+1))\right]. \tag{11.38}$$

Note that in both off-line VI and online scheduling, we always need to find the optimal solution for (11.38), which can be solved as follows. From the principle of water-filling method, it can be derived that with a given total transmission power in the tth frame, the optimal power allocation on each subcarrier can be written as

$$p_i(t) = \max\left\{0, \frac{1}{\beta_t} - \frac{\sigma_z^2}{\|h_i(t)\|^2}\right\}, \quad \forall i = 1, 2, \ldots, N_F, \tag{11.39}$$

where β_t is determined by the total transmission power on all the subcarriers. Therefore, the key of solution is to find the optimal total transmission power (or β_t) for the

tth frame such that the right-hand-side of (11.38) is minimized. Notice that the number of information bits delivered in the tth frame is given by (11.32), the optimization problem on the right-hand-side of (11.38) can be rewritten as

$$\min_{\beta_t} \sum_{i=1}^{N_F} \max\left\{0, \frac{1}{\beta_t} - \frac{\sigma_z^2}{\|h_i(t)\|^2}\right\} + \overline{V}_{t+1}(\max\{0, q(t) - d(t)\}), \tag{11.40}$$

which can be solved by one-dimensional search on β_t.

From the above solution, the difference between this problem and Example 11.1 can be observed. In both problems, the power allocation follows the expression of (11.39). Their difference lies in the choice of β_t. In Example 11.1, β_t is determined by the peak transmission power constraint, whereas in this problem, β_t should be optimized via (11.40). In other words, if the MAC layer queue dynamics are considered in the power allocation, different system states or stages will result in different water levels. This insight is intuitive. For example, if there are still a lot of bits waiting to be delivered, the transmitter tends to use high-transmission power (small β_t), and vice versa. It is worth to mention that the value function $\overline{V}_1(B)$ is the minimum average system cost to deliver all the B information bits within T frames.

Finally, a numerical simulation result is provided in Figure 11.3 to demonstrate the performance gain of the finite-horizon MDP formulation, where the baseline is the power allocation via conventional physical layer water-filling method with constant power constraint in each frame. It can be observed that the MDP approach always has

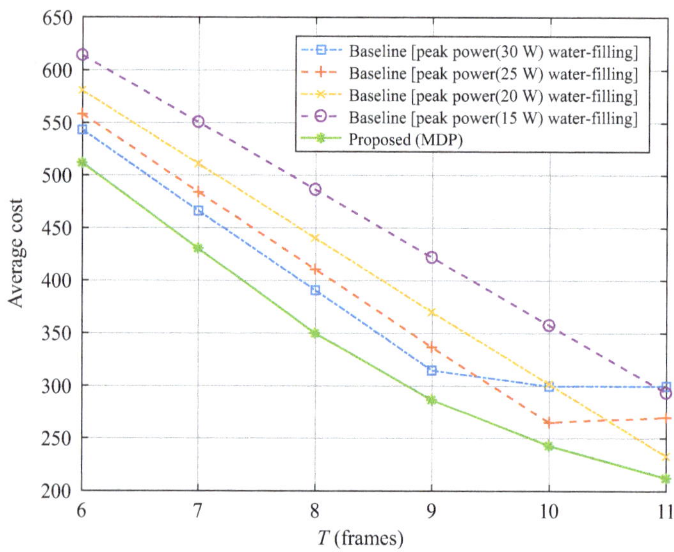

Figure 11.3 Performance comparison between conventional water-filling algorithm (baseline algorithm) and finite-horizon MDP algorithm (proposed algorithm)

less cost than the physical layer approach with various peak power levels. This gain mainly comes from the cross-frame power scheduling, which exploits the channel temporal diversity.

11.2.3 Infinite-horizon MDP with discounted cost

When the scope of optimization is extended to infinite time horizon, the MDP formulation and solution would become quite different from the case of finite time horizon. In the example of Section 11.2.2.1, the system cost is the summation of all transmission powers in all the subcarriers and frames. It can be imagined that if it is extended to infinite time horizon with the possibility of new packet arrival at the transmitter, the system cost will tend to infinity, i.e., it cannot be measured. In order to handle this issue, in infinite-horizon MDP, two measurements on the system cost are considered, namely, discounted cost and average cost. This section will introduce the formulation and solution for discounted system cost, and the case of average cost is left to the next one.

In infinite-horizon MDP, it is usually assumed that the cost function and the control policy are the same for all stages. Hence, let $\mathbf{s}$ and $\mathbf{a}$ be the system state and control action in certain stage, the corresponding cost can be denoted as $g(\mathbf{s}, \mathbf{a})$ and the overall discounted cost for one certain control policy Ω and initial system state at the first stage $\mathbf{s}_1$ can be written as

$$\overline{G}(\Omega) = \lim_{T\to+\infty} \mathbb{E}\left[\sum_{t=1}^{T} \gamma^{t-1} g(\mathbf{s}_t, \mathbf{a}_t)\Big|\mathbf{s}_1\right], \tag{11.41}$$

where $\mathbf{s}_t$ and $\mathbf{a}_t = \Omega(\mathbf{s}_t)$ are the system state and control action of the tth stage, respectively. The expectation is taken over all possible state transition, and the infinite summation usually converges due to the discount factor $\gamma \in (0, 1)$. As a result, the infinite-horizon MDP can be mathematically described as follows:

Problem 11.4 (Infinite-horizon MDP with discounted cost). *Find the optimal control polices, denoted as Ω^*, such that the overall cost $\overline{G}$ is minimized, i.e.:*

$$\Omega^* = \arg\min_{\Omega} \overline{G}(\Omega) = \arg\min_{\Omega} \lim_{T\to+\infty} \mathbb{E}\left[\sum_{t=1}^{T} \gamma^{t-1} g\Big(\mathbf{s}_t, \Omega(\mathbf{s}_t)\Big)\right]. \tag{11.42}$$

In order to derive the solution of the above problem, the following cost-to-go function (value function) is first defined for one arbitrary system state $\mathbf{s}_1$ at the first stage:

$$\begin{aligned} V(\mathbf{s}_1) &= \min_{\Omega} \lim_{T\to+\infty} \mathbb{E}\left[\sum_{t=1}^{T} \gamma^{t-1} g\Big(\mathbf{s}_t, \Omega(\mathbf{s}_t)\Big)\Big|\mathbf{s}_1\right] \\ &= \min_{\Omega}\left\{g(\mathbf{s}_1, \mathbf{a}_1) + \lim_{T\to+\infty} \mathbb{E}\left[\sum_{t=2}^{T} \gamma^{t-1} g(\mathbf{s}_t, \mathbf{a}_t)\Big|\mathbf{s}_1\right]\right\}. \end{aligned} \tag{11.43}$$

With the definition of V, the system cost starting from the tth stage given the state $\mathbf{s}_t$ at the tth stage can be written as

$$\begin{aligned}
&\min_{\Omega} \lim_{T\to+\infty} \mathbb{E}\left[\sum_{n=t}^{T} \gamma^{n-1} g(\mathbf{s}_n, \mathbf{a}_n)\Big|\mathbf{s}_t\right] \\
&= \gamma^{t-1} \min_{\Omega} \lim_{T\to+\infty} \mathbb{E}\left[\sum_{n=t}^{T} \gamma^{n-t} g(\mathbf{s}_n, \mathbf{a}_n)\Big|\mathbf{s}_t\right] \\
&= \gamma^{t-1} \min_{\Omega} \lim_{T'\to+\infty} \mathbb{E}\left[\sum_{k=1}^{T'} \gamma^{k-1} g(\mathbf{s}_{k+t-1}, \mathbf{a}_{k+t-1})\Big|\mathbf{s}_t\right],
\end{aligned} \tag{11.44}$$

where the second equality is due to $k = n - t + 1$ and $T' = T - t + 1$. If we define the new notation for system state by letting $\mathbf{s}'_k = \mathbf{s}_{k+t-1}$ and $\mathbf{a}'_k = \mathbf{a}_{k+t-1}$, the minimization of the above equation can be written as

$$\begin{aligned}
&\min_{\Omega} \lim_{T'\to+\infty} \mathbb{E}\left[\sum_{k=1}^{T'} \gamma^{k-1} g(\mathbf{s}_{k+t-1}, \mathbf{a}_{k+t-1})\Big|\mathbf{s}_t\right] \\
&= \min_{\Omega} \lim_{T'\to+\infty} \mathbb{E}\left[\sum_{k=1}^{T'} \gamma^{k-1} g(\mathbf{s}'_k, \mathbf{a}'_k)\Big|\mathbf{s}'_1\right] \\
&= V(\mathbf{s}'_1) = V(\mathbf{s}_t).
\end{aligned} \tag{11.45}$$

Hence, it can be derived that

$$\min_{\Omega} \lim_{T\to+\infty} \mathbb{E}\left[\sum_{n=t}^{T} \gamma^{n-1} g(\mathbf{s}_n, \mathbf{a}_n)\Big|\mathbf{s}_t\right] = \gamma^{t-1} V(\mathbf{s}_t). \tag{11.46}$$

Since the time horizon is infinite, the optimal policy minimizing the system cost since the first stage also minimizes the system cost since any arbitrary stage. Hence, (11.43) can be written as

$$\begin{aligned}
V(\mathbf{s}_1) &= \min_{\Omega} \left\{ g(\mathbf{s}_1, \mathbf{a}_1) + \mathbb{E}_{\mathbf{s}_2} \lim_{T\to+\infty} \mathbb{E}_{\{\mathbf{s}_i|i=3,4,\ldots\}} \left[\sum_{t=2}^{T} \gamma^{t-1} g(\mathbf{s}_t, \mathbf{a}_t)\Big|\mathbf{s}_2\right]\right\} \\
&= \min_{\Omega(\mathbf{s}_1)} \left\{ g(\mathbf{s}_1, \mathbf{a}_1) + \mathbb{E}_{\mathbf{s}_2} \min_{\Omega} \lim_{T\to+\infty} \mathbb{E}_{\{\mathbf{s}_i|i=3,4,\ldots\}} \left[\sum_{t=2}^{T} \gamma^{t-1} g(\mathbf{s}_t, \mathbf{a}_t)\Big|\mathbf{s}_2\right]\right\} \\
&= \min_{\Omega(\mathbf{s}_1)} \left\{ g(\mathbf{s}_1, \mathbf{a}_1) + \gamma \mathbb{E}_{\mathbf{s}_2} V(\mathbf{s}_2) \right\},
\end{aligned} \tag{11.47}$$

where the last equality is due to (11.46). Similarly, for arbitrary system state at the arbitrary tth stage $\mathbf{s}_t$, the Bellman equation for infinite-horizon MDP with discounted cost can be written as follows:

$$V(\mathbf{s}_t) = \min_{\Omega(\mathbf{s}_t)} \left\{ g(\mathbf{s}_t, \mathbf{a}_t) + \gamma \mathbb{E}_{\mathbf{s}_{t+1}} V(\mathbf{s}_{t+1}) \right\}. \tag{11.48}$$

Regarding to the solution, if the value function has already been calculated, it is straightforward to see that the optimal control action for arbitrary stage is

$$\Omega^*(\mathbf{s}_t) = \mathbf{a}_t^* = \arg\min_{\mathbf{a}_t} \left\{ g(\mathbf{s}_t, \mathbf{a}_t) + \gamma \mathbb{E}_{\mathbf{s}_{t+1}} V(\mathbf{s}_{t+1}) \right\}, \quad \forall t, \mathbf{s}_t. \tag{11.49}$$

On the other hand, the value function should satisfy the Bellman equation in (11.48). This is a fixed-point problem with minimization on the right-hand-side, and we have to rely on the iterative algorithm, which is named as VI. The detail steps of VI is elaborated below, and please refer to [3] for the proof of convergence.

VI algorithm for infinite-horizon MDP with discounted cost

The value functions defined in (11.48) can be evaluated by the following steps:

1. Let $i = 0$ and initialize the value function $V(\mathbf{s})$ for all possible $\mathbf{s} \in \mathcal{S}$, which is denoted as $V^i(\mathbf{s})$.
2. In the ith iteration, update the value function as

$$V^{i+1}(\mathbf{s}_t) = \min_{\Omega} \left\{ g(\mathbf{s}_t, \mathbf{a}_t) + \gamma \mathbb{E}_{\mathbf{s}_{t+1}} V^i(\mathbf{s}_{t+1}) \right\}, \tag{11.50}$$

   for all possible $\mathbf{s}_t \in \mathcal{S}$.
3. If the update from V^i to V^{i+1} for any system state is negligible (or less than one predetermined threshold), the iteration terminates. Otherwise, let $i = i + 1$ and jump to Step 2.

In the following section, we still use the case of multi-carrier power allocation to demonstrate the formulation via infinite-horizon MDP with the discounted cost. It takes the packet arrival at the MAC layer into considered, which is not addressed in Section 11.2.2.1.

11.2.3.1 Case study: multi-carrier power allocation with random packet arrival

The resource allocation problem introduced in Example 11.2 can also be addressed with discounted cost, which will be elaborated in this example. Specifically, a point-to-point OFDM communication link with N_F subcarriers and random packet arrival at the transmitter is considered. It is assumed that each packet consists of B information bits, and one packet should be transmitted within one frame. The key elements of MDP formulation are elaborated below:

- **System state:** In the tth frame ($t = 1, 2, 3, \ldots$), the system state $\mathbf{s}_t$ is uniquely specified by the CSI of all the subcarriers $\{h_i(t)|i = 1, 2, \ldots, N_F\}$ and the QSI $q(t)$. The latter denotes the number of remaining packets waiting at the transmitter. Thus:

$$\mathbf{s}_t = \left\{ \{h_i(t)|i = 1, 2, \ldots, N_F\}, q(t) \right\}. \tag{11.51}$$

- **Control policy:** The control action in the tth frame ($t = 1, 2, 3, \ldots$) is the power allocation on all the subcarriers, i.e., $\mathbf{a}_t = \{p_i(t) | i = 1, 2, \ldots, N_F\}$. Then the control policy in the tth frame, denoted as Ω, can be written as

$$\Omega(\mathbf{s}_t) = \mathbf{a}_t, \quad \forall t, \mathbf{s}_t. \tag{11.52}$$

- **Transition kernel:** The block fading channel model is considered, and the CSI in each frame is i.i.d. distributed. Therefore, the transition kernel can be written as

$$\Pr(\mathbf{s}_{t+1} | \mathbf{s}_t, \mathbf{a}_t) = \Pr\left(\{h_i(t+1) | i = 1, 2, \ldots, N_F\}\right) \Pr\left(q(t+1) \middle| \mathbf{s}_t, \mathbf{a}_t\right), \tag{11.53}$$

where

$$q(t+1) = \max\{0, q(t) - d(t)\} + c(t), \tag{11.54}$$

and

$$d(t) = \left\lfloor \frac{1}{B} \sum_{i=1}^{N_F} \log_2\left(1 + \frac{p_i(t) \|h_i(t)\|^2}{\sigma_z^2}\right) \right\rfloor \tag{11.55}$$

are the number of packets transmitted in the tth frame, $c(t)$ is the number of arrival packets in the tth frame. It is usually assumed that $c(t)$ follows the Poisson arrival with expectation λ, as in Example 11.2. Thus, there are λ arrival packets in one frame on average.
- **Cost:** The average power consumption at the transmitter is

$$\overline{P} = \lim_{T \to +\infty} \mathbb{E}\left[\frac{1}{T} \sum_{t=1}^{T} \sum_{i=1}^{N_F} p_i(t)\right]. \tag{11.56}$$

According to the Little's Law, the average transmission delay of one packet is

$$\overline{W} = \frac{\overline{Q}}{\lambda} = \lim_{T \to +\infty} \mathbb{E}\left[\frac{1}{T} \sum_{t=1}^{T} \frac{q(t)}{\lambda}\right], \tag{11.57}$$

where $\overline{Q}$ is the average number of packets waiting at the transmitter. The weighted sum of average power and delay is

$$\overline{P} + \eta \overline{W} = \lim_{T \to +\infty} \mathbb{E}\left\{\frac{1}{T} \sum_{t=1}^{T} \left[\eta \frac{q(t)}{\lambda} + \sum_{i=1}^{N_F} p_i(t)\right]\right\}, \tag{11.58}$$

where η is the weight on the average transmission delay. The problem of minimizing $\overline{P} + \eta \overline{W}$ is an infinite horizon MDP with average cost, whose solution will be introduced in the next section. Usually, people prefer to consider the discount approximation of $\overline{P} + \eta \overline{W}$ as follows:

$$\overline{G} = \lim_{T \to +\infty} \mathbb{E}\left\{\sum_{t=1}^{T} \gamma^{t-1} \left[\eta \frac{q(t)}{\lambda} + \sum_{i=1}^{N_F} p_i(t)\right]\right\}. \tag{11.59}$$

The main reason for approximating average cost via discounted cost is that it has better converge rate in VI.

Hence, the resource allocation problem can be formulated as

$$\min_{\Omega} G = \min_{\Omega} \lim_{T\to+\infty} \mathbb{E}\left[\sum_{t=1}^{T} \gamma^{t-1} \underbrace{\left(\eta\frac{q(t)}{\lambda} + \sum_{i=1}^{N_F} p_i(t)\right)}_{g(\mathbf{s}_t,\mathbf{a}_t)}\right], \tag{11.60}$$

which is an infinite-horizon MDP with discounted cost. The Bellman equation for this problem is

$$V(\mathbf{s}_t) = \min_{\Omega}\left\{\eta\frac{q(t)}{\lambda} + \sum_{i=1}^{N_F} p_i(t) + \gamma \mathbb{E}_{\mathbf{s}_{t+1}} V(\mathbf{s}_{t+1})\right\}. \tag{11.61}$$

Note that the space of the system state includes all possible values of CSI, and it is actually impossible to evaluate value function. Similar to Section 11.2.2.1, since the CSI is i.i.d. distributed in each frame, the expectation with respect to the CSI can be taken on both side of the above Bellman equation, i.e.:

$$\begin{aligned}
\overline{V}\big(q(t)\big) &= \mathbb{E}_h \min_{\Omega}\left\{\eta\frac{q(t)}{\lambda} + \sum_{i=1}^{N_F} p_i(t) + \gamma \mathbb{E}_{\mathbf{s}_{t+1}} V(\mathbf{s}_{t+1})\right\} \\
&= \mathbb{E}_h \min_{\Omega}\left\{\eta\frac{q(t)}{\lambda} + \sum_{i=1}^{N_F} p_i(t) + \gamma \sum_{q(t+1)} \Pr\big(q(t+1)|\mathbf{s}_t,\mathbf{a}_t\big)\overline{V}\big(q(t+1)\big)\right\} \\
&= \eta\frac{q(t)}{\lambda} + \mathbb{E}_h \min_{\Omega}\left\{\sum_{i=1}^{N_F} p_i(t) + \gamma \sum_{c(t)} \frac{\lambda^{c(t)}e^{-\lambda}}{c(t)!}\overline{V}\big(q(t+1)\big)\right\} \\
&= \eta\frac{q(t)}{\lambda} + \mathbb{E}_{h,c} \min_{\Omega}\left\{\sum_{i=1}^{N_F} p_i(t) + \gamma \overline{V}\big(q(t+1)\big)\right\},
\end{aligned} \tag{11.62}$$

where $\mathbb{E}_h$ is the expectation over CSI, $\mathbb{E}_{\mathbf{s}_{t+1}}$ is the expectation over next system state, and $\mathbb{E}_{h,c}$ is the expectation over random packet arrival.

The off-line VI can be applied to compute the value function $\overline{V}$ for all possible queue length. In order to avoid infinite transmission queue, we can set a buffer size. Thus, the overflow packets will be dropped. With the value function, the optimal scheduling action can be calculated via:

$$\Omega^*(\mathbf{s}_t) = \mathbf{a}_t^* = \arg\min_{\Omega}\left\{\sum_{i=1}^{N_F} p_i(t) + \gamma \sum_{c(t)} \frac{\lambda^{c(t)}e^{-\lambda}}{c(t)!}\overline{V}\big(q(t+1)\big)\right\}. \tag{11.63}$$

Note that in both off-line VI and online scheduling, we always need to find the optimal solution for (11.63), which can be solved with the approach introduced in Section 11.2.2.1 (i.e., water-filling with optimized water level).

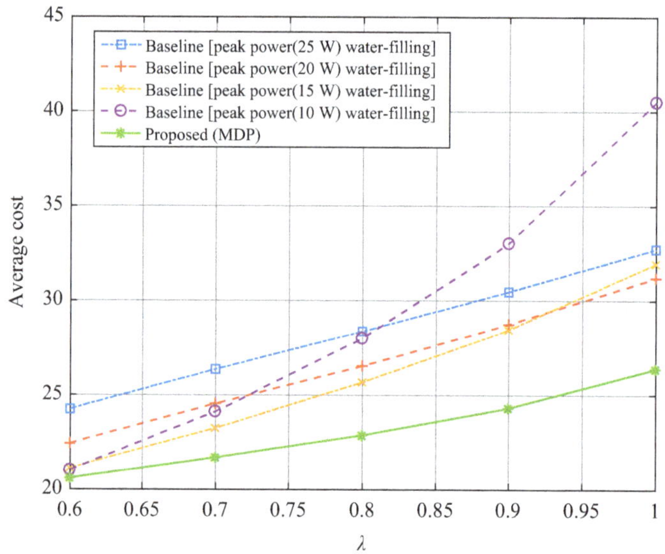

Figure 11.4 Performance comparison between conventional water-filling algorithm (baseline algorithm) and infinite-horizon MDP algorithm (proposed algorithm)

Finally, a numerical simulation result is provided in Figure 11.4 to demonstrate the performance gain of the infinite-horizon MDP formulation, where the baseline is the power allocation via conventional physical layer water-filling method with constant power constraint in each frame. It can be observed that the MDP approach always has less cost than the physical layer approach with various peak power levels. Particularly, the performance gain of the MDP formulation is more significant in the region of heavier traffic (larger λ).

11.2.4 Infinite-horizon MDP with average cost

In the problem formulation of Section 11.2.3.1, the average cost, a weighted summation of average power and average delay, is approximated as discounted cost so that the solution for infinite-horizon MDP with discounted cost can be applied. In this section, we shall show how to handle the exact average cost via infinite-horizon MDP.

Let $\mathbf{s}$ and $\mathbf{a}$ be the system state and control action in certain stage, the corresponding cost is $g(\mathbf{s}, \mathbf{a})$ and the overall average cost for one certain control policy Ω can be written as

$$\overline{G}(\Omega) = \lim_{T \to +\infty} \mathbb{E}\left[\frac{1}{T}\sum_{t=1}^{T} g(\mathbf{s}_t, \mathbf{a}_t)\right], \tag{11.64}$$

where $\mathbf{s}_t$ and $\mathbf{a}_t = \Omega(\mathbf{s}_t)$ are the system state and control action of the tth stage, respectively. The expectation is taken over all possible state transition. Therefore, the infinite-horizon MDP with average cost can be mathematically described as follows:

Problem 11.5 (Infinite-horizon MDP with discounted cost). *Find the optimal control policy, denoted as* Ω^**, such that the overall cost* $\overline{G}$ *is minimized, i.e.:*

$$\Omega^* = \arg\min_{\Omega} \overline{G}(\Omega) = \arg\min_{\Omega} \lim_{T \to +\infty} \mathbb{E}\left[\frac{1}{T}\sum_{t=1}^{T} g\Big(\mathbf{s}_t, \Omega(\mathbf{s}_t)\Big)\right]. \tag{11.65}$$

Comparing with Problem 11.4, it can be observed that the discounted cost MDP values the current cost more than the future cost (due to discount factor γ), but the average cost MDP values them equally. Moreover, when the discount factor γ of Problem 11.4 is close to 1, the discounted cost MDP becomes closer to the average cost MDP.

Unlike the case of discounted cost, the value function for the case of average cost does not have straightforward meaning. Instead, the value function is defined via the following Bellman equation:

$$\theta + V(\mathbf{s}_t) = \min_{\Omega(\mathbf{s}_t)}\Big\{g(\mathbf{s}_t, \mathbf{a}_t) + \mathbb{E}_{\mathbf{s}_{t+1}} V(\mathbf{s}_{t+1})\Big\}, \quad \forall \mathbf{s}_t, \mathbf{a}_t, \tag{11.66}$$

where $V(\mathbf{s})$ is the value function for system state $\mathbf{s}$. As proved in [3], this Bellman equation could bring the following insights on Problem 11.5:

- θ is the minimized average system cost, i.e.:

$$\theta = \min_{\Omega} \lim_{T \to +\infty} \mathbb{E}\left[\frac{1}{T}\sum_{t=1}^{T} g\Big(\mathbf{s}_t, \Omega(\mathbf{s}_t)\Big)\right]. \tag{11.67}$$

- The optimal control action for arbitrary system state $\mathbf{s}_t$ at arbitrary tth stage can be obtained by solving the right-hand-side of (11.66), i.e.:

$$\Omega^*(\mathbf{s}_t) = \mathbf{a}_t^* = \arg\min_{\mathbf{a}_t}\Big\{g(\mathbf{s}_t, \mathbf{a}_t) + \mathbb{E}_{\mathbf{s}_{t+1}} V(\mathbf{s}_{t+1})\Big\}. \tag{11.68}$$

Moreover, the VI to calculate the value function is elaborated below.

VI algorithm for infinite-horizon MDP with average cost

The value functions for infinite-horizon MDP can be evaluated by the following steps:

1. Let $i = 0$ and initialize the value function $V^i(\mathbf{s})$ for all possible $\mathbf{s} \in \mathcal{S}$. Moreover, arbitrarily choose one system state as the reference state, which is denoted as $\mathbf{s}_{ref}$.
2. In the ith iteration, update the value function for all system states as follows:

$$V^{i+1}(\mathbf{s}_t) = \min_{\Omega(\mathbf{s}_t)}\Big\{g(\mathbf{s}_t, \mathbf{a}_t) + \mathbb{E}_{\mathbf{s}_{t+1}} V^i(\mathbf{s}_{t+1})\Big\} - V^i(\mathbf{s}_{ref}), \forall \mathbf{s}_t. \tag{11.69}$$

3. If the update from V^i to V^{i+1} for all system states is negligible, the iteration terminates. Otherwise, let $i = i + 1$ and jump to Step 2.

11.2.4.1 Case study: multi-carrier power allocation with average cost

Following the definition of system state, transition kernel and control policy in Section 11.2.3.1, the average cost minimization of the point-to-point OFDM transmission is given by

$$\min_{\Omega} \overline{G} = \min_{\Omega} \lim_{T \to +\infty} \mathbb{E} \left[\sum_{t=1}^{T} \frac{1}{T} \underbrace{\left(\eta \frac{q(t)}{\lambda} + \sum_{i=1}^{N_F} p_i(t) \right)}_{g(\mathbf{s}_t, \mathbf{a}_t)} \right]. \tag{11.70}$$

Its Bellman equation after taking expectation on the CSI can be written as

$$\theta + \overline{V}\Big(q(t)\Big) = \eta \frac{q(t)}{\lambda} + \mathbb{E} \min_{\Omega} \left\{ \sum_{i=1}^{N_F} p_i(t) + \sum_{c(t)} \frac{\lambda^{c(t)} e^{-\lambda}}{c(t)!} \overline{V}\Big(q(t+1)\Big) \right\}, \tag{11.71}$$

where $\overline{V}(q(t))$ is the value function with $q(t)$ packets at the transmitter. The VI can be used to evaluate the value function for all the possible queue lengths (a maximum queue length can be assumed to avoid infinite queue). Moreover, with the value function, the optimal scheduling action at arbitrary one stage (say the tth stage) with arbitrary system state $\mathbf{s}_t$ can be derived via:

$$\Omega^*(\mathbf{s}_t) = \mathbf{a}_t^* = \arg\min_{\mathbf{a}_t} \left\{ \sum_{i=1}^{N_F} p_i(t) + \sum_{c(t)} \frac{\lambda^{c(t)} e^{-\lambda}}{c(t)!} \overline{V}(q(t+1)) \right\}. \tag{11.72}$$

Note that $q(t+1)$ depends on both $p_i(t)$ $(\forall i)$ and $c(t)$. This problem can be solved with the approach introduced in Section 11.2.2.1 (i.e., water-filling with optimized water level).

11.3 Reinforcement learning

In the previous section, when introducing the solution of MDP, we actually assume that the state transition kernel and the system cost function of each stage are precisely known. If this knowledge is unknown, which might happen in practice, the methods of reinforcement learning can be used to collect the information in an online way and finally yield the desired solution.

In this section, we shall use the case of infinite-horizon MDP with discounted cost as an example to explain some methods of reinforcement learning. The approaches can be similarly applied on the other forms of MDPs. It has been introduced in the previous section that the solution of MDP includes the off-line VI and online scheduling. Now, consider the following VI, which is supposed to be finished before running the system:

$$
\begin{aligned}
V^{i+1}(\mathbf{s}_t) &= \min_{\mathbf{a}_t} \left\{ g(\mathbf{s}_t, \mathbf{a}_t) + \gamma \mathbb{E}_{\mathbf{s}_{t+1}} V^i(\mathbf{s}_{t+1}) \right\} \\
&= \min_{\mathbf{a}_t} \left\{ g(\mathbf{s}_t, \mathbf{a}_t) + \gamma \sum_{\mathbf{s}_{t+1}} \Pr(\mathbf{s}_{t+1} | \mathbf{s}_t, \mathbf{a}_t) V^i(\mathbf{s}_{t+1}) \right\}. \qquad (11.73)
\end{aligned}
$$

It can be observed that the VI relies on the knowledge of cost function $g(\mathbf{s}_t, \mathbf{a}_t)$ and the state transition probability $\Pr(\mathbf{s}_{t+1} | \mathbf{s}_t, \mathbf{a}_t)$. In other words, the VI is infeasible if they are unknown. In the following example, we extend the power allocation example of Section 11.2.3.1 from ideal mathematical model to practical implementation and show that the cost function or the transition kernel (state transition probability) may be unknown in some situation.

Example 11.3 (Multi-carrier power allocation with unknown statistics). *In the example of Section 11.2.3.1, after taking the expectation on CSI, the equivalent Bellman equation is given by*

$$
\overline{V}\Big(q(t)\Big) = \eta \frac{q(t)}{\lambda} + \mathbb{E}_{h,c} \min_{\Omega} \left\{ \sum_{i=1}^{N_F} p_i(t) + \gamma \overline{V}\Big(q(t+1)\Big) \right\}.
$$

The expectation on the right-hand-side is with respect to the distributions of CSI and packet arrival. It is usually to assume that they are Rayleigh fading and Poisson arrival, respectively. In practice, the BS may be lack of their statistics, e.g., mean or variance, or they may even not follow the assumed distributions. Hence, the VI based on the above equation cannot be carried on off-line.

In order to match the above equations with the standard form of Bellman equation (11.73), we can define the following control policy with respect to the queue length:

$$
\overline{\Omega}\Big(q(t)\Big) = \Big\{ \Omega(\mathbf{s}_t) | \forall h_i(t), i = 1, 2, \ldots, N_F \Big\} = \mathbf{a}_t. \qquad (11.74)
$$

Thus, $\overline{\Omega}$ is a mapping from the queue length to the power allocations for all possible CSI. With the definition of $\overline{\Omega}$, the Bellman equation for the example of Section 11.2.3.1 can be written as

$$
\overline{V}\Big(q(t)\Big) = \min_{\overline{\Omega}(q(t))} \left\{ \eta \frac{q(t)}{\lambda} + \mathbb{E}_h \left[\sum_{i=1}^{N_F} p_i(t) \right] + \gamma\, \mathbb{E}_{h,c} \Big[\overline{V}\Big(q(t+1)\Big) \Big] \right\},
$$

where cost function and state transition probability of (11.73) are given by

$$g(\mathbf{s}_t, \mathbf{a}_t) = \eta \frac{q(t)}{\lambda} + \mathbb{E}\left[\sum_{i=1}^{N_F} p_i(t)\right], \tag{11.75}$$

and

$$\Pr(\mathbf{s}_{t+1}|\mathbf{s}_t, \mathbf{a}_t) = \mathbb{E}_h \Pr\left(q(t+1)\middle|q(t), \overline{\Omega}(q(t)), \{h_i(t)\}\right), \tag{11.76}$$

respectively. Hence, the cost function requires the knowledge of CSI distribution, and the state transition probability depends on the distributions of both CSI and packet arrival. Without the knowledge of both distribution, the off-line VI is infeasible.

In order to find the optimal policy without the *a priori* knowledge on the statistics of the system, we have to perform VI in an online way, which is usually referred to as reinforcement learning. In the remaining of this section, we shall introduce two learning approaches. The first approach can be applied on the example of Section 11.2.3.1 without any knowledge on CSI distribution, and the second one, which is call as *Q-learning*, is more general to handle unknown statistics of packet arrival.

11.3.1 Online solution via stochastic approximation

In this section, we shall focus on the particular type of MDPs as elaborated in Example 11.3. For the elaboration convenience, we extend the formulation of infinite-horizon MDP with discounted cost in Section 11.2.3 by including an independent random variable in each stage. Specifically, suppose that ξ_t is a random variable (or a set of random variables) at the tth stage, and its distribution is i.i.d. with respect to t. Now, consider an infinite-horizon MDP with discounted cost, where the cost function at the tth stage is $g(\mathbf{s}_t, \mathbf{a}_t, \xi_t)$ and the transition kernel is given by the distribution $\Pr(\mathbf{s}_{t+1}|\mathbf{s}_t, \mathbf{a}_t, \xi_t)$. $\mathbf{s}_t$ is the system state at the tth stage, and ξ_t (which is not included in $\mathbf{s}_t$) can also be observed at the beginning of the tth stage. In a policy Ω, the control action $\mathbf{a}_t$ is determined according to both $\mathbf{s}_t$ and ξ_t, i.e., $\mathbf{a}_t = \Omega(\mathbf{s}_t, \xi_t)$. The MDP problem can be described as

$$\min_{\Omega} \lim_{T\to+\infty} \mathbb{E}\left[\sum_{t=1}^{T} \gamma^{t-1} g(\mathbf{s}_t, \mathbf{a}_t, \xi_t)\middle|\mathbf{s}_1\right], \tag{11.77}$$

where the value function is defined as

$$V(\mathbf{s}) = \min_{\Omega} \lim_{T\to+\infty} \mathbb{E}\left[\sum_{t=1}^{T} \gamma^{t-1} g(\mathbf{s}_t, \mathbf{a}_t, \xi_t)\middle|\mathbf{s}_1 = \mathbf{s}\right]. \tag{11.78}$$

Note that this is the exact MDP problem discussed in Section 11.2.3.1, where ξ_t and system state $\mathbf{s}_t$ refer to the CSI and queue length in the tth frame, respectively. Its Bellman equation can be written as

$$V(\mathbf{s}_t) = \mathbb{E}_{\xi_t} \min_{\mathbf{a}_t} \left\{ g(\mathbf{s}_t, \mathbf{a}_t, \xi_t) + \gamma \Pr(\mathbf{s}_{t+1}|\mathbf{s}_t, \mathbf{a}_t, \xi_t) V(\mathbf{s}_{t+1}) \right\}. \tag{11.79}$$

In this section, we assume that ξ_t $(\forall t)$, $g(\mathbf{s}_t, \mathbf{a}_t, \xi_t)$, and $\Pr(\mathbf{s}_{t+1}|\mathbf{s}_t, \mathbf{a}_t, \xi_t)$ can be observed or measured at each stage, but the distribution of ξ_t is unknown. This refers to the situation that the CSI distribution in the example of Section 11.2.3.1 is unknown (the distribution of packet arrival is known). Hence, the off-line VI is infeasible as the right-hand-side of (11.79) cannot be calculated. Instead, we can first initialize a control policy, evaluate the value function corresponding to this policy via stochastic approximation in an online way, and then update the policy and reevaluate the value function again. By such iteration, it can be proved that the Bellman equation of (11.79) can be finally solved. The algorithm is elaborated below.

Online value and policy iteration

1. Let $i = 0$, and initialize a control policy Ω^i.
2. Run the system with the policy Ω^i and evaluate the corresponding value function V^i which satisfies:

   $$V^i(\mathbf{s}_t) = \mathbb{E}_{\xi_t} \left[g(\mathbf{s}_t, \mathbf{a}_t, \xi_t) + \gamma \Pr(\mathbf{s}_{t+1}|\mathbf{s}_t, \mathbf{a}_t, \xi_t) V^i(\mathbf{s}_{t+1}) \right], \quad \forall \mathbf{s}_t, \tag{11.80}$$

   where $\mathbf{s}_{t+1}$ denotes the next stage system state given the current stage system state $\mathbf{s}_t$.
3. Update the control policy from Ω^i to Ω^{i+1} via:

   $$\Omega^{i+1}(\mathbf{s}_t, \xi_t) = \arg\min_{\mathbf{a}_t} \left\{ g(\mathbf{s}_t, \mathbf{a}_t, \xi_t) + \gamma \Pr(\mathbf{s}_{t+1}|\mathbf{s}_t, \mathbf{a}_t, \xi_t) V^i(\mathbf{s}_{t+1}) \right\}. \tag{11.81}$$

   Note that ξ_t can be observed at the tth stage (e.g., the CSI can be estimated at the beginning of each frame in the example of Section 11.2.3.1), the above optimization problem can be solved.
4. If the update on the control policy is negligible, terminate the algorithm. Otherwise, let $i = i + 1$ and jump to Step 2.

It can be proved that the policy and value function obtained by the above iterative algorithm, denoted as V^∞ and Ω^∞, can satisfy the Bellman equation in (11.79). Thus, Ω^∞ is the optimal control policy and V^∞ represents the minimum discounted cost for each initial system state. Notice that in the second step of the above algorithm, we should solve a fixed-point problem with unknown statistics. The stochastic approximation introduced in Section 11.1 can be applied. Specifically, the procedure is elaborated below.

Stochastic approximation algorithm for value function

1. Let $j = 1$. Initialize the value function V^i, and denote it as $V^{i,j}$.
2. At the jth stage, denote $\mathbf{s}_j$ as the system state and update the value function as follows:

$$V^{i,j+1}(\mathbf{s}_j) = \frac{j}{j+1}V^{i,j}(\mathbf{s}_j) + \frac{1}{j+1}\left[g(\mathbf{s}_j, \mathbf{a}_j, \xi_j) + \gamma V^{i,j}(\mathbf{s}_{j+1})\right], \tag{11.82}$$

and

$$V^{i,j+1}(\mathbf{s}) = V^{i,j}(\mathbf{s}), \quad \forall \mathbf{s} \neq \mathbf{s}_j. \tag{11.83}$$

Thus, the value for the current system state $\mathbf{s}_j$ is updated, and others remain the same. As a remark, notice that $g(\mathbf{s}_j, \mathbf{a}_j, \xi_j) + \gamma V^{i,j}(\mathbf{s}_{j+1})$ is an unbiased estimation of $\mathbb{E}_{\xi_t}\left[g(\mathbf{s}_t, \mathbf{a}_t, \xi_t) + \gamma \Pr(\mathbf{s}_{t+1}|\mathbf{s}_t, \mathbf{a}_t, \xi_t)V^i(\mathbf{s}_{t+1})\right]$. Moreover, since the knowledge on $\mathbf{s}_{j+1}$ is required, the above update should be calculated after observing the next system state.
3. If the update on the value function is negligible, terminate the algorithm. Otherwise, let $j = j + 1$ and jump to Step 2.

11.3.1.1 Case study: multi-carrier power allocation without channel statistics

In this example, we shall continue the optimization of power allocation without channel statistics, as initiated in Example 11.3, by the online value and policy iteration described in this section (Section 11.3.1). The notations will follow the definitions in Example 11.3 and Section 11.2.3.1. In order to match the MDP formulation of this section, we shall treat the QSI only as the system state (i.e., $\mathbf{s}_t$), and the CSI as the independent variables in each optimization stage (i.e., ξ_t). Specifically, the problem formulation of Example 11.3 is established below.

- **System state:** Since the distribution of CSI is i.i.d. in each frame, we can treat the CSI as the independent random variables ξ_t, instead of the system state. Thus:

$$\xi_t = \{h_i(t)|\forall i\}, \tag{11.84}$$

whose distribution is unknown. The system state becomes:

$$\mathbf{s}_t = \{q(t)\}. \tag{11.85}$$

- **Control policy:** As elaborated in Example 11.3, when the CSI is removed from the system state, the control action becomes the power allocation for all possible CSI given the QSI. Thus, the control action of the tth frame is

$$\mathbf{a}_t = \overline{\Omega}(\mathbf{s}_t) = \left\{\Omega\Big(q(t), \{h_i(t)\}\Big)\Big|\forall i, h_i(t)\right\}. \tag{11.86}$$

  As a remark note that Ω and $\overline{\Omega}$ represent the same scheduling behavior, however their mathematical meanings are different: Ω is a policy with respect to QSI and CSI, and $\overline{\Omega}$ is a policy with respect to QSI only. Hence, one action in $\overline{\Omega}$ consists of a number of actions in Ω with the same QSI.
- **Transition kernel:** Given the system state, CSI and control action of the tth frame, the transition kernel can be written as

$$\Pr(\mathbf{s}_{t+1}|\mathbf{s}_t, \mathbf{a}_t, \xi_t) = \Pr\Big(q(t+1)\Big|q(t), \{h_i(t)|\forall i\}, \{p_i(t)|\forall i\}\Big), \tag{11.87}$$

  where

$$q(t+1) = \max\{0, q(t) - d(t)\} + c(t), \tag{11.88}$$

$$d(t) = \left\lfloor \frac{1}{B}\sum_{i=1}^{N_F} \log_2\left(1 + \frac{p_i(t)\|h_i(t)\|^2}{\sigma_z^2}\right)\right\rfloor \tag{11.89}$$

  is the number of packets transmitted in the tth frame, $c(t)$ is the number of arrival packets in the tth frame. Note that the randomness of $q(t+1)$ comes from random packet arrival $c(t)$.
- **Cost:** The overall cost function as defined in Section 11.2.3.1 is

$$\overline{G} = \lim_{T\to+\infty} \mathbb{E}\left\{\sum_{t=1}^{T}\gamma^{t-1}\left[\eta\frac{q(t)}{\lambda} + \sum_{i=1}^{N_F} p_i(t)\right]\right\}. \tag{11.90}$$

As elaborated in Example 11.3, the Bellman equation for the above MDP problem is

$$\overline{V}\Big(q(t)\Big) = \min_{\overline{\Omega}(q(t))}\left\{\eta\frac{q(t)}{\lambda} + \mathbb{E}_{\xi_t}\left[\sum_{i=1}^{N_F} p_i(t)\right] + \gamma\, \mathbb{E}_{\xi_t, c(t)}\left[\overline{V}\Big(q(t+1)\Big)\right]\right\}, \tag{11.91}$$

or equivalently:

$$\overline{V}\Big(q(t)\Big) = \mathbb{E}_{\xi_t}\min_{\overline{\Omega}(q(t))}\left\{\eta\frac{q(t)}{\lambda} + \sum_{i=1}^{N_F} p_i(t) + \gamma\sum_{c(t)}\Pr\Big(c(t)\Big)\overline{V}\Big(q(t+1)\Big)\right\}. \tag{11.92}$$

Note that without the distribution knowledge of CSI ξ_t, the expectations in the above Bellman equation cannot be calculated directly. Hence, we have to rely on the online value and policy iteration introduced in this section, which consists of two levels of iteration. The outer iteration is for updating the policy, and the inner one is to find the value function corresponding to the policy. The procedure is elaborated below.

Step 1 (Initialize a policy): In order to obtain an initial policy, we can first initialize the value function, which is denoted as $\overline{V}^0$, and derive a power allocation policy by solving the right-hand-side of the above Bellman equation (11.92) with the initialized value function. Specifically, from the principle of water-filling method, it is known that given a total transmission power in the tth frame, the optimal power allocation on each subcarrier can be written as

$$p_i(t) = \max\left\{0, \frac{1}{\beta_t} - \frac{\sigma_z^2}{\|h_i(t)\|^2}\right\}, \quad \forall i = 1, 2, \ldots, N_F, \tag{11.93}$$

where β_t depends on the total transmission power of the tth frame. β_t is usually referred to as the Lagrange multiplier as the above power allocation is derived via convex optimization [5]. Moreover, given $\mathbf{s}_t$ and ξ_t, the β_t with respect to the initialized value function $\overline{V}^0$, denoted as β_t^1, can be calculated according to the right-hand-side of (11.92), i.e.,

$$\beta_t^1 = \arg\min\left\{\eta\frac{q(t)}{\lambda} + \sum_{i=1}^{N_F}\left\{0, \frac{1}{\beta_t} - \frac{\sigma_z^2}{\|h_i(t)\|^2}\right\} + \gamma\sum_{c(t)}\Pr\left(c(t)\right)\overline{V}^0\left(q(t+1)\right)\right\}.$$

As a result, the initial power allocation policy is then given by

$$p_i^1(t) = \max\left\{0, \frac{1}{\beta_t^1} - \frac{\sigma_z^2}{\|h_i(t)\|^2}\right\}, \quad \forall i = 1, 2, \ldots, N_F. \tag{11.94}$$

Step 2 (Value function evaluation): Given the power allocation policy derived based on β_t^i $(i = 1, 2, \ldots)$, the corresponding value function can be calculated as follows:

- Let $j = 1$. Initialize the value function by $\overline{V}^{i,j} = \overline{V}^{i-1}$.
- At the jth stage, denote $\mathbf{s}_j$ as the system state and update the value function as follows:

$$\overline{V}^{i,j+1}(\mathbf{s}_j) = \frac{j}{j+1}\overline{V}^{i,j}(\mathbf{s}_j) + \frac{1}{j+1}\left[g(\mathbf{s}_j, \mathbf{a}_j, \xi_j) + \gamma V^{i,j}(\mathbf{s}_{j+1})\right], \tag{11.95}$$

and

$$\overline{V}^{i,j+1}(\mathbf{s}) = \overline{V}^{i,j}(\mathbf{s}), \quad \forall \mathbf{s} \neq \mathbf{s}_j. \tag{11.96}$$

- Let $j = j + 1$ and repeat the above step until the iteration converged. Let $\overline{V}^i$ be the converged value function.

Step 3 (Policy evaluation): Update the β_t as

$$\beta_t^{i+1} = \arg\min \left\{ \eta \frac{q(t)}{\lambda} + \sum_{i=1}^{N_F} \left\{ 0, \frac{1}{\beta_t} - \frac{\sigma_z^2}{\|h_i(t)\|^2} \right\} + \gamma \sum_{c(t)} \Pr\left(c(t)\right) \overline{V}^i\left(q(t+1)\right) \right\}.$$

As a result, the updated power allocation policy is then given by

$$p_i^{i+1}(t) = \max\left\{0, \frac{1}{\beta_t^{i+1}} - \frac{\sigma_z^2}{\|h_i(t)\|^2}\right\}, \quad \forall i = 1, 2, \ldots, N_F. \tag{11.97}$$

If the update on the power allocation policy is negligible, terminate the algorithm; otherwise, let $i = i + 1$ and jump to the Step 2.

The above online algorithm will converge to the optimal power allocation as in Section 11.2.3.1, and the performance illustrated in Figure 11.4 also applies on it. Moreover, it can be observed from the above algorithm that the calculation of β_t in each iteration requires the distribution knowledge to packet arrival. We may learn it from the history of packet arrival, if it is unknown at the very beginning. In fact, Q-learning is one more elegant way to handle this situation, which is elaborated in the below section.

11.3.2 Q-learning

The stochastic-approximation-based learning approach in the previous section is able to handle the situation that the controller knows the transition kernel $\Pr(\mathbf{s}_{t+1}|\mathbf{s}_t, \mathbf{a}_t, \xi_t)$ but does not know its expectation with respect to ξ_t, i.e., $\Pr(\mathbf{s}_{t+1}|\mathbf{s}_t, \mathbf{a}_t) = \mathbb{E}_{\xi_t}[\Pr(\mathbf{s}_{t+1}|\mathbf{s}_t, \mathbf{a}_t, \xi_t)]$. Regarding to the example in Section 11.3.1.1, it refers to the circumstance that the transmitter knows the distribution of packet arrival in each frame, but not the CSI distribution. Q-learning is a more powerful tool to solve MDP problems with unknown transition kernel $\Pr(\mathbf{s}_{t+1}|\mathbf{s}_t, \mathbf{a}_t)$. In other words, it can handle the power allocation even without the statistics of packet arrival.

We use the infinite-horizon MDP with discounted cost in Problem 11.4 as the example to demonstrate the method of Q-learning. First of all, the Q function is defined as

$$Q(\mathbf{s}, \mathbf{a}) = \min_{\Omega} \lim_{T\to+\infty} \mathbb{E}\left[\sum_{t=1}^{T} \gamma^{t-1} g(\mathbf{s}_t, \mathbf{a}_t) \middle| \mathbf{s}_1 = \mathbf{s}, \mathbf{a}_1 = \mathbf{a}\right]. \tag{11.98}$$

Hence, the relation between value function and Q function is

$$V(\mathbf{s}) = \min_{\mathbf{a}} Q(\mathbf{s}, \mathbf{a}), \tag{11.99}$$

and the optimal control action for system state $\mathbf{s}$ is

$$\Omega^*(\mathbf{s}) = \arg\min_{\mathbf{a}} Q(\mathbf{s}, \mathbf{a}). \tag{11.100}$$

In other words, the optimal control policy can be easily obtained with the Q function of the MDP. Moreover, the Bellman equation in (11.48) can be written in the form of Q function, i.e.:

$$Q(\mathbf{s}_t, \mathbf{a}_t) = g(\mathbf{s}_t, \mathbf{a}_t) + \gamma \Pr(\mathbf{s}_{t+1}|\mathbf{s}_t, \mathbf{a}_t) V(\mathbf{s}_{t+1}), \tag{11.101}$$

$$V(\mathbf{s}_t) = \min_{\mathbf{a}_t} \left\{ g(\mathbf{s}_t, \mathbf{a}_t) + \gamma \Pr(\mathbf{s}_{t+1}|\mathbf{s}_t, \mathbf{a}_t) \min_{\mathbf{a}_{t+1}} Q(\mathbf{s}_{t+1}, \mathbf{a}_{t+1}) \right\}, \tag{11.102}$$

or

$$Q(\mathbf{s}_t, \mathbf{a}_t) = g(\mathbf{s}_t, \mathbf{a}_t) + \gamma \Pr(\mathbf{s}_{t+1}|\mathbf{s}_t, \mathbf{a}_t) \min_{\mathbf{a}_{t+1}} Q(\mathbf{s}_{t+1}, \mathbf{a}_{t+1}). \tag{11.103}$$

In order to compute and store the values of Q function, it is actually required that both the state and action spaces should be finite. Regarding the example of Section 11.3.1.1, we should quantize the transmission power into finite levels.

The Bellman equation (11.103) provides an iterative way to evaluate the Q function. The procedure is described below.

VI algorithm for *Q* function

The Q function defined in (11.103) can be evaluated by the following steps:

1. Let $i = 0$ and initialize the Q function $Q(\mathbf{s}, \mathbf{a})$ for all possible $\mathbf{s} \in \mathcal{S}$ and $\mathbf{a} \in \mathcal{A}$, which is denoted as $Q^i(\mathbf{s}, \mathbf{a})$.
2. In the ith iteration, update the Q function as

   $$Q^{i+1}(\mathbf{s}_t, \mathbf{a}_t) = g(\mathbf{s}_t, \mathbf{a}_t) + \gamma \Pr(\mathbf{s}_{t+1}|\mathbf{s}_t, \mathbf{a}_t) \min_{\mathbf{a}_{t+1}} Q^i(\mathbf{s}_{t+1}, \mathbf{a}_{t+1}),$$

   for all possible $\mathbf{s}_t \in \mathcal{S}$ and $\mathbf{a}_t \in \mathcal{A}$.
3. If the update from Q^i to Q^{i+1} for any system state and action is negligible (or less than one predetermined threshold), the iteration terminates. Otherwise, let $i = i + 1$ and jump to Step 2.

The above VI require the knowledge on the distribution of $\Pr(\mathbf{s}_{t+1}|\mathbf{s}_t, \mathbf{a}_t)$. If they are not available at the controller, the Q-learning algorithm is provided below.

***Q*-learning algorithm**

1. Let $j = 1$, initialize the Q function, denoted as Q^j.
2. At the jth stage, denote $\mathbf{s}_j$ as the system state and $\mathbf{a}_j$ as the action, update the value function as follows:

   $$Q^{i+1}(\mathbf{s}_i, \mathbf{a}_i) = \frac{j}{j+1} Q^i(\mathbf{s}_i, \mathbf{a}_i) + \frac{1}{j+1} \left[g(\mathbf{s}_j, \mathbf{a}_j) + \gamma \min_{\mathbf{a}_{i+1}} Q^i(\mathbf{s}_{i+1}, \mathbf{a}_{i+1}) \right],$$

   and

   $$Q^{i+1}(\mathbf{s}, \mathbf{a}) = Q^i(\mathbf{s}, \mathbf{a}), \quad \forall(\mathbf{s}, \mathbf{a}) \neq (\mathbf{s}_j, \mathbf{a}_j). \tag{11.104}$$

Since the knowledge on $\mathbf{s}_{j+1}$ is required, the above update should be calculated after observing the next system state.

3. If the update on the Q function is negligible, terminate the algorithm. Otherwise, let $j = j + 1$ and jump to Step 2.

In the above algorithm, the control action for each stage should be chosen to guarantee that the Q function for all pairs of system state and control action can be well trained.

11.3.2.1 Case study: multi-carrier power allocation via Q-learning

In this example, we shall still consider the power-allocation problem introduced in Section 11.3.1.1, however, with more practical assumption that not only the distribution of CSI but also the distribution of packet arrival is unknown. We shall show that the Q-learning method could provide an online optimization algorithm. The definitions of system state, control policy, transition kernel, and cost function follow that in Section 11.3.1.1. Note that Q function is defined in terms of system state and control action both with finite space, we can quantize the choice of Lagrange multiplier β_t into a finite space $\mathscr{B}$. Thus, the Q function can be written as

$$Q\Big(q(t), \beta_t\Big), \text{ where } \beta_t \in \mathscr{B}. \tag{11.105}$$

Hence, the Bellman equation in terms of Q function becomes:

$$\begin{aligned} Q\Big(q(t), \beta_t\Big) &= \eta \frac{q(t)}{\lambda} + \mathbb{E}_{\xi_t}\left[\sum_{i=1}^{N_F} p_i(t)\right] + \gamma \mathbb{E}_{\xi_t} \Pr\Big(q(t+1)|q(t), \beta_t, \xi_t\Big) \\ &\min_{\beta} Q\Big(q(t+1), \beta\Big). \end{aligned} \tag{11.106}$$

Without the statistics knowledge of ξ_t and packet arrival $c(t)$, the Q-learning algorithm is provided below:

1. Let $i = 1$. Initialize the Q function, denoted as Q^i.
2. In the ith frame, let $q(i)$ and β_i be the system state (QSI) and the Lagrange multiplier for power allocation, update the value function as follows:

$$\begin{aligned} Q^{i+1}\Big(q(i), \beta_i\Big) &= \frac{i}{i+1} Q^i\Big(q(i), \beta_i\Big) \\ &+ \frac{1}{i+1}\left[\eta \frac{q(i)}{\lambda} + \sum_{j=1}^{N_F} p_j(i) + \gamma \min_{\beta} Q\Big(q(i+1), \beta\Big)\right], \end{aligned}$$

and

$$Q^{i+1}(\mathbf{s}, \beta) = Q^i(\mathbf{s}, \beta), \quad \forall (\mathbf{s}, \beta) \neq \Big(q(i), \beta_i\Big). \tag{11.107}$$

Since the knowledge on $q(i+1)$ is required, the above update should be calculated after observing the next system state.

3. If the update on the Q function is negligible, terminate the algorithm. Otherwise, let $i = i + 1$ and jump to Step 2.

With the Q function, the power allocation for the tth frame can be written as

$$p_i(t) = \max\left\{0, \frac{1}{\beta_t} - \frac{\sigma_z^2}{\|h_i(t)\|^2}\right\}, \quad \forall i = 1, 2, \ldots, N_F, \tag{11.108}$$

where $\beta_t = \min_\beta Q(q(t), \beta)$. Note that in this solution, the Lagrange multiplier β_t is determined according to the QSI only. A better solution may be obtained if we treat both CSI and QSI as the system state in the Q-learning algorithm (β_t is then determined according to both CSI and QSI). However, it requires the quantization of CSI and larger system complexity.

Comparing with the method introduced in Section 11.3.1, it can be observed that the Q-learning approach is more general in the sense that it can be applied on the situation without knowledge of transition kernel. However, the price to pay is that the Q function depends on both system state and control action. Thus, the storage and computation complexities for evaluating the Q function is higher.

11.4 Summary and discussion

In this chapter, we focus on the wireless resource allocation along a number of frames, where the MDP is used to formulate the scheduling as an stochastic optimization problem. As the foundation of stochastic learning, we first elaborate on the basics of stochastic approximation. Then we introduce the MDP with three different formulations, and one example of power allocation is provided for each formulation. As we can see, MDP is powerful to handle the multistage optimization problem with random future. Moreover, it is common that some system statistics are unknown before running; we continue to introduce the reinforcement learning to construct online algorithms, which collect the system information and drive the scheduling to optimal.

In order to simplify the elaboration, we have ignored the proofs of some mathematical statements in this chapter. For the readers who are interested in a more rigorous treatment on mathematical derivation, please refer to [2,3] for the discussions on the MDP and [6] for the discussions on the reinforcement learning. Moreover, the application of MDP and reinforcement learning in wireless resource allocation has drawn a number of research interests. For example, the infinite-horizon MDP has been used to optimize the point-to-point link [7], cellular uplink [8,9], cellular downlink [10], relay networks [11], and wireless cache systems [12], where the average transmission delay is either minimized or constrained. Moreover, the low-complexity algorithm design via approximate MDP can be considered to avoid the curse of dimensionality [13].

References

[1] Robbins H, and Monro S. A Stochastic Approximation Method. The Annals of Mathematical Statistics. 1951;22(3):400–407.
[2] Bertsekas D. Dynamic Programming and Optimal Control: Volume I. 3rd ed. Belmont: Athena Scientific; 2005.
[3] Bertsekas D. Dynamic Programming and Optimal Control: Volume II. 3rd ed. Belmont: Athena Scientific; 2005.
[4] Kleinrock L. Queueing Systems. Volume 1: Theory. 1st ed. New York: Wiley-Interscience; 1975.
[5] Boyd S, and Vandenberghe L. Convex Optimization. 1st ed. Cambridge: Cambridge University Press; 2004.
[6] Sutton R, and Barto A. Reinforcement Learning: An Introduction. 2nd ed. Cambridge: MIT Press; 2018.
[7] Bettesh I, and Shamai S. Optimal Power and Rate Control for Minimal Average Delay: The Single-User Case. IEEE Transactions on Information Theory. 2006;52:4115–4141.
[8] Moghadari M, Hossain E, and Le LB. Delay-Optimal Distributed Scheduling in Multi-User Multi-Relay Cellular Wireless Networks. IEEE Transactions on Communications. 2013;61(4):1349–1360.
[9] Cui Y, and Lau VKN. Distributive Stochastic Learning for Delay-Optimal OFDMA Power and Subband Allocation. IEEE Transactions on Signal Processing. 2010;58(9):4848–4858.
[10] Cui Y, and Jiang D. Analysis and Optimization of Caching and Multicasting in Large-Scale Cache-Enabled Heterogeneous Wireless Networks. IEEE Transactions on Wireless Communications. 2017;16(1):250–264.
[11] Wang R, and Lau VKN. Delay-Aware Two-Hop Cooperative Relay Communications via Approximate MDP and Stochastic Learning. IEEE Transactions on Information Theory. 2013;59(11):7645–7670.
[12] Zhou B, Cui Y, and Tao M. Stochastic Content-Centric Multicast Scheduling for Cache-Enabled Heterogeneous Cellular Networks. IEEE Transactions on Wireless Communications. 2016;15(9):6284–6297.
[13] Powell WB. Approximate Dynamic Programming: Solving the Curses of Dimensionality. 2nd ed. New Jersey:John Wiley & Sons; 2011.

Chapter 12
Q-learning-based power control in small-cell networks

Zhicai Zhang[1], Zhengfu Li[2], Jianmin Zhang[3], and Haijun Zhang[3]

Because of the time-varying nature of wireless channels, it is difficult to guarantee the deterministic quality of service (QoS) in wireless networks. In this chapter, by combining information theory with the effective capacity (EC) principle, the energy-efficiency optimization problem with statistical QoS guarantee is formulated in the uplink of a two-tier femtocell network. To solve the problem, we introduce a Q-learning mechanism based on Stackelberg game framework. The macro users act as leaders and know the emission power strategy of all femtocell users (FUS). The femtocell user is the follower and only communicates with the macrocell base station (MBS) without communicating with other femtocell base stations (FBSs). In Stackelberg game studying procedure, the macro user chooses the transmit power level first according to the best response of the femtocell, and the micro users interact directly with the environment, i.e., leader's transmit power strategies, and find their best responses. Then, the optimization problem is modeled as a noncooperative game, and the existence of Nash equilibriums (NEs) is studied. Finally, in order to improve the self-organizing ability of femtocell, we adopt Q-learning framework based on noncooperative game, in which all the FBS are regarded as agents to achieve power allocation. Numerical results show that the algorithm cannot only meet the delay requirements of delay-sensitive traffic but also has good convergence.

12.1 Introduction

In recent years, most voice and data services have occurred in indoor environments. However, due to long-distance transmission and high penetration loss, the indoor coverage of macrocell may not be so good. As a result, FBS has gained wide attention in wireless industry [1,2]. With the exponential growth of mobile data traffics, wireless

[1]College of Physics and Electronic Engineering, Shanxi University, China
[2]Beijing Key Laboratory of Network System Architecture and Convergence, Beijing University of Posts and Telecommunications, China
[3]School of Computer & Communication Engineering, University of Science and Technology Beijing, China

communication networks play a more and more important role in the global emissions of carbon dioxide [3]. Obviously, the increasing energy cost will bring significant operational cost to mobile operators. On the other hand, limited battery resources cannot meet the requirement of mass data rate. In this chapter, the concept of green communication is proposed to develop environmentally friendly and energy saving technologies for future wireless communications. Therefore, the use of energy aware communication technology is the trend of the next generation wireless network design.

In a two-tier network with shared spectrum, due to cross-layer interference, the target user and the femtocell user of each signal-to-interference-plus-noise ratio (SINR) sampling macrocell are coupled. The SINR target concept establishes application-related minimum QoS requirements for each user. It is reasonably expected that since home users deploy femtocell for their own benefit and because they are close to their BS, femtocell users and cellular users seek different SINRs (data rates)—usually higher data rates using femtocell. However, QoS improvements from femtocell should be at the expense of reducing cellular coverage.

In practice, a reliable delay guarantee is provided for delay sensitivity. High data rate services, such as video calling and video conferencing, are the key issues of wireless communication network. However, due to the time-varying nature of wireless channel, it is difficult and unrealistic to apply the traditional fixed delay QoS guarantee. To solve the problem, the statistical QoS metric with delay-bound violation probability have been widely adopted to guarantee the statistical delay QoS [4–6]. In [5], for delay-sensitive traffic in single-cell downlink Orthogonal Frequency Division Multiple Access (OFDMA) networks, the effective spectrum design based on EC delay allocation is studied. In [6], a joint power and subchannel allocation algorithm in vehicular infrastructure communication network is proposed. It has the requirement of delayed QoS. However, as far as we know, EC-based delay provision in two-tier femtocell cellular networks has not been widely studied.

In addition, due to the scarcity of spectrum, the microcell and macrocell usually share the same frequency band. However, in the case of co-channel operation, intensive and unplanned deployment will lead to serious cross-tier and co-tier interference, which will greatly limit the performance of the network. Microcell base stations are low-power, low-cost, and user-deployed wireless access points that use local broadband connections as backhauls. Not only users but also operators benefit from femtocell. On the one hand, users enjoy high-quality links; on the other hand, operators reduce operating expenses and capital expenditure due to service uninstallation and user deployment of FBS.

Therefore, it is necessary to design effective interference suppression mechanism in the two-tier femtocell networks to reduce cross-tier and co-tier interference. In [7,8], the author reviews the interference management in two-level microcellular networks and small cellular networks. In [9], the authors have proposed a novel interference coordination scheme using downlink multicell chunk allocation with dynamic inter-cell coordination to reduce co-tier interference. In [10], based on cooperative Nash bargaining game theory, this chapter proposes a cognitive cell joint uplink subchannel and power-allocation algorithm to reduce cross-layer interference. In [11], in order to maximize the total capacity of all femtocell users under the constraints of co-layer/cross-layer interference and given minimum capacity QoS, a resource allocation scheme for cognitive nano-cellular is proposed. However, the delay QoS provisioning was not taken into consideration in [9–11]. Also, as a randomly deployed base station, the traditional centralized network scheduling is difficult to optimize its network performance. Therefore, the reinforcement learning which can provide agents with self-organization capability has attracted considerable interest in academy and industry [12].

In [13], the author has studied the self-optimization, self-configuration and self-optimization of small cell network. In [14], aiming at the power control problem in ad-hoc networks, an enhanced learning algorithm based on random virtual game theory is proposed. In [15,16], aiming at the utility maximization problem of two-tier femtocell networks, an enhanced learning algorithm based on hierarchical Stackelberg game is proposed. However, the algorithms in [15,16] require frequent routing information exchanges between macrocells and micro cells, which greatly increases network load.

In recent years, there have been many researches on energy-efficient resource management [17,18]. Energy efficiency was first proposed by Goodman *et al.*, which is defined as the number of error-free delivered bits for each energy-unit used in transmission and is measured in bit/joule [19]. FBS is a low-power, low-cost base station that can enhance indoor environment coverage and unload traffic from macrocell. A low complexity energy-efficient subchannel allocation scheme is proposed in [17], but the method does not consider interference caused by neighbors. In [18], joint subchannel allocation and power control are modeled as a potential game to maximize energy efficiency of multicell uplink OFDMA systems, but QoS guarantees are without consideration.

In addition to energy saving management of radio resources, femtocell network is another promising technology for energy saving. Because of this type of deployment strategy, the transmitter is closer to the receiver and reduces penetration and path loss. As we know, FBS is installed by end users, who have not enough professional skills to configure parameters of FBS. On this account, FBS should have self-learning ability to automatically configure and optimize its operating information, e.g., transmit power assignment. In recent years, reinforcement learning mechanism, such as *Q*-learning, is widely used in radio resource allocation of wireless network [20–22]; however, most of the existing works are focusing in cognitive radio networks.

In addition, providing delay QoS guarantees while minimizing energy consumption is a key problem in green communication systems. For example, in real-time services, such as multimedia video conferencing and live sports events, latency time is a key QoS metric. Since the time-varying channel, deterministic delay QoS guarantee mechanisms used in wired networks cannot take affect in wireless networks [4]. To address this issue, statistical QoS provisioning, in terms of delay exponent and EC, has become an effective method to support real-time service in wireless networks [23–25].

Machine learning can be widely used in modeling various technical problems of next-generation systems, such as large-scale Multiple-Input Multiple-Output (MIMO), device-to-device networks, heterogeneous networks constituted by femtocells and small cells [26]. Therefore there are some existing works about the application of machine learning to small cell networks. In [27], a heterogeneous fully distributed multi-objective strategy based on a reinforcement learning model are proposed to be built for self-configuration and optimization of femtocells. In [28], the state of the system consists of the user's specific allocation of small cell resource blocks and channel quality, and the action consists of downlink power control actions. The reward is quantified based on an improvement in SINR. The results show that the compensation strategy based on the reinforcement learning model has achieved excellent performance improvement.

Machine learning is a discipline that specializes in algorithms that can be learned from data. In general, these algorithms are run by generating models built from observational data and then using the generated models to predict and make decisions. Most problems in machine learning can be translated into multi-objective optimization problems where multiple targets must be optimized simultaneously in the presence of two or more conflicting targets. Mapping multiple optimization problems to game theory can provide a stable solution [29]. Game theory focuses on the nature of equilibrium states. For example, an in-depth study of the concept of algorithmic game theory is the concept of anarchy price. The anarchy of certain problems (such as routing in a crowded network) is the biggest difference between the NE configuration (the best way each participant routes in the case of other people's behavior) and the global optimal solution. However, NE is a subtle object. In large systems with multiple entities with limited information, it is more natural to assume that each entity self-adjusts its behavior based on past experience, producing results that may be stable or unstable. Therefore, it is desirable to use the understanding of the characteristics of such an adaptive algorithm to draw conclusions about the behavior of the overall system [30].

In this chapter, we will study energy-efficient power control in uplink two-tier femtocell networks with delayed QoS guarantees. Based on the concept of EC, we formulate an energy-efficiency optimization problem with statistical QoS guarantee. To solve the problem, a transmit power learning mechanism based on Stackelberg game is proposed. In the learning process, macro users are leaders and can communicate with micro users. Femto-users act as followers and only know the power strategy of leader rather than other followers. Besides, leader knows followers' best responses of transmit power and selects strategy first; followers move subsequently. We use EC as a network performance metric to provide statistical delay QoS.

Then we adopt pricing mechanism to protect macrocell users (MU) from severe cross-layer interference. The optimization problem is modeled as a noncooperative game. Then we study the existence of NEs. Specifically, considering that femtocells are deployed by end users who have not enough professional skills to configure and optimize FBSs's parameters, such as transmit power, we use Q-learning theory to enable femtocells to achieve self-organizing capability in terms of transmission power and other parameters. And we propose a distributed Q-learning procedure based on Stackelberg game. Simulation results show the proposed algorithm has a better performance in terms of convergence compared with a conjecture-based multi-agent Q-learning (CMAQL) algorithm with no information exchange between each player [31]. Based on the noncooperative game framework, a Boltzmann distribution-based weighted filter Q-learning algorithm (BDb-WFQA) is proposed to realize power allocation. The simulation results show the proposed BDb-WFQA algorithm can meets the large-scale delay requirements, has a better convergence performance, and a small EC loss compared with the Noncooperative Game-based Power Control Algorithm (NGb-PCA). This algorithm can meet the large-scale delay requirements, and has better convergence performance, and has a small EC loss.

The rest of the chapter is organized as follows. In Section 12.2, we briefly discuss EC and formulate an energy-efficiency optimization problem with statistical delay provisioning. A noncooperative game theoretic solution is proposed in Section 12.3. A *Q*-learning mechanism based on Stackelberg game framework and a WFQA based on Boltzmann distribution are proposed in Section 12.4. Simulation results are shown in Section 12.5. In Section 12.6, we conclude the chapter.

12.2 System model

12.2.1 System description

The scenario considered in this chapter is shown in Figure 12.1, where N femtocells are overlaid in a macrocell, which constitutes a two-tier femtocell network. FBSs are in closed subscriber group (CSG) mode, i.e., mobile stations (MSs) that are not the members of the CSG, are not allowed to access the CSG FBSs.

As Figure 12.2, the representative macrocell is covered by several femtocell. In each femtocell, FBS provides services for its FUS. To analyze traceability, we assume that only one active MU/FU is scheduled in each MBS/FBS in each signaling slot. It is worth pointing out that the algorithm obtained under this assumption can be easily extended to each MBS/FBS scenario with multiple active users.

However, in the two-tier network, cross-layer interference significantly hinders the performance of traditional power-control schemes. For example, signal-strength-based power control (channel inversion) adopted by cellular users results in unacceptable deterioration of cellular SINR. Because users carry out high-power transmissions

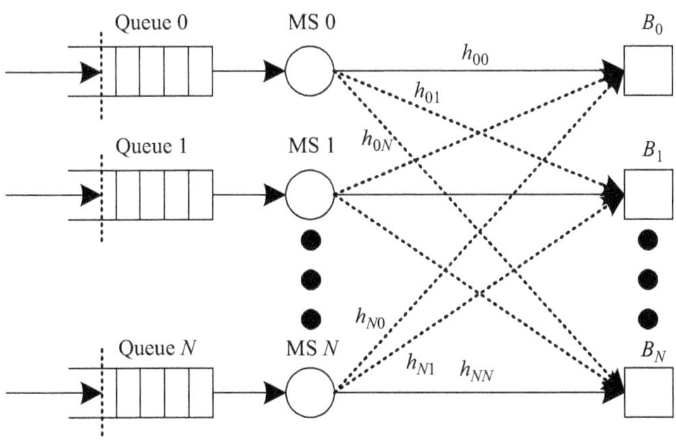

Figure 12.1 System model of two-tier femtocell networks

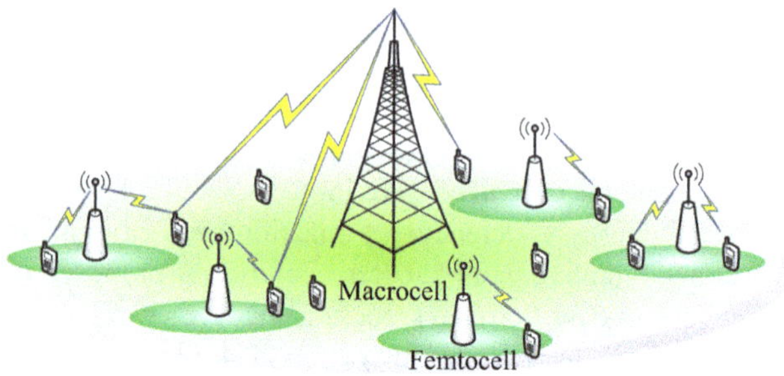

Figure 12.2 The scenario of two-tier femtocell networks

at the edges of their cell to meet their received power targets and cause excessive cross-layer interference in nearby microcellular networks. Due to scalability, security, and limited availability of backhaul bandwidth, base station (BS) and femtocells base station APS.

Let $i \in N = \{0, 1, \ldots, N\}$ denote the index of active users, where $i = 0$ indicates the scheduled user in macrocell B_0 and $i \in \{1, 2, \ldots, N\}$ denotes the scheduled users in femtocell B_i.

Let B_i $(i \in \mathcal{N})$ denote the base station (BS), where $\mathcal{N} = \{0, 1, 2, \ldots, N\}$. B_0 denotes the MBS, and B_i $(i \in \mathcal{N}, i \neq 0)$ is FBS. We assume that each MS will be allocated only a subchannel, and in order to avoid intra cell interference during each frame time slot, only one active MS in each cell can occupy the same frequency. Let $i \in \mathcal{N}$ denote the index of scheduled user in B_i.

The received SINR of MS i in B_i can be expressed as

$$\gamma_i(p_i, \mathbf{p}_{-i}) = \frac{p_i h_{ii}}{\sum_{j \neq i} p_j h_{ij} + \sigma_i^2}, \quad \forall i \in \mathcal{N}, \tag{12.1}$$

where p_i denotes the transmit power of MS i, and $\mathbf{p}_{-i}, (-i \in \mathcal{N})$ denotes the transmit power of other MSs except MS i. h_{ii} and h_{ij} are the channel gains from MS i to BS B_i, B_j respectively, σ_i^2 is the variance of additive white Gaussian noise (AWGN) of MS i.

Similarly, the received SINR of MU is

$$\gamma_0 = \frac{h_{0,0} p_0}{\sum_{i=1}^{N} h_{i,0} p_i + \sigma_0^2}, \tag{12.2}$$

where $h_{i,0}$ is the channel gain from FBS B_i to the active MU and $h_{0,0}$ denotes the channel gain from MBS to its active MU.

According to the Shannon's capacity formula, the ideal achievable data rate of MS i is

$$R_i(p_i, \mathbf{p}_{-i}) = w\log_2(1 + \gamma_i(p_i, \mathbf{p}_{-i})), \tag{12.3}$$

where w is the bandwidth of each subchannel.

12.2.2 Effective capacity

The concept of statistical delay guarantee has been extensively studied in the effective bandwidth theory [32]. Delay provisioning is an important and challenging problem in wireless networks for delay-sensitive services such as video calls, video conferencing, and online games. However, owing to the time-varying nature of wireless channels, it is difficult and unrealistic to have deterministic delay guarantees for mobile services.

According to Shannon's law, the capacity potential of a femtocell can be quickly verified by associating the wireless link capacity (in bits per second) in the bandwidth with the SINR. SINR is a function of the desired transmitter's transmit power, path loss, and shadows. Path losses cause the transmitted signal to decay as $Ad^{-\alpha}$, where A is a fixed loss, d is the distance between the transmitter and receiver, and α is the path loss exponent. The key to increasing capacity is to enhance reception between intended transmitter receiver pairs by minimizing d and α.

To solve this problem, the concept of EC is proposed in [4], which is defined as the maximum constant arrival rate guaranteed by a statistical delay specified by the QoS index of θ on a time-varying channel.

Based on large deviation principle, Chang [32] has pointed out that with sufficient condition, for a dynamic queueing system with stationary ergodic arrival and service processes, the queue length process $Q(t)$ converges to a random variable $Q(\infty)$:

$$\lim_{Q_{th}\to\infty} \frac{\log(\Pr\{Q(\infty) > Q_{th}\})}{Q_{th}} = -\theta, \tag{12.4}$$

where Q_{th} is queue length bound and $\theta > 0$ is the decay rate of the tail distribution of the queue length $Q(\infty)$.

If $Q_{th} \to \infty$, we get the approximation of the buffer violation probability, $\Pr\{Q(\infty) > Q_{th}\} \approx e^{-\theta Q_{th}}$.

We can find that the larger θ corresponds to the faster fading rate, which means more stringent QoS constraints, while the smaller θ leads to a slower fading rate, which means a looser QoS requirement. Similarly, the delay-outage probability can be approximated by [4], $\Pr\{Delay > D_{th}\} \approx \xi e^{-\theta\delta D_{th}}$, where D_{th} is the maximum tolerable delay, ξ is the probability of a non-empty buffer, and δ is the maximum constant arrival rate.

The concept of EC is proposed by Wu *et al.*, in [4], it is defined as the maximum constant arrival rate that can be supported by the time-varying channel, while ensuring the statistical delay requirement specified by the QoS exponent θ. The EC is formulated as

$$E^c(\theta) = -\lim_{K\to\infty} \frac{1}{K\theta} \ln(\mathbb{E}\{e^{-\theta\sum_{k=1}^{K} S[k]}\}), \tag{12.5}$$

where $\{S[k] | k = 1, 2, \ldots, K\}$ denotes the discrete-time, stationary, and ergodic stochastic service process. $\mathbb{E}\{\cdot\}$ is the expectation over the channel state.

We assume that the channel fading coefficients remain unchanged over the frame duration T and vary independently for each frame and each MS. From (12.5), $S_i[k] = TR_i[k]$ is obtained. Based on the above analysis, the EC of MS i can be simplified as

$$E_i^c(\theta_i) = -\frac{1}{\theta_i T} \ln(\mathbb{E}\{e^{-\theta_i T R_i[k]}\}). \tag{12.6}$$

12.2.3 Problem formulation

The energy efficiency under statistical delay guarantees of MS i is defined as the ratio of the EC to the totally consumed energy as follows:

$$\eta_i(p_i, \mathbf{p}_{-i}) = \frac{E_i^c(\theta_i)}{p_i + p_c}. \tag{12.7}$$

In (12.7), p_c represents the average energy consumption of device electronics, including mixers, filters, and digital-to-analog converters, and excludes that of the power amplifier. Femtocell is deployed randomly by end users, so cross-layer interference against MU is uncertain. When the cross-tier interference exceeds MUs' threshold, the communication of MUs is seriously affected or even interrupted.

Given the minimum SINR guarantee γ_i^*, FU i's utility function can be expressed as

$$u_i(p_i, \mathrm{p}_{-i}) = \begin{cases} E_i^C(\theta_i), & \text{if } \gamma_i(p_i, \mathrm{p}_{-i}) \geq \gamma_i^* \\ 0, & \text{otherwise} \end{cases}, \tag{12.8}$$

where p_{-i} denotes the transmit power of other FBSs except FBS i.

Our goal is to maximize the energy efficiency of each MS while meeting the delay QoS guarantee. Therefore, the corresponding problem is

$$\max \frac{-\ln(\mathbb{E}\{e^{-\theta_i T R_i[k]}\})}{\theta_i T (p_i + p_c)}, \tag{12.9a}$$

$$p_i \geq p_{\min}, \forall i \in \mathcal{N}, \tag{12.9b}$$

$$p_i \leq p_{\max}, \forall i \in \mathcal{N}, \tag{12.9c}$$

$$\theta_i > 0, \forall i \in \mathcal{N}, \tag{12.9d}$$

where $p_{\min}$ and $p_{\max}$ are the lower and upper bounds of each MS's transmit power, respectively.

12.3 Noncooperative game theoretic solution

In this section, we formulate the FBSs' selfish behavior as a noncooperative game. Let $G = \{N', \{\mathrm{P}_i\}, u_i(p_i, \mathrm{p}_{-i})\}$ denote the noncooperative power control game (NPCG), where $N' = \{1, 2, \ldots N\}$ is the set of FBSs, $\{\mathrm{P}_i\}$ is the strategy set of all players, and $u_i(p_i, \mathrm{p}_{-i})$ is the utility function. It is obvious that the level of each FU's utility depends on its FBS's transmit power and other FBSs' strategies. We assume that each

FBS is rational. Each player pursues the maximization of its own utility, which can be denoted as

$$\max_{p_i \in P_i} u_i(p_i{}^*, \mathrm{p}^*{}_{-i}), \forall i \in N', \tag{12.10a}$$

$$subjectto : p_i \geq p_i{}^{\min}, \tag{12.10b}$$

$$p_i \leq p_i{}^{\max}, \tag{12.10c}$$

$$\theta_i > 0. \tag{12.10d}$$

Definition 12.1. *A given power control strategy $(p_i{}^*, \mathrm{p}^*{}_{-i})$ is an NE point of NPCG, if for $\forall i \in N', \forall p_i \in P_i$, the following inequality is satisfied:*

$$u_i(p_i{}^*, \mathrm{p}^*{}_{-i}) \geq u_i(p_i, \mathrm{p}^*{}_{-i}). \tag{12.11}$$

On the NE point, no player can improve their utility by changing its strategy unilaterally [33]. Generally speaking, we can prove the existence of NE by the following Theorem 12.1.

Theorem 12.1. *An NE exits in the NPCG $G = \{N', \{\mathrm{P}_i\}, u_i(p_i, \mathrm{p}_{-i})\}$, if for all $i \in N'$, the following two conditions are satisfied:*

1. *In Euclidean space R^N, the strategy set $\{P_i\}$ is a non-empty, convex, and compact subset.*
2. *The utility function $u_i(p_i, \mathrm{p}_{-i})$ is continuous in (p_i, p_{-i}) and quasi-concave in p_i.*

Proof. For condition (1), it is obvious that $\{\mathrm{P}_i\}$ is a non-empty, convex, and compact subset. We prove condition (2) in the following:

For fixed p_{-i}, let $h_i = (g_{i,i}/\sum_{j\neq 0,i} g_{j,i}p_j + g_{0,i}p_0 + \sigma^2)$ denote the channel gain-to-interference-plus-noise ratio of FU i and $f(h_i)$ is the probability density of h_i. For almost all practical environment, we assume $f(h_i)$ is continuous and differentiable in h_i:

$$\tilde{E}_{C_i}(\theta_i) = -\frac{1}{\theta_i} \ln \left[\int_0^\infty e^{-\theta_i R_i(p_i)} f(h_i) dh_i \right] - ug_{i,0}p_i. \tag{12.12}$$

It is apparent that $u_i(p_i, \mathrm{p}_{-i})$ is continuous in (p_i, p_{-i}). In addition, $ug_{i,0}p_i$ is linear about p_i, which does not affect the concavity of the equation. Based on $(\int_a^b f(p,h)dh)'_p = \int_a^b f'_p(p,h)dh$, it is easy to prove that $(\partial^2 \tilde{E}_{C_i}(\theta_i)/\partial p_i{}^2) \leq 0$. Thus, $\tilde{E}_{C_i}(\theta_i)$ is concave and the condition (2) is proved. □

Therefore, the NPCG $G = \{N', \{\mathrm{P}_i\}, u_i(p_i, \mathrm{p}_{-i})\}$ admits an NE point.

12.4 *Q*-learning algorithm

As far as we know, FBS is installed by end users, who have not enough professional skill to configure parameters of FBS. On this account, FBS should have self-learning ability to automatically configure and optimize the FBS's operating information. In Stackelberg learning game, every user in the network behaves as an intelligent agent, whose goal is to maximize its expected utility. The game is repeated to achieve the best strategy. Stackelberg learning framework has two hierarchies: (1) MU maximizes its expected utility by knowing the response of all FUs to each possible game and (2) given an MU strategy, FU performs a noncooperative game.

In this section, we will adopt the reinforcement learning mechanism based on the Stackelberg game framework to achieve the energy-saving transmission power allocation while ensuring the delay of QoS requirements.

To be compatible with reinforcement learning mechanism [13], the transmit power of MS i is discretized as $\mathcal{P}_i = (p_{i,v_i}|v_i = 1, 2, \ldots, V_i)$. The probability of MS i choosing transmit power p_{i,v_i} at time slot t is π^t_{i,v_i} ($\pi^t_{i,v_i} \in \pi^t_i$), and $\pi^t_i = (\pi^t_{i,v_i}|v_i = 1, 2, \ldots, V_i)$, which satisfies $\sum_{v_i=1}^{V_i} \pi^t_{i,v_i} = 1$.

Then, the expected utility of MS i is given by

$$u_i(\pi^t_i, \boldsymbol{\pi}^t_{-i}) = \mathbb{E}\{\eta_i(\mathbf{p})|\pi^t_i, \boldsymbol{\pi}^t_{-i}\} = \sum_{\mathbf{p}\in\mathbf{P}} \eta_i(\mathbf{p}) \prod_{j\in\mathcal{N}} \pi^t_{j,v_j}, \tag{12.13}$$

where $\mathbf{p} = (p_{0,v_0}, \ldots p_{i,v_i} \ldots, p_{N,v_N}) \in \mathbf{P}$ is the actions of all MSs at time slot t, and $\mathbf{P} = \times_{i\in\mathcal{N}} \mathcal{P}_i$.

12.4.1 Stackelberg game framework

The Stackelberg game model [33] is very suitable for two-tier femtocell networks, where MS 0 is formulated as a leader, and MSs i ($i \in \mathcal{N}, i \neq 0$) are modeled as followers. In Stackelberg game framework, the leader can first know the strategy information of all followers, then choose the action, and followers can receive the leader's strategy and then act.

Based on above analysis, it is easy to find that the goal of MS 0 is to maximize its revenue as

$$\max u_0(\pi_0, \boldsymbol{\pi}_{-0}), \tag{12.14}$$

and the objective of MS i, ($i \in \mathcal{N}, i \neq 0$) is

$$\max u_i(\pi_i, \boldsymbol{\pi}_{-i}). \tag{12.15}$$

Because of this fact, FBS is deployed by end users randomly. There is no communication or coordination between femtocells. They pursue their profits selfishly. Equation (12.10a)–(12.10d) can be modeled as a noncooperative power allocation sub-game $G = [\{i\}, \{\mathcal{P}_i\}, \{u_i\}]$ ($i \in \mathcal{N}, i \neq 0$).

Theorem 12.2. *Given MS 0's strategy π_0, there exists a mixed strategy $\{\pi^*_i, \boldsymbol{\pi}^*_{-i}\}$ satisfies:*

$$u_i(\pi^*_i, \boldsymbol{\pi}^*_{-i}) \geq u_i(\pi_i, \boldsymbol{\pi}^*_{-i}), \tag{12.16}$$

which is an NE point.

Proof. As it has been shown in [33], every limited strategic game has a mixed strategy equilibrium, i.e., there exists $NE(\pi_0)$ for given π_0. □

Lemma 12.1. *The problem exists a Stackelberg equilibrium (SE) point* $\{\pi_0^*, \pi_i^*, \boldsymbol{\pi}_{-i}^*\}$ $(\forall i \in \mathcal{N}, i \neq 0)$*, which is a mixed strategy.*

The proof of the existence of SE point is omitted here for brevity. We will employ reinforcement learning mechanism, called Q-learning, to find SE point.

12.4.2 Q-learning

Based on reinforcement learning, each femtocell can be an intelligent agent with self-organization and self-learning ability, and its operation parameters can be optimized according to the environment. Q-learning is a common reinforcement learning method, which is widely used in self-organizing femtocell networks. It does not need teachers' signals. It can optimize its operation parameters through experiments and errors. Each BS acts as an intelligent agent, maximizing its profit by interacting directly with the environment.

We define $p_{i,v_i} \in \mathcal{P}_i$ $(\forall i \in \mathcal{N})$ as actions of Q-learning model, and $\boldsymbol{\pi}_{-i}^t$ $(-i \in \mathcal{N})$ are environment states. In a standard Q-learning model, an agent interacts with its environment to optimize its operation parameters. First, the agent perceives the environment and observes its current state $s \in S$. Then, the agent selects and performs an action $a \in A$ according to a decision policy $\pi : s \rightarrow a$ and the environment will change to the next state $s+1$. Meanwhile, the agent receives a reward W from the environment.

In each state, there is a Q-value associated with each action. The definition of a Q-value is the sum of the received reward (possibly discounted) when an agent performs an associated action and then follows a given policy thereafter [34]. Similarly, the optimal Q-value is the sum of the received reward when the optimal strategy is followed. Therefore, the Q-value can be expressed as

$$Q_\pi^t(a,s) = W^t(a,s) + \lambda \max_{a \in A} Q_\pi^{t-1}(a,s+1), \tag{12.17}$$

where $W^t(a,s)$ is the received reward when an agent performs an action a at the state s in the time slot t and λ denotes a discount factor, $0 \leq \lambda < 1$. However, at the beginning of the learning, the (12.17) has not been established. The deviation between the optimal value and the realistic value is

$$\Delta Q_\pi^t(a,s) = W^t(a,s) + \lambda \max_{a \in A} Q_\pi^{t-1}(a,s+1) - Q_\pi^{t-1}(a,s), \tag{12.18}$$

Therefore, the Q-value is updated as the following rule:

$$\begin{aligned} Q_\pi^t(a,s) &= Q_\pi^{t-1}(a,s) + \rho_t \Delta Q_\pi^t(a,s) \\ &= (1-\rho_t) Q_\pi^{t-1}(a,s) + \rho_t [W^t(a,s) + \lambda \max_{a \in A} Q_\pi^{t-1}(a,s+1)], \end{aligned} \tag{12.19}$$

where ρ_t is a learning factor.

Q-learning represents the knowledge by means of a Q-function, whose Q-value is defined as $Q_i^{t+1}(p_{i,v_i}, \boldsymbol{\pi}_{-i}^{t+1})$ and is updated according to

$$Q_i^{t+1}(p_{i,v_i}, \boldsymbol{\pi}_{-i}^{t+1}) = Q_i^t(p_{i,v_i}, \boldsymbol{\pi}_{-i}^{t+1}) + \alpha^t(r_i(p_{i,v_i}, \boldsymbol{\pi}_{-i}^{t+1}) - Q_i^t(p_{i,v_i}, \boldsymbol{\pi}_{-i}^{t+1})), \quad (12.20)$$

where $\alpha^t \in [0, 1)$ is the learning rate. In (12.20), $r_i(p_{i,v_i}, \boldsymbol{\pi}_{-i}^{t+1})$ is the reward function of MS i when selecting p_{i,v_i} and other MSs' strategies are $\boldsymbol{\pi}_{-i}^{t+1}$. The relationship between reward and utility function of MS i is

$$u_i(\pi_i^t, \boldsymbol{\pi}_{-i}^t) = \sum_{v_i=1}^{V_i} \pi_{i,v_i} r_i(p_{i,v_i}, \boldsymbol{\pi}_{-i}^t). \quad (12.21)$$

Each BS updates its strategy based on Boltzmann distribution [14], which is formally described as

$$\pi_{i,v_i}^t = \frac{\exp(Q_i^t(p_{i,v_i}, \boldsymbol{\pi}_{-i}^{t+1})/\tau)}{\sum_{v_i=1}^{V_i} \exp(Q_i^t(p_{i,v_i}, \boldsymbol{\pi}_{-i}^{t+1})/\tau)}, \quad (12.22)$$

where τ ($\tau > 0$) is temperature parameter. Higher value of τ causes the probabilities of all actions of MS i to be nearly equal; lower value of τ leads to the probability of actions' bigger difference with respect to their Q-values.

12.4.3 Q-learning procedure

In this section, we will study the QoS aware power allocation in sparse and dense deployment of femtocell networks. The Q-learning mechanism based on Stackelberg game framework is adopted.

12.4.3.1 Sparsely deployed scenario

In sparsely deployed femtocell networks, for example, in rural areas, the interference between FBS is negligible due to path loss and penetration loss.

As we have assumed before, MBS knows complete strategies of all FBSs and updates its Q-value by (12.20). The reward function of MS 0 is the following:

$$r_0(p_{0,v_0}, \boldsymbol{\pi}_{-0}^{t+1}) = \sum_{\mathbf{p} \in \mathbf{P}} \{\eta_0(\mathbf{p})\delta_{-(0,v_0)}^{t+1}\}, \quad (12.23)$$

where $\delta_{-(0,v_0)}^{t+1} = \prod_{j \in \mathcal{N}, j \neq 0} \pi_{j,v_j}^{t+1}$ denotes the probability of actions vector $\mathbf{p}_{-(0,v_0)} = (p_{1,v_1}, \ldots p_{i,v_i} \ldots, p_{N,v_N})$.

For MS i ($\forall i \in \mathcal{N}, i \neq 0$), due to the fact that FBSs can receive MBS's transmit power strategy, and there is no interference between FBSs, the reward function of MS i is

$$r_i(p_{i,v_i}, \pi_0^{t+1}) = \sum_{v_0=1}^{V_0} \delta_{-(i,v_i)}^{t+1} \eta_i(p_{i,v_i}, p_{0,v_0}), \quad (12.24)$$

where $\delta_{-(i,v_i)}^{t+1} = \pi_{0,v_0}^{t+1}$.

12.4.3.2 Densely deployed scenario

In densely deployed femtocell networks, such as in urban areas, FBS is close to each other and the interference between with each other cannot be ignored.

In this scenario, the reward function of MS 0 is according to (12.13). The reward function of MS i ($\forall i \in \mathcal{N}, i \neq 0$) in this scenario is

$$r_i(p_{i,v_i}, \pi_0^{t+1}) = \sum_{v_0=1}^{V_0} \delta_{-(i,v_i)}^{t+1} \hat{\eta}_i(p_{i,v_i}, p_{0,v_0}). \tag{12.25}$$

Since there is no communication or cooperation between FBSs, if the selected power level at time shot $t+1$ satisfies $p_{i,v_i}^{t+1} = p_{i,v_i}$, $\hat{\eta}_i^{t+1}(p_{i,v_i}, p_{0,v_0})$ is estimated by (12.20), else $\hat{\eta}_i^{t+1}(p_{i,v_i}, p_{0,v_0}) = \hat{\eta}_i^t(p_{i,v_i}, p_{0,v_0})$:

$$\hat{\eta}_i^{t+1}(p_{i,v_i}, p_{0,v_0}) = \frac{\eta_i(p_{i,v_i}, \mathbf{p}_{-i}) - \hat{\eta}_i^t(p_{i,v_i}, p_{0,v_0})}{\rho^t(p_{i,v_i}, p_{0,v_0}) + 1} + \hat{\eta}_i^t(p_{i,v_i}, p_{0,v_0}). \tag{12.26}$$

In (12.20), $\eta_i(p_{i,v_i}, \mathbf{p}_{-i})$ is the real value when $p_{i,v_i}^{t+1} = p_{i,v_i}$, which can be calculated by the feedback information from FBS B_i. $\rho^t(p_{i,v_i}, p_{0,v_0})$ is the times number when the MS 0s transmit power is p_{0,v_0}, and MS i selects power level p_{i,v_i} until time shot t [14].

12.4.3.3 Distributed Q-learning algorithm

Theorem 12.3. *The proposed algorithm can discover a SE mixed strategy.*

Due to the limited space, the convergence of the proposed algorithm can be found in [35]. As the Algorithm 12.1, a distributed Q-learning algorithm is proposed.

Algorithm 12.1: Distributed Q-learning algorithm

Step 1: Initialization: for $t = 0$, $Q_i^t(p_{i,v_i}, \boldsymbol{\pi}'_{-i})$, $\forall i \in \mathcal{N}$;
power discretization: $\mathbf{p}_i = (p_{i,1}, \ldots, p_{i,v_i}, \ldots, p_{i,V_i})$;
Learning:
Step 2: Update $t = t + 1$;
Step 3: Update π_i^t according to (12.22);
Step 4: Update MS 0s transmit power according to $p_{0,v_0^*} = \arg\max(Q_0^t(p_{0,v_0}, \boldsymbol{\pi}'_{-0}))$, and send the value of π_0^t to FBS.
Step 5: Update MS i's ($i \neq 0$) transmit power according to
$p_{i,v_i^*} = \arg\max(Q_i^t(p_{i,v_i}, \boldsymbol{\pi}'_{-i}))$, and send the value of π_i^t to MBS.
Step 6: Calculate MS 0s reward according to (12.23), calculate MS i's ($i \neq 0$) reward by (12.25).
Step 7: Update MS i's Q-value by (12.20).
Step 8: Back to **Step 2**.
End learning

12.4.4 The proposed BDb-WFQA based on NPCG

In the NPCG $G = \{N', \{\mathrm{P}_i\}, u_i(p_i,\mathrm{p}_{-i})\}$, the strategy set $\mathrm{P}_i = [p_i^{\min}, p_i^{\max}]$ is continuous, which is not applicable in the Q-learning method. To be compatible with the Q-learning method, we discretize the continuous power set $p_i \in \mathrm{P}_i = [p_i^{\min}, p_i^{\max}]$ [14] as following:

$$p_i(a_i) = \left(1 - \frac{a_i}{M_i}\right) p_i^{\min} + \frac{a_i}{M_i} p_i^{\max}, \tag{12.27}$$

where $a_i \in A_i = \{0, 1, \ldots, M_i\}$ and A_i is the set of FBS Bi's action space. The number of action space is $M_i + 1$.

Thus, the NPCG $G = \{N', \{\mathrm{P}_i\}, u_i(p_i,\mathrm{p}_{-i})\}$ transforms to the discrete game $G_d = \{N', \{A_i\}, u_i(p_i(a_i),\mathrm{p}_{-i})\}$. Based on the discrete game G_d, we design an appropriate Q-learning algorithm to achieve the EC-based power allocation for FBSs.

According to the Q-learning theory, agent, state, and action can be defined as follows:

Agent: All of the FBSs B_i. As stated in Section 12.2, there is only one scheduled active FU in each FBS during each signaling slot. Therefore, $i \in N' = \{1, 2, \ldots N\}$.

State: FBS Bi's policy π_i^{t-1} and the received interference of FU i $I_i^t = \sum_{j \neq i} g_{j,i} p_j + g_{0,i} p_0 + \sigma^2$. $\pi_i^{t-1} = (\pi_{i,0}^{t-1}, \ldots, \pi_{i,a_i}^{t-1}, \ldots, \pi_{i,M_i}^{t-1})$ is a probability vector, where π_{i,a_i}^{t-1} is the probability with which FBS B_i chooses action a_i at time $t-1$.

Action: Each discrete transmit power can be denoted by each action a_i. Therefore, we use action $a_i \in A_i$ to replace the FBS Bi's transmit power. According to policy π_i^{t-1}, FBS B_i selects transmit power a_i with probability π_{i,a_i}^{t-1}.

The Q-value can be formulated according to the utility function of discrete game G_d:

$$Q_{\pi_i}^t(a_i, s_i) = W^t(a_i, s_i) + \lambda \max_{a \in A} Q_{\pi_i}^{t-1}(a_i, s_i + 1) = \pi_{i,a_i}^{t-1} u_i^t(p_i(a_i), \mathrm{p}_{-i}). \tag{12.28}$$

Therefore, we adopt the following rule to update Q-value:

$$Q_{\pi_i}^t(a_i, s_i) = Q_{\pi_i}^{t-1}(a_i, s_i) + \rho_t(A)[\pi_{i,a_i}^{t-1} u_i^t(p_i(a_i), \mathrm{p}_{-i}) - Q_{\pi_i}^{t-1}(a_i, s_i)], \tag{12.29}$$

where $\rho_t(A)$ is the learning factor. In practice, FBS B_i knows neither the opponents' strategy π_{-i}^{t-1} nor the true utility before running the action a_i. But the FBS B_i can compute the attainable utility $u_i^t(p_i(a_i), \mathrm{p}_{-i})$ through the feedback information of the receiver; thus, we design the following learning factor to estimate the utility:

$$\rho_t(A) = \begin{cases} \dfrac{1}{t\pi_{i,a_i}^{t-1}} & \text{if } \mathbf{A} = a_i, \\[2ex] \dfrac{\alpha}{t + \alpha} & \text{otherwise.} \end{cases} \tag{12.30}$$

where α is the filter parameter. Notice that the $t\pi_{i,a_i}^{t-1}$ is approximately equal to the frequency of FBS B_i selecting the action a_i until time t. Therefore, the $Q_{\pi_i}^t(a_i, s_i)$

is the approximation of FU i's expected utility when FBS B_i adopts the action a_i. Additionally, $\alpha/(t+\alpha)$ decreases with the increase of the time slot and

$$\frac{\alpha}{t+\alpha}=\begin{cases}\geq 0.5 & \text{if t} \leq \alpha,\\ < 0.5 & \text{if t} > \alpha.\end{cases} \tag{12.31}$$

Therefore, we can believe that the α represents the weight of the historical learning process and can speed up learning. Moreover, in order to ensure fast convergence, we propose a weighted filter algorithm-based Boltzmann distribution [31] to update the policy π_i^t:

$$\pi_{i,a_i}^t=\frac{\alpha^2}{t^2+\alpha^2}\frac{\exp(Q_{\pi_i}^t(a_i,s_i)/T)}{\sum_{j=0}^{M_i}\exp(Q_{\pi_i}^t(j,s_i)/T)}+\frac{t^2\pi_{i,a_i}^{t-1}}{t^2+\alpha^2}, \tag{12.32}$$

where T is the temperature parameter.

The convergence of the proposed algorithm is proved as follows. Because the NEs exit in the NPCG and the action set of discrete game G_d is the discretized strategy set of NPCG, there is at least one action $a_i{}^*$ at which the maximum Q-value $Q_{\pi_i}^*$ is attained [13]. Although there may be more than one optimal action $a_i{}^*$, the maximum Q-value $Q_{\pi_i}^*$ is unique. Additionally, we can prove the $\sum_{t=1}^{\infty}(\alpha/(t+\alpha))=\infty$ and $\sum_{t=1}^{\infty}(\alpha/(t+\alpha))^2<\infty$ easily. According to [16], we achieve $Q_{\pi_i}^t(a_i,s_i)\to Q_{\pi_i}^*(a_i{}^*,s_i)$ as $t\to\infty$ with probability 1, where $Q_{\pi_i}^*(a_i{}^*,s_i)$ denotes the optimal Q-value for optimal action $a_i{}^*$ at state s_i.

The proposed BDb-WFQA algorithm is given in Algorithm 12.2.

Algorithm 12.2: The proposed BDb-WFQA algorithm

Step 1: Initialization: for $t=0$;
Step 2: Select $a_i^0=rand(0,M_i)$;
Step 3: Compute $p_i(a_i^0)$ using (12.27);
Step 4: Compute the received interference I_i^0;
Step 5: Compute $u_i{}^0(p_i(a_i{}^0),\mathrm{p}_{-i})$ using (12.8);
Step 6: Initialize $Q_{\pi_i}^0(a_i{}^0,s_i)=u_i{}^0(p_i(a_i{}^0),\mathrm{p}_{-i})$;
Step 7: Initialize policy $\pi_i{}^0$,for the action $j\in A_i$,$\pi_{i,j}^0=\frac{\exp(Q_{\pi_i}^0(j,s_i)/T)}{\sum_{k=0}^{M_i}\exp(Q_{\pi_i}^0(k,s_i)/T)}$
End Initialization
Step 8: Learning: for $t=t+1$.
Step 9: Select $a_i^t=l$ according to π_i^{t-1}, $l\in A_i$;
Step 10: Compute $p_i(a_i^t)$ using (12.27);
Step 11: Compute the received interference I_i^t of FU i;
Step 12: Compute $u_i^t(p_i(a_i{}^t),\mathrm{p}_{-i})$ using (12.8);
Step 13: Update the Q-value using (12.19);
Step 14: Update the π_i^t using (12.22);
End learning

12.5 Simulation and analysis

12.5.1 Simulation for Q-learning based on Stackelberg game

In this section, we will introduce the simulation of the proposed algorithm and simulate a CMAQL algorithm to compare with the proposed algorithm [31]. Macro-users and micro-users are distributed randomly in the two-tier femtocell networks and share the same spectrum with $w = 200$ kHz. The channel-fading is modeled as Rayleigh block-fading channels, the fading-block duration $T = 1$ ms. Noise spectral density is $N_0 = -174$ dBm/Hz. The channel gain for macro-user and femto-users are λL^{-3} and λL^{-4}, respectively, where L is the transmitter–receiver separation in meters, and $\lambda = 2 \times 10^{-4}$ [36].

The additional circuit power p_c is 10 dBm for all users, the lower bound of transmit power for each user is $p_{\min} = 10$ dB m, and upper bounds for femto-users and macro-user are $p_{\max} = 20$ dB m and $p_{\max} = 30$ dB m, respectively. The transmit power region $[p_{\min}, p_{\max}]$ is divided into d parts equally in the Q-learning procedure, and we consider $d = 3, 10, 20$, respectively, in the simulation.

Figure 12.3 shows expected utilities with respect to the QoS exponent. When the value of θ is small, i.e., $\theta \leq 10^{-4}$, there is no significant expected utility change. This is because the smaller the QoS index, the looser the delay requirements, and the EC is close to Shannon capacity, regardless of the arrival rate and delay requirements. Instead, when the value of θ is larger, and the delay requirement is tighter, EC and expected utility decrease correspondingly. On the other hand, the discretization of transmit power results in the best transmit power error, while the smaller of d will lead to a higher expected utility loss.

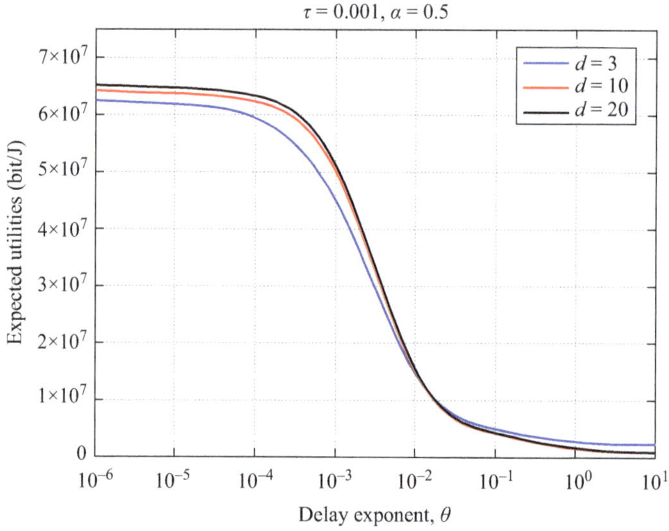

Figure 12.3 Expected utilities versus different QoS exponent

Figures 12.4 and 12.5 show the convergence of the proposed algorithm. From these figures, we can see that the proposed algorithm has faster convergence speed than CMAQL algorithm. The reason is that micro-users in the proposed *Q*-learning mechanism can share transmit power strategy with macro-user, while the value of $\delta^{t+1}_{-(i,v_i)}$ is estimated by only the past experiences in CMAQL algorithm.

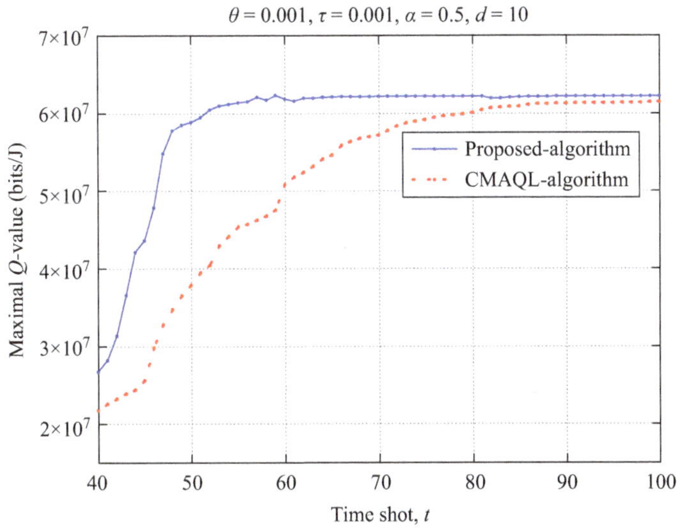

Figure 12.4 The convergence of Q-learning mechanism

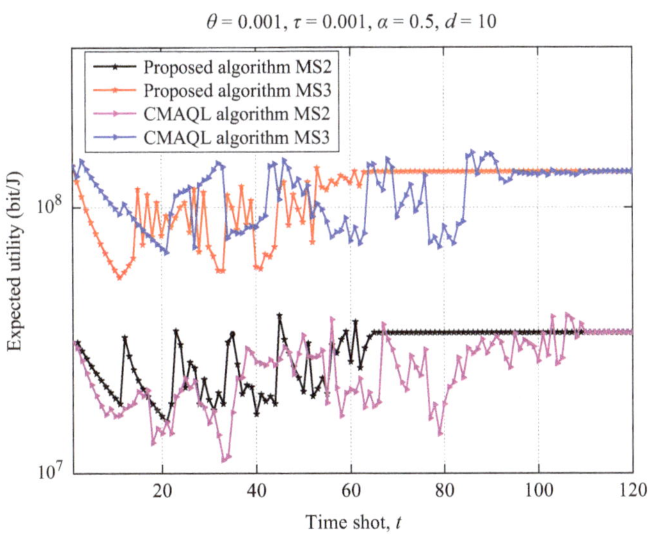

Figure 12.5 The convergence of expected utilities

12.5.2 Simulation for BDb-WFQA algorithm

This section shows the performance of the proposed algorithm through numerical simulations. We consider a two-tier femtocell network in which a macrocell is overlaid by three cochannel deployed femtocells, which is similar to the scenario in [15]. The related simulation parameters, as shown in Table 12.1, note that channel fading is considered to be independent. And the Rayleigh block fading is also allocated. The path loss of MUs and FUs is kd^{-3} and kd^{-4}, respectively, where d is the distance from the transmitter to receiver and $k = 2 \times 10^{-4}$ [36].

The convergence of the proposed BDb-WFQA is shown in Figure 12.6. For the comparison purpose, two other algorithms are also simulated. The first one is the NGb-PCA. The second one is the hierarchical reinforcement learning algorithm (HRLA) in [15], which employs the discrete power as action profile and chooses

Table 12.1 Simulation parameters

Parameter	Value
The channel bandwidth w	100 kHz
Macrocell radius	500 m
Femtocell radius	20 m
FBS transmit power $p_i^{\min}$	10 dB m
FBS transmit power $p_i^{\max}$	20 dB m
The number of discrete power value	3
Power of AWGN σ^2	−110 dB m
The minimum SINR γ_i^*	5 dB

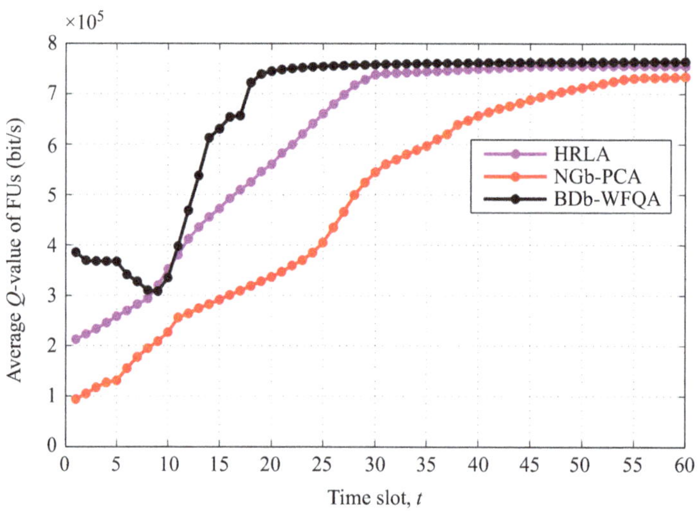

Figure 12.6 The convergence of algorithms

action through the Boltzmann distribution. From Figure 12.6, after 20 iterations, the BDb-WFQA Algorithm is stable, which guarantees the convergence of the algorithm. In addition, we find that compared with NGB-PCA and HLA, the proposed BDB-WFQA has faster convergence. This is because the proposed BDb-WFQA employs the discrete power as action profile and uses the weighted filter way to update the policy where the filter parameter α can be considered as a believable parameter to accelerate learning.

The average EC of FUs is shown in Figure 12.7. It is can be observed that the average EC of FUs reduces with the increase of delay QoS exponent θ for both the NGb-PCA and the proposed BDb-WFQA. This is because a larger θ means a more stringent delay requirement. In addition, we find that the performance of the proposed BDb-WFQA is slightly lower than that of NGB-PCA. This is because the proposed BDb-WFQA uses discrete action contours, but it may lose correct power values. However, as mentioned earlier, we know that the proposed BDb-WFQA converges faster than NGb-PCA.

The average EC of MUs is shown in Figure 12.8. From the five curves in the Figure 12.8, it can be observed that the average EC of MUs increases with the increase of μ. Besides, we can see that when the pricing factor $\mu = 0$, the average EC of MUs is the smallest. This is because that $\mu = 0$ means there is not interference constraint at FBSs' side, the FBSs will choose the optimal transmit power to selfishly increase their EC, which will cause severe cross-tier interference to macrocell. When the $\mu \geq 170$ dB, MUs gain the largest average EC. This is because the sufficiently large pricing will make the FBSs choose the smallest transmit power; thus, the cross-tier interference each MU received is smallest, and the achievable average

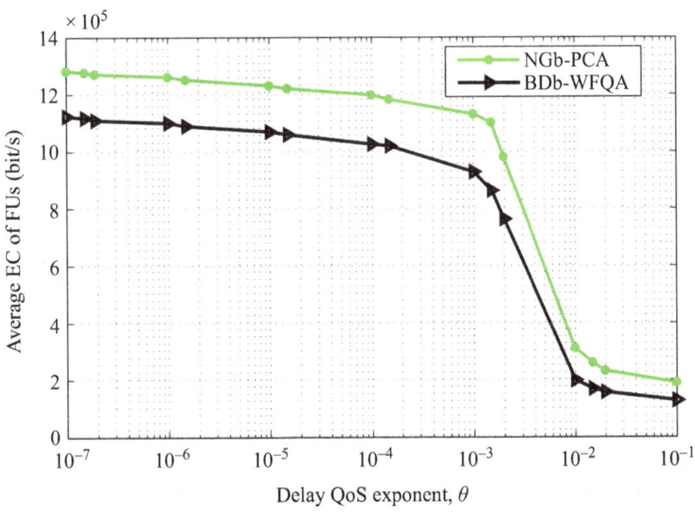

Figure 12.7 The average EC of FUs versus delay QoS parameter θ

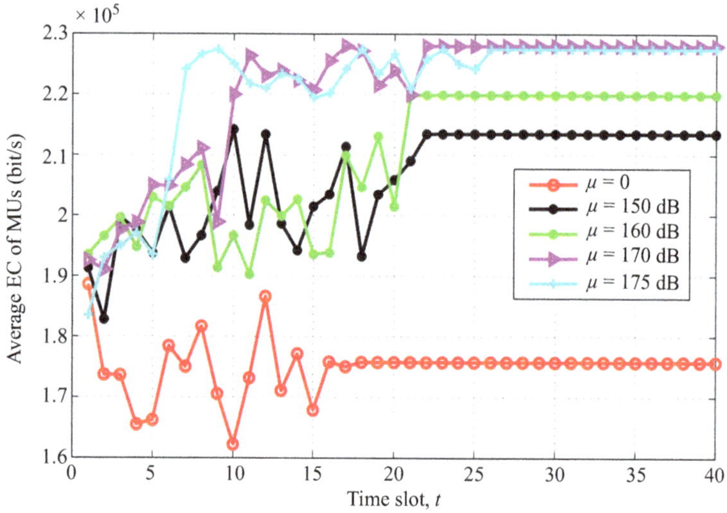

Figure 12.8 The average EC of MUs

EC is largest. Therefore, we can choose a pricing factor which can guarantee that the received cross-tier interference of MUs is acceptable, and the FBSs can achieve a good EC performance.

12.6 Conclusion

We investigate the energy efficient power control in two-tier femtocell networks with considering delay-QoS guarantee. In order to enhance FBS's ability of self-configuration and self-optimization, we propose a Q-learning mechanism based on Stackelberg game framework. In the learning procedure, macro-user is a leader, who knows transmit power strategies of all femto-users and chooses power level first; while femto-users acting as followers can communicate with only the leader and move subsequently. Finally, a distributed Q-learning algorithm based on Stackelberg game is proposed to study the downlink power control problem in two-layer femto-cellular networks with statistical delay QoS constraints and interlayer interference constraints. We also design a network performance measure with the statistical delay QoS provisioning based on the concept of EC. Then we model the power allocation problem as a noncooperative game and verify the existence of NEs. In particular, we adopt a Q-learning theory to achieve self-organizing ability of femtocells and propose BDb-WFQA to realize power allocation of FBS. Simulation results show that compared with CMAQL algorithm, the Q-learning mechanism based on Stackelberg game framework has faster convergence speed. Simulation results also show that the proposed BDb-WFQA increases the achievable EC of MUs through the pricing method and provides a delay QoS provision for MUs and FUs. Furthermore, the

BDb-WFQA has a better performance in the convergence compared with NGb-PCA and HRLA. In the future, we will continue to study wireless resource optimization issues and further use game theory to ensure the QoS in wireless networks.

References

[1] Chandrasekhar V, Andrews JG, and Gatherer A. Femtocell networks: A survey: In IEEE Commun Mag. 2008;46(9):59–67.
[2] Zhang H, Jiang C, Beaulieu NC, *et al.* Resource allocation in spectrum-sharing OFDMA femtocells with heterogeneous services. IEEE Trans Commun. 2014;62(7):2366–2377.
[3] Li GY, Xu Z, Xiong C, *et al.* Energy-efficient wireless communications: tutorial, survey, and open issues. IEEE Wireless Commun. 2011;18(6):28–35.
[4] Wu D, and Negi R. Effective capacity: a wireless link model for support of quality of service. IEEE Trans Wireless Commun. 2003;2(4):630–643.
[5] Xiong C, Li GY, Liu Y, *et al.* Energy-efficient design for downlink OFDMA with delay-sensitive traffic. IEEE Trans Wireless Commun. 2013;12(6): 3085–3095.
[6] Zhang H, Ma Y, Yuan D, *et al.* Quality-of-service driven power and sub-carrier allocation policy for vehicular communication networks. IEEE J Sel Areas Commun. 2011;29(1):197–206.
[7] Palanisamy P, and Nirmala S. Downlink interference management in femtocell networks-a comprehensive study and survey. In: Proc. IEEE ICICES; 2013. p. 747–754.
[8] Zhang H, Jiang C, Cheng J, *et al.* Cooperative interference mitigation and handover management for heterogeneous cloud small cell networks. IEEE Wireless Commun. 2015;22(3):92–99.
[9] Rahman M, and Yanikomeroglu H. Enhancing cell-edge performance: a downlink dynamic interference avoidance scheme with inter-cell coordination. IEEE Trans Wireless Commun. 2010;9(4):1414–1425.
[10] Zhang H, Jiang C, Beaulieu NC, *et al.* Resource allocation for cognitive small cell networks: a cooperative bargaining game theoretic approach. IEEE Trans Wireless Commun. 2015;14(6):3481–3493.
[11] Zhang H, Jiang C, Mao X, *et al.* Interference-limited resource optimization in cognitive femtocells with fairness and imperfect spectrum sensing. IEEE Trans Veh Technol. 2016;65(3):1761–1771.
[12] Li Z, Lu Z, Wen X, *et al.* Distributed power control for two-tier femtocell networks with QoS provisioning based on *Q*-learning. In: Vehicular Technology Conference IEEE; 2015. p. 1–6.
[13] Zhang H, Jiang C, Hu RQ, *et al.* Self-organization in disaster resilient heterogeneous small cell networks. IEEE Network. 2016;30(2):116–121.
[14] Long C, Zhang Q, Li B, *et al.* Non-cooperative power control for wireless ad hoc networks with repeated games. IEEE J Sel Areas Commun. 2007;25(6): 1101–1112.

[15] Chen X, Zhang H, Chen T, *et al.* Improving energy efficiency in green femtocell networks: a hierarchical reinforcement learning framework. In: Proc. IEEE ICC, Budapest, Hungary; 2013. p. 2241–2245.
[16] Zhang Z, Wen X, Li Z, *et al.* QoS-aware energy-efficient power control in two-tier femtocell networks based on *Q*-learning. In: Proc. ICT; 2014. p. 313–317.
[17] Miao G, Himayat N, Li GY, *et al.* Low-complexity energy-efficient scheduling for uplink OFDMA. IEEE Trans Commun. 2012;60(1):112–120.
[18] Zappone A, Alfano G, Buzzi S, *et al.* Energy-efficient non-cooperative resource allocation in multi-cell OFDMA systems with multiple base station antennas. In: IEEE GreenCom; 2011. p. 82–87.
[19] Saraydar CU, Mandayam NB, and Goodman DJ. Pareto efficiency of pricing-based power control in wireless data networks. In: IEEE Wireless Communications and Networking Conference (WCNC); 1999. p. 231–235 vol. 1.
[20] Wang L, Chen X, Zhao Z, *et al.* Exploration vs exploitation for distributed channel access in cognitive radio networks: a multi-user case study. In: 11th International Symposium on Communications and Information Technologies (ISCIT); 2011. p. 360–365.
[21] van den Biggelaar O, Dricot JM, Doncker PD, *et al.* A new distributed algorithm for the allocation of cognitive radio sensing times. In: IEEE International Symposium on Personal Indoor and Mobile Radio Communications (PIMRC); 2012. p. 1208–1213.
[22] Panahi FH, and Ohtsuki T. Optimal channel-sensing policy based on Fuzzy *Q*-learning process over cognitive radio systems. In: IEEE International Conference on Communications (ICC); 2013. p. 2677–2682.
[23] Qiao D, Gursoy MC, and Velipasalar S. Energy efficiency in multiaccess fading channels under QoS constraints. EURASIP J Wireless Commun Networking. 2012;2012(1):136.
[24] Musavian L, and Le-Ngoc T. Energy-efficient power allocation for delay-constrained systems. In: IEEE Global Communications Conference (GLOBE-COM); 2012. p. 3554–3559.
[25] Xiong C, Li GY, Liu Y, *et al.* QoS driven energy-efficient design for downlink OFDMA networks. In: IEEE Global Communications Conference (GLOBECOM); 2012. p. 4320–4325.
[26] Jiang C, Zhang H, Ren Y, *et al.* Machine learning paradigms for next-generation wireless networks. IEEE Wireless Commun. 2017;24(2):98–105.
[27] Alnwaimi G, Vahid S, and Moessner K. Dynamic heterogeneous learning games for opportunistic access in LTE-based macro/femtocell deployments. IEEE Trans Wireless Commun. 2015;14(4):2294–2308.
[28] Onireti O, Zoha A, Moysen J, *et al.* A cell outage management framework for dense heterogeneous networks. IEEE Trans Veh Technol. 2016;65(4):2097–2113.
[29] Rekha JU, Chatrapati KS, and Babu AV. Game Theory and Its Applications in Machine Learning. In: Satapathy SC, Mandal JK, Udgata SK, Bhateja V.

(eds) Information Systems Design and Intelligent Applications. Advances in Intelligent Systems and Computing. 2016;435. New Delhi:Springer.

[30] Blum A, Blum M, Kearns M, Sandholm T, and Hajiaghayi MT. Machine Learning, Game Theory, and Mechanism Design for a Networked World.

[31] Cheng X, Zhao Z, Zhang H, *et al.* Conjectural variations in multi-agent reinforcement learning for energy-efficient cognitive wireless mesh networks. In: IEEE Wireless Communication and Networking Conference (WCNC); 2012. p. 820–825.

[32] Chang CS. Stability, queue length, and delay of deterministic and stochastic queueing networks. IEEE Trans Autom Control. 1994;39(5):913–931.

[33] Fudenburg D, and Tirole J. Game Theory. In: The Cambridge:MIT Press; 1991.

[34] Watkins CJ, and Dayan P. Technical note *Q*-learning. Mach Learn. 1992;8(3–4): 279–292.

[35] Sastry PS, Phansalkar VV, and Thathachar MAL. Decentralized learning of Nash equilibria in multi-person stochastic games with incomplete information. IEEE Trans Syst Man Cybern. 1994;24(5):769–777.

[36] Chandrasekhar V, Andrews JG, Muharemovic T, *et al.* Power control in two-tier femtocell networks. IEEE Trans Wireless Commun. 2009;8(8):4316–4328.

Chapter 13
Data-driven vehicular mobility modeling and prediction

Yong Li[1]*, Fengli Xu*[1]*, and Manzoor Ahmed*[2]

Vehicular networks have been recently attracting an increasing attention from both the industry and research communities. One of the challenges in this area is the understanding of vehicular mobility and further propose accurate and realistic mobility models to aid the vehicular communication and networks design and evaluation. In this chapter, different from the current works focusing on designing microscopic level models that are describing the individual mobility behaviors, we are exploring the use of open Jackson queuing network frameworks to model the macroscopic level vehicular mobility. The proposed intuitive model can accurately describe the vehicular mobility, and further predict various measures of network-level performance. These measures include the vehicular distribution and vehicular-level performance, such as average sojourn time in each area and the number of sojourned areas in the vehicular networks. Model validation based on two large-scale urban vehicular motion traces reveals that such a simple model can accurately predict a number of system measure concerned with the vehicular network performance. Moreover, we develop two applications to illustrate the proposed model's effectiveness in the analysis of system-level performance and dimensioning of vehicular networks.

13.1 Introduction

Recently, as more and more vehicles are equipped with multiple sensors and heterogeneous communication access devices to enable wireless connectivity, interests on vehicular communications and networks have grown tremendously [1]. It is seen as the key technology for improving road safety and building intelligent transportation system (ITS) [2]. Many applications of vehicular networks are also emerging, including automatic collision warning, remote vehicle diagnostics, emergency management and assistance for safely driving, vehicle tracking, automobile high speed Internet access, and multimedia content sharing. In USA, Federal Communications Commission has

[1]Department of Electronic Engineering, Tsinghua University, China
[2]Department of Computer Science, Qingdao University, China

allocated 75 MHz of spectrum for dedicated short-range communications in vehicular networks, and IEEE is also working on the related standard specifications. The aim of these leading consortia and standardization bodies is to develop technologies and protocols for information transmission between vehicles and roadside units (RSU), infrastructures equipment, known as vehicles to infrastructures (V2I), and between vehicles, known as vehicles to vehicles (V2V).

Urban vehicular ad hoc networks (VANETs) [3] are recognized as an important component of the next generation ITS to alleviate serious problems, such as traffic jams and accidents, as well as to enable new mobile applications to the public [1]. Since urban VANETs are highly mobile, it is difficult to maintain a connected and stable network for communication. Thus, they are usually distributed, self-organized by the mobile vehicles, characterized by very high velocity, and limited degrees of freedom in nodes mobility patterns. This brings a strong interaction between the vehicular mobility and network protocol design, which is the main focus of current development of VANETs [3]. First, mobility in the macroscopic means the flows of vehicular traffic directed from one region to another, which influences the spatial distribution of vehicles, and the data traffic may also be altered by mobility. Thus, a specific relationship between the mobility and wireless communication exists in VANETs. Second, mobility in the microscope means the individual vehicular mobility, which influences the position of each vehicle. Then, the communication rate changes when the vehicles communicate with the RSUs [4] or vehicles [5] via V2I or V2V.

In terms of VANETs' design, since the development of VANETs' technologies has huge impact on the automotive market, we should put a growing effort in the development of communication protocols and mobility models by efficiently utilizing their relationship and the influences of mobility on the communications, specific to the vehicular networks. In terms of protocol and vehicular network system performance evaluation, economic issues and technology limitations make theoretical analysis and simulation as the prime choices in the validation of VANETs, and also as the widely adopted first step in the development of real world technologies [5]. A critical aspect in the theoretical analysis and simulation of VANETs is the need for a realistic mobility model reflecting the real behaviors of vehicles in terms of both large-scale vehicular traffic and microscope level of individual mobility. In conclusion, mobility models are significant to the development of vehicular networks and related works have become an important part on vehicular networks [5].

After a few years of exciting work, a large variety of mobility models are available, which can be categorized in three different classes known as synthetic, survey-based model and trace-based models. The synthetic models as their name implies are obtained by mathematical models, while the survey models extract mobility patterns through surveys, and finally the trace-based models generate mobility patterns from real mobility traces [5]. These models vary from the most trivial to the most realistic ones, or from freely available models to commercial vehicular simulator. However, these models consider each vehicle as a distinct entity, and they are in the microscopic level [6]. Although microscopic level models describe the individual mobility behaviors precisely, unfortunately, they fail to capture the overall mobility in the whole network. In contrast, macroscopic level description can lead to gross quantities of metrics like vehicular distribution, density and means of velocity, by treating vehicular traffic according to fluid dynamics, and then large-scale overall vehicular behaviors and traffic can be easily revealed. Further, such models are indispensable for network dimensioning, answering the "what if" questions like how the network performance changes or the deployed network evolves as the number of vehicles or communication demands scale-up [7]. Thus, macroscopic level vehicular mobility models are crucial for the development of vehicular networking protocols and algorithms.

In this chapter, against this background, we consider the problem of modeling the macroscopic level vehicular mobility. Specifically, we explore the use of an open Jackson queueing network to model the vehicular mobility among areas divided by the intersections of the city road. In the model, vehicles arrive in the system according to a random process, move from one area to another area by making independent probabilistic transitions, and finally depart the system. The question we address is can this simple queueing network model accurately describe the vehicular mobility and further predict various measures of network-level performance like the vehicular distribution, and vehicular-level performance like average sojourn time in each area and the number of sojourned areas in the vehicular networks. Our novel contributions are summarized as follows:

1. We model the macroscopic level vehicular mobility as an open Jackson queueing network. Under this model, we obtain three important metrics related to vehicular mobility and system performance, which are vehicular area distribution, average sojourn time in each area, and average mobility length.
2. Using two large-scale urban vehicular motion traces, we validate the accuracy of the proposed queueing network model by comparing the model-predicted results with the observations in the traces. The results reveal that such a simple model can accurately predict a number of system metrics concerned with the vehicular network performance.
3. Under the proposed open Jackson queueing network for vehicular mobility, we introduce two specific applications. One is the decision of how much the capacity of the RSUs should provide with the increase of the communication demand coming with the increasing of city vehicles. The second is investigating the performance of the combined communications of V2I and V2V. The applied applications illustrate that the proposed model is effective in the analysis of system-level performance and dimensioning of vehicular networks.

The rest of this chapter is organized as follows. After introducing related work in Section 13.2, we give the model motivation and describe the system model in Section 13.3. While in Section 13.4, we derive related system performance metrics based on the proposed model. Moreover, in Section 13.5, we introduce the vehicular mobility trace for model simulation and provide the validation results, and followed by two specific applications of vehicular network performance analysis in Section 13.6. Finally, we conclude the chapter in Section 13.7.

13.2 Related work

In terms of individual mobility models in the microscopic level, after a few years of exciting developments, a large variety of models are available. Different from the synthetic models [8,9] and survey-based models [10,11], the trace-based model try to extract mobility patterns from existing mobility trace by approximating the movements based on observed movement patterns [12,13]. Even both the synthetic and survey-based models are very complex, still they are not able to come close to realistic modeling of motion patterns. All these microscopic level individual mobility modeling approaches have the limitation of obtaining global mobility patterns instead of precise movements; and they also sometimes are too complex to solve by mathematical equations. On the contrary to modeling the individual mobility, our work focuses on the macroscopic mobility modeling. To our knowledge, this is the first work that gives a simple queueing model of mobility with large-scale urban vehicular mobility empirical data validation.

Recent works [14] focus on studying the metric of *inter-contact time*, which denotes the time between two successive communication contacts of two vehicles, and it finds that the inter-contact time exhibits the exponential distribution over a large range of timescales. Poisson distributed contact rate has been validated to fit well to real vehicular traces and is widely used to model opportunistic vehicular systems [15,16]. Instead of studying inter-contact time, Li *et al.* [17] puts forward another key metric known as *contact duration*, which is how long a contact lasts. In contrast to these works, revealing the vehicular contact patterns that indirectly reflect the macroscopic mobility, we directly model the vehicular mobility among areas and reveal the vehicular mobility flow and its spatial distribution in direct manner.

Previous works on modeling and performance analysis with queueing network model studied mostly the wired network and applications like peer-to-peer live streaming systems [18,19]. The most closely related works are theoretical analysis for cellular and Wi-Fi networks [7,20,21]. Ashtiani *et al.* [20] used a closed queueing network with fixed nodes to model the users and traffic in the cellular network, while Kim *et al.* [21] utilized M/M/c/c queues to model cellular network mobile users. Similarly, Chen *et al.* [7] proposed a mixed queueing network model to describe the user mobility among access points in the campus wireless network environment. All these models for wireless networks are proposed under different assumptions of mathematical properties. In contrast, our work focuses on modeling the large-scale urban vehicular mobility. Rather than giving complex mathematical deviation, we justify that using the simplest open Jackson queueing model can capture the essential properties for vehicular mobility which are validated by two empirical traces. Moreover, we introduce two typical applications, which show our proposed model is useful in the vehicular network performance analysis and design.

13.3 Model

13.3.1 Data sets and preprocessing

Shanghai trace [14] was collected in SG project [22], in which 2,019 operational taxis continuously covered the whole month of February 2007 without any interruptions in Shanghai city. In this trace, a taxi sends its location coordinates by GPRS to the central database every 1 min when it has passengers onboard but every 15 s when it is vacant for the reason of real-time scheduling. However, the different intervals of reporting may distort the records of the physical movements of the taxis, since most of the taxis are not vacant most of the time. Another drawback of this trace is that the number of taxis is limited. Indeed, 2,000 taxis and 1 min duration may not be sufficient to record the statistical features of contact duration in a high-speed large urban environment.

In collecting *Beijing* trace, we used the mobility track logs obtained from 27,000 participating Beijing taxis carrying GPS receivers during May 2010. The reason behind choosing taxis as vehicular devices is that taxis are more sensitive to urban environments in terms of underlying road topology, traffic control, and urban planning, and they have broader coverage in terms of space and operation time than that of buses and private cars. Specifically, we utilized the GPS devices to collect the taxis' locations and timestamps, and further GPRS modules report the records every 15 s for moving taxis. The specific information contained in such a report includes the taxi's ID, the longitude and latitude coordinates of the taxi's location, timestamps, instant speed, and heading direction. *Beijing* trace is the largest vehicular data trace available.

By collecting the GPS information of longitude and latitude coordinates, we obtain the taxi's moving traces that are locations varying with the time. Since these locations are measured by the GPS devices, the noise may exist in the collected data due to the inaccuracy of the GPS device. At the same time, since the taxies may not report their location time at the same timeslots at the fixed frequency like in Shanghai trace. Therefore, we need to process the data trace to get accurate locations of all the taxies in the same frequency and timeslots. In order to achieve these goals, we first use the city map of Shanghai and Beijing to correct the taxi's locations based on the coordinates of city road. Then, we use the method of linear interpolation (LI) to insert location points to let all the taxies have a location information at every 15 s. For the LI method, we first select any location that is near to each other in the original trace at two time points. If their interval is larger than 15 s, we use the selected two time points and map the information to estimate the unknown ones. For example, suppose we have the location information of one taxi in the original trace at time $t_1 < t_2 < \cdots < t_n$, and their corresponding locations are $l_1, l_2, \ldots, l_n$. If we want to insert the location information of time t that are calculated according to the time interval, we find m that satisfies $t_m \leq t < t_{m+1}$. Then, we calculate the location by the following expression through LI:

$$l_t = l_m \cdot \frac{t_{m+1} - t}{t_{m+1} - t_m} + l_{m+1} \cdot \frac{t - t_m}{t_{m+1} - t_m}. \tag{13.1}$$

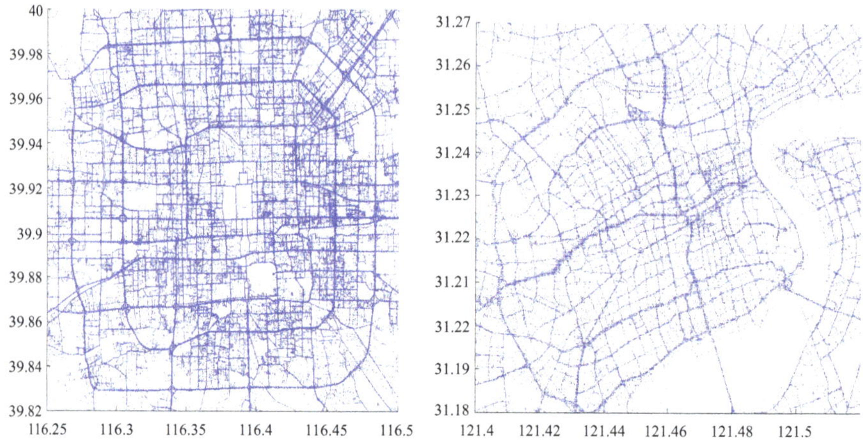

(a) (b)

Figure 13.1 City maps recovered from one data taxis mobility trace of (a) Beijing and (b) Shanghai

In order to verify the above data preprocess approach does not introduce inaccurate information to the original data trace, we use the obtained data of locations of 1 day to plot the trajectory of all taxies, which are shown in Figure 13.1(a) and (b) for Beijing and Shanghai data, respectively. From these two figures, we can see that our data set is fine-grained that even using 1 day's data can recover the map of the whole city. In order to further show the accuracy of our data processing, we compare the obtained figures with the original Beijing and Shanghai Map and find that all the trajectories are in the city road and thus demonstrate that the map drawn by these trajectories are very similar with the original city map.

13.3.2 Model motivation

Consider a vehicle moving in the city roads. It will travel along a road and then come across an intersection. It may then wait for traffic signal for a while in the intersection to choose the direction and travel to another road to drive on. In the downtown of a city, the road is usually very crowded, and the intersections may be very dense, which lead to very long waiting time at the intersections and short driving time along the road. Therefore, the intersection is an important factor to model the urban vehicular mobility. From the viewpoint of the whole city, we observe a large group of vehicles waiting at the area of each intersection, and the streams of traffic moving from one area to another area. That is to say, in order to describe the vehicular distribution, we need to pay attention to the areas around the intersections and understand vehicular behaviors of transition from one area to another from the system viewpoint.

Thus, if we divide the whole urban area into different areas that include at least one intersection, we can model the vehicles moving from one area to another adjacent area

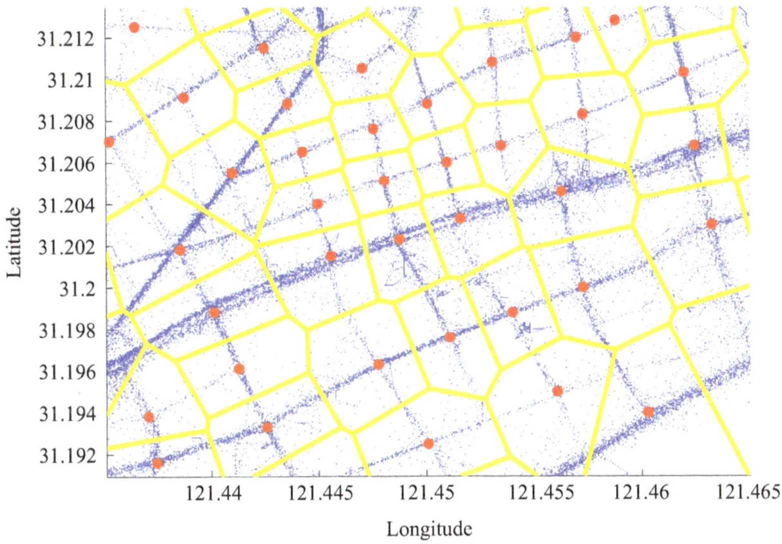

Figure 13.2 Illustration of area partition algorithm for a part of Shanghai city

and model the vehicular traffic transiting from one area to another. Taking Figure 13.2 as an example. We first marked the important intersections surrounded by a large number of vehicles as the red point and then divide the whole area around the selected intersections. The methods for the area partition can be chosen flexibly by the specific applications. For example, if we want to adapt the model to the use in the vehicular network design, i.e., deploying the RSU system, then the Voronoi diagram can be used, where each point is divided into the intersection that it is nearest to and then obtain the boundary of each area.

Let us consider the two-dimensional vehicular mobility defined by a sequence of steps that a vehicle travels in the city road, which is modeled by the above areas formed by the intersections. A step is denoted by a tuple (t^1, t^2, A), where A is the area, t^1 is the time entering area A, and t^2 is the time that it departs the area. In the first step, the vehicle enters the modeled region from the entering area, and after some step, it moves out of the modeled region. Every vehicle moves in this way by transiting from one area to another area. In this way, we can depict one vehicle's mobility and overall describe the traffic flows of the whole system by combining all the vehicles and intersections together as a system. Now, we are ready to introduce our queue network to model the above vehicle-mobility scenario.

13.3.3 Queue modeling

By the method of area dividing described above, the whole vehicular system can be described by the partitioned areas. The number of areas is denoted by N, across which the vehicles transit, and the vehicles move into the system, move from one area to another, and finally moving out the system. We use a queueing network to model the above system, which is shown in Figure 13.3. It includes N server nodes with infinite queue size, which models the N partitioned areas in the system. The servers are denoted by set $\mathcal{N} = \{A_1, A_2, A_N\}$. The vehicular movement into the system and moving from one area to another are modeled by the entrance into the queueing network and the transition from one server to another.

Now, we describe the dynamic behaviors of vehicles moving in different areas of the system. In such a vehicular mobility system with multi-areas, the vehicle dynamic behaviors occur on two different timescales. One is on the long timescale, in which the vehicle may enter and depart the system. The other is on the short timescale, where a vehicle changes areas, which means it switches from one area to another. In the viewpoint of queueing network model, the vehicles enter into the system with certain rate, stay in the server's queue, and then transfer into another server. For the long timescale dynamics, we assume that vehicles arrive server $n, n \in \mathcal{N}$ with rates of λ_n. When a vehicle moves to area n, it will continually stay in this area for a period of time. We assume that the average amount of time that a vehicle stay in area n is μ_n. The distribution of staying time is arbitrary. For the short timescale dynamics, we consider that after the vehicle staying in area n for a random period of time, it switches to another area m with probability p_{nm} or leaves the system with probability p_{n0}. In this way, the vehicles move from one area to another, depart or enter the system.

We have modeled the vehicular mobility system including N areas as an open network of N servers with infinite queues. In such an open system, vehicles freely join and leave the system. The exogenous arrival rate for server n is λ_n. After staying in the queue of server n for time period of $1/\mu_n$, it will leave the queueing network with probability p_{n0} or switch to other server like m with probability p_{nm} and denote the switching matrix as $\mathbf{P}$. Therefore, for server n, its load is denoted by $\rho_n = \lambda_n/\mu_n$.

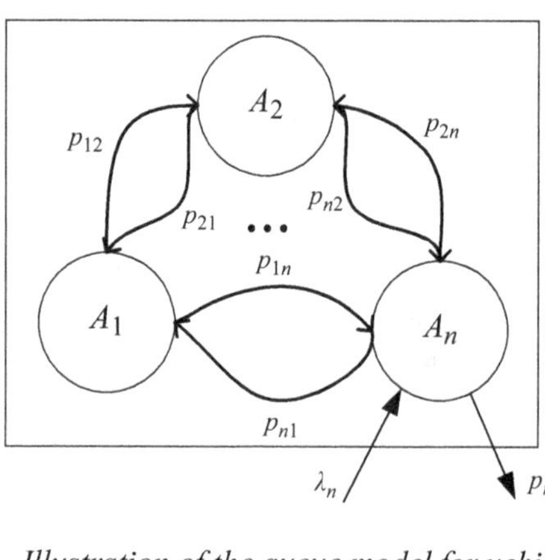

Figure 13.3 Illustration of the queue model for vehicular mobility

Table 13.1 Parameters and notations used in the queueing network model

Notations	Meaning
$\mathcal{N}$	The set of areas, or the servers set
N	The number of servers in the system
λ_n	The exogenous arrival rate for server n
$1/\mu_n$	The expected vehicle residence time at area n
ρ_n	The load of server n
M	The number of vehicles in the system
p_{nm}	Probability of the vehicles moving from area n to area m
W_n	The number of vehicles in area n

Since each server's queue is considered as an infinite queue, and each user is served immediately if we assume there are an infinite number of servers, and each vehicle is independent of each other. Now, we summarize the key parameters and their notations in our model in Table 13.1.

As we model the vehicular mobility system as the aforementioned open queueing system, it would be very easy to get further results if it is an open Jackson network, which has well-known results about the user distribution and waiting time. Related to the Jackson network, we need to demonstrate that the exogenous arrival to each server follows Poisson process. If it holds, the queueing network can be modeled by a network with infinite server queue (i.e., $M/G/\infty$). Thus, we need to study the property of exogenous arrival rate in the system. By leveraging the Beijing and Shanghai traces, we find that the actual exogenous arrival process of the vehicular mobility matches well with the exponential distribution. Thus, the vehicular mobility system can be modeled as an open Jackson network. Based on this model, we will derive some important metrics to depict the system performance.

13.4 Performance derivation

Based on the open Jackson-queueing network model for the vehicular mobility, we will investigate three important metrics related to the system performance, which are vehicular area distribution, average sojourn time, and average mobility length. First, we give formal definitions of above three metrics:

- Vehicular area distribution: The vehicular area distribution is defined as the steady population probability distribution of all areas in the system. That is how the vehicles of whole system are distributed in the areas. Basically, we can define it as the separate distribution of each area and joint area distribution of all the areas. Specifically, the vehicular distribution of separated area n is the probability that there are w_n vehicles in area n, which can be expressed as $P(W_n = w_n)$, where

$w_n = 0, 1, \ldots, M, 1 \leq n \leq N$, and W_n is the random variable denoting the number of vehicles in an area n. Then, the joint area vehicular distribution, denoted by $\pi(\vec{w_n})$, can be expressed as follows:

$$\pi(\vec{w_n}) = P(W_1 = w_1, \ldots, W_N = w_N). \tag{13.2}$$

- Average sojourn time: Average sojourn time is defined in the viewpoint of vehicles, which is how long a vehicle will stay in the system. That is the time period between that the vehicles entering into the region by one of the areas and that it finally leaves the region. By averaging when all the vehicles in the system is in a steady state, the average sojourn time. This parameter is related to the session time of a vehicle visiting in the system.
- Average mobility length: Average mobility length is also defined in terms of vehicles. That is the average number of areas that a vehicle will travel along when it is in the system during the sojourn time. This parameter is related to the average number of transitions of a vehicle during a session.

13.4.1 Vehicular distribution

Considering area n, the exogenous arrival rate for area n is λ_n. The vehicles switch from one area to another area according to the matrix $\mathbf{P}$. We treat $\mathcal{N}$ areas as an open Jackson network of N nodes with infinite servers, and view each vehicle as a customer that sojourns at node n for a random period of time with mean $1/\mu_n$, which is the server time in node n. Then, we let $\gamma = (\gamma_1, \gamma_2, \ldots, \gamma_N)$ be the effective arrival rate vector for all areas in $\mathcal{N}$. According to Figure 13.4, for any area, say n, we have

$$\gamma_n = \lambda_n + \sum_{j \neq n} \gamma_n p_{jn}, \quad 1 \leq n \leq N. \tag{13.3}$$

We express above expression in matrix form accordingly as

$$\gamma = \lambda + \gamma \mathbf{P}, \tag{13.4}$$

where $\mathbf{P}$ is the $N \times N$ area-switching matrix.

For area n, let $\rho_n = \pi_n/\mu_n$, and ρ_n is the load of area n. Actually, ρ_n is the average number of vehicles in area n. That is to say, ρ_n is the expected number of vehicles in

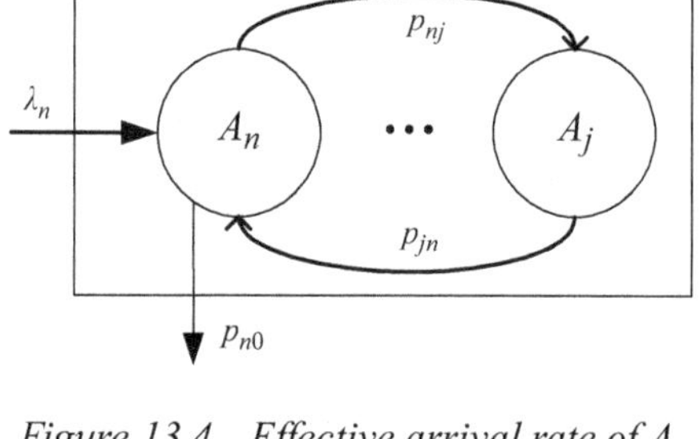

Figure 13.4 Effective arrival rate of A_i

area n. Furthermore, we get the separated area vehicular distribution and joint area vehicular distribution expressions by the following lemma:

Lemma 13.1. *The vehicular distribution of the separated area n is*

$$P(W_n = w_n) = \frac{{\rho_n}^{w_n} e^{-\rho_n}}{w_n!}, \tag{13.5}$$

and the joint area vehicular distribution is

$$\pi(\vec{w_n}) = \prod_{j=1}^{N} \frac{{\rho_j}^{w_j} e^{-\rho_j}}{w_j!}, \tag{13.6}$$

and the expected number of vehicles that stay in the area n in dynamic vehicular mobility system is ρ_n.

Proof. We consider one area, say n. The user arrival rate is γ_n and the serve time is $1/\mu_n$. We view this area as an infinite serve node. Therefore, according to the Jackson networks, for the vehicular mobility system, we have

$$\pi(\vec{w_n}) = P(W_1 = w_1, \ldots, W_N = w_N) = \prod_{j=1}^{N} \frac{{\rho_j}^{w_j} e^{-\rho_j}}{w_j!}. \tag{13.7}$$

The marginal distribution for individual area is expressed as

$$P(W_n = w_n) = \frac{{\rho_n}^{w_n} e^{-\rho_n}}{w_n!}. \tag{13.8}$$

We note that the distribution of vehicles at the queue of node n follows Poisson with mean ρ_n. Therefore, the expected number of vehicles that stay in area n is ρ_n, which proves the lemma. □

13.4.2 Average sojourn time

Using the queueing model for the vehicular mobility, we can further analyze important system metrics related to the mobility behaviors and properties. As introduced above, now, we derive two metrics of average sojourn time and mobility length based on the proposed queueing model of mobility.

First, recall the queueing server set is $\mathscr{N} = \{A_1, A_2, A_N\}$, which are the system states that vehicles transfer from one to another, and $\mathbf{P} = \{p_{nm}, 1 \leq n \leq N, 1 \leq m \leq N\}$ is the switching matrix of the probability that a vehicle moves into area m when it is in area n. Note that we already define the probability that a vehicle leaves the modeled region from area n with probability p_{n0}. In order to define a complete system state, we add another system state A_0 into $\mathscr{N}$, namely, the state when the vehicle is out of the region and denote the new system states set as $\mathscr{M}$, which is $\mathscr{M} = \mathscr{N} \bigcup A_0$.

We define p_{0n} as the probability that a vehicle move into the system through area n. From the definition of p_{n0} and p_{0n}, we can obtain the following expression for them:

$$p_{n0} = 1 - \sum_{m=1}^{N} p_{nm}, \quad n = 1, \ldots, N, \tag{13.9}$$

$$p_{0n} = \frac{\gamma_n}{\sum_{m=1}^{N} \gamma_m}, \quad n = 1, \ldots, N. \tag{13.10}$$

In order to better distinguish the transitions on states $\mathscr{N}$ and $\mathscr{M}$, we refer $\mathbf{P} = \{p_{nm}\}, 0 \leq n, m \leq N$ as the system transition on states $\mathscr{M}$ and denote the sub-matrix $\mathbf{R}$ as the transition among the areas on state $\mathscr{N}$, where $\mathbf{R} = P\{p_{nm}\}, 1 \leq n, m \leq N$.

Now, we obtain the average vehicular sojourn time by the following theorem:

Theorem 13.1. *Denoting the vehicular sojourn time in the system by S, we can obtain the average sojourn time designated as E[S] by the following expression:*

$$E[S] = \sum_{n \in \mathscr{N}} p_{0n} E[T_n], \tag{13.11}$$

where $E[T_n]$ is obtained by $[E[T_1], \ldots, E[T_N]] = \boldsymbol{T} = (\boldsymbol{I} - \boldsymbol{R})^{-1}\boldsymbol{U}$, $\boldsymbol{I}$ is an all-one vector, and $\boldsymbol{U} = [1/\mu_1, \ldots, 1/\mu_N]$.

Proof. We denote T_n as the vehicular sojourn time in the system on the condition that it is always in the system. That is to say, the vehicle will not move out of the region when it stays in any area of $n, n \in \mathscr{N}$. Considering the staying duration in area n, and using the Jackson queue network model, we have

$$E[T_n] = \frac{1}{\mu_n} + \sum_{m \in \mathscr{N}} p_{nm} E[T_m]. \tag{13.12}$$

Define $\mathbf{U} = [1/\mu_1, \ldots, 1/\mu_N]$ and $\mathbf{T} = [E[T_1], \ldots, E[T_N]]$, we can derive as the following matrix form:

$$\mathbf{T} = \mathbf{U} + \mathbf{RT}. \tag{13.13}$$

By [23], we can obtain that $(\mathbf{I} - \mathbf{R})$ is reversible. Thus, we can have

$$\mathbf{T} = (\mathbf{I} - \mathbf{R})^{-1}\mathbf{U}. \tag{13.14}$$

Therefore, the average sojourn time $E[s]$ is

$$E[S] = \sum_{n \in \mathscr{N}} p_{0i} E[T_n], \tag{13.15}$$

which proves the theorem. □

13.4.3 Average mobility length

Based on the average sojourn time of Theorem 13.1, we can obtain the average vehicular mobility length in the following theorem:

Theorem 13.2. *The average vehicular mobility length, denoted by* $E[L]$*, can be expressed by*

$$E[L] = \sum_{n \in \mathcal{N}} p_{0n} E[T_n], \tag{13.16}$$

where $E[T_n]$ *is obtained by* $[E[T_1], \ldots, E[T_N]] = \boldsymbol{T} = (\boldsymbol{I} - \boldsymbol{R})^{-1}\boldsymbol{I}$

Proof. Note that in the obtained sojourn time that a vehicle stays in the system when it is already in the system. In (13.13), we can change the sojourn time to mobility length just by setting the staying time in each area by 1. That is, $\mathbf{U} = [1, \ldots, 1]$. Then, we obtain:

$$\mathbf{T} = (\mathbf{I} - \mathbf{R})^{-1}\mathbf{1}. \tag{13.17}$$

Therefore, the average sojourn time $E[L]$ is

$$E[L] = \sum_{n \in \mathcal{N}} p_{0n} E[T_n], \tag{13.18}$$

which proves the theorem. □

13.5 Model validation

By leveraging the two most largest urban vehicular mobility traces as introduced earlier, we validate our proposed open Jackson queueing network model. The traces record the location information, i.e., longitude and latitude of the vehicles during the trace-collection period. Therefore, we first need to per-process the traces to fit our model. Then, we validate our model by using the empirical trace data in terms of following metrics: vehicular arrival rate, vehicular area distribution, average sojourn time, and mobility length.

13.5.1 Time selection and area partition

In order to use the GPS vehicular trace to validate our model, we need to preprocess the data on two different dimensions, i.e., time and location. In terms of time dimension, since the traces record the continuous vehicular mobility trajectory of the whole day, we need to select the period of time that is more stable in terms of numbers of the vehicles in the system to observe. While in the location dimension, we need to partition the urban map into areas according to the intersections and further decide which vehicles belong to which area considering their longitude and latitude information.

13.5.1.1 Area partition

In order to divide the vehicular mobility system into areas that they transit on, we need to take the road and city structure into consideration. As mentioned before, the intersection is the most important factor to model the urban vehicular mobility and distributions. Therefore, we divide the system according to the position of the intersections in the city roads. More specifically, we take the intersections as the center of each area and then partition the roads into area according to the distance to the intersections. We use Voronoi diagram to achieve this, which is a frequently used method of decomposition of a given space [24]. In Voronoi diagram, we are given a finite set of sites $\{p_1, \ldots, p_n\}$ in the Euclidean plane. The site of p_i's corresponding Voronoi cell, denoted by V_i, consists of all points whose distance to p_i is not greater than their distance to any other sites. While using the Voronoi diagram to partition the system region, we take all the intersections as the set of sites, denoted by $\mathscr{I} = \{p_1, p_2, \ldots, p_N\}$, and refer to all the points in the system region as $\mathscr{L}$. We denote the d_{li} as the distance between point $l, l \in \mathscr{L}$ and site $i, i \in \mathscr{I}$, which is the geometric distance between point l and the point of intersection in site i. Then, the area for site, i, is designated as V_i and expressed as $V_i = \{l | d_{li} \leq d_{lj}, \forall j \in \mathscr{I} \setminus \{i\}, \forall l \in \mathscr{L}\}$. According to the regulation introduced above, we obtain the boundary and partition of the desired system region into different areas.

For example, we consider a part of Shanghai city, the area partitioned results depicted via Voronoi diagram is shown in Figure 13.2. The blue points in the figure are the records of vehicular trajectory, the marked red points are the intersections, and the cells are the partition areas distinguished by the yellow curves. We observe that the region is divided by different areas, and according to the coordinate information of each vehicle, we can decide which area it belongs to. Consequently, the vehicular mobility is modeled by the transitions from one area to another.

13.5.1.2 Observation period selection

Since we are interested in the time period when vehicles are most active and stable, we look into the trace to investigate the time period when there are enough vehicles in the system, and the system is relatively stable and stationary. In the urban city, the vehicle traffic appears almost the same patterns varying from one day to another, except the little difference between the working day and weekend. Thus, we can select the suitable observation period by just investigating the aggregate vehicular arrival rate in the timescale of 1 day.

We plot the aggregated vehicular arrival rate into the system of both Beijing and Shanghai trace in Figure 13.5. By observing the curve of Beijing trace, we find the arrival rate is very low from the midnight to 5 AM, and it increases quickly during 6–9 AM. Then, the arrival rate almost keeps in the same level during the daytime from 9 AM to 7 PM. Following comes a rate-decreasing period. Similar arrival rate patterns can be observed in the Shanghai trace, except the traffic also keeps in the very active state during the period of 7 PM to the midnight. Combining the Shanghai and Beijing traces' results, our goals of finding the most active period for the vehicular mobility system, we select the period of 9 AM to 7 PM as the observation period as the data to validate our proposed model. When processing the vehicular mobility trajectory, if a

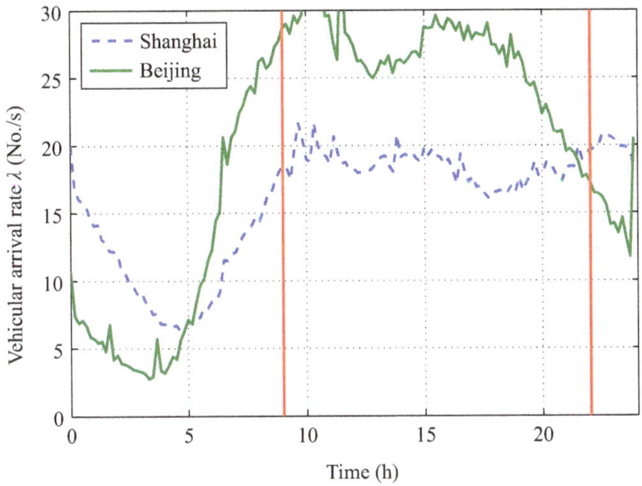

Figure 13.5 Average vehicular arrival rate to the system in timescale of 1 day for Shanghai and Beijing trace

vehicle does not have records in any of the areas in the period of 10 min, we assume that it has departed from the system and will take it as a new vehicle moving into the system when it appears.

13.5.2 Arrival rate validation

We note that in the open Jackson network model, an important requirement is that the exogenous arrival to each area should follow Poisson process. Therefore, we need to validate this. In the open Jackson model, the arrival to each server is the aggregated customer arrival process. Consequently, the aggregated arrivals is the metric to validate the rationality of our proposed model. We validate the vehicular arrival rate in the timescales of not only 1 day but also all the days. First, we investigate the aggregated arrivals of all vehicles in the system. Based on the introduced method of area partition, we count the exogenous arrivals to each area. To validate that the aggregated arrival rate follows Poisson process, we need to justify the arrival time that can be fitted well by the exponential distribution. We plot the Complementary Cumulative Distribution Function (CCDF) of the arrival time distribution obtained from Shanghai and Beijing traces in Figure 13.6. Since we plot the results in linear-log scale, where the curves of exponential distribution will become a straight line, we can observe that the arrival time may match the exponential distribution well. Furthermore, we use exponential distribution to fit the 90% of the distribution, where the red curves are the exponential distribution and the blue curves are the empirical curves. The goodness of fit is measured quantitatively by the *R*-square statistics [25], which is defined as the percentage of the variation between the empirical CCDF and the fitted distribution. We obtain that the average adjusted *R*-square statistics is over 98% for both Shanghai and Beijing traces. This confirms

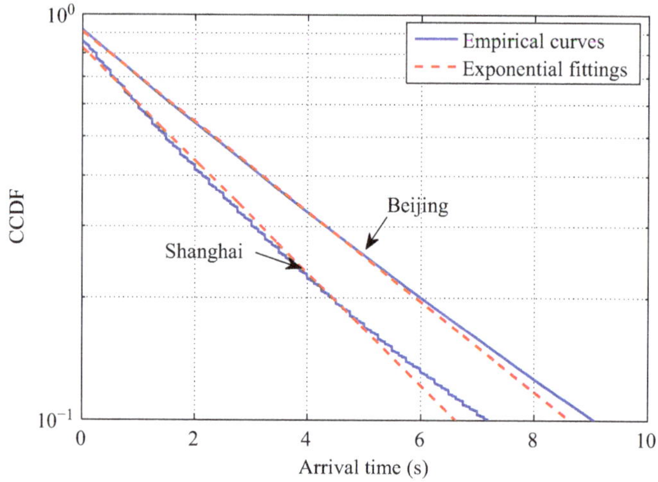

Figure 13.6 Exponential fittings for aggregated distribution of vehicular arrival time of Shanghai and Beijing trace

the accuracy of the exponential distribution of the aggregated arrival time, which indicates the aggregated arrivals follows Poisson process.

Second, in order to further validate the exogenous arrivals to each area following Poisson distribution, we investigate the distribution of the exogenous arrival time of each area in the timescale of 1 day and select the first 15 days for studying. To measure the closeness of the Poisson distribution and empirical ones, we use the Kolmogorov–Smirnov test (KS test) instead of CCDF fitting due to the large amount of curves in each area of 15 days. The KS statistic can quantify the distance between the empirical distribution function of the sample and the cumulative distribution function of the theoretical distribution [26]. The smaller the KS statistic, the closer the two distributions are. In our study, we set the significance level [26] of KS test to 0.01, which means the confidence level is 99%. Figure 13.7 shows the goodness-of-fit measured by the acceptance ratio of KS tests of each day by averaging the results of all areas. From the results, we can observe that the acceptance ratio of Beijing trace is above 90% except the second day, which have a relatively smaller vehicle mobility records. With regards to the Shanghai trace, we note a good match between the model distribution and empirical results, and the average acceptance ratios are around 80%, which means the overall accuracy of the Poisson model is about 80%. Combing the results of Shanghai and Beijing, we come to the conclusion that the exogenous arrivals to each area can be accurately modeled by the Poisson distribution.

Now, we have completed the validation of Poisson-based exogenous arrivals' accuracy in our model. Next, we focus on validating the results of vehicular distribution, sojourn time, and mobility length, which are important metrics obtained from the open Jackson queuing network-based vehicular mobility model.

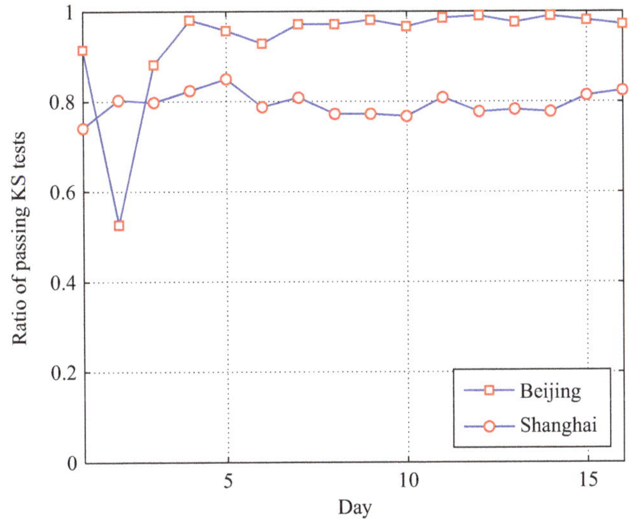

Figure 13.7 Results of KS tests passing ratio for the vehicular arrival rate in timescale of 1 day for Shanghai and Beijing traces

13.5.3 Vehicular distribution

At the first glance, we would like to know how well the proposed queueing network model matches the empirical results. We note that the direct metrics obtained from the model is the vehicular distribution at each area. Therefore, we first investigate the vehicular distribution by comparing the results obtained from the theoretical model and empirical data.

By selecting the six most busiest areas from Beijing and Shanghai traces, we plot the distribution of the number of vehicles in these areas by the empirical results and also plot the model results obtained from Lemma 13.1. The results of Beijing trace are shown in Figure 13.8, and that of Shanghai trace are shown in Figure 13.9, where the blue solid lines are the model results and the red dotted lines are the empirical results. From the results, it is quite obvious that our model results match the empirical results well, which shows the accuracy of our proposed queuing network-based vehicular mobility model.

In order to measure the closeness of the predicted results obtained from the model and the empirical results of all areas, we use the distribution obtained in Lemma 13.1 to fit the empirical curves of all the areas in both Shanghai and Beijing traces. The goodness of fit is also measured quantitatively by the R-square statistics [25], which is defined as the percentage of the variation between the empirical CCDF and the fitted distribution. Figure 13.10 shows the adjusted R-square statistics of the model distribution fittings, where the adjusted R-square statistics are computed with

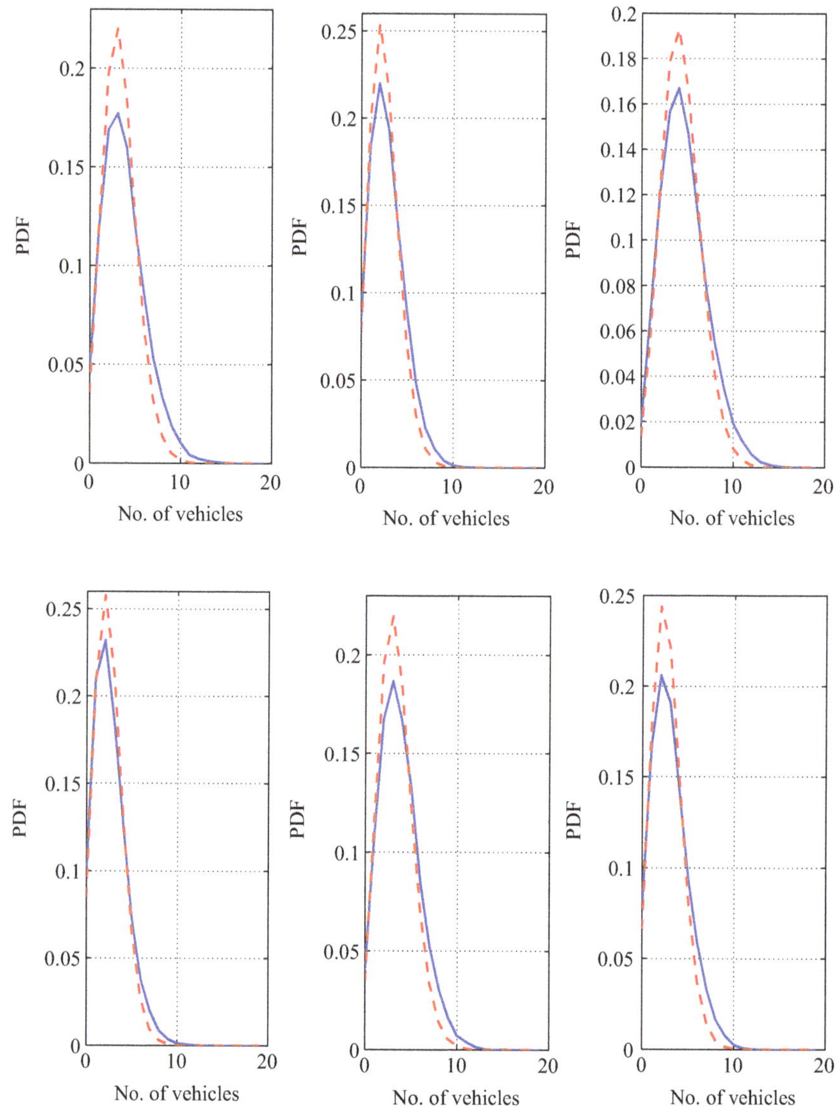

Figure 13.8 Vehicular distribution of six intersections in Beijing trace, where the red and dotted curves are the empirical results obtained from the trace, and the blue and solid curves are the theoretical results obtained by our proposed model

MATLAB® Curve Fitting Toolbox. It can be seen from Figure 13.10 that the average adjusted R-square statistics of over 90% areas for the Shanghai trace is larger than 98%, and that of over 90% areas for the Shanghai trace is larger than 95%. This confirms the accuracy of the model-based prediction for vehicular distribution.

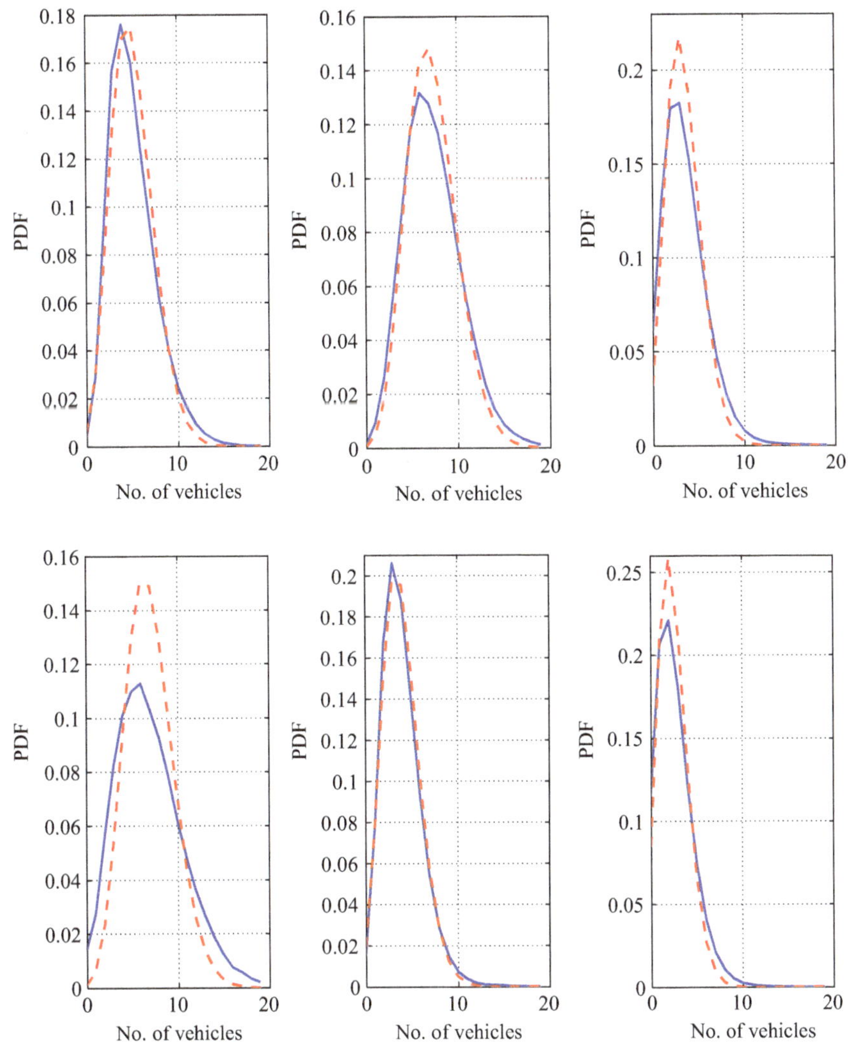

Figure 13.9 Vehicular distribution of six intersections in Shanghai trace, where the red and dotted curves are the empirical results obtained from the trace, and the blue and solid curves are the theoretical results obtained by our proposed model

13.5.4 Average sojourn time and mobility length

In this subsection, we compare the theoretical results of mean sojourn time and mobility length predicted from our proposed model against the empirical results obtained by the mobility traces, where the theoretical results is calculate by Theorems 13.1 and 13.2.

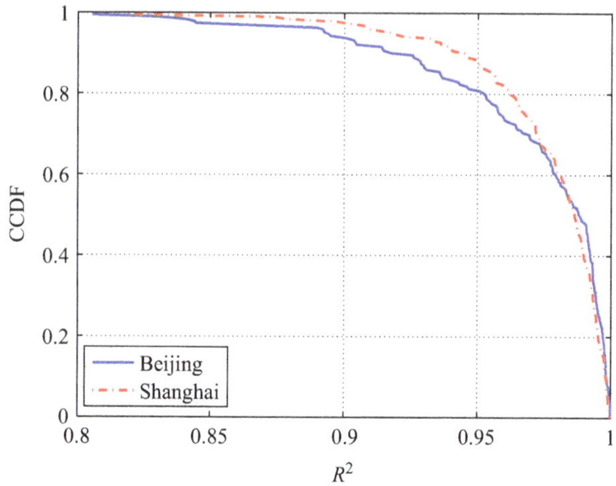

Figure 13.10 Distribution of the adjusted R-square statistics of the theoretical results and empirical results

Table 13.2 Predicted and empirical results of average sojourn time and mobility length of Shanghai trace

Mobility Number of vehicles	Average sojourn time		Average mobility length Predicted and empirical results
	Predicted results	Empirical results	
1,000	1,498.0	1,412.6	16.69
3,000	1,462.6	1,386.3	16.28
4,441	1,475.4	1,374.2	16.10

For testing the scalability of the accuracy of the proposed model, we vary the number of evolved vehicles. Both for the Shanghai and Beijing trace, we sort the vehicles according to their number of positions recorded in the GPS trace. We first select vehicles that have at least 80% of record during the trace collection time, and then put more and more vehicles into the system for testing. For Shanghai trace, we selected different number of vehicles as 1,000, which have 80% of record, 3,000 and 4,441, which is the total number of vehicles in the trace. For Beijing trace, the set of the number of vehicles as 3,000, 6,000, 10,000, and 28,590.

The results of average sojourn time and mobility length of predicted and empirical results under the Shanghai and Beijing traces are shown in Tables 13.2 and 13.3, respectively. From the results of average sojourn time, we find that the predicted results match the empirical results very well for both Shanghai and Beijing traces. Especially, when we only use vehicles with more completed record, the predicted results are very near to the empirical results. For example, in Shanghai trace,

Table 13.3 Predicted and empirical results of average sojourn time and mobility length of Beijing trace

Mobility Number of vehicles	Average sojourn time		Average mobility length Predicted and empirical results
	Predicted results	Empirical results	
3,000	1,689.8	1,720.5	9.86
6,000	1,662.5	1,699.7	9.62
10,000	1,613.4	1,658.1	9.23
28,590	1,450.4	1,500.1	8.37

the average deviation between the predicted and empirical results is only 5.7% when the number of vehicular is $1,000$; while in Beijing trace, it is only 1.8% when the number of vehicular is $3,000$. With the increasing number of vehicles, although more and more vehicles are with imperfect records, which will induce some errors into the system and model, the accuracy of our model is also accepted. For example, the prediction accuracy are higher than 6.9% and 3.4% when the number of vehicles are $4,441$ and $28,590$ in Shanghai and Beijing trace, respectively. In terms of average mobility length, the predicted results comply with the theoretical results completely. Consequently, we validate that our model is accurate enough to model the vehicular mobility to obtain the average and stable system performance.

13.6 Applications of networking

In vehicular networks, regarding with random and bursty data traffic initiated from vehicles, RSUs play as the gateways to the internet and to the infrastructure of other systems, such as ITS. Vehicles transmit their Internet access requests and information to RSUs, and RSUs then send responses to the Internet for querying the data and information needed by vehicles. Therefore, deploying RSUs appropriately is significant to the performance of vehicular networks. On one hand, the capacity and the number of deployed RSUs determine the capacity and service that can be provided to the vehicular network. On the other hand, a large number of RSUs deployed with large capacity mean more infrastructure cost. Therefore, the decision for RSU deployment should depend on the demands of vehicles. Basically, in the large urban city, it is very difficult to make such decisions due to the dynamics of vehicular traffic and randomness of vehicle mobility. However, based on our proposed vehicular model, we can obtain some fundamental results of the relationship with the RSUs capacity and network performance. Using the proposed queuing network-based vehicular mobility model, we will analyze how much RSUs' capacity should be provided with the rise in communication demands resulting from the increase of urban vehicles.

In reality, it is difficult to cover roads with enough RSUs so that each vehicle on the road can always be connected to nearby RSU in terms of infrastructure cost.

Instead, vehicle-to-vehicle communications by opportunistic contacts offer higher bandwidth communication capacity for data transmission, which can be utilized to form what is known as the opportunistic vehicular network or vehicular delay tolerant network (VDTN) [27]. By exploiting the delay-tolerant nature of non-real-time applications, service providers can delay and even shift the data transmission to VDTN. Therefore, in vehicular networking, vehicle-to-RSUs communications (V2I) and vehicle-to-vehicle communications (V2V) usually are combined to offer services. In this situation, investigating the performance of the networks considering the vehicular mobility and different approaches of communications is a very difficult problem. As the second part of this section, we will study the application of using the queueing model to investigate the performance of the combined communications of V2I and V2V.

13.6.1 RSU capacity decision

Now, we investigate the problem that if the number of vehicles increases with the development of economy and human demands, can the deployed RSUs handle the emerging growth of the communication demand. In this case, we define the capacity of the RSU as the maximum number of vehicles that can be served. Since the vehicular system in terms of both vehicular traffic and communication traffic is dynamic, we further define the RSU is overloaded if the time period, in which the number of served vehicles is more than its capacity, exceeds 95% of the whole measurement time. That is if the probability that the served vehicles, more than the maximum number, is below 95%, the RSU is not overloaded. By increasing the exogenous arrival rate λ, we look at the case that what fraction of RSUs will be overloaded.

From both the Shanghai and Beijing traces, we select those vehicles having almost completely recorded during the whole trace collection time, which is 3,000 taxis, and obtain the original λ from the trace, increase it from λ to 5λ, and plot the results in Figures 13.11 and 13.12 for Beijing and Shanghai traces, respectively. From the results, we can observe that with the increase in RSUs' capacity, the fraction of the overloaded RSU decrease very fast, while with the increase of the vehicular exogenous arrival rate, it also increases. For quantitative analysis, we examine the requirements of the capacity of RSU needed when at least 95% RSUs are not overloaded. In Beijing trace, we note that when the vehicular arrival rate is λ, the required capacity is about 10, on the other hand, when the arrival rate is increased by five times, i.e., 5λ, the required RSU capacity is about 35, which is only 3.5 times compared with the case when the arrival rate is λ. With regard to the Shanghai trace, we can obtain similar results when we increase the arrival rate by five times from λ to 5λ, the required RSU capacity increases about three times. This result is not obvious since the capacity does not need to increase linearly with the arrival rate, which means the number of vehicles in the system. Moreover, according to the obtained results shown in the figures, we can decide the RSU deploying policy according to the network performance requirements and investigation of deploying RSU equipment in terms of the cost.

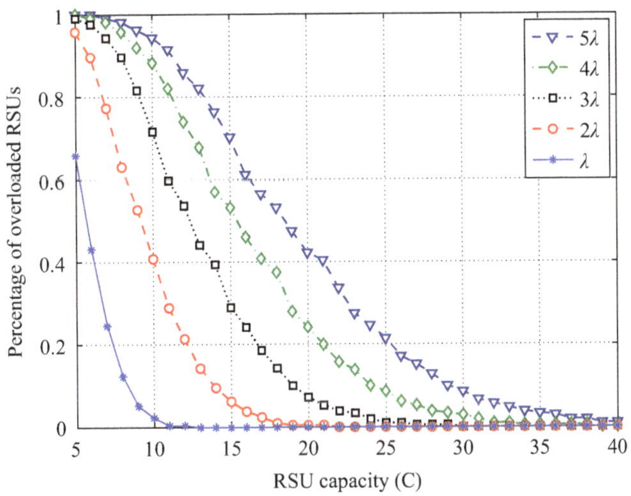

Figure 13.11 Fraction of overloaded RSU according to its capacity when scaling-up the vehicular arrival rate for Beijing trace

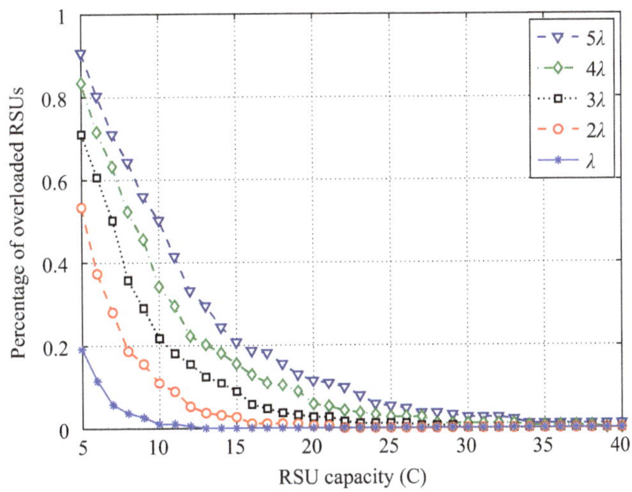

Figure 13.12 Fraction of overloaded RSU according to its capacity when scaling-up the vehicular arrival rate for Shanghai trace

13.6.2 V2I and V2V combined performance analysis

Based on the queueing network model, we derive closed-form expressions for performance metrics, such as the probability of network-wide satisfied areas and the average number of areas that satisfy the communication demands with V2I and V2V communications.

We say satisfied area occurs when every vehicle in this area can receive the demanded data rate. When the network is in the state of satisfying communication, all vehicles in the network are satisfied. For a given vehicular network, it is hard for the network to enjoy the satisfying communication all the time. Then, the metric of the steady-state probability that the network is in satisfying communication is an important parameter to evaluate the performance of the vehicular network. Another important metric is the expected number of areas that are enjoying the satisfying communication. Now we give more precise definitions for these two metrics.

The communication capacity index for area n, denoted by $\Phi_n(W_n)$, is defined as

$$\Phi_n(W_n) = \frac{c_n + p_n}{d_n\,(W_n)}, \tag{13.19}$$

where W_n is the number of vehicles in the area of n, c_n is the communication capacity of the deployed RSU in area n, p_n is the capacity of the V2V communications in this area, and $d_n\,(W_n)$ is the demand of communication of vehicles in the area. Based on $\Phi_n(W_n)$, the probability that area n is enjoying the satisfying communication can be defined as

$$AS_n = P(\Phi_n(W_n) \geq 1)\,. \tag{13.20}$$

The probability of all the areas that are satisfying is defined as

$$PS = P(\Phi_n(W_n) \geq 1, n = 1, \ldots, N) \tag{13.21}$$

Similarly, the average number of satisfied areas is defined as

$$NS = \sum_{n=1}^{N} P(\Phi_n(W_n) \geq 1). \tag{13.22}$$

Now, we consider area n and calculate the area satisfying probability AS_n. In this case, c_n is the capacity of the deployed RSU located at the center of this area, p_n is the V2V communication capacity, which depends on the number of vehicles in this area. We assume each vehicle can offer capacity of u_i, then $p_n = \sum_{i \in W_n} u_i$. Assume vehicles in area n need communication capacity of r_n, then $d_n(W_n) = W_n r_n$. Hence, AS_n can be expressed as

$$AS_n = P\left(c_n + \sum_{i \in W_n} u_i \geq W_n r_n\right). \tag{13.23}$$

In the reality vehicular system, the V2V communication capacity depends on the wireless interface. Suppose there are two classes of vehicles, i.e., one class uses Bluetooth and the other uses Wi-Fi to achieve the short range peer-to-peer communications. Therefore, one class of vehicles has low capacity of u^j, and the other class has large capacity of u^k; then, according to Theorem 13.1, the satisfying probability of area n, AS_n, is given by

$$
\begin{aligned}
AS_n &= P\left(c_n + \sum_{i\in W_n} u_i \geq W_n r_n\right) \\
&= P\left(c_n + u^j W_n^j + u^k W_n^k \geq (W_n^j + W_n^k) r_n\right) \\
&= \sum_{w_n^j=0}^{\infty} \sum_{w_n^k=0}^{\infty} \begin{array}{l} P\left(W_n^j = w_n^j\right) P\left(W_n^k = w_n^k\right) \times \\ P\left(\frac{c_n + u^j w_n^j + u^k w_n^k}{(w_n^j + w_n^k)} \geq r_n \,\middle|\, W_n^j = w_n^j, W_n^k = w_n^k\right) \end{array} \\
&= \sum_{w_n^j=0}^{\infty} \sum_{w_n^k=0}^{\infty} \begin{array}{l} P\left(W_n^j = w_n^j\right) P\left(W_n^k = w_n^k\right) \times \\ 1\left(c_n + u^j w_n^j + u^k w_n^k \geq (w_n^j + w_n^k) r_n\right) \end{array} \\
&= \sum_{w_n^j=0}^{\infty} \sum_{w_n^k=0}^{\infty} \begin{array}{l} P(W_n^j = w_n^j) P(W_n^k = w_n^k) \times \\ 1\left((r_n - u^j) w_n^j + (r_n - u^k) m_j^k \leq c_n\right) \end{array} \\
&= \sum_{0 \leq (r_n - u^j) w_n^j + (r_n - u^k) m_j^k \leq c_n} \frac{(\rho_n^j)^{w_n^j}}{w_n^j!} \frac{(\rho_n^l)^{w_n^k}}{w_n^k!} e^{-\rho_n^j - \rho_n^k}
\end{aligned}
\tag{13.24}
$$

Consequently, we can obtain the probability of the network-wide satisfying PS and the number of satisfying areas NS as follows:

$$
\begin{aligned}
PS &= P(\Phi_n(W_n) \geq 1, \quad n = 1, \ldots, N) \\
&= \prod_{n=1}^{N} P(\Phi_n(W_n) \geq 1) = \prod_{n=1}^{N} AS_n;
\end{aligned}
\tag{13.25}
$$

$$
NS = \sum_{n=1}^{N} P(\Phi_n(W_n) \geq 1) = \sum_{n=1}^{N} AS_n.
\tag{13.26}
$$

Based on the above derivation, we set a vehicular network environment to observe the performance. For the considered system, we assume there are 20 areas. The RSU's communication capacity of each area is set to be uniformly distributed, i.e., [500, 12,000] bps by considering the capacity provided by usually deployed Wi-Fi or 3G/4G base stations. The capacity of V2V communication of each vehicle is divide into two class, one class having [750, 850] bps, and [50, 150] bps of low capacity vehicles. Related to the load of each area, we use the data obtained from Beijing trace, select most 20 areas with largest number of vehicles in the system, let half of vehicles are with high capacity, and the other half are with low capacity. We set this area load settings as ρ and increase the load by 3, 5, 7, and 9 times. In the vehicular side, we set their communication by an exponential distribution with parameter of ϑ. Under the above settings, we can obtain the vehicular networking system performance of area satisfying probability, network-wide satisfying probability, and the number of satisfying areas. By varying the mean of the communication demand of vehicles $1/\vartheta$, the results of satisfying probability of the most loaded areas are shown in Figure 13.13. From the results, we can see that the satisfying probability is near 100% when the average demand is less than 300 bps. With the increase in demand, the satisfying probability

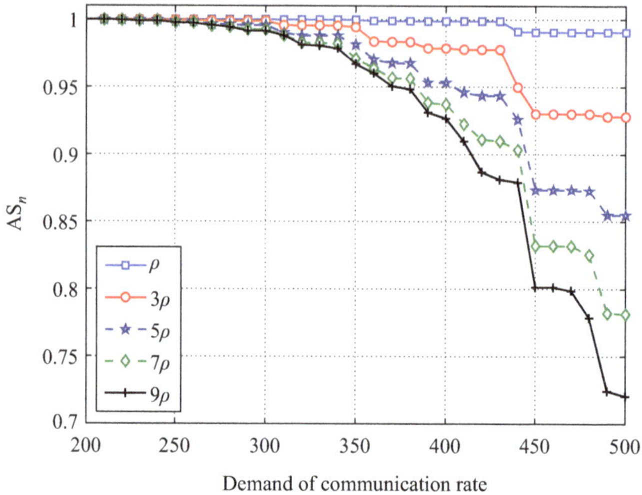

Figure 13.13 Area satisfying probability according to the demand of communication rate

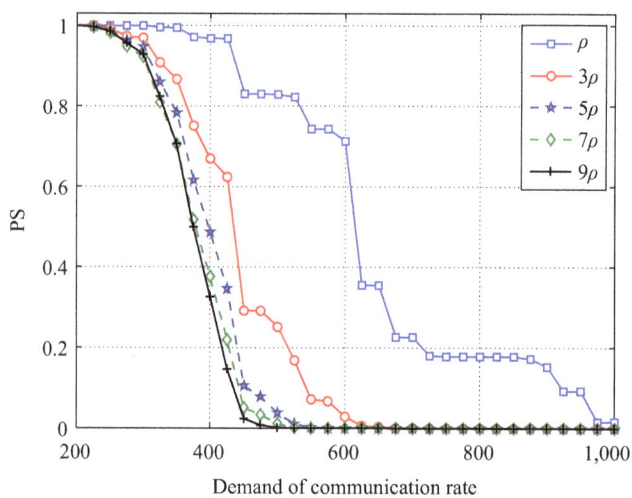

Figure 13.14 Network-wide satisfying probability according to the demand of communication rate

decreases. The larger the load, the sharper the decreasing rate. Under these results, we can decide how to deploy the RSU equipment according to the performance curve and specific requirements. In terms of the network-wide performance, Figures 13.14 and 13.15 show the results of *PS* and *NS*. With the increase in average demand and

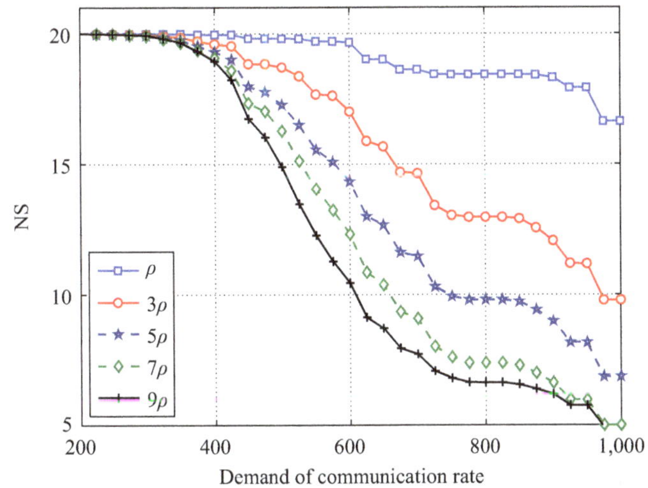

Figure 13.15 The number of satisfying areas according to the demand of communication rate

area load, we can see both *PS* and *NS* decrease. In this case, we can use the related results to design the network system according to the requirements and decide how to deploy infrastructure and RSU devices supporting V2V communications.

13.7 Conclusions

In this chapter, we used the open Jackson queueing network to model the macroscopic level vehicular mobility. Our proposed simple model can accurately describe the vehicular mobility and predict various measures of network-level and vehicular-level performance. Based on two large-scale urban city vehicular motion traces, we validated the accuracy of our proposed model. Finally, we proposed two applications as an example to illustrate our proposed model effectiveness in the analysis of system-level performance and dimensioning of vehicular networks.

References

[1] Khabazian M, Aissa S, and Mehmet-Ali M. Performance modeling of message dissemination in vehicular ad hoc networks with priority. IEEE Journal on Selected Areas in Communications. 2011;29(1):61–71.

[2] Dimitrakopoulos G, and Demestichas P. Intelligent transportation systems. IEEE Vehicular Technology Magazine. 2010;5(1):77–84.

[3] Li F, and Wang Y. Routing in vehicular ad hoc networks: A survey. IEEE Vehicular Technology Magazine. 2007;2(2):12–22.

[4] Abdrabou A, and Zhuang W. Probabilistic delay control and road side unit placement for vehicular ad hoc networks with disrupted connectivity. IEEE Journal on Selected Areas in Communications. 2011;29(1):129–139.

[5] Harri J, Filali F, and Bonnet C. Mobility models for vehicular ad hoc networks: A survey and taxonomy. IEEE Communications Surveys & Tutorials. 2009;11(4):19–41.

[6] Helbing D. Traffic and related self-driven many-particle systems. Reviews of Modern Physics. 2001;73(4):1067.

[7] Chen YC, Kurose J, and Towsley D. A simple queueing network model of mobility in a campus wireless network. In: Proceedings of the 3rd ACM Workshop on Wireless of the Students, by the Students, for the Students. ACM; 2011. p. 5–8.

[8] Rojas A, Branch P, and Armitage G. Experimental validation of the random waypoint mobility model through a real world mobility trace for large geographical areas. In: Proceedings of the 8th ACM International Symposium on Modeling, Analysis and Simulation of Wireless and Mobile Systems. ACM; 2005. p. 174–177.

[9] Hsu Wj, Merchant K, Shu Hw, *et al.* Weighted waypoint mobility model and its impact on ad hoc networks. ACM SIGMOBILE Mobile Computing and Communications Review. 2005;9(1):59–63.

[10] Zheng Q, Hong X, and Liu J. An agenda based mobility model. In: Simulation Symposium, 2006. 39th Annual. IEEE; 2006. pp. 8.

[11] Musolesi M, and Mascolo C. A community based mobility model for ad hoc network research. In: Proceedings of the 2nd International Workshop on Multi-Hop Ad Hoc Networks: From Theory to Reality. ACM; 2006. p. 31–38.

[12] Yoon J, Noble BD, Liu M, *et al.* Building realistic mobility models from coarse-grained traces. In: Proceedings of the 4th International Conference on Mobile Systems, Applications and Services. ACM; 2006. p. 177–190.

[13] Kim M, Kotz D, and Kim S. Extracting a mobility model from real user traces. In: INFOCOM 2006. 25th IEEE International Conference on Computer Communications. Proceedings. IEEE; 2006. p. 1–13.

[14] Zhu H, Li M, Fu L, *et al.* Impact of traffic influxes: Revealing exponential inter-contact time in urban VANETs. IEEE Transactions on Parallel and Distributed Systems. 2011;22(8):1258–1266.

[15] Lee K, Yi Y, Jeong J, *et al.* Max-contribution: On optimal resource allocation in delay tolerant networks. In: INFOCOM, 2010 Proceedings IEEE. IEEE; 2010. p. 1–9.

[16] Zhu H, Fu L, Xue G, *et al.* Recognizing exponential inter-contact time in VANETs. In: INFOCOM, 2010 Proceedings IEEE. IEEE; 2010. p. 1–5.

[17] Li Y, Jin D, Wang Z, *et al.* Exponential and power law distribution of contact duration in urban vehicular ad hoc networks. IEEE Signal Processing Letters. 2013;20(1):110–113.

[18] Kelly FP. Networks of queues with customers of different types. Journal of Applied Probability. 1975;12(3):542–554.

[19] Menasche DS, Rocha AA, Li B, *et al.* Content availability and bundling in swarming systems. In: Proceedings of the 5th International Conference on Emerging Networking Experiments and Technologies. ACM; 2009. p. 121–132.

[20] Ashtiani F, Salehi JA, and Aref MR. Mobility modeling and analytical solution for spatial traffic distribution in wireless multimedia networks. IEEE Journal on Selected Areas in Communications. 2003;21(10):1699–1709.

[21] Kim K, and Choi H. A mobility model and performance analysis in wireless cellular network with general distribution and multi-cell model. Wireless Personal Communications. 2010;53(2):179–198.

[22] Li M, Zhu H, Zhu Y, *et al.* ANTS: Efficient vehicle locating based on ant search in ShanghaiGrid. IEEE Transactions on Vehicular Technology. 2009;58(8):4088–4097.

[23] Kemeny J G, and Snell J L. Markov Chains[M]. New York: Springer-Verlag, 1976.

[24] Kise K, Sato A, and Iwata M. Segmentation of page images using the area Voronoi diagram. Computer Vision and Image Understanding. 1998; 70(3):370–382.

[25] Schermelleh-Engel K, Moosbrugger H, and Müller H. Evaluating the fit of structural equation models: Tests of significance and descriptive goodness-of-fit measures[J]. Methods of psychological research online, 2003;8(2):23–74.

[26] Zhang G, Wang X, Liang YC, *et al.* Fast and robust spectrum sensing via Kolmogorov–Smirnov test. IEEE Transactions on Communications. 2010;58(12):3410–3416.

[27] Câmara D, Frangiadakis N, Filali F, *et al.* Vehicular Delay Tolerant Networks. Handbook of Research on Mobility and Computing: Evolving Technologies and Ubiquitous Impacts. IGI Global; 2011. p. 356–367.

Index